Genetics and Genomics: Concepts and Applications

Genetics and Genomics: Concepts and Applications

Edited by Rosanna Mann

SYRAWOOD
PUBLISHING HOUSE

New York

Published by Syrawood Publishing House,
750 Third Avenue, 9ᵗʰ Floor,
New York, NY 10017, USA
www.syrawoodpublishinghouse.com

Genetics and Genomics: Concepts and Applications
Edited by Rosanna Mann

© 2019 Syrawood Publishing House

International Standard Book Number: 978-1-68286-711-2 (Hardback)

Cataloging-in-Publication Data

Genetics and genomics : concepts and applications / edited by Rosanna Mann.
    p. cm.
Includes bibliographical references and index.
ISBN 978-1-68286-711-2
1. Genetics. 2. Genomics. 3. Genomes. I. Mann, Rosanna.
QH430 .G46 2019
576.5--dc23

TABLE OF CONTENTS

PREFACE

The world is advancing at a fast pace like never before. Therefore, the need is to keep up with the latest developments. This book was an idea that came to fruition when the specialists in the area realized the need to coordinate together and document essential themes in the subject. That's when I was requested to be the editor. Editing this book has been an honour as it brings together diverse authors researching on different streams of the field. The book collates essential materials contributed by veterans in the area which can be utilized by students and researchers alike.

Genomics is the science that studies the structural, functional and evolutionary aspects of the entire set of genes of an organism. The interdisciplinary study of genetics and genomics investigates the interrelationship between genotypes and phenotypes, and between genes and diseases. It also strives to understand the different genetic variations created by the processes of mutation, natural selection and evolution. The application areas of these fields are spread across a range of industries like pharmaceuticals, crop production, pest control, etc. The various fields and subfields of gene studies along with the recent technological advances have been covered in exhaustive details in this book. Different approaches, evaluations, methodologies and advanced studies on genetics and genomics have been included. Geneticists, students and researchers working in this domain will find it a valuable source of reference.

Each chapter is a sole-standing publication that reflects each author's interpretation. Thus, the book displays a multi-facetted picture of our current understanding of application, resources and aspects of the field. I would like to thank the contributors of this book and my family for their endless support.

Editor

COL5A1 gene variants previously associated with reduced soft tissue injury risk are associated with elite athlete status in rugby

Shane M. Heffernan[1,9*], Liam P. Kilduff[2], Robert M. Erskine[3,4], Stephen H. Day[1], Georgina K. Stebbings[1], Christian J. Cook[2,5], Stuart M. Raleigh[6], Mark A. Bennett[2], Guan Wang[7], Malcolm Collins[8], Yannis P. Pitsiladis[7] and Alun G. Williams[1,4]

Abstract

Background: Two common single nucleotide polymorphisms within the *COL5A1* gene (SNPs; rs12722 C/T and rs3196378 C/A) have previously been associated with tendon and ligament pathologies. Given the high incidence of tendon and ligament injuries in elite rugby athletes, we hypothesised that both SNPs would be associated with career success.

Results: In 1105 participants (RugbyGene project), comprising 460 elite rugby union (RU), 88 elite rugby league athletes and 565 non-athlete controls, DNA was collected and genotyped for the *COL5A1* rs12722 and rs3196378 variants using real-time PCR. For rs12722, the injury-protective CC genotype and C allele were more common in all athletes (21% and 47%, respectively) and RU athletes (22% and 48%) than in controls (16% and 41%, $P \leq 0.01$). For rs3196378, the CC genotype and C allele were overrepresented in all athletes (23% and 48%) and RU athletes (24% and 49%) compared with controls (16% and 41%, $P \leq 0.02$). The CC genotype in particular was overrepresented in the back and centres (24%) compared with controls, with more than twice the odds (OR = 2.25, $P = 0.006$) of possessing the injury-protective CC genotype. Furthermore, when considering both SNPs simultaneously, the CC–CC SNP-SNP combination and C–C inferred allele combination were higher in all the athlete groups ($\geq$18% and $\geq$43%) compared with controls (13% and 40%; $P = 0.01$). However, no genotype differences were identified for either SNP when RU playing positions were compared directly with each other.

Conclusion: It appears that the C alleles, CC genotypes and resulting combinations of both rs12722 and rs3196378 are beneficial for rugby athletes to achieve elite status and carriage of these variants may impart an inherited resistance against soft tissue injury, despite exposure to the high-risk environment of elite rugby. These data have implications for the management of inter-individual differences in injury risk amongst elite athletes.

Keywords: Rugby union, Rugby league, Tendon, Ligament, Genetics

* Correspondence: shane.heffernan@ucd.ie
[1]MMU Sports Genomics Laboratory, Manchester Metropolitan University, Crewe, Manchester, UK
[9]School of Public Health, Physiotherapy and Sports Science, University College Dublin, Dublin 4, Ireland
Full list of author information is available at the end of the article

Background

Elite rugby athletes regularly experience high velocity collisions that lead to increasingly high injury occurrence rates that are likely to be a consequence of the increasing size and strength of the athletes [1–4]. This increased size and strength is likely to result in greater changes in momentum during player collisions, as well as during voluntary accelerations and decelerations. This has resulted in rugby union (RU) having one of the highest reported injury incidence rates in professional team sports [5]. Meta-analyses have shown that for every 1000 h, an elite RU athlete will experience approximately 81 injuries during match play and three during training, with the majority being ligament, tendon and muscle injuries of the lower limbs [6]. Indeed, in the most recent Rugby World Cup (2015) this rate of incidence was more than 90 injuries per 1000 h [7]. Furthermore, pooled data from 10 studies of elite rugby league (RL) athletes show that injury incidence is approximately twice (172 per 1000 h) that of RU [8]. Similar to RU, the majority of injuries in RL occur on the lower limbs, consisting mainly of sprains and strains [8]. Injury incidence differs across RU playing position, with elite back row players showing the highest rate among forwards and centres the highest among backs [7]. Therefore, investigating the molecular genetic components of these injuries, including in the context of playing positions that differ in terms of physiological characteristics [9, 10], match play demands [11] as well as genetically [12], may progress understanding towards greater individualisation of match play exposure and training load and mode, in order to reduce injury risk [13].

The collagen fibril, which consists predominately of type I collagen, is the primary structural component of tendons, ligaments and other non-cartilaginous connective tissues [14]. The formation and diameter of the collagen fibril is regulated by, amongst other molecules, the minor fibrillar type V collagen protein [15–18]. The type V collagen isoform comprises two α1(V) and one α2(V) chains, encoded by the *COL5A1* and *COL5A2* genes respectively [16, 19] and forms between 1 and 5% of total collagen content [18, 20]. The *COL5A1* gene is the most explored genetic locus in relation to tendon and ligament injuries [21–25], while mutations in the *COL5A1* gene have been identified in Ehlers-Danlos syndrome, a disease characterised by joint hypermobility, laxity and muscle hypotonia [26]. This results in irregularly large collagen fibrils within connective tissue [27] and is attributed to a reduced synthesis of collagen type V [17, 28].

Two common single nucleotide polymorphisms (SNPs, rs12722 C/T and rs3196378 C/A) located in the functional 3′ untranslated region (3′ UTR) *COL5A1* gene on chromosome 9 have been associated with tendon [23] and ligament ([rs12722] [29]) injuries. Specifically, the

CC genotype of the more extensively investigated rs12722 polymorphism has been previously associated with reduced risk of chronic Achilles tendinopathy (odds ratio (OR) = 0.42-AUS and OR = 0.38-SA, respectively; [23, 25]), anterior cruciate ligament injury in females (OR = 6.6; [22]) and lateral epicondylitis (OR = 1.4; [21]), suggesting a protective role of the C allele against injury. Although there are conflicting results [23, 30], our current understanding also suggests that the CC genotype and/or the C allele of rs3196378 would also have a protective role [31, 32]. Considering the high frequencies of tendon and ligament inquiries in elite rugby [6–8, 33], assessing these specific genetic variants may be of use in helping improve the management of injury risk in individual players.

Given the association of the two *COL5A1* gene variants with injury risk, it is possible that possession of the risk alleles might reduce an individual's ability to withstand exposure to the environment of competitive rugby without suffering more frequent injuries. Consequently, those individuals would be forced to miss training, selection and competitive events important for their career progression. Thus, athletes carrying the C allele at either or both rs12722 and rs3196378 might be at an advantage in terms of their ability to achieve success in elite competitive rugby and at a disadvantage in terms of their shorter-term and longer-term musculoskeletal health. Therefore, the objective of the present study was to investigate if *COL5A1* rs12722 and rs3196378 genotype and allele frequencies differed between elite rugby athletes and a control population, and/or between playing positions. It was hypothesised that the *COL5A1* rs12722 and rs3196378 injury-protective C alleles and/or CC genotypes would be overrepresented in elite rugby athletes compared with controls.

Methods

Participants

As part of the ongoing RugbyGene project [12, 13, 34], a total of 1105 individuals were recruited and gave written informed consent to participate in the present study. An a priori calculation for 80% power to detect a small effect size (w) of 0.1 required a sample of >785 participants. The sample comprised elite Caucasian male rugby athletes (*n* = 540; mean (standard deviation) height 1.85 (0.07) m, mass 101 (14) kg, age 29 (7) years) including 72% British, 16% South African, 7% Irish and 5% of other nationalities. Caucasian controls (68% male; *n* = 565; height 1.75 (0.10) m, mass 75 (13) kg, age 26 (11) years) included 86% British, 12% South African, 1% Irish and 1% of other nationalities recruited mainly during 2012–2016. Eight athletes competed in both elite RL and RU and were included in both groups that were analysed separately. Athletes were considered elite if they had competed regularly (> 5 matches) since 1995 in the highest professional league in the UK, Ireland or South Africa for RU and the highest professional league in the UK for RL. Of the RU athletes, 51.7% had competed at international level for a "High Performance Union" (Regulation 16, worldrugby.org) and 43.2% of RL had competed at international level. All data for the athlete group's international status were confirmed as of 1st January 2017. Most participants in the current study were also participants in previous publications regarding variations in the *ACTN3*, *ACE* and *FTO* genes [12, 34].

Sample collection and genotyping

Description of all molecular procedures have previously been described in detail [12]. Briefly, blood (~70% of all samples), saliva (~25%) or buccal swab samples (~5%) were obtained via the following protocols. Blood from a superficial forearm vein was put into an EDTA tube and stored in sterile tubes at −20 °C until processing. Saliva samples were collected using Oragene DNA OG-500 collection tubes (DNA Genotek Inc., Ontario, Canada) according to the manufacturer's protocol and stored at room temperature until processing. Sterile buccal swabs (Omni swab, Whatman, Springfield Mill, UK) were rubbed against the buccal mucosa of the cheek for approximately 30 s and the tips stored at −20 °C until processing. At MMU and Glasgow, DNA was stored at 4 °C following isolation performed using the QIAamp DNA Blood Mini kit and standard spin column protocol (Qiagen, West Sussex, UK). In Cape Town, DNA was isolated from whole blood [35] and samples stored at −20 °C. At Northampton, DNA was isolated from whole blood using Flexigene kits (Qiagen), with the resulting samples stored at −20 °C.

Genotyping at all three centres was performed using TaqMan assays (Applied Biosystems, Paisley, UK) for both the *COL5A1* rs12722 and rs3196378 variants. Our genotyping methods and quality control procedures have been fully described in our earlier study [12]. Minor adaptations were made to the volumes used in each assay mix depending on whether the DNA was obtained from buccal swabs or saliva/blood. PCR was performed on either a Chromo4 (Bio-Rad, Hertfordshire, UK) or StepOnePlus thermal cycler (Applied Biosystems). Genotypes were called based on reporter dye intensity and visualized using cluster plots. The TaqMan assays included VIC and FAM dyes that for rs12722 indicated C and T alleles on the forward DNA strand, respectively. Thus, VIC/FAM were interpreted as: 5′-CACACCCA[C/T]GCGCCCCG-3′. For rs3196378, VIC and FAM dyes indicated C and A alleles on the forward DNA strand, respectively and were interpreted as: 5′-CCCACCCC[A/C]GCCCTGGC-3′. Genotype calling was 100% successful for both polymorphisms

in the athlete samples. For rs12722, one of the 566 control samples was unsuccessful and, for rs3196378, 10 of the 566 control samples were unsuccessful. There was 100% agreement among reference samples genotyped in the three genotyping centres, i.e. Glasgow, Northampton and MMU laboratories.

Data analysis

SPSS for Windows version 22 (SPSS Inc., Chicago, IL) software was used to conduct Pearson's Chi-square (χ^2) tests to compare genotype (using three analysis models; additive, recessive and dominant), allele and inferred haplotype frequencies between athletes and controls, and between RU subgroups based on playing position and controls. With 80% statistical power, analyses of all genetic models in positional subgroups compared with controls (forwards, backs and back 3-centres) were able to detect a small-to-medium effect size (w) of 0.12. Multifactor Dimensionality Reduction (MDR; www.multifactordimensionalityreduction.org) software was used to calculate SNP-SNP epistasis interactions [36]. Haplotypes were inferred using SNPStats [37]. Sixty five tests were subjected to Benjamini-Hochberg (BH; [38]) corrections to control false discovery rate and corrected probability values are reported. Odds ratios (OR) were calculated to estimate effect size. CubeX online software (www.oege.org/software/cubex) was used to determine linkage disequilibrium statistics [39]. Alpha was set at 0.05.

Results

Genotype frequencies were in Hardy-Weinberg equilibrium for both rs12722 and rs3196378 in the control ($P \geq 0.21$) and athlete groups (RL, $P \geq 0.11$; RU, $P \geq 0.754$). There was no sexual dimorphism of genotype frequency for either rs12722 ($P = 0.279$) or rs3196378 ($P = 0.374$) within the control group. *COL5A1* rs12722 and rs3196378 were in tight linkage disequilibrium for both controls (D′ = 0.902, r^2 = 0.784) and all athletes (D′ = 0.877, r^2 = 0.738). Athletes were taller and heavier ($P < 0.05$) but not older ($P > 0.05$) than controls.

rs12722

The CC genotype, proportion of C allele carriers and C allele were overrepresented in all athletes (21.1%, 73.3% and 47.2%, respectively) and RU athletes (22.0%, 73.3% and 47.6%) compared with controls (15.6%, 66.5% and 41.1%, Table 1 and Fig. 1, $P \leq 0.01$). Furthermore, the CC genotype, proportion of C allele carriers (Table 1) and C allele (Fig. 1) were overrepresented in the subgroups of RU forwards (22.1%, 71.5% and 46.8%) and backs (21.8%, 75.6% and 48.7%) compared with controls (15.6%, 66.5% and 41.1%). Additionally, of the RU subgroups, the back three and centres differed from controls and showed the greatest C allele and CC genotype

Table 1 *COL5A1* rs12722 and rs3196378 genotype and allele distribution of controls and athletes separated by code (RL and RU) and into positional sub-groups for RU, presented as genotype/allele counts followed by percentage in parentheses

Genotype	Controls	All athletes	RL athletes	RU athletes	Forwards	Backs	Back 3 and Centres
rs12722							
TT	189 (33.4)	144 (26.7)	23 (26.1)	123 (26.7)	75 (28.5)	48 (24.4)	28 (22.8)
CT	288 (51.0)	282 (52.2)	51 (58.0)	236 (51.3)	130 (49.4)	106 (53.8)	65 (52.8)
CC	88 (15.6)	111 (21.1)*	14 (15.9)	101 (22.0)*	58 (22.1)*	43 (21.8)*	30 (24.4)*
Total	565	540	88	460	263	197	123
T allele carriers	477 (84.4)	426 (78.9)*	74 (84.1)	359 (78.0)*	205 (77.9)*	154 (78.2)*	93 (75.6)*
C allele carriers	379 (66.5)	396 (73.3)*	65 (73.9)	337 (73.3)*	188 (71.5)	149 (75.6)*	95 (77.2)*
rs3196378							
AA	183 (32.9)	144 (26.7)	23 (26.1)	123 (26.7)	74 (28.1)	49 (24.9)	28 (22.8)
CA	286 (51.4)	271 (50.2)	49 (55.7)	227 (49.3)	122 (46.4)	105 (53.3)	65 (52.8)
CC	87 (15.6)	125 (23.1)*	16 (18.2)	110 (24.0)*	67 (25.5)*	43 (21.8) *	30 (24.4)*
Total	556	540	88	460	263	197	123
A allele carriers	469 (84.4)	415 (76.9)*	72 (81.8)	350 (76.1)*	196 (74.5)*	154 (78.2)*	93 (75.6)*
C allele carriers	373 (67.1)	396 (73.3)*	65 (73.9)	337 (73.3)*	189 (71.9)	148 (75.1)*	95 (77.2)*

RU, rugby union, RL rugby league. The genotype, T allele carrier and C allele carrier data represent the additive, dominant model and recessive models, respectively. Eight athletes competed in both elite RL and RU and were included in both groups that were analysed separately
Asterisks (*) indicate Chi-square differences from controls ($P \leq 0.03$)

frequency (50.8% versus 41.1% and 24.4% versus 15.6%, respectively, Table 1 and Fig. 1, $P \leq 0.02$). Compared with controls, those back three and centre players had 2.3 times the odds of possessing the CC genotype and 1.5 times the odds of possessing the C allele (Table 2).

rs3196378

The CC genotype, proportion of C allele carriers and C allele were overrepresented in all athletes (23.1%, 73.3% and 48.4%) and RU athletes (23.9%, 73.3% and 48.6%) compared with controls (15.6%, 67.7% and 41.4%, Table 1 and

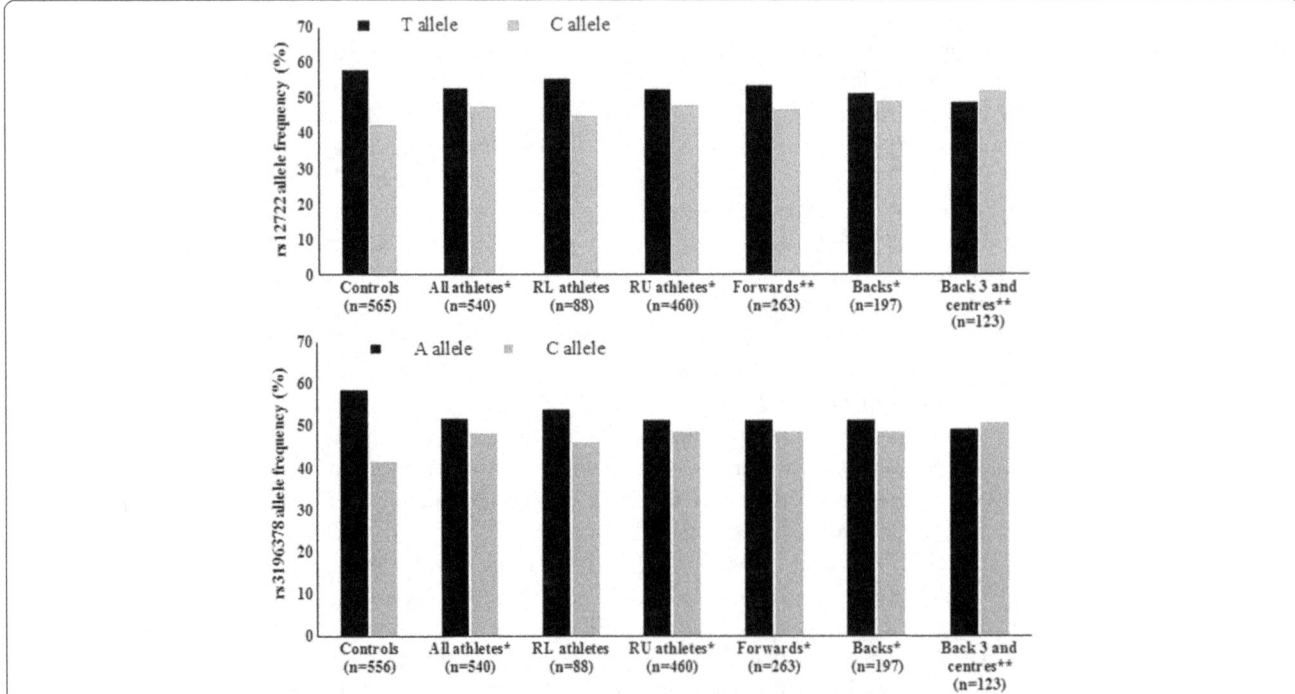

Fig. 1 Allele frequency of *COL5A1* rs12722 (**a**) and rs3196378 (**b**) for control and athlete groups. Asterisks indicate a difference in allele frequency between the particular athlete group and controls. A single asterisk (*) designates $P = 0.01$ and a double asterisk (**) designates $P = 0.02$. RU, rugby union, RL rugby league. Eight athletes competed in both elite RL and RU and were included in both groups that were analysed separately

Table 2 Odds ratio statistics for RU player status of *COL5A1* gene variants (rs12722 and rs3196378)

Positional comparison	Model	[b]Odds Ratio	95% Confidence Interval
rs12722			
All athletes v Controls	CC/TT	1.70	1.19–2.42
	CC/T$_{carriers}$	1.45	1.07–1.97
	C$_{allele}$/T$_{allele}$	1.28	1.08–1.52
RU athletes v Controls	CC/TT	1.76	1.22–2.54
	CC/T$_{carriers}$	1.53	1.11–2.09
	C$_{allele}$/T$_{allele}$	1.30	1.09–1.55
Forwards v Controls	CC/TT	1.66	1.08–2.54
	CC/T$_{carriers}$	1.53	1.06–2.22
	C$_{allele}$/T$_{allele}$	1.26	1.02–1.55
Backs v Controls	CC/TT	1.92	1.19–3.12
	CC/T$_{carriers}$	1.51	1.01–2.27
	C$_{allele}$/T$_{allele}$	1.36	1.08–1.72
Back 3 and centres v Controls	CC/TT	2.30	1.29–4.08
	CC/T$_{carriers}$	1.76	1.09–2.78
	C$_{allele}$/T$_{allele}$	1.48	1.12–1.96
rs3196378			
All athletes v Controls	CC/AA	1.83	1.29–2.59
	CC/A$_{carriers}$	1.66	1.22–2.25
	C$_{allele}$/A$_{allele}$	1.32	1.12–1.56
RU athletes v Controls	CC/AA	1.88	1.31–2.70
	CC/A$_{carriers}$	1.73	1.27–2.37
	C$_{allele}$/A$_{allele}$	1.34	1.12–1.59
Forwards v Controls	CC/AA	1.90	1.25–2.89
	CC/A$_{carriers}$	1.88	1.31–2.69
	C$_{allele}$/A$_{allele}$	1.34	1.09–1.66
Backs v Controls	CC/AA	1.85	1.14–2.99
	CC/A$_{carriers}$	1.54	1.02–2.31
	C$_{allele}$/A$_{allele}$	1.33	1.06–1.68
Back 3 and centres v Controls	CC/AA	2.25	1.27–4.00
	CC/A$_{carriers}$	1.78	1.11–2.84
	C$_{allele}$/A$_{allele}$	1.46	1.11–1.93
Haplotypes and SNP epistasis			
All athletes v Controls	C–C/T–A	1.31	1.10–1.56
	[a]CC–CC	1.97	1.44–2.69
RU athletes v Controls	C–C/T–A	1.32	1.11–1.59
	[a]CC–CC	2.11	1.53–2.91
Forwards v Controls	C–C/T–A	1.31	1.05–1.62
	[a]CC–CC	2.19	1.51–3.16
Backs v Controls	C–C/T–A	1.36	1.07–1.72
	[a]CC–CC	1.86	1.29–2.67
Back 3 and centres v Controls	C–C/T–A	1.49	1.12–1.99
	[a]CC–CC	2.35	1.47–3.77

[a]MDR best model versus all other genotype combinations for rs12722 and rs3196378
[b]All odds ratios were statistically significant ($P < 0.05$)

Fig. 1, $P \leq 0.02$). Furthermore, CC genotype, proportion of C allele carriers (Table 1) and C allele (Fig. 1) were over-represented in backs (21.8%, 75.1% and 48.5%) compared with controls (15.6%, 67.7% and 41.4%, $P \leq 0.02$). Forwards also had higher CC genotype and C allele frequencies (25.5% and 48.7%; Table 1 and Fig. 1) and showed almost twice the odds of being CC genotype than carrying an A allele, compared with controls (Table 2). For the back three and centres group, 24.4% were CC genotype, 77.2%

were C allele carriers and C allele frequency was 47.8% - all of which were greater than controls ($P \leq 0.02$; Table 1 and Fig. 1, OR = 2.25, Table 2).

Haplotype and SNP epistasis analysis
There was a greater frequency of the CC–CC SNP-SNP combination in all athletes (18.3%), RU athletes (18.9%), RU forwards (19.4%), RU backs (18.3%) and to the greatest extent the back three and centre group (20.3%;

OR = 2.35; Table 2), compared with control (12.8%; all athlete comparisons with the control group were *P* = 0.01). Furthermore, C–C inferred haplotype frequencies were higher in all the athlete groups compared with controls, reflected by a greater frequency of the T–A inferred haplotype in the control group (*P* = 0.01; Fig. 2).

Discussion

The present observations are the first to identify associations between *COL5A1* rs12722 and rs3196378 polymorphisms and athlete status in a large cohort of elite rugby athletes. As hypothesised, the apparent injury-protective C allele and CC genotype, of both SNPs [23], were overrepresented in elite rugby athletes compared with controls. This association persisted across playing position, with the C allele being overrepresented in RU forwards and backs including the back three and centres group, compared with controls. Furthermore, when the two SNPs are combined, the CC–CC combination and C–C inferred allele combination showed similar overrepresentation in elite rugby athletes compared with controls.

The results provide an insight into the potential injury susceptibility of some elite rugby athletes. September et al. [23] identified a higher frequency of the CC genotype in asymptomatic controls for both rs12722 and rs3196378 compared with tendinopathy patients [23, 32]. Moreover, the rs12722 T allele has been associated with ligament injury [21, 22, 24, 40] and Achilles tendinopathy [23, 25] with the C allele again identified as protective in these studies, despite a lack of replication in another study [30]. Greater joint laxity has almost a 3-fold increase in risk of knee ligament rupture [41] and greater joint laxity has recently been associated with the rs12722 T allele in non-white females [40]. Collectively, these data suggest that the C allele of rs12722 and/or

rs3196378 (or other variant(s) in strong LD with these SNPs) might be beneficial in protecting against tendon and ligament injuries. This is reflected in the present study showing greater C allele frequency in elite rugby athletes compared with controls. Based on these data, we propose that when exposed to the high-risk environment of rugby during training and especially during competitive matches, ceteris paribus, carriage of the C alleles at the rs12722 and rs3196378 loci provides both a shorter-term and longer-term advantage to rugby athletes in the form of reduced injury risk. Athletes with fewer and/or less severe injuries, all else being equal, will miss fewer matches, training and selection events and thus be more likely to progress towards elite status in their athletic careers compared with their peers.

The rs12722 CC genotype has also been related to a lower incidence of exercise-associated muscle cramping (EAMC) in Caucasian ironman and ultra-marathon athletes [42]. The authors hypothesised that this was due to similar mechanisms of reduced tendon injury susceptibility, in that rs12722 alters soft tissue structural and mechanical properties (tissue stiffness and thickening). Some recent findings might support this hypothesis, as greater tendon stiffness was associated with the rs12722 T allele in one study [43], however another study reported no association of rs12722 with tendon structural or mechanical properties [44]. These data suggest that in addition to the apparent protection from tendon and ligament injury, the greater frequencies of the C allele in elite rugby athletes might be protective against muscle cramping and possibly reduced tendon stiffness. Indeed, recent evidence from elite RL shows that over 70% of athletes experience EAMC per season and that history of cramping is the strongest predictor of future EAMC [45]. In contrast, the TT genotype has been associated

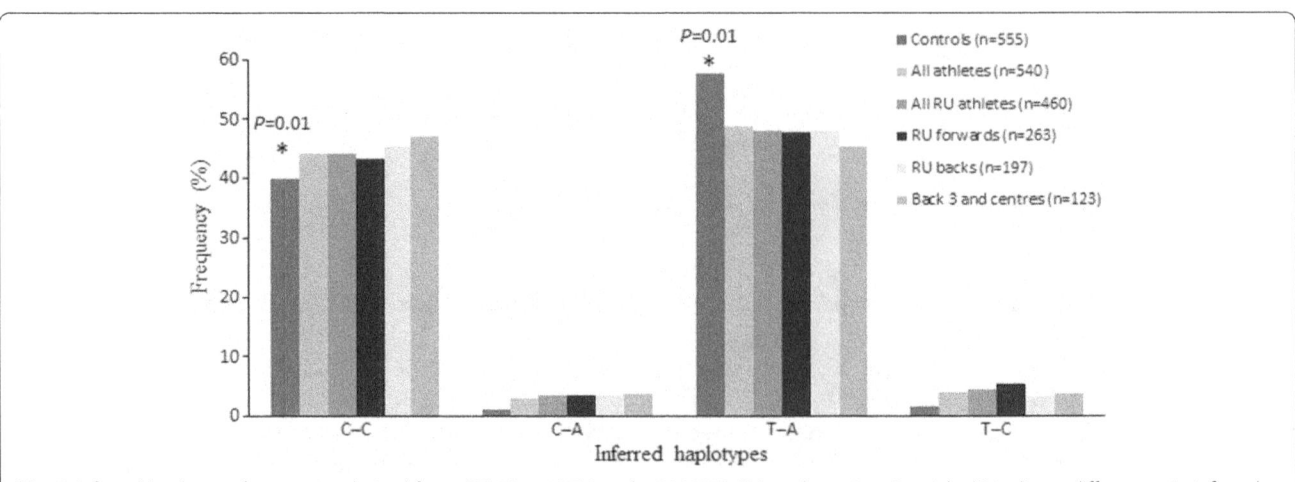

Fig. 2 Inferred haplotype frequencies derived from *COL5A1* rs12722 and rs3196378. RU, rugby union. Asterisks (*) indicate differences in inferred haplotype frequencies between the controls and each athlete group (*P* = 0.01). Eight athletes competed in both elite RL and RU and were included in both groups that were analysed separately

with greater endurance running ability of Caucasian ironman triathletes (TT = 294 min, CC = 307 min; [46]). However, recent data show no association of rs12722 with either running economy or V $_{2max}$ [47]. While endurance capacity is of value in elite rugby, the predominant focus of player selection and training programs is towards power, speed and strength - i.e. short-term, anaerobic performance [with notable differences between playing positions; 9, 10].

Limited data exist regarding *COL5A1* genetic variation and team sport athletes. In a study of 73 soccer athletes, including some elite players, no rs12722 TT genotype individuals were identified (a potentially interesting observation but difficult to interpret because of the varied geographic ancestry of the athletes), but there was a tendency for more severe muscle injuries in the TC genotype group ($P = 0.08$), compared with CC [48]. Here, consistent with those observations, we show an overrepresentation of the protective C allele and CC genotype of both rs12722 and rs3196378, in addition to CC–CC SNP-SNP combination and C–C inferred allele combination in elite rugby athletes.

Some possible mechanisms have been proposed to explain the association of *COL5A1* gene variants and soft tissue injury [31, 32]. Laguett et al. [32] have shown that the *COL5A1* 3′ UTR – where both rs12722 and rs3196378 are situated - affects mRNA stability. For both SNPs, the alleles associated with greater soft-tissue injury risk were associated with greater Hsa-miR-608 stability, which in turn may alter the Col5α1 protein secondary structure - proposed to play a role in type V collagen production [31]. This would suggest that C/T allele differences at rs12722 might alter the co-polymerisation of collagen type V and type I fibrils. However, to date, this has not been demonstrated experimentally and exactly how this may translate into functional properties is currently unknown. Nevertheless, it appears that the C allele and CC genotype of rs12722 and rs3196378 appear beneficial for rugby athletes to achieve elite status, probably through greater resistance to soft tissue injury. Interestingly, while most relevant investigations have focussed on rs12722, we show in a large cohort (total $n = 1090$) that strong linkage disequilibrium exists in both controls and athletes between rs12722 and rs3196378. As such, it is likely that the associations of rs12722 with tendon and ligament injuries would be similar for rs3196378. It is possible that combining genetic data from multiple gene variants associated with injury susceptibility, such as those presented here, with other indicators of injury risk and recovery during rehabilitation could be used to better manage the prevention and recovery from elite player injury in the future.

Conclusion

In conclusion, we have presented the first associations between *COL5A1* 3′ UTR rs12722 and rs3196378 and elite status within a large cohort of rugby athletes. The C alleles of both polymorphisms, separately and in combination, were overrepresented in all athletes, RU forwards, backs and in RU back three and centre players versus controls. We propose that rugby athletes possessing more C alleles at these two genetic loci are probably at a lower risk of injury, given their exposure to the high-risk environment of elite rugby. However, these data pertain to only two SNPs of many that may be relevant to soft tissue injury and interpretation of the present results should be in that context. Future investigations should seek to combine elite rugby genotype data such as these with injury incidence data during rugby matches and training. It will be important to establish whether inter-individual differences in injury risk, within a population that we demonstrate here appear to be at an overall lower genetic risk of tendon and ligament injury compared with non-athletes, nevertheless are associated with those same genetic loci.

Abbreviations

3′ UTR: 3′ untranslated region; *COL5A1*: collagen type V α1; *COL5A2*: collagen type V α2; DNA: deoxyribonucleic acid; EAMC: exercise-associated muscle cramping; EDTA: ethylenediaminetetraacetic acid; MDR: multifactor dimensionality reduction; miRNA: micro ribonucleic acid; PCR: Polymerase chain reaction; RL: rugby league; RU: rugby union; SNP: single nucleotide polymorphisms

Acknowledgements

The authors wish to thank all athletes, their respective scientific support staff and the control participants for their time and willingness to participate in this research. We would like to thank Louis El Khoury for his technical assistance at the University of Northampton and Ben Davies, Craig Roberts, Bill Ribbans and Beth Vance for assistance in participant recruitment.

Funding

The present study was funded by Manchester Metropolitan University. Publication of this manuscript was supported by Manchester Metropolitan University.

Twitter

Follow the RugbyGene project at @RugbyGeneStudy.
Follow Shane Heffernan @Dr_Heff56.

About this supplement

This article has been published as part of *BMC Genomics* Volume 18 Supplement 8, 2017: Proceedings of the 34th FIMS World Sports Medicine Congress. The full contents of the supplement are available online at https://bmcgenomics.biomedcentral.com/articles/supplements/volume-18-supplement-8.

Authors' contributions

Listed alphabetically: MC, SD, RE, SH, LK, YP, AW and GW conceived and designed the study. MB, CC, MC, RE, SH, LK, SD, SR, GS, AW and GW contributed to data collection. SH and AW analysed data and drafted the manuscript. All authors contributed to interpretation of data, revised the article critically for important intellectual content and approved the final version of the manuscript.

Ethics approval and consent to participant

Ethical approval was granted by the ethics committees of Manchester Metropolitan University, Glasgow University, University of Cape Town and University of Northampton and all experimental procedures complied with the Declaration of Helsinki (2003). All participates gave written informed consent to take part in the present study.

Competing interests

The authors declare that they have no competing interests.

Author details

[1]MMU Sports Genomics Laboratory, Manchester Metropolitan University, Crewe, Manchester, UK. [2]A-STEM, College of Engineering, Swansea University, Swansea, UK. [3]Research Institute for Sport & Exercise Sciences, Liverpool John Moores University, Liverpool, UK. [4]Institute of Sport, Exercise and Health, University College London, London, UK. [5]School of Sport, Health and Exercise Sciences, Bangor University, Bangor, UK. [6]Centre for Physical Activity and Chronic Disease, Institute of Health and Wellbeing, University of Northampton, Northampton, UK. [7]FIMS Reference Collaborating Centre of Sports Medicine for Anti-Doping Research, University of Brighton, Brighton, UK. [8]Division of Exercise Science and Sports Medicine, Department of Human Biology, University of Cape Town (UCT), Cape Town, South Africa. [9]School of Public Health, Physiotherapy and Sports Science, University College Dublin, Dublin 4, Ireland.

References

1. Bradley WJ, Cavanagh BP, Douglas W, Donovan TF, Morton JP, Close GL. Quantification of training load, energy intake, and physiological adaptations during a rugby preseason: a case study from an elite eropean rugby union squad. J Stength Cond Res. 2015;29(2):534–44.
2. Sedeaud A, Vidalin H, Tafflet M, Marc A, Toussaint J. Rugby morphologies: "bigger and taller", reflects an early directional selection. J Sports Med Phys Fitness. 2013;53(2):185–91.
3. Cunniffe B, Proctor W, Baker JS, Davies B. An evaluation of the physiological demands of elite rugby union using global positioning system tracking software. J Strength Cond Res. 2009;23(4):1195–203.
4. Owen NJ, Watkins J, Kilduff LP, Bevan HR, Bennett MA. Development of a criterion method to determine peak mechanical power output in a countermovement jump. J Strength Cond Res. 2014;28(6):1552–8.
5. Brooks JH, Kemp SP. Recent trends in rugby union injuries. Clin Sports Med. 2008;27(1):51–73.
6. Williams S, Trewartha G, Kemp S, Stokes KA. Meta-analysis of injuries in senior men's professional rugby union. Sports Med. 2013;43(10):1043–55.
7. Fuller CW, Taylor A, Kemp SP, Raftery M. Rugby World Cup 2015: World Rugby injury surveillance study. Brit J Sports Med. 2016:In press.
8. King D, Gissane C, Clark T, Marshall SW. The incidence of match and training injuries in rugby league: a pooled data analysis of published studies. Int J Sports Sci Coach. 2014;9(2):417–32.
9. Smart D, Hopkins WG, Quarrie KL, Gill N. The relationship between physical fitness and game behaviours in rugby union players. Eur J Sports Sci. 2014;14(supp 1):S8–S17.
10. Smart DJ, Hopkins WG, Gill ND. Differences and changes in the physical characteristics of professional and amateur rugby union players. J Strength Cond Res. 2013;27(11):3033–44.
11. Jones MR, West DJ, Crewther BT, Cook CJ, Kilduff LP. Quantifying positional and temporal movement patterns in professional rugby union using global positioning system. Eur J Sport Sci. 2015;15(6):488–96.
12. Heffernan SM, Kilduff LP, Erskine RM, Day SH, McPhee JS, McMahon GE, et al. Association of ACTN3 R577X but not ACE I/D gene variants with elite rugby union player status and playing position. Physiol Genomics. 2016;48(3):196–201.
13. Heffernan SM, Kilduff LP, Day SH, Pitsiladis YP, Williams AG. Genomics in rugby union: a review and future prospects. Eur J Sport Sci. 2015;15(6):460–8.
14. Imamura Y, Scott IC, Greenspan DS. The pro-α3 (V) collagen chain complete primary structure, expression domains in adult and developing tissues, and comparison to the structures and expression domains of the other types V and XI procollagen chains. J Biol Chem. 2000;275(12):8749–59.
15. Birk DE, Fitch J, Babiarz J, Doane K, Linsenmayer T. Collagen fibrillogenesis in vitro: interaction of types I and V collagen regulates fibril diameter. J Cell Sci. 1990;95(4):649–57.
16. Wenstrup RJ, Florer JB, Brunskill EW, Bell SM, Chervoneva I, Birk DE, Type V. Collagen controls the initiation of collagen fibril assembly. J Biol Chem. 2004;279(51):53331–7.
17. Sun M, Chen S, Adams SM, Florer JB, Liu H, Kao WW-Y, et al. Collagen V is a dominant regulator of collagen fibrillogenesis: dysfunctional regulation of structure and function in a corneal-stroma-specific Col5a1-null mouse model. J Cell Sci. 2011;124(23):4096–105.
18. Chanut-Delalande H, Bonod-Bidaud C, Cogne S, Malbouyres M, Ramirez F, Fichard A, et al. Development of a functional skin matrix requires deposition of collagen V heterotrimers. Mol Cell Biol. 2004;24(13):6049–57.
19. Malfait F, Wenstrup RJ, De Paepe A. Clinical and genetic aspects of Ehlers-Danlos syndrome, classic type. Genet Med. 2010;12(10):597–605.
20. McLaughlin J, Linsenmayer T, Birk D, Type V. Collagen synthesis and deposition by chicken embryo corneal fibroblasts in vitro. J Cell Sci. 1989;94(2):371–9.
21. Altinisik J, Meric G, Erduran M, Ates O, Ulusal AE, Akseki D. The BstUI and DpnII variants of the COL5A1 Gene are associated with Tennis Elbow. 2015;43(7):1784–9.
22. Posthumus M, September AV, O'Cuinneagain D, van der Merwe W, Schwellnus MP, Collins M. The COL5A1 gene is associated with increased risk of anterior cruciate ligament ruptures in female participants. Am J Sports Med. 2009;37(11):2234–40.
23. September AV, Cook J, Handley CJ, van der Merwe L, Schwellnus MP, Collins M. Variants within the COL5A1 gene are associated with Achilles tendinopathy in two populations. Brit J Sports Med. 2009;43(5):357–65.
24. O'Connell K, Knight H, Ficek K, Leonska-Duniec A, Maciejewska-Karlowska A, Sawczuk M, et al. Interactions between collagen gene variants and risk of anterior cruciate ligament rupture. Eur J Sports Sci. 2015;15(4):341–50.
25. Mokone G, Schwellnus M, Noakes T, Collins M. The COL5A1 gene and Achilles tendon pathology. Scand J Med Sci Sports. 2006;16(1):19–26.
26. Beighton P, Paepe AD, Steinmann B, Tsipouras P, Wenstrup RJ. Ehlers-Danlos syndromes: revised nosology, Villefranche, 1997. Am J Med Genet. 1998;77(1):31–7.
27. Vogel A, Holbrook K, Steinmann B, Gitzelmann R, Byers P. Abnormal collagen fibril structure in the gravis form (type I) of Ehlers-Danlos syndrome. Lab Investig. 1979;40(2):201–6.
28. Malfait F, De Paepe A. Molecular genetics in classic Ehlers–Danlos syndrome. Am J Med Genet. 2005; 139 C(1):17–23.
29. Posthumus M, September AV, Keegan M, O'Cuinneagain D, Van der Merwe W, Schwellnus MP, et al. Genetic risk factors for anterior cruciate ligament ruptures: COL1A1 gene variant. Brit J Sports Med. 2009;43(5):352–6.
30. Brown KL, Seale KB, El Khoury LY, Posthumus M, Ribbans WJ, Raleigh SM, et al. Polymorphisms within the COL5A1 gene and regulators of the extracellular matrix modify the risk of Achilles tendon pathology in a British case-control study. Ahead of print: J Sports Sci; 2016.
31. Abrahams Y, Laguette MJ, Prince S, Collins M. Polymorphisms within the COL5A1 3'-UTR that alters mRNA structure and the MIR608 gene are associated with Achilles Tendinopathy. Ann Hum Genet. 2013;77(3):204–14.
32. Laguette M-J, Abrahams Y, Prince S, Collins M. Sequence variants within the 3'-UTR of the COL5A1 gene alters mRNA stability: implications for musculoskeletal soft tissue injuries. Matrix Biol. 2011;30(5):338–45.
33. Fuller CW, Taylor A, Raftery M. Epidemiology of concussion in men's elite rugby-7s (sevens world series) and rugby-15s (rugby world cup, junior world championship and rugby trophy, Pacific nations cup and English premiership). Brit J Sports Med. 2015;49(7):478–83.
34. Heffernan SM, Stebbings GK, Kilduff LP, Erskine RM, Day SH, Morse CI, et al. Fat mass and obesity associated (FTO) gene influences skeletal muscle phenotypes in non-resistance trained males and elite rugby playing position. BMC Genet. 2017;18(1):4.

35. Lahiri DK, Nurnberger JIA. Rapid non-enzymatic method for the preparation of HMW DNA from blood for RFLP studies. Nucleic Acids Res. 1991;19(19):5444.

36. Moore JH, Gilbert JC, Tsai CT, Chiang FT, Holden T, Barney N, et al. A flexible computational framework for detecting, characterizing, and interpreting statistical patterns of epistasis in genetic studies of human disease susceptibility. J Theor Biol. 2006;241(2):252–61.

37. Sole X, Guino E, Valls J, Iniesta R, Moreno V. SNPStats: a web tool for the analysis of association studies. Bioinformatics. 2006;22(15):1928–9.

38. Benjamini Y, Hochberg Y. Controlling the false discovery rate: a practical and powerful approach to multiple testing. J R Stat Soc. 1995;57(1):289–300.

39. Gaunt TR, Rodríguez S, Day IN. Cubic exact solutions for the estimation of pairwise haplotype frequencies: implications for linkage disequilibrium analyses and a web tool'CubeX. BMC Bioinformatics. 2007;8(1):428.

40. Bell RD, Shultz SJ, Wideman L, Henrich VC. Collagen gene variants previously associated with anterior cruciate ligament injury risk are also associated with joint laxity. Sports Health: A Multidisciplinary Approach. 2012;4(4):312–8.

41. Uhorchak JM, Scoville CR, Williams GN, Arciero RA, Pierre PS, Taylor DC. Risk factors associated with noncontact injury of the anterior cruciate ligament a prospective four-year evaluation of 859 west point cadets. Am J Sports Med. 2003;31(6):831–42.

42. O'Connell K, Posthumus M, Schwellnus MP, Collins M. Collagen genes and exercise-associated muscle cramping. Clin J Sport Med. 2013;23(1):64–9.

43. Kirk E, Moore C, Chater-Diehl E, Singh S, Rice C. Human COL5A1 polymorphisms and quadriceps muscle-tendon mechanical stiffness in vivo. Ahead of print: Exp Physiol; 2016.

44. Foster BP, Morse CI, Onambele GL, Williams AG. Human COL5A1 rs12722 gene polymorphism and tendon properties in vivo in an asymptomatic population. Eur J Appl Physiol. 2014;114(7):1393–402.

45. Summers KM, Snodgrass SJ, Callister R. Predictors of calf cramping in rugby league. J Strength Cond Res. 2014;28(3):774–83.

46. Posthumus M, Schwellnus MP, Collins M. The COL5A1 gene: a novel marker of endurance running performance. Med Sci Sports Exerc. 2011;43(4):584–9.

47. Bertuzzi R, Pasqua LA, Bueno S, Lima-Silva AE, Matsuda M, Marquezini M, et al. Is the COL5A1 rs12722 gene polymorphism associated with running economy? PLoS One. 2014;9(9):e106581.

48. Pruna R, Artells R, Ribas J, Montoro B, Cos F, Muñoz C, et al. Single nucleotide polymorphisms associated with non-contact soft tissue injuries in elite professional soccer players: influence on degree of injury and recovery time. BMC Musculoskelet Dis. 2013;14(1):1.

2

Early nutritional programming affects liver transcriptome in diploid and triploid Atlantic salmon, *Salmo salar*

L. M. Vera[1*], C. Metochis[1], J. F. Taylor[1], M. Clarkson[1], K. H. Skjærven[2], H. Migaud[1] and D. R. Tocher[1]

Abstract

Background: To ensure sustainability of aquaculture, plant-based ingredients are being used in feeds to replace marine-derived products. However, plants contain secondary metabolites which can affect food intake and nutrient utilisation of fish. The application of nutritional stimuli during early development can induce long-term changes in animal physiology. Recently, we successfully used this approach to improve the utilisation of plant-based diets in diploid and triploid Atlantic salmon. In the present study we explored the molecular mechanisms occurring in the liver of salmon when challenged with a plant-based diet in order to determine the metabolic processes affected, and the effect of ploidy.

Results: Microarray analysis revealed that nutritional history had a major impact on the expression of genes. Key pathways of intermediary metabolism were up-regulated, including oxidative phosphorylation, pyruvate metabolism, TCA cycle, glycolysis and fatty acid metabolism. Other differentially expressed pathways affected by diet included protein processing in endoplasmic reticulum, RNA transport, endocytosis and purine metabolism. The interaction between diet and ploidy also had an effect on the hepatic transcriptome of salmon. The biological pathways with the highest number of genes affected by this interaction were related to gene transcription and translation, and cell processes such as proliferation, differentiation, communication and membrane trafficking.

Conclusions: The present study revealed that nutritional programming induced changes in a large number of metabolic processes in Atlantic salmon, which may be associated with the improved fish performance and nutrient utilisation demonstrated previously. In addition, differences between diploid and triploid salmon were found, supporting recent data that indicate nutritional requirements of triploid salmon may differ from those of their diploid counterparts.

Keywords: Atlantic salmon, nutritional programming, aquaculture, microarray, liver transcriptome, plant-based feeds

Background

Aquaculture is the fastest growing animal food-production sector, providing around 50% of fish and seafood consumed by humans worldwide [1]. However, to achieve sustainability of the aquaculture industry, the amount of fish meal (FM) and fish oil (FO) in traditional feed formulations must to be significantly reduced by alternative ingredients, such as those of plant origin. During the last decade or so, a considerable number of scientific studies have focused on the effects of these alternative plant-based ingredients on fish physiology, showing that high levels of replacement

may have negative effects on fish growth, health, welfare or disease resistance [2]. Plant proteins such as soybean meal (SBM), soybean protein concentrate (SPC), corn gluten, sunflower meal and pea protein concentrate (PPC) are increasingly being used in commercial fish feeds. However, plants contain various secondary metabolites that can act as anti-nutritional factors (ANFs) which can reduce feed intake and nutrient utilisation of vegetable-based feeds by fish [3]. These ANFs include lectins, saponins, phytic acid and proteinase inhibitors amongst others. In addition, high levels of plant proteins can also affect the bioavailability of amino acids, minerals or vitamins [4]. On the other hand, replacement of FO with vegetable oils (VO) reduces the content of long-chain polyunsaturated fatty acids (LC-PUFA) in the feeds. All VOs lack the n-3 LC-PUFAs,

* Correspondence: luisa.veraandujar@stir.ac.uk; vera.andujar@gmail.com
[1]Institute of Aquaculture, Faculty of Natural Sciences, University of Stirling, FK94LA, Stirling, Scotland, UK
Full list of author information is available at the end of the article

namely eicosapentaenoic acid (EPA; 20:5n-3) and docosahexaenoic acid (DHA, 22:6n-3), which are essential for marine fish due to their limited ability to synthesise these from their C_{18} PUFA precursors [5]. In humans, EPA and DHA are essential for normal growth and development and play an important role in nervous system, as well as preventing cardiovascular and inflammatory diseases [6]. Therefore, to avoid compromising the nutritional value of aquaculture products, it is crucial to ensure an adequate level of n-3 LC-PUFA in the fish fillet.

In mammals, the concept and application of nutritional programming has been widely studied and reported. The basis of this concept is that nutritional changes during critical periods in early development can permanently induce changes in animal metabolism and physiology, as a result of adaptive changes at the cellular, molecular and biochemical levels [7, 8]. Early nutritional cues during plastic developmental windows prepare the animal to better adapt to the same nutritional environment when it corresponds with that experienced during early stages of their life cycle [9]. The underlying mechanisms responsible for this imprinting include changes in gene expression and also permanent epigenetic changes in DNA methylation or histone modifications at regulatory regions of key genes involved in nutrient-sensitive pathways [10, 11]. In fish, a few studies have explored the potential of nutritional intervention to improve the acceptance of plant-based diets and the ability of fish to utilise them [12–19]. Recently, Geurden et al. [20] reported that early exposure of rainbow trout (*Onchorhynchus mykiss*) to a plant-based diet improved growth and feed utilisation when these fish were fed the same diet later in life. Furthermore, these positive effects were accompanied by modifications in brain and liver transcriptome that included changes in hepatic pathways mediating intermediary and xenobiotic metabolism, proteolysis and cytoskeletal regulation of cell cycle [21]. Other research in fish species has shown that nutritional programming can have impacts on PUFA metabolism in European seabass (*Dicentrarchus labrax*) [12], carbohydrate utilisation in zebrafish (*Danio rerio*) [13, 14], Siberian sturgeon (*Acipenser baerii*) [15] and gilthead seabream (*Sparus aurata*) [16], acceptance of plant-derived protein in zebrafish [17], muscle catabolic capacities in rainbow trout [18], and gut health of seabass fed a plant-based diet in which FM and FO were partially replaced with terrestrial plant meals and VO [19]. In addition, early nutritional programming has been showed to be effective in improving performance of the progeny through conditioning the broodstock to low FM and FO [22].

We recently investigated the concept of nutritional programming in Atlantic salmon (*Salmo salar*) as a strategy to improve the acceptance and utilisation of plant-based feeds. Specifically, an early nutritional stimulus was applied for 3 weeks from first feeding. Then, when juveniles were fed the same plant-based diet 4 months later, fish that had received the nutritional stimulus during early development showed increased growth performance and feed conversion, in comparison with salmon that had been fed exclusively a marine diet throughout their life [23]. However, the molecular mechanisms underlying these differences remained unclear. Nutrigenomics is a recent discipline that is being increasingly applied to provide greater understanding of the molecular actions of nutrients and other dietary components. Very often, nutrigenomic investigations involve studies in which differences between alternative dietary conditions are explored using genome-wide analyses at transcriptomic, proteomic, metabolomic and/or epigenomic levels [24]. In fish nutrition, these studies have provided new insights into the molecular pathways affected by the replacement of marine FM and FO with plant-based ingredients [25–28].

In the present study, a microarray analysis approach was used to explore the effects of nutritional programming on molecular pathways in the liver of Atlantic salmon challenged with a plant-based diet, in order to determine the metabolic processes affected and, therefore, potentially responsible for the differences in growth performance and nutrient utilisation reported previously [23].

Methods
Nutritional trial and sampling
Eggs and milt from unrelated Atlantic salmon two sea-winter broodstock (20 dams and 5 sires) (Landcatch Natural Selection Ltd., Ormsary, Scotland) were collected and transferred to the Institute of Aquaculture (University of Stirling, Scotland) where the feeding trial took place in the temperate freshwater recirculation facility. Eggs were divided into two groups (~1680 eggs each) and triploidy induced in one subgroup according to Johnstone and Stet [29]. Towards the end of the alevin stage (~ 950 degree days, dd), fish were transferred to 0.3 m^2 tanks under 24 h light. Water temperature, initially at 5.6 ± 0.1 °C, was increased over 11 days prior to first feeding and maintained at 12.7 ± 0.5 °C for the duration of the feeding trial.

Three diets formulated and manufactured by BioMar Ltd. (TechCentre, Brande, Denmark) were used in the present study (Table 1). The marine diet (Diet M) was a commercial-like formulation containing FM (80%) as protein source and FO (4%) as the added lipid. The vegetable-based diet (Diets V1 and V2) used during the nutritional stimulus and subsequent challenge phases contained only a low proportion of FM (10%) and a mixture of plant proteins (SPC, PPC and wheat gluten) as the protein sources, whilst rapeseed oil (RO) was the

Table 1 Formulations of the experimental diets

	Stimulus phase		Challenge phase
	M	V1	V2
Feed ingredients (%)			
Fish meal	64.8	5.0	5.0
Crustacean and fish peptones	14.6	5.0	5.0
Soya protein concentrate	0.0	16.4	9.0
Wheat gluten	0.0	21.4	18.2
Pea protein concentrate	0.0	21.0	24.6
Wheat	13.6	14.0	13.4
Fish oil	4.0	0.0	0.0
Rapeseed oil	0.0	6.0	17.1
Vitamins/Minerals Premix	2.3	5.5	5.2
AminoAcid Mix	0.7	5.8	2.5
Proximate composition			
Moisture (%)	8.1	7.8	7.1
Crude lipid (%)	13.3	11.3	21.6
Crude protein (%)	57.1	56.6	49.6
Ash (%)	11.5	7.6	7.5
Crude energy (MJ/kg)	20.5	20.6	22.7
Total phosphorus (%)	1.8	1.4	1.4
LC-PUFA			
n-3 LC-PUFA (%)	26.4	2.9	1.17
EPA (%)	13.0	1.4	0.57
DHA (%)	12.1	1.4	0.57

DHA docosahexaenoic acid, *EPA* eicosapentaenoic acid, *LC-PUFA* long-chain polyunsaturated fatty acids

sole added lipid source (i.e. 0% FO). During the first 3-weeks of exogenous feeding, termed the "stimulus" phase, diploid (2n) and triploid (3n) salmon were divided into triplicate groups (260 fish/tank, $n = 3$) and either fed Diet M or Diet V1 to produce four treatment groups (2 nM, 3 nM, 2 nV, 3 nV). All fish were then fed Diet M for 15-weeks (marine phase) before all groups were then

challenged with Diet V2 for 6-weeks (challenge phase) as described in detail previously [23]. At the end of the challenge phase, fish were randomly selected and euthanised with an overdose (1000 ppm) of buffered tricaine methanesulfonate (MS-222, Pharmaq, Norway), followed by a blow to the head. Samples from dissected livers were collected in RNALater® (Sigma-Aldrich, Poole, UK) for transcriptomic analysis ($n = 6$ fish/treatment group) or snap frozen in liquid nitrogen for DNA methylation analysis ($n = 3$/treatment group) and stored according to manufacturer's instructions until processing (Fig. 1).

Transcriptome analysis
Liver samples were homogenised in 1 mL of TRIzol® (Invitrogen, UK) and total RNA extracted in accordance with the manufacturer's instructions. RNA pellets were rehydrated in MilliQ water and total RNA concentration determined using an ND-1000 Nanodrop spectrophotometer (Labtech Int., East Sussex, UK). RNA integrity was assessed by agarose gel electrophoresis.

Analysis of liver transcriptome was performed using an Atlantic salmon custom-made oligoarray (ArrayExpress accession number A-MEXP-2065) with 44 k features per array in a four-array-per-slide format (Agilent Technologies UK Ltd., Wokingham, UK), described in detail previously [30]. To reduce the risk of not being able to identify between paralogues of duplicated genes, the probes used in this oligoarray were designed in the 3'-end of each sequence. A dual-label experimental design was used for the microarray hybridisations with Cy3-labelled test samples competitively hybridised to a common Cy5-labelled pooled-reference per array. A total of 24 arrays were employed, one array per individual fish ($n = 6$). The common reference was a pool of equal amounts of amplified RNA from all test samples. Indirect labelling was employed in preparing the microarray targets. Amplified antisense RNA (aRNA) was produced from each RNA sample using TargetAmpTM 1-Round Aminoallyl-aRNA Amplification Kit

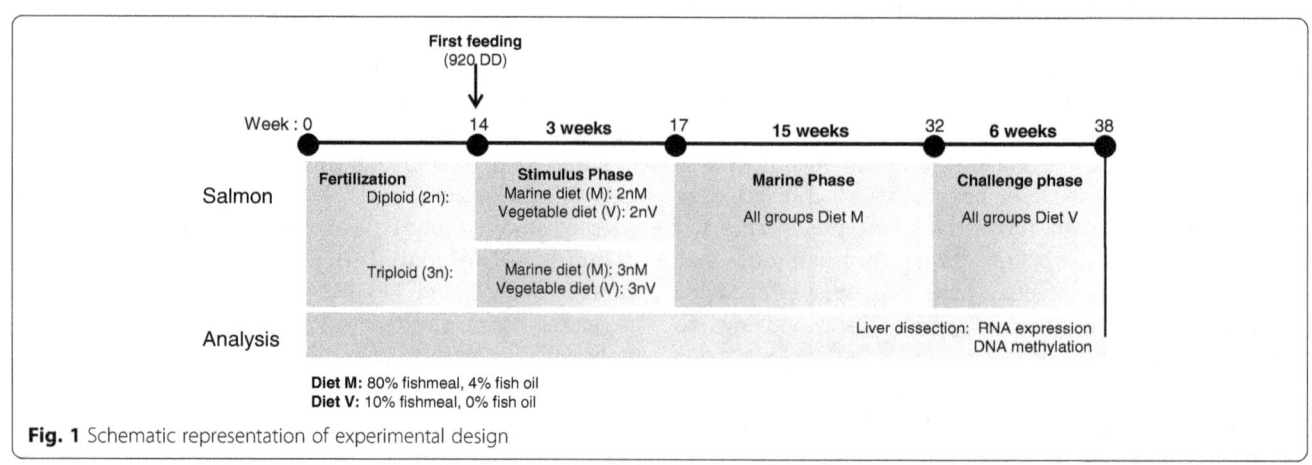

Fig. 1 Schematic representation of experimental design

101 (Epicentre, Madison, Wisconsin, USA), as per manufacturer's methodology, followed by Cy3 or Cy5 fluor (PA23001 or PA25001, GE HealthCare) incorporation through a dye-coupling reaction, as described by Betancor et al. [31]. The hybridisations were performed using SureHyb hybridisation chambers (Agilent) in a DNA Microarray Hybridisation Oven (Agilent). Sample order was semi-randomised, with one replicate per experimental group being loaded onto each slide. For each hybridisation, 825 ng of Cy3-labelled experimental biological replicate and Cy5-labelled reference pool were combined, following the protocol described by Morais et al. [32]. Details of the microarray experiment were submitted to ArrayExpress under accession number E-MTAB-5813.

RT-qPCR validation

Validation of microarray expression data was performed by reverse transcriptase RT-qPCR. The expression of 8 genes was determined in liver from fish of all treatments (Table 2). cDNA was reverse transcribed from 1 µg of total RNA using QuantiTect Reverse Transcription kit (Qiagen Ltd., Manchester, UK). The resulting cDNA was diluted 20-fold with milliQ water. Real-time PCR was performed using Luminaris Color Higreen qPCR Master mix (Thermo Fisher Scientific, MA, USA) and a Mastercycler RealPlex 2 thermocycler (Eppendorf, UK), which was programmed to perform the following protocol:

50 °C for 2 min, 95 °C for 10 min, followed by 40 cycles at 95 °C for 15 s, annealing temperature Tm for 15 s and 72 °C for 30 s. This was followed by a temperature ramp from 70 to 90 °C for melt-curve analysis to verify that no primer-dimer artefacts were present and only one product was generated from each qPCR assay. qPCR was performed in 96-well plates in duplicate. The final volume of the PCR reaction was 10 µL: 2.5 µL of cDNA, 5 µL of the qPCR Master Mix and 2.5 µL of forward and reverse primers. The efficiency of the primers was verified and validated by performing standard curves for all genes investigated. The primers used were designed using the software PRIMER3 [33]. Target specificity was checked in silico using Blast (NCBI). Only primer pairs with no unintended targets were selected. The relative expression of target genes was calculated by the ΔCt method [34] using β-actin, ef1a and rpl1 as the reference genes, which were chosen as the most stable according to RefFinder [35].

Table 2 Primers used for validation of microarray analysis by RT-qPCR

Genes	Primer sequence (5'-3')	Amplicon	Ta	Accession number	Reference
cpn2	F: AAGGATGGAGAGCTGCTGTT	154 bp	59 °C	XM_014180427.1	New design
	R: CTATCCCTCCGGTGAAGTCC				
gsta3	F: ACCTCCCTGTGTTCGAGAAG	231 bp	59 °C	NM_001140755.1	New design
	R: CGTCATCAGGTTGAGGCTTC				
hspa4	F: CTACGCTGTGGAAATCGTGG	242 bp	59 °C	XM_014136794.1	New design
	R: CACTTAACCCCTCCTCTGCA				
hspa5	F: CTACGCCTACTCGCTCAAGA	166 bp	59 °C	XM_014136127.1	New design
	R: CTCCTTCTTCTTGGCCTGGA				
tryp	F: GATACATGGACGGAGGCAGA	210 bp	59 °C	NM_001140895.2	New design
	R: ACAGGGAGGAAAGCAGCTAG				
elovl5b	F: ACAAAAAGCCATGTTTATCTGAAAGA	141 bp	60 °C	NM_001136552.1	Betancor et al. [63]
	R: AAGTGGGTCTCTCTGGGGCTGTG				
elovl6	F: ATCTGAGGAAACCGCTGGTG	177 bp	56 °C	XM_014199191.1	New design
	R: CAAAGGCGTAGGCCCAAAAC				
fads2d6a	F: GCTGGCCCATCTAGCAGAAA	119 bp	59 °C	NM_001123575.2	New design
	R: TGTCTGAGCCAAGTCACACC				
βactin	F: ATCCTGACAGAGCGCGGTTACAGT	112 bp	60 °C	AF012125	McStay et al. [64]
	R: TGCCCATCTCCTGCTCAAAGTCCA				
ef1a	F: CACCACCGGCCATCTGATCTACAA	78 bp	60 °C	DQ834870	Ytteborg et al. [65]
	R: TCAGCAGCCTCCTTCTCGAACTTC				
rpl1	F: ACTATGGCTGTCGAGAAGGTGCT	120 bp	60 °C	NM_001140826.1	Carmona-Antoñanzas et al. [66]
	R: TGTACTCGAACAGTCGTGGGTCA				

cpn2 calpain-2, gsta3 glutathione S-transferase alpha 3, hspa4 heat shock protein 4-like, hspa5 heat shock protein 5-like, tryp trypsin, elovl5b fatty acyl elongase 5 isoform b, elovl6 fatty acyl elongase 6, fads2d6a delta-6 fatty acyl desaturase isoform a, βactin β-actin, ef1a elongation factor 1 alpha, rpl1 ribosomic protein L1

DNA methylation (5-methylcytosine) level

Approximately 20 mg liver tissue of 6 fish from two different time points, after the marine and challenge phases, and of both ploidies were defrosted in ATL lysis buffer (Qiagen) and homogenised using a Precellys 24 homogeniser at 3 × 15 s at 6000 rpm with intervals of 10 s (Bertin Instruments, France). DNA extractions were performed according to the DNeasy Blood and Tissue kit (Qiagen), and the quantity of DNA measured using Qubit Fluorometric Quantification (Thermo Fisher Scientific).

DNA methylation level was measured using HPLC as described in detail previously [36]. Extracted DNA was digested to single nucleotides using DNA Degradase according to manufacturer's instructions (Zymo Research, Irvine, CA, USA). After enzymatic digestion, samples were diluted to a volume of 60 μL with the appropriate concentration of 30 ng/μL using 1xTE buffer and stored at –20 °C until HPLC analysis. A dilution curve of known adenine, guanine, cytosine, thymine, methyl-cytosine and uracil nucleotide standard mix was analysed prior to and after the experimental DNA samples. Uracil was included in the standard mix as a reference for RNA free-DNA. Chromeleon software (Thermo Fisher Scientific) was used for data processing from the HPLC results. Percentage DNA methylation was calculated using molar equivalents for both cytosine (dCMP) and methyl-cytosine (5mdCMP), where the molar equivalents were the peak areas divided by the extinction coefficients, 9300 and 11,800 for dCMP and 5mdCMP, respectively.

Statistical analysis

Microarray data were analysed in GeneSpring GX version 12.6.1 (Agilent) by two-way analysis of variance (ANOVA) with the statistical cut-off at $p < 0.05$, which tested the explanatory power of the variables "diet" and "ploidy" as well as "diet x ploidy" interaction. Data were submitted to the Kyoto Encyclopedia of Genes and Genomes (KEGG) [37] for biological function analysis. To this end, features of the array were annotated using the KEGG Automatic Annotation Server (KAAS) to obtain functional annotations, which returned a total of 60.5% of all features with a functional annotation. The significance of differences in RT-qPCR data between dietary groups was determined using a Mann-Whitney test ($p < 0.05$). Statistical analysis for DNA methylation data were conducted using two-way ANOVA, followed by Sidak's multiple comparisons test in GraphPad Prism7 (GraphPad Software, USA).

Results

In the present study, the only time the salmon were fed different diets was during the initial 3-week "stimulus" phase when fish were fed either Diet M or Diet V1. After that, the fish were fed identically and thus they were all fed the marine diet (Diet M) during the "marine" phase, and the vegetable diet (Diet V2) during the "challenge" phase. Therefore, for simplicity, the terms "V-fish" and "M-fish" are used to denote fish that were fed Diet V1 and Diet M, respectively, during the stimulus phase.

Liver transcriptome response

Two-way ANOVA of the cDNA microarray data returned a high number of differentially expressed gene features (DEG) ($p < 0.05$). A total of 3877 probes were affected by nutritional history (diet), with 3355 probes exhibiting differential expression only in response to nutritional history and 390 affected by both nutritional history and ploidy (Fig. 2). Ploidy affected the expression of 1522 probes, with 1089 of these showing differential expression exclusively in response to ploidy. Finally, there was a significant interaction effect between nutritional history and ploidy in 604 probes. The lists of genes affected by each factor (diet and ploidy) and diet x ploidy interaction were further analysed by assigning them KEGG orthology (KO) numbers and mapping them to a known database of categories (KEGG), excluding non-annotated features (38–43%). Regarding annotation, it is recognised that microarray techniques might not allow clear identification between paralogues of duplicated genes, which can be of interest in Atlantic salmon. However, to minimise this risk, the probes used in our oligoarray were designed in the 3′-end of each target sequence.

Effects of nutritional history

The Diet V1 had a profound effect on the expression of genes involved in metabolism when measured at the end of the challenge phase. The full list of annotated genes

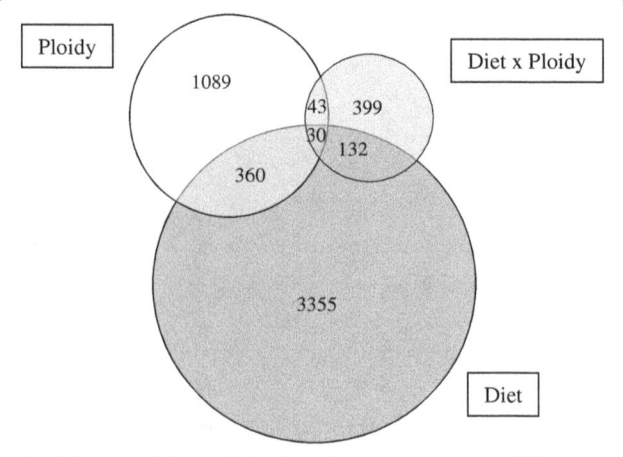

Fig. 2 Impact of diet and ploidy on liver transcriptome of Atlantic salmon at the end of the challenge. Venn diagram shows differentially expressed mRNA transcripts. The area of the circles is scaled to the number of transcripts (Two-way ANOVA, $p < 0.05$)

affected by nutritional history is presented in Additional file 1. The functional categories most affected by nutritional history were metabolism (28%) (mainly carbohydrate, amino acid and lipid metabolism) followed by signalling (15%), immune system, endocrine system and translation processes (each 7%) (Fig. 3). More than 38% of all DEG showed a fold-change (FC) above 1.5. Pathways analysis showed that the top differentially expressed pathways affected by nutritional history were oxidative phosphorylation (55 DEG), RNA transport (48 DEG), protein processing in endoplasmic reticulum (46 DEG), endocytosis (45 DEG) and purine metabolism (44 DEG) (Additional file 2). Analysis of the top 100 most significant DEG according to p-value showed highest representation of metabolism (35%), followed by translation (13%), and folding, sorting and degradation (12%) (Additional file 3). Within metabolism, the most represented categories were energy metabolism, carbohydrate metabolism and amino acid metabolism. The genes showing the highest FC within the top 100 DEG list were: *guanine deaminase* (purine metabolism), *enabled* (regulation of actin cytoskeleton), *glucosamine-fructose-6-phosphatase aminotransferase* (alanine, aspartate and glutamate metabolism), *DNA replication licensing factor MCM4* (DNA replication and cell cycle) and *ATP-binding cassette, subfamily A, member 1* (ABC transporters, involved in lipid digestion and absorption). *Guanine deaminase* (FC = +3.4) was up-regulated in salmon fed Diet V (V-fish), whereas the other 4 genes were downregulated in this group (FCs between −2.3 and −3.0).

Protein metabolism

Most genes involved in proteasome, phagosome, lysosome, endocytosis and phagocytosis pathways were up-regulated in V-fish (Additional file 2). Oxidative phosphorylation and endocytosis were also up-regulated in V-fish (87% and 71%, respectively), whereas protein processing in endoplasmic reticulum and RNA transport were downregulated in these fish. In particular, six genes belonging to the DnaJ/Hsp40 family were down-regulated in V-fish (FCs between −1.2 and −2.6), this family of molecular chaperones being involved in protein translation, folding, unfolding, translocation and degradation.

Intermediate metabolism

KEGG pathway analysis of genes that were affected by nutritional history and belonging to the metabolism category revealed that 71% of these DEG were up-regulated in V-fish. In particular, key pathways of intermediary metabolism were up-regulated, including oxidative phosphorylation, pyruvate metabolism, TCA cycle, glycolysis and fatty acid metabolism (Fig. 4). Indeed, 90% and 89% of genes found differentially expressed in this study and involved in the biosynthesis of unsaturated fatty acids and fatty acid elongation were upregulated, respectively, including *elongation of very long chain fatty acids protein 4, elongation of very long chain fatty acids protein 5, elongation of very long chain fatty acids protein 6, acyl-CoA 6-desaturase* and *stearoyl-CoA desaturase* (Fig. 5). However, the response of genes involved in cholesterol and phospholipid efflux was less clear. Thus, the cellular

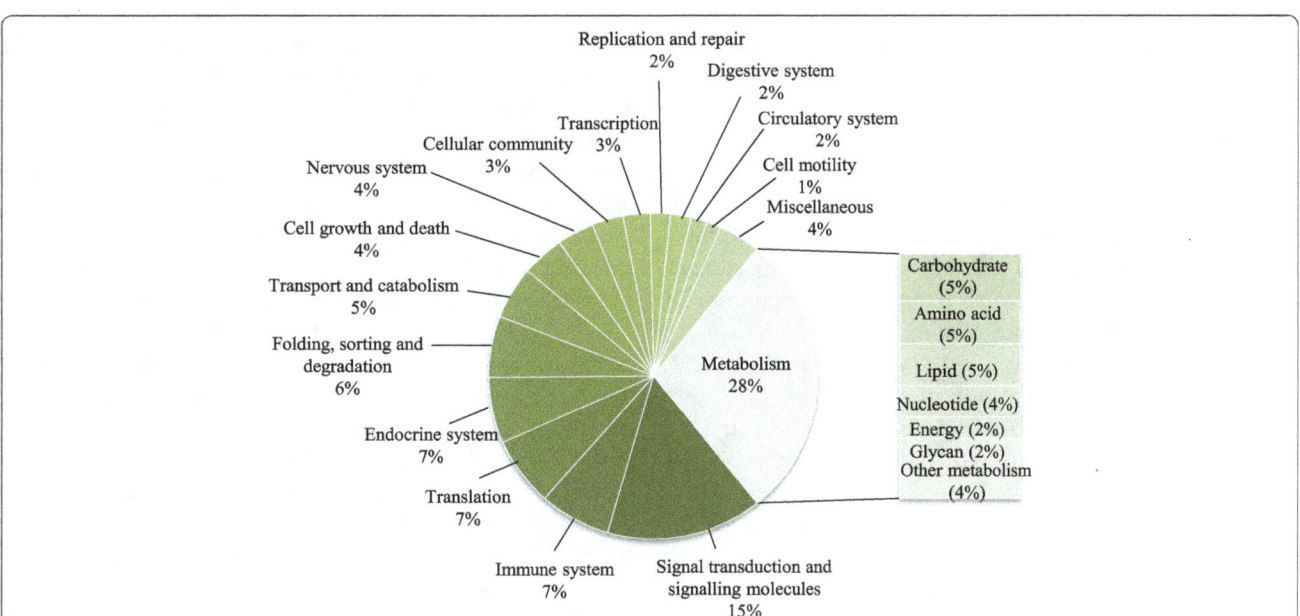

Fig. 3 Functional categories of genes differentially expressed in liver of Atlantic salmon and affected by diet. Non-annotated genes and features corresponding to the same gene are not represented

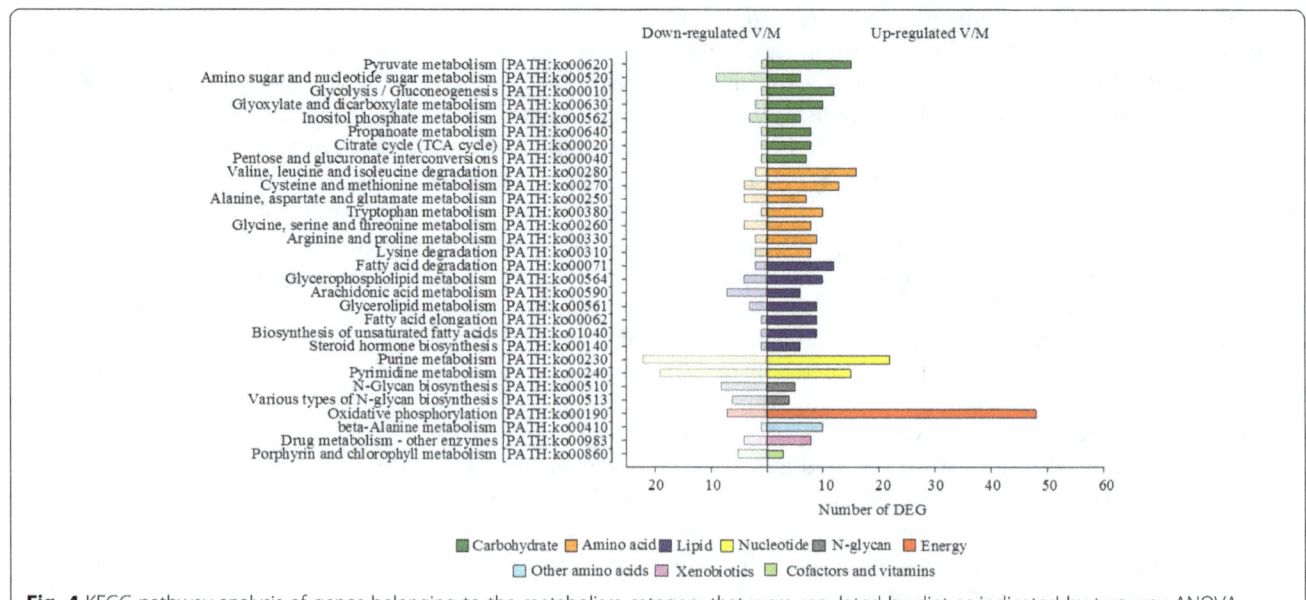

Fig. 4 KEGG pathway analysis of genes belonging to the metabolism category that were regulated by diet as indicated by two-way ANOVA analysis. Bars represent number of up- and down-regulated genes in fish fed diet V versus diet M during early development. Different colours indicate different nutrient groups

transporter *ATP-binding cassette, subfamily A, member 1* and *apolipoprotein A-IV* genes were down-regulated in V-fish, whereas *apolipropotein B* was up-regulated.

Glutathione metabolism was also up-regulated in V-fish (Additional file 1). Genes coding for enzymes involved in key reactions of this pathway that were up-regulated in V-fish including *glutathione synthase*, *glutathione peroxidase*, *glutathione-S-transferase* and *protein-disulfide reductase (glutathione)* although, in these cases, FCs were only moderate (between +1.2 and +1.6). In addition, other conjugating enzymes involved in phase II detoxification metabolism were also up-regulated (e.g. *glucuronosyltransferase*, FC = +1.5).

Immune system
Nutritional history also affected the expression of genes related to immune processes (Additional file 1). In fact, *major histocompatibility complex, class I* showed a FC of +6.2 in salmon that were fed Diet V1. Consistent with this, some genes involved in B cell and T cell receptor signalling were upregulated in V-fish including, for example, *T-cell receptor CD4*, *interferon gamma* (*ifnγ*), *interleukin-10* (*il10*), *CD81 antigen* and *CD4 antigen*. Finally, *CD59 antigen* was also up-regulated in V-fish. However, the *mannose-binding lectin* (*mbl*) gene, involved in innate immunity, was down-regulated and an inhibitor of the complement cascade was up-regulated (*complement component 4 binding protein, alpha*). Finally, the *B-cell linker* gene was also up-regulated in M-fish.

Cell cycle
The cell cycle pathway was generally down-regulated in V-fish whereas four different caspase genes inducing

apoptosis and inflammation processes were up-regulated (*caspase 1*, *caspase 3*, *caspase 7* and *caspase 9*) (Additional file 1). In addition, genes involved in mRNA surveillance, spliceosome, RNA degradation pathways, ribosome biogenesis and post-translational repair and/or degradation of misfolded proteins were down-regulated in the V-fish.

Ploidy effects
Microarray analysis revealed that ploidy also had an effect on the hepatic transcriptome of salmon. In particular, ploidy significantly influenced the expression of 1522 probes in total, with 1089 being affected only by this factor, whereas 28% of the gene features differentially expressed between ploidies were also affected by nutritional history (Fig. 2). The functional categories most affected by ploidy were signalling (25%), followed by metabolism (20%) (lipid, carbohydrate and amino acid metabolism), immune (9%) and endocrine (8%) responses (Additional file 4). A total of 604 probes showed a significant interaction between diet x ploidy, which corresponded to 302 annotated genes (Additional file 5). The functional categories showing most diet x ploidy interactions were metabolism (16%) (mainly amino acid and lipid) and signalling (15%), followed by endocrine system, immune system and folding, sorting and degradation (each 8%) (Fig. 6). Pathways analysis showed that the top differentially expressed pathways affected by diet x ploidy were spliceosome (11 DEG), PI3K-Akt signalling pathway (11 DEG), RNA transport (10 DEG) and apoptosis (10 DEG), followed by MAPK signalling pathway, cell cycle and platelet activation (9 DEG each).

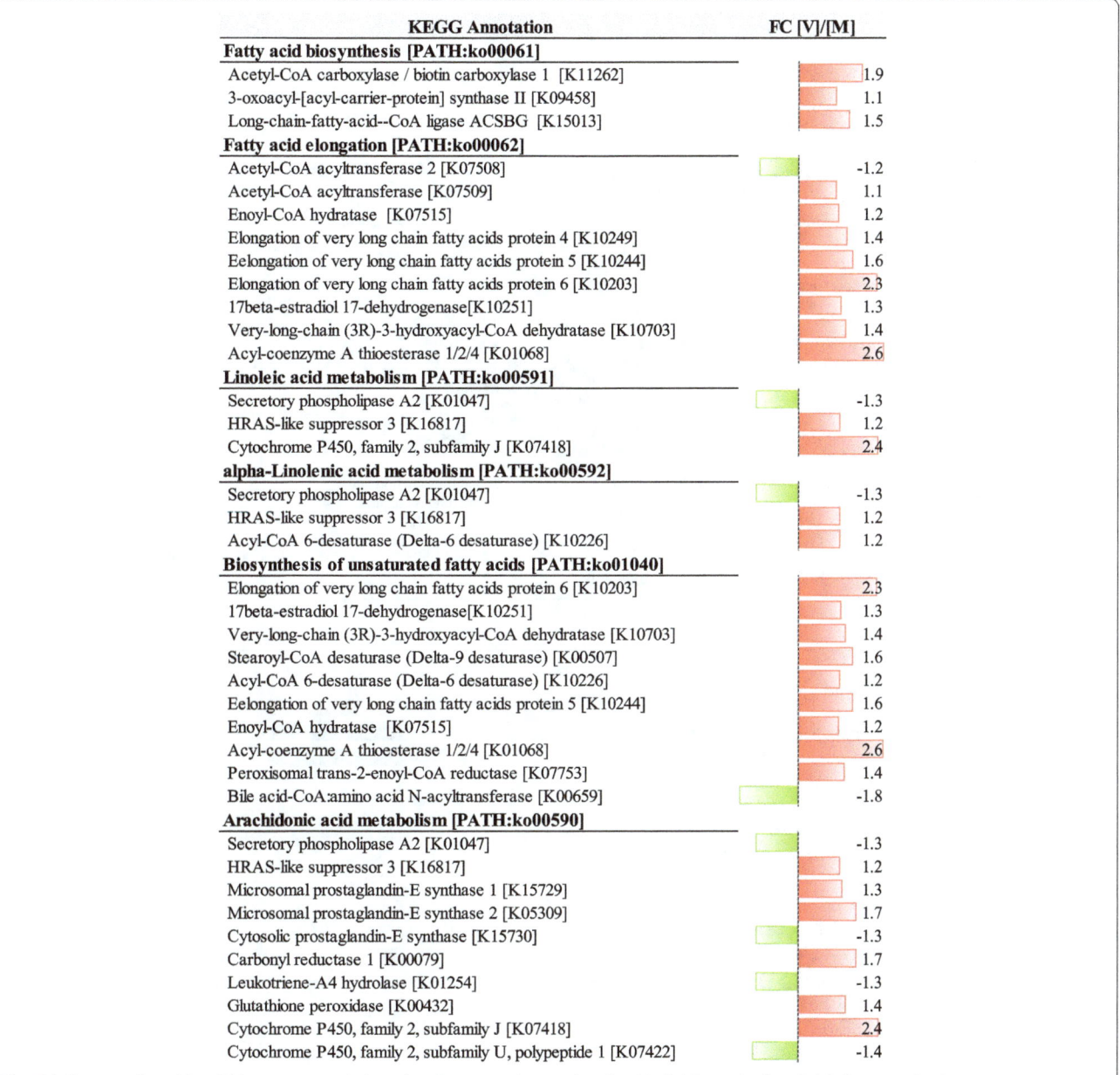

Fig. 5 Influence of nutritional history on regulation of pathways and genes involved in lipid metabolism. Fold-change ratios between gene expression in salmon fed diet V and diet M are shown (FC [V]/[M]). Pathway analysis was performed using the Kyoto Encyclopedia of Genes and Genome (KEGG)

Importantly, the effect of nutritional programming of genes involved in these pathways depended on ploidy. Thus, in diploids, 39% of these genes were up-regulated in the V-fish, whereas in triploids, 61% were up-regulated in this dietary group (Additional file 6 and Additional file 7). Within metabolism, the most represented categories were amino acid, lipid, carbohydrate and nucleotide metabolism (Fig. 7). However, in the metabolic pathways, the effect of diet x ploidy interaction was different than that observed for other functional categories. Hence, in triploids fed the V-diet during the stimulus phase, there was a general down-regulation of the genes involved in metabolism affected by the interaction of both factors (63% down-regulated). In contrast, in diploids, 52% and 48% of the DEG were up-regulated and down-regulated in V-fish, respectively. When considering only the top 100 most significant DEG according to *p* value, the genes showing the highest FCs in diploids were: *nuclear pore complex protein Nup85* (FC = +5.1 in the V-fish, RNA transport) *suppressor of cytokine signalling 1* (FC = +4.3, involved in negative regulation of cytokines and protein ubiquitination),

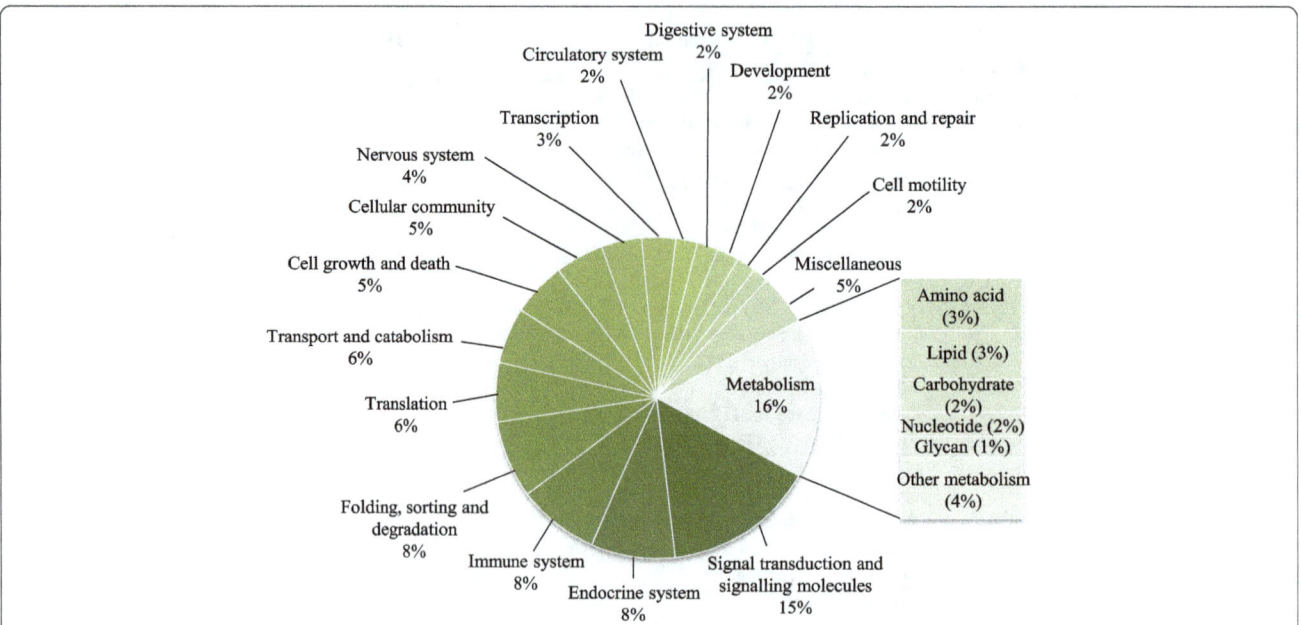

Fig. 6 Functional categories of genes differentially expressed in liver of Atlantic salmon and affected by the interaction diet x ploidy. Non-annotated genes and features corresponding to the same gene are not represented

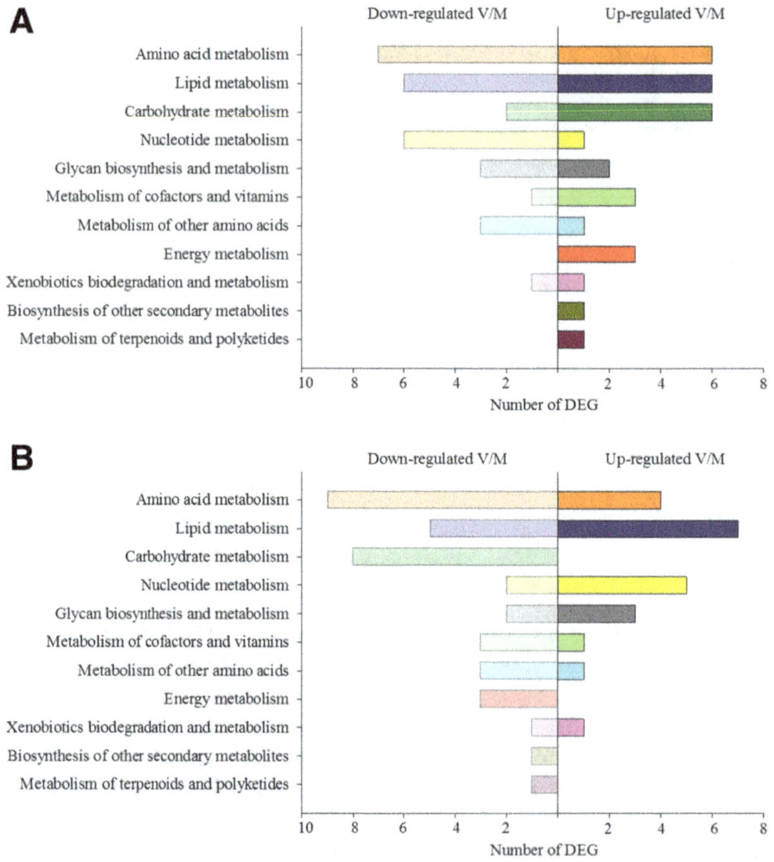

Fig. 7 Analysis of genes belonging to the metabolism category that were regulated by diet x ploidy in diploid (**a**) and triploid (**b**) salmon, as indicated by two-way ANOVA analysis. Bars represent number of up- and down-regulated genes in fish fed diet V versus diet M during the nutritional programming phase

Ca2+ transporting ATPase (FC = −4.2, calcium signalling pathway), *chondroitin sulfate proteoglycan 2* (FC = +2.9, cell adhesion) and *dynein heavy chain 1*(FC = +2.5, phagosome). In triploids, the genes showing the highest FCs were: *zinc finger CCHC domain-containing protein 9* (FC = −4.5 in the V-fish, involved in suppression of the MAPK signalling pathway), *splicing factor, arginine/serine-rich 4/5/6* (FC = −4.0, spliceosome), *chondroitin sulfate proteoglycan 2* (FC = −2.9, cell adhesion), *8-oxo-dGDP phosphatase* (FC = +2.9, involved in the hydrolysis of oxidised nucleoside diphosphate derivatives) and *beta-arrestin* (FC = +2.6, involved in sequestration of G-protein-coupled receptors).

Genetic information processing and cellular processes

The expression of a number of genes related to protein synthesis and metabolism were affected by the interaction between diet and ploidy. In particular, spliceosome and RNA transport pathways presented a higher number of up-regulated DEG in triploid V-fish (Additional file 6, Additional file 7). In addition, purine metabolism was also up-regulated in this group. Pathways involved in cell proliferation (cell cycle, MAPK signalling pathway), differentiation and survival of cells (neutrophin signalling pathway), cell communication (cAMP-dependent pathway and gap junction), cytokinesis and membrane trafficking (regulation of actin cytoskeleton) showed a higher number of up-regulated genes in the triploid V-fish (Additional file 6, Additional file 7).

Metabolism

KEGG pathway analysis of genes belonging to the metabolism category and affected by diet x ploidy interaction revealed down-regulation of a number of genes involved in amino acid, carbohydrate and energy metabolism in triploid V-fish (Fig. 7). In particular, there was down-regulation of genes involved in glycolysis, TCA cycle, pentose phosphate pathway and oxidative phosphorylation. However, fold-changes in the genes affected by diet x ploidy interaction and involved in metabolic pathways were moderate and the number of probes affected by this interaction (604) was only 11% of all the probes differentially expressed in our analysis (5408), which clearly showed that, irrespective of ploidy, diet had the stronger effect on the liver transcriptome in the present study, affecting the expression of a total of 3877 probes (Additional file 8).

DNA methylation level in liver

The mean DNA methylation level was found to be stable for both dietary groups and both ploidy with 5-methylcytosine representing a mean of 3.06 ± 0.16% of total cytosine. The global liver DNA methylation level of

diploid salmon was not affected by the different early nutritional exposure to either V- or M-diets. However, examining only triploid salmon, using a two way ANOVA (Sidak's multiple comparisons tests), revealed a significant increase in DNA methylation in both V- and M-fish after the challenge phase ($p < 0.03$) (Fig. 8).

RT-qPCR validation of array data

The expression of 8 selected genes was measured by RT-qPCR to validate the microarray results (Table 3). The candidate genes chosen represented different pathways showing differential expression between V- and M-fish. Good correlation between results obtained with both methodologies was obtained for most genes in terms of FC and direction of change (up- or down-regulated). Indeed in diploids, 100% of all genes showed the same response in both analyses, whereas in triploids 80% of genes showed the same response. For most genes, FCs were not high which likely explains why only one gene (*gsta3*) showed statistical significance after RT-qPCR analysis (Mann-Whitney, $p < 0.05$).

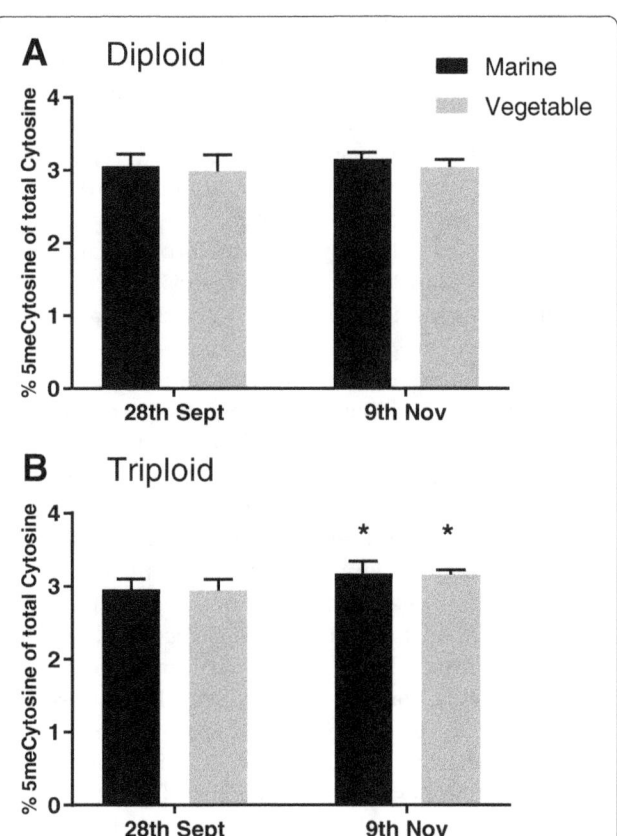

Fig. 8 Significant higher global DNA methylation levels in livers ($n = 6$) from triploid salmon after the challenge period. DNA methylation quantified as % 5mdCytosine of total cytosine measured by HPLC. Data are presented as means ± SD. Significant differences between first feed feeding groups are marked by asterisks (* = $p < 0.03$)

Table 3 Validation of microarray analysis by RT-qPCR

Genes	FC Diploids V/M			FC Triploids V/M		
	Microarray	RT-qPCR	p-value	Microarray	RT-qPCR	p-value
Calpain2 (cpn2)	+2.76	+1.78	>0.05	+1.44	−1.34	>0.05
Glutathione S-transferase alpha 3 (gsta3)	+1.27	+1.19	>0.05	+1.99	+1.74[*]	0.012
Heat shock protein 4 (hspa4)	−1.68	−1.45	>0.05	−1.47	−1.35	>0.05
Heat shock protein 5 (hspa5)	−1.68	−1.44	>0.05	−2.49	−2.69	>0.05
Trypsin (tryp)	+4.32	+3.46	>0.05	+2.39	+1.80	>0.05
Fatty acyl elongase 5 isoform b (elovl5b)	+1.84	+1.53	>0.05	+1.71	+1.23	>0.05
Fatty acyl elongase 6 (elovl6)	+1.96	+1.29	>0.05	+2.70	+1.73	>0.05
Delta-6 fatty acyl desaturase isoform a (fads2d6a)	+1.15	+1.98	>0.05	+1.29	−1.22	>0.05

Data are presented as the fold change (FC) between expression levels in diploid and triploid salmon fed the vegetable (V) diet versus the marine (M) diet. Asterisks indicate fold changes that are statistically significant (Mann-Whitney test, $p < 0.05$)

Discussion

Nutritional history

Up-regulation in V-fish of genes involved in proteasome, phagosome, lysosome, endocytosis and phagocytosis pathways suggests increased protein turnover and high dietary protein absorption and utilisation in this group. Up-regulation of protein catabolism seems to be part of a pruning process, eliminating cellular proteins that are not required by hepatic cells at that precise time [38]. Tacchi et al. [26] also reported a similar stimulation of hepatic protein metabolism in salmon fed on a low marine protein diet in which FM was partially replaced by plant-derived proteins. In fact, the up-regulation of intermediary metabolism observed in the present study may be partly explained by increased protein catabolism as this process has high energy demands [39]. However, protein processing in endoplasmic reticulum and RNA transport were downregulated in the V-fish, including the expression of molecular chaperones involved in protein translation, folding, unfolding, translocation and degradation. Higher expression of molecular chaperones in M-fish during the challenge period possibly indicated that these fish were subjected to higher stress due to their first exposure to Diet V, which suggested that nutritional intervention during early development induced physiological changes in the V-fish and increased their ability to respond to the dietary challenge at the end of the trial. Increased levels of anti-nutritional compounds present in many commercially applied plant ingredients can induce mild to severe inflammation in the intestine and hinder dietary nutrient absorption interfering with proper digestion in salmonid and marine fish [40]. Hence, anti-nutritional factors may promote transcriptional changes that characterise the so-called diet-induced stress condition [27]. In fact, up-regulation of most genes involved in the PI3K-Akt signalling pathway in M-fish also suggested the presence of dietary components promoting diet-induced stress in this group [41, 42].

In the V-fish, nutritional programming resulted in the up-regulation of most pathways involved in intermediary metabolism, including oxidative phosphorylation, pyruvate metabolism, TCA cycle, glycolysis and fatty acid metabolism. These are interconnected pathways all involved in the conversion of dietary nutrients to cellular components, thus, at least partly explaining the improved performance of V-fish during the challenge period. This is more likely associated with a metabolic adaptation and enhancement promoting improved nutrient utilisation in V-fish, rather than "compensatory" metabolism observed after periods of reduced performance, as suggested by higher feed efficiency values [23] and lower expression of *growth hormone receptor* (*ghr*) in V-fish. In addition, up-regulation of genes involved in the biosynthesis of unsaturated fatty acids and fatty acid elongation is consistent with increased biosynthesis of n-3 LC-PUFA (18:3n-3 to EPA, as well as EPA to DHA conversions) in V-fish exposed to very low levels of these fatty acids in the stimulus period compared to M-fish that received high dietary levels. Moreover, these data are in agreement with previous studies reporting enhanced *delta 6-desaturase* mRNA levels in juvenile European seabass that had been exposed to a dietary deficiency of n-3 LC-PUFA at the larval stage [43]. On the other hand, the response of genes involved in cholesterol and phospholipid efflux showed no clear trends, in accordance with findings by Balasubramanian et al. [21] in rainbow trout previously fed marine or vegetable diets after a later vegetable diet challenge.

The V-fish also showed up-regulation of genes involved in the antioxidant defence of cells, including key genes involved in glutathione metabolism, which may indicate higher oxidative stress in V-fish, possibly due to higher production of reactive oxygen species (ROS) as a consequence of the upregulation of oxidative phosphorylation [44]. This was consistent with previous research showing up-regulation of genes involved in oxidative stress response when salmon were fed diets with a high content of vegetable protein [26].

Microarray analysis also showed effects of nutritional history on the immune system of salmon. In fact, *major histocompatibility complex class I* (*mhcI*) was upregulated over 6-fold in V-fish. The main function of MHCI is to display peptide fragments of non-self-proteins, mainly generated from the degradation of cytosolic proteins by the proteasome within cells, and bind CD8 receptors expressed on cytotoxic T cells (CTLs), which triggers an immediate response of the immune system against these non-self-antigens [45]. Normal cells display peptides from endogenous cellular protein turnover on class I MHC, although CTLs are not activated in response to them due to central and peripheral tolerance mechanisms. When a cell expresses non-self-proteins, such as after infection, a fraction of the class I MHC will display these peptides on the cell surface. Consequently, CTLs specific for the MHC peptide complex will recognize and kill antigen-presenting cells [46]. Therefore, upregulation of the *mhcI* gene in V-fish is likely associated with the observed increase in protein catabolism in these salmon. Nonetheless, taking into account the role that the MHC complex has in the elimination of viral infections, the nutritional programming induced in the present study may also increase robustness of V-fish to resist viral diseases. In fact, upregulation of genes involved in B cell and T cell receptor signalling in V-fish supports this hypothesis and may suggest an increased presence of T cells having a central role in cell-mediated immunity in these fish [45, 47]. CD4 carrying T-cells are involved in several immunological processes, including maturation to plasma cells and memory B cells, and the activation of cytotoxic T cells and macrophages (T helper cells), as well as the maintenance of immunological tolerance, a state of immunological unresponsiveness to substances having the potential to elicit an immune response (T regulatory cells) [47]. The latter is particularly important for the normal physiology of salmon as it is the primary mechanism by which the immune system "learns" to discriminate self- from non-self-epitopes. Therefore, results of the present study suggested the development of immune tolerance in V-fish to the vegetable-based diet. Finally, up-regulation of *CD59 antigen* in V-fish could be an additional indication of adaptation and the development of tolerance to the vegetable-based diet, since this protein is responsible for the inhibition of the formation of the membrane attack complex (MAC) during complement activation and thus protects cells from complement-mediated lysis [48, 49]. However, downregulation of *mbl* and up-regulation of the *complement component 4 binding protein, alpha* gene, suggest that innate immunity was down-regulated in the V-fish. MBL

recognizes carbohydrate patterns, found on the surface of many pathogens including bacteria, viruses, protozoa and fungi, resulting in activation of the lectin pathway of the complement system [49, 50]. In contrast, up-regulation of *B-cell linker* gene in M-fish could suggest the induction of adaptive immune responses in these fish when exposed to Diet V for the first time during the challenge phase.

Up-regulation of caspase genes in the V-fish suggested that apoptosis may be increased in liver of these fish, probably related to the central role played by this biological process in the differentiation and maintenance of liver [51]. In addition, down-regulation of genes involved in RNA degradation, ribosome biogenesis and post-translational repair and/or degradation of misfolded proteins could indicate lower stress levels in V-fish during the challenge period, in agreement with down-regulation of molecular chaperones observed in this group.

Ploidy effects

While triploid fish were included in the study due to the increasing interest in triploidy in aquaculture, the primary aim of the present research was to investigate nutritional programming in salmon, and so our interest was in determining whether nutritional programming had different effects in diploids and triploids rather than the effect of ploidy itself. Therefore, our analysis with respect to ploidy focused on the interaction between the diet (nutritional history) and ploidy, since these data provide information that may reveal differences between diploids and triploids in response to nutritional programming that could underpin and thus help to elucidate differences in nutritional requirements between ploidy.

Several pathways involved in protein synthesis and metabolism were affected by the interaction between diet and ploidy. In general, nutritional programming with the V-diet resulted in up-regulation of protein synthesis in triploid salmon, when compared to their diploid counterparts. These differences could be related to cellular differences between ploidies. Indeed, Shrimpton et al. [52] suggested previously that triploids possess lower numbers of cells which can, in combination with altered surface to volume ratios, potentially modify signal transmission and transport of RNA, proteins and other materials. Protein degradation also seemed to be up-regulated in triploid V-fish, increasing protein turnover in order to support the higher growth of these fish observed during the final challenge phase of the feeding trial [23]. The effect of nutritional programming in key genes involved in cellular processes also differed between ploidy which also seemed to be up-regulated in the triploid V-fish. In fact, the up-regulation of the MAPK signalling pathway can be linked to the down-regulation of *zinc finger CCHC domain-containing protein 9* in this group. Altogether these data suggested enhancement of signal transduction in the triploid V-salmon, perhaps related to the larger size of triploid cells, which may affect the amount of each signalling protein required for effective signal transmission in these fish [52].

Pathway analysis revealed down-regulation of key genes involved in intermediary metabolism in triploid V-fish. These data may suggest a lower capacity of triploid salmon to generate metabolic energy when fed a vegetable-based diet. In general, triploid salmon have been reported to have increased oxygen demand in comparison with their diploid counterparts [53], as well as reduced capacity to transfer oxygen from water to tissues [54–56], which could affect the capacity to produce energy by oxidative phosphorylation [57]. Given the increasing evidence that farmed triploid salmon have higher requirements for certain nutrients, such as histidine and phosphorus, in order to support normal and healthy growth, it is likely that the use of plant-based products in triploid feeds should be combined with increased supplementation of certain nutrients [58–60].

Early nutritional exposure did not change the global DNA methylation level in liver

Global DNA methylation measures the total percentage of methylated cytosine compared to unmethylated cytosine, as previously described [61], and an increase in DNA methylation as observed in the triploid salmon after the challenge phase might be inversely correlated to gene expression levels [62]. In the present study, DNA methylation was investigated as it is known to be the most permanent epigenetic mark. However, further studies using a nucleotide specific sequencing approach, such as reduced representation bisulfite sequencing (RRBS) [11], could clarify where in the genome and which genes are regulated by DNA methylation in triploids.

Conclusions

The present study demonstrated that an early nutritional stimulus in Atlantic salmon affected the expression of genes involved in a large number of metabolic processes, suggesting the development of physiological adaptations in V-fish that enabled improved nutrient utilisation and, possibly, the ability to cope with the presence of ANFs in plant-based diets. Indeed, the increased expression of genes related to anti-inflammatory processes, apoptosis, acquired immune leukocyte receptors and regulators of essential immune responses appeared to be key for the development of immune tolerance to certain components and the avoidance and/or mitigation of autoimmune damage. The present transcriptional

data were also consistent with, and do not exclude, the possibility of programming triploid Atlantic salmon to efficiently utilise plant-based diets. Nonetheless, some differences between diploid and triploid salmon in response to nutritional programming were evident. These differences, however, might have been influenced by not only epigenetic regulatory processes, and metabolic and physiological differences between the two ploidies, but also the fact that the experimental diets were formulated according to the nutritional standards of diploid salmon while recent research suggests that nutritional requirements of triploid salmon can be different.

Additional files

Additional file 1: List of genes differentially expressed according to two-way ANOVA ($p < 0.05$) and affected by nutritional history. FC: Fold Change; KO: KEGG Orthology; M: marine V: vegetable. (XLSX 68 kb)

Additional file 2: Pathways significantly enriched based on early nutritional history, as indicated by two-way ANOVA ($p < 0.05$). Bars represent the number of up-regulated (red) and down-regulated (green) genes in salmon fed diet V versus diet M. Pathway analysis was performed using the Kyoto Encyclopedia of Genes and Genome (KEGG). DEG: Differentially Expressed Genes. (PPTX 160 kb)

Additional file 3: List of the top 100 most significant differentially expressed genes affected by diet (Two-way ANOVA, $p < 0.05$). FC: Fold Change; KO: KEGG Orthology; M: marine; V: vegetable. (XLSX 16 kb)

Additional file 4: Functional categories of genes differentially expressed in liver of Atlantic salmon and affected by ploidy. Non-annotated genes and features corresponding to the same gene are not represented (PPTX 1443 kb)

Additional file 5: List of genes differentially expressed according to two-way ANOVA ($p < 0.05$) and affected by the interaction diet x ploidy. FC: Fold Change; KO: KEGG Orthology; M: marine V: vegetable. (XLSX 30 kb)

Additional file 6: Pathways significantly enriched based on diet x ploidy, as indicated by two-way ANOVA ($p < 0.05$). Bars represent the number of up-regulated (red) and down-regulated (green) genes in diploid salmon fed diet V versus diet M. Pathway analysis was performed using the Kyoto Encyclopedia of Genes and Genome (KEGG). DEG: Differentially Expressed Genes. (PPTX 167 kb)

Additional file 7: Pathways significantly enriched based on diet x ploidy, as indicated by two-way ANOVA ($p < 0.05$). Bars represent the number of up-regulated (red) and down-regulated (green) genes in triploid salmon fed diet V versus diet M. Pathway analysis was performed using the Kyoto Encyclopedia of Genes and Genome (KEGG). DEG: Differentially Expressed Genes. (PPTX 168 kb)

Additional file 8: List of the top 100 most significant differentially expressed genes affected by diet x ploidy (Two-way ANOVA, $p < 0.05$). FC: Fold Change; KO: KEGG Orthology; M: marine; V: vegetable. (XLSX 17 kb)

Abbreviations

5mdCMP: 5-Methyl-2´-Deoxycytidine-5´-Monophosphate; ANFs: Anti-nutritional factors; aRNA: Antisense RNA; CTLs: Cytotoxic T Cells; dCMP: Deoxycytidine monophosphate; DEG: Differentially expressed genes; DHA: Docosahexaenoic acid; Diet M: Marine diet; Diet V: Vegetable diet; EPA: Eicosapentaenoic acid; FC: Fold change; FM: Fish meal; FO: Fish oil; KEGG: Kyoto encyclopedia of genes and genomes; KO: KEGG orthology; LC-PUFA: Long-Chain polyunsaturated fatty acids; MAC: Membrane attack complex; MBL: Manose-Binding Lectin; MHCl: Major histocompatibility complex I; PPC: Pea protein concentrate; RO: Rapeseed oil; ROS: Reactive oxygen species; RRBS: Reduced representation bisulfite sequencing; SBM: Soybean meal; SPC: Soybean protein concentrate; TCA: Tricarboxylic acid; VO: Vegetable oil

Acknowledgements

The authors thank Drs. Mónica Betancor and John Taggart for their assistance with microarray analysis.

Funding

This study, LMV, CM and JFT received funding from the European Union Seventh Framework Programme (FP7/2007–2013) under the grant agreement No. 288925, Advanced Research Initiatives for Nutrition & Aquaculture (ARRAINA).

Authors' contributions

Conceived and designed the experiments: DRT, JFT and HM. Performed the experiment: MC, CM, JFT and LMV. Analysed and interpreted microarray and qPCR data: LMV and CM. Analysed and interpreted DNA methylation data: KHS. Wrote the paper: LMV, KHS and DRT. Revised and edited the paper: DRT, JFT, MC, CM, HM, KHS and LMV. All authors read and approved the final version of the manuscript.

Ethics approval

All experimental procedures and husbandry practices were conducted in compliance with the Animals Scientific Procedures Act 1986 (Home Office Code of Practice) in accordance with EU regulation (EC Directive 86/609/EEC) and approved by the Animal Welfare and Ethical Review Board of the University of Stirling. All fish were monitored daily by the Named Animal Care and Welfare Officer (NACWO).

Competing interests

The authors declare that they have no competing interests.

Author details

[1]Institute of Aquaculture, Faculty of Natural Sciences, University of Stirling, FK94LA, Stirling, Scotland, UK. [2]National Institute of Nutrition and Seafood Research (NIFES), Nordnes, 5817 Bergen, Norway.

References

1. Tacon AGJ. Aquaculture: a catch for all? OECD Obs. 2010;278:32.
2. Hardy RW. Utilization of plant proteins in fish diets: effects of global demand and supplies of fishmeal. Aquaculture Res. 2010;41:770–6.
3. Francis G, Makkar HPS, Becker K. Antinutritional factors present in plant-derived alternate fish feed ingredients and their effects in fish. Aquaculture. 2001;199:197–227.
4. Kaushik S. Use of alternative protein sources for the intensive rearing of carnivorous fishes. In: Flos R, Tort L, Torres P, editors. Mediterranean Aquaculture. Chichester: Ellis Horwood Ltd; 1990. p. 125–38.
5. Tocher DR. Metabolism and functions of lipids and fatty acids in teleost fish. Rev Fish Sci. 2003;11:107–84.
6. Simopoulos AP. Essential fatty acids in health and chronic disease. Am J Clin Nutr. 1999;70:560–9.
7. Lucas A. Programming by early nutrition: an experimental approach. J Nutr. 1998;128:401S–6S.
8. Patel MS, Srinivasan M. Metabolic programming: causes and consequences. J Biol Chem. 2002;277:1629–32.
9. Gluckman PD, Hanson MA, Spencer HG. Predictive adaptive responses and human evolution. Trends Ecol Evol. 2005;20:527–33.
10. Symonds ME, Sebert SP, Hyatt MA, Budge H. Nutritional programming of the metabolic syndrome. Nat Rev Endocrinol. 2009;5:604–10.
11. Gut P, Verdin E. The nexus of chromatin regulation and intermediary metabolism. Nature. 2013;502:489–98.
12. Vagner M, Zambonino Infante JL, Robin JH, Person-Le Ruyet JI. It possible to influence European sea bass (*Dicentrarchus labrax*) juvenile metabolism by a nutritional conditioning during larval stage? Aquaculture. 2007;267:165–74.
13. Fang L, Liang XF, Zhou Y, Guo XZ, He Y, Yi TL, Liu LW, Yuan XC, Tao YX. Programming effects of high-carbohydrate feeding of larvae on adult glucose metabolism in zebrafish, *Danio rerio*. Br J Nutr. 2014;111:808–18.
14. Rocha F, Dias J, Engrola S, Gavaia P, Geurden I, Dinis MT, Panserat S. Glucose metabolism and gene expression in juvenile zebrafish (*Danio rerio*) challenged with a high carbohydrate diet: effects of an acute glucose stimulus during late embryonic life. Br J Nutr. 2015;113:403–13.
15. Gong G, Xue M, Wang J, XF W, Zheng YH, Han F, Liang XF, XO S. The regulation of gluconeogenesis in the Siberian sturgeon (*Acipenser baerii*) affected later in life by a short-term high-glucose programming during early life. Aquaculture. 2015;436:127–36.
16. Rocha F, Dias J, Geurden I, Dinis MT, Panserat S, Engrola S. High-glucose feeding of gilthead seabream (*Sparus aurata*) larvae: effects on molecular and metabolic pathways. Aquaculture. 2016;451:241–53.
17. Perera E, Yufera M. Soybean meal and soy protein concentrate in early diet elicit different nutritional programming effects on juvenile zebrafish. Zebrafish. 2016;13:61–9.
18. Panserat S, Marandel L, Geurden I, Veron V, Dias K, Plagnes-Juan E, Pegourié G, Arbenoits E, Santigosa E, Weber G, Verlhac Trichet V. Muscle catabolic capacities and global hepatic epigenome are modified in juvenile rainbow trout fed different vitamin levels at first feeding. Aquaculture. 2017;468:515–23.
19. Torrecillas S, Mompel D, Caballero MJ, Montero D, Merrifield D, Rodiles A, Robaina L, Zamorano MJ, Karalazos V, Kaushik S, Izquierdo M. Effect of fishmeal and fish oil replacement by vegetable meals and oils on gut health of European sea bass (*Dicentrarchus labrax*). Aquaculture. 2017;468:386–98.
20. Geurden I, Borchert P, Balasubramanian MN, Schrama JW, Dupont-Nivet M, Quillet E, Kaushik SJ, Panserat S, Medale F. The positive impact of the early feeding of a plant-based diet on its future acceptance and utilisation in rainbow trout. PLoS One. 2013;8(12):e83162.
21. Balasubramanian MN, Panserat S, Dupont-Nivet M, Quillet E, Montfort J, Le Cam A, Medale F, Kaushik SJ, Geurden I. Molecular pathways associated with the nutritional programming of plant-based diet acceptance in rainbow trout following an early feeding exposure. BMC Genomics. 2016;17:449.
22. Izquierdo MS, Turkmen S, Montero D, Zamorano MJ, Alfonso JM, Karalazos V, Fernández-Palacios H. Nutritional programming through broodstock diets to improve utilization of very low fishmeal and fish oil diets in gilthead sea bream. Aquaculture. 2015;449:18–26.
23. Clarkson M, Metochis C, Vera LM, Leeming D, Migaud H, Tocher DR, Taylor JF. Early nutritional intervention can improve utilisation of vegetable-based diets in diploid and triploid Atlantic salmon (Salmo salar). Br J Nutr. 2017;118(1):17–29.
24. Mathers JC. Nutrigenomics in the modern era. Proc Nutr Soc. 2016. doi:10.1017/S002966511600080X.
25. Morais S, Pratoomyot J, Taggart JB, Bron JE, Guy DR, Bell G, Tocher DR. Genotype-specific responses in Atlantic salmon (*Salmo salar*) subject to dietary fish oil replacement by vegetable oil: a liver transcriptomic analysis. BMC Genomics. 2011;12:255.
26. Tacchi L, Secombes CJ, Bickerdike R, Adler MA, Venegas C, Takle H, Martin SAM. Transcriptomic and physiological responses to fishmeal substitution with plant proteins in formulated feed in farmed Atlantic salmon (*Salmo salar*). BMC Genomics. 2012;13:363.
27. De Santis C, Crampton VO, Bicskei B, Tocher DR. Replacement of dietary soy- with air classified faba bean protein concentrate alters the hepatic transcriptome in Atlantic salmon (*Salmo salar*) parr. Comp Biochem Physiol Part D 2015;16:48-58.
28. De Santis C, Martin SAM, Dehler CE, Iannetta PPM, Leeming D, Tocher DR. Influence of dietary inclusion of a wet processed faba bean protein isolate on post-smolt Atlantic salmon (*Salmo salar*). Aquaculture. 2016;465:124–33.
29. Johnstone R, Stet RJM. The production of gynogenetic Atlantic salmon, Salmo Salar L. Theor Appl Genet. 1995;90:819–26.
30. Tacchi L, Bron JE, Taggart JB, Secombes CJ, Bickerdike R, Adler MA, Takle H, Martin SAM. Multiple tissue transcriptomic responses to Piscirickettsia salmonis in Atlantic salmon (*Salmo salar*). Physiol Genomics. 2011;43:1241-54.
31. Betancor MB, Olsen RE, Solstorm D, Skulstad OF, Tocher DR. Assessment of a land-locked Atlantic salmon (*Salmo salar* L.) population as a potential genetic resource with a focus on long-chain polyunsaturated fatty acid biosynthesis. Biochim Biophys Acta. 2016;1861:227-38.

32. Morais S, Taggart JB, Guy DR, Bell G, Tocher DR. Hepatic transcriptome analysis of inter-family variability in flesh n-3 long-chain polyunsaturated fatty acid content in Atlantic salmon. BMC Genomics. 2012;13:410.

33. Untergasser A, Cutcutache I, Koressaar T, Ye J, Faircloth BC, Remm M, Rozen SG. Primer3-new capabilities and interfaces. Nucleic Acids Res. 2012;40:e115.

34. Pfaffl MW. A new mathematical model for relative quantification in real-time RT-PCR. Nucleic Acids Res. 2001;29:e45.

35. Xie F, Xiao P, Chen D, Xu L, Zhang B. miRDeepFinder: a miRNA analysis tool for deep sequencing of plant small RNAs. Plant Mol Biol. 2012;80:75-84.

36. Skjaerven KH, Hamre K, Penglase S, Finn RN, Olsvik PA. Thermal stress alters expression of genes involved in one carbon and DNA methylation pathways in Atlantic cod embryos. Comp Biochem Physiol A Mol Integr Physiol. 2014;173:17-27.

37. Kanehisa M, Goto S. KEGG: Kyoto Encyclopedia of Genes and Genomes. Nucleic Acid Res. 2000;28:27-30.

38. Lecker SH, Goldberg AL, Mitch WE. Protein degradation by the ubiquitin-proteasome pathway in normal and disease states. J Am Soc Nephrol. 2006; 17:1807-19.

39. Gottesman S, Maurizi MR, Wickner S. Regulatory subunits of energy-dependent proteases. Cell. 1997;91:435-8.

40. Krogdahl Å, Penn M, Thorsen J, Refstie S, Bakke AM. Important antinutrients in plant feedstuffs for aquaculture: an update on recent findings regarding responses in salmonids. Aquacult Res. 2010;41:333-44.

41. Kim D, Chung J. Akt: versatile mediator of cell survival and beyond. J Biochem Mol Biol. 2001;35:106-15.

42. Hers I, Vincent EE, Tavare JM. Akt signalling in health and disease. Cell Signal. 2011;23: 1515-27.

43. Vagner M, Robin JH, Zambonino-Infante JL, Tocher DR, Person-Le Ruyet J. Ontogenic effects of early feeding of sea bass (Dicentrarchus labrax) larvae with a range of dietary n-3 highly unsaturated fatty acid levels on the functioning of polyunsaturated fatty acid desaturation pathways. Br J Nutr. 2009;101:1452-62.

44. Pompella A, Visvikis A, Paolicchi A, Tata V, Casini AF. The changing faces of glutathione, a cellular protagonist. Biochem Pharmacol. 2003;66:1499-1503.

45. Janeway CA, Travers PJr, Walport M, Shlomchik MJ. Immunobiology. 5th ed. New York: Garland Science Inc.; 2001.

46. Nakanishi T, Fischer U, Dijkstra JM, Hasegawa S, Somamoto T, Okamoto N, Ototake M. Cytotoxic T cell function in fish. Dev Comp Immunol. 2002;26: 131-9.

47. Guyton CG, Hall JE. Textbook of Medical Physiology. 11th ed. Philadelphia: Elsevier Saunders Inc.; 2010.

48. Maio M, Brasoveanu LI, Coral S, Sigalotti L, Lamaj E, Gasparollo A, Visintin A, Altomonte M, Fonsatti E. Structure, distribution, and functional role of protectin (CD59) in complement-susceptibility and in immunotherapy of human malignancies. Int J Oncol. 1998;13:305-18.

49. Holland MC, Lambris JD. The complement system in teleosts. Fish Shellfish Immunol. 2002;12:399-420.

50. Boshra H, Li J, Sunyer JO. Recent advances on the complement system of teleost fish. Fish Shellfish Immunol. 2006;20:239-62.

51. Schuster N, Krieglstein K. Mechanisms of TGF-β-mediated apoptosis. Cell Tissue Res. 2002;307:1-14.

52. Shrimpton JM, Sentlinger AMC, Heath JW, Devlin RH, Heath DD. 2007. Biochemical and molecular differences in diploid and triploid ocean-type chinook salmon (Oncorhynchus tshawytscha) smolts. Fish Physiol Biochem. 2007;33:259-68.

53. Atkins ME, Benfey TJ. Effect of acclimation temperature on routine metabolic rate in triploid salmonids. Comp Biochem Physiol. 2008;149A:157-61.

54. Farrell AP. 2009. Environment, antecedents and climate change: lessons from the study of temperature physiology and river migration of salmonids. J Exp Biol. 2009;212:3771-80.

55. Pörtner HO, Farrell AP. Physiology and climate change. Science. 2008;322:690-2.

56. Pörtner HO, Knust R. 2007. Climate change affects marine fishes through the oxygen limitation of thermal tolerance. Science. 2007;315:95-7.

57. Hansen TJ, Olsen RE, Stien L, Oppedal F, Torgersen T, Breck O, Remen M, Vågseth T, Fjelldal PG. Effect of water oxygen level on performance of diploid and triploid Atlantic salmon post-smolts reared at high temperature. Aquaculture. 2015;435:354-60.

58. Fjelldal PG, Hansen TJ, Lock E-J, Wargelius A, Fraser TWK, Sambraus F, El-Mowafi A, Alberktsen S, Waagbø R, Ørnsrud R. Increased dietary phosphorous prevents vertebral deformities in triploid Atlantic salmon (Salmo salar L.). Aquacult Nutr. 2016;22:72-90.

59. Taylor JF, Waagbø R, Díez-Padrisa M, Campbell P, Walton J, Hunter D, Matthew C, Migaud H. Adult triploid Atlantic salmon (Salmo salar) have higher dietary histidine requirements to prevent cataract development in seawater. Aquacult Nutr. 2015;21:18-32.

60. Smedley MA, Clokie BGJ, Migaud H, Campbell P, Walton J, Hunter D, Corrigan D, Taylor JF. Dietary phosphorus and protein supplementation enhances seawater growth and reduces severity of vertebral malformation in triploid Atlantic salmon (Salmo salar L.). Aquaculture. 2016;456:357-68.

61. Skjærven KH, Hamre K, Penglase S, Finn RN, Olsvik PA. Thermal stress alters expression of genes involced in one carbon and DNA methylation pathways in Atlantic cod embryos. Comp Biochem Phys A. 2014;173:17-27.

62. Potok ME, Nix DA, Parnell TJ, Cairns BR. Reprogramming the maternal zebrafish genome after fertilization to match the paternal methylation pattern. Cell. 2013;153:759-72.

63. Betancor MB, Howarth FJE, Glencross BD, Tocher DR. Influence of dietary docosahexaenoic acid in combination with other long-chain polyunsaturated fatty acids on expression of biosynthesis genes and phospholipid fatty acid compositions in tissues of post-smolt Atlantic salmon (Salmo salar). Comp Biochem Physiol B Biochem Mol Biol. 2014;172-173:74-89.

64. McStay E, Migaud H, Vera LM, Sánchez-Vázquez FJ, Davie A. Comparative study of pineal clock gene and AANAT2 expression in relation to melatonin synthesis in Atlantic salmon (Salmo salar) and European seabass (Dicentrarchus labrax). Comp Biochem Physiol A Mol Integr Physiol. 2014; 169:77-89.

65. Ytteborg E, Vegusdal A, Witten PE, Berge GM, Takle H, Ostbye TK, Ruyter B. Atlantic salmon (Salmo salar) muscle precursor cells differentiate into osteoblasts in vitro: Polyunsaturated fatty acids and hyperthermia influence gene expression and differentiation. BBA-Mol Cell Biol L. 2010;1801:127-37.

66 Carmona-Antoñanzas G, Taylor JF, Martinez-Rubio L, Tocher DR. Molecular mechanism of dietary phospholipid requirement of Atlantic salmon, Salmo salar, fry. Mol Cell Biol Lipids. 2015;1851(11):1428-41.

3

Independent impacts of aging on mitochondrial DNA quantity and quality in humans

Ruoyu Zhang[1] ⓘD, Yiqin Wang[1], Kaixiong Ye[2], Martin Picard[3] and Zhenglong Gu[1*]

Abstract

Background: The accumulation of mitochondrial DNA (mtDNA) mutations, and the reduction of mtDNA copy number, both disrupt mitochondrial energetics, and may contribute to aging and age-associated phenotypes. However, there are few genetic and epidemiological studies on the spectra of blood mtDNA heteroplasmies, and the distribution of mtDNA copy numbers in different age groups and their impact on age-related phenotypes. In this work, we used whole-genome sequencing data of isolated peripheral blood mononuclear cells (PBMCs) from the UK10K project to investigate in parallel mtDNA heteroplasmy and copy number in 1511 women, between 17 and 85 years old, recruited in the TwinsUK cohorts.

Results: We report a high prevalence of pathogenic mtDNA heteroplasmies in this population. We also find an increase in mtDNA heteroplasmies with age ($\beta = 0.011$, $P = 5.77e\text{-}6$), and showed that, on average, individuals aged 70-years or older had 58.5% more mtDNA heteroplasmies than those under 40-years old. Conversely, mtDNA copy number decreased by an average of 0.4 copies per year ($\beta = -0.395$, $P = 0.0097$). Multiple regression analyses also showed that age had independent effects on mtDNA copy number decrease and heteroplasmy accumulation. Finally, mtDNA copy number was positively associated with serum bicarbonate level ($P = 4.46e\text{-}5$), and inversely correlated with white blood cell count ($P = 0.0006$). Moreover, the aggregated heteroplasmy load was associated with blood apolipoprotein B level ($P = 1.33e\text{-}5$), linking the accumulation of mtDNA mutations to age-related physiological markers.

Conclusions: Our population-based study indicates that both mtDNA quality and quantity are influenced by age. An open question for the future is whether interventions that would contribute to maintain optimal mtDNA copy number and prevent the expansion of heteroplasmy could promote healthy aging.

Keywords: Aging, Heteroplasmy, mtDNA copy number, Whole genome sequencing

Background

Mitochondria play a central role in cellular energy metabolism, as well as in a range of other cellular activities, such as calcium signaling, iron homeostasis, hormone synthesis, and programmed cell death [1–3]. Mitochondria differ from all other organelles in animals in having their own DNA (mitochondrial DNA, mtDNA), which in humans encodes 37 genes: 22 tRNAs, 2 rRNAs and 13 protein subunits of the electron transport chain and Complex V/ATP synthase. Although they contribute only ~1% of the mitochondrial proteome, the 13 mtDNA-encoded proteins are nevertheless essential for mitochondrial oxidative phosphorylation and cellular energetics [4]. A single mammalian cell hosts hundreds to thousands of copies of mtDNA, which are thought to have played a critical role in the evolution of mammalian genomic complexity [5]. Because of its multi-copy nature, spontaneous mtDNA mutations often affect only a small proportion of the cell's mtDNA, a state termed *heteroplasmy*. In contrast, if all mtDNA molecules harbor a specific mutation, it is said to be in a state of *homoplasmy*. mtDNA heteroplasmy is implicated in several human diseases, in which the ratio of mutated to wild-type mtDNA is critical in determining whether a specific mutation is deleterious [2, 6, 7]. In previous studies, we demonstrated that even in healthy

* Correspondence: zg27@cornell.edu
[1]Division of Nutritional Sciences, Cornell University, Ithaca, NY 14853, USA
Full list of author information is available at the end of the article

adults, low-frequency heteroplasmies with high pathogenic potential were common [8].

In addition to mtDNA mutation burden, the number of mtDNA molecules per cell, or "mtDNA copy number", is also strictly regulated, ensuring that mitochondria can generate appropriate levels of energy and intracellular signals to maintain normal cellular functions. Altered mtDNA copy number has been shown to be involved in age-related diseases, including cancer, neurodegeneration disorders, and diabetes [9–11]. In the general population, mtDNA copy number measured in peripheral blood has also been shown to be associated with a variety of physiological phenotypes, and to be linked with aging and mortality [12, 13]. For example, higher mtDNA copy number was linked with better physical and mental health status in aged populations [12]. It has been speculated that both mtDNA heteroplasmy and copy number may contribute to the aging process, but the effects of mtDNA heteroplasmy and copy number were only discussed separately, thus remaining inconclusive in humans [14].

Aging is commonly characterized as a time-dependent progressive loss of physiological integrity, leading to impaired function and increased vulnerability to death [14]. One important factor in aging is the accumulation of DNA damage over time [15]. mtDNA has been considered a major target of aging-associated mutation accumulation, possibly because it experiences higher oxidative damages, more turnover, and has lower replication fidelity compared to nuclear DNA (nDNA) [16–18]. Mice carrying elevated mtDNA mutation burden present premature signs of aging including hair loss, kyphosis, and premature death (lifespan shortened by up to 50%) [19, 20]. In human studies, mtDNA heteroplasmy incidence increases with age [21–23], while lower mtDNA copy number has been reported in aged populations [12, 24]. Ding et al. reported an trend of increased heteroplasmies and decreased mtDNA copy number with age in their study population [25]. However, previous studies were limited in one or more ways: i) limited power in detecting low-to-medium frequency heteroplasmies in blood due to low sequencing depth; ii) relatively small sample sizes, limiting statistical power; iii) small age range; iv) whole blood as the source of DNA, which contains several sources of contaminants for mtDNA analysis; and/or v) assessing either mtDNA mutation or copy number, but not both in the same biological samples. Thus, it is largely unknown whether the impacts of age on mtDNA mutation burden and on copy number are independent from each other.

Whole genome sequencing (WGS) data allows us to study mtDNA heteroplasmy and copy number simultaneously. Previous large-scale studies of mtDNA heteroplasmy or copy number mostly used sequencing data of total genomic DNA extracted from transformed cell lines or whole blood. It is possible that during cell line transformation, both mtDNA heteroplasmy and copy number could undergo marked changes [26]. Moreover, estimating mtDNA copy number from WGS data relies on the ratio of sequencing reads for the mitochondrial and nuclear genomes extracted from the biological samples. There are numerous factors in the whole blood that can bias the estimation of mtDNA copy number. For example, platelets have high mtDNA content, but lack nuclear DNA; mtDNA from platelets therefore artificially raises the estimated mtDNA copy number from the whole blood [27, 28]. In the current study, we focused our analysis on WGS data of isolated platelet-free peripheral blood mononuclear cells (PBMCs) DNA obtained from the UK10K project TwinsUK cohort, which includes individuals ranging from 17 to 85 years of age [29]. The TwinsUK is a cohort with WGS data for more than 1500 generally healthy female individuals with phenotypic data. The resulting mtDNA-phenotypic dataset is one of the largest available in a general human population for analysis on the relationship between age and mtDNA heteroplasmy and copy number. Our analyses reveal that these two mtDNA properties are significantly correlated with age, and that these age effects were independent. Our results further indicate that mtDNA copy number and heteroplasmy load were significantly associated with age-related physiological parameters in this population, suggesting potential pathways by which age-related mtDNA alterations may impact the aging process.

Methods
Data access permission
Data used in this study was obtained from UK10K project, "UK10K Data Access Agreement" was approved by UK10K Data Access Officer.

mtDNA variation identification and haplogroup assignment
Whole genome sequencing and subsequent read mapping of the TwinsUK cohorts were accomplished by the UK10K project [29]. Briefly, DNA (1–3 µg) extracted from PBMCs was sheared to 100–1000 bp, and sheared DNA was sent to Illumina paired-end DNA library preparation. After size selection (300–500 bp insert size), the DNA library was sequenced by the Illumina HiSeq platform with paired-end read lengths of 100 bp. Sequencing reads mapping to the mitochondrial genome were extracted from indexed bam files to identify heteroplasmy in each individual. Retrieved reads were re-mapped to the combined human genome, hg19 for the nuclear genome and the revised Cambridge Reference Sequence (rCRS) for the mitochondrial genome, using

bowtie2 [30]. Read pairs with proper orientation and less than 5 mismatches were retained from the mapping results. The nuclear genome contains some regions with high similarity to part of the mtDNA (nuclear mitochondrial DNA, abbreviated as NUMTs). To minimize the effect of NUMTs for heteroplasmy calling, we further required the retained reads to be uniquely mapped to the mitochondrial genome. Filtered reads were further processed following the GATK best practice workflow, including Mark duplicates (duplicated reads were removed), Indel realignment, and Base quality score recalibration steps. Homoplasmies were identified using GATK HaplotypeCaller and GenotypeGVCFs [31]. Haplogroups were assigned using homoplasmic variants identified from each sample by HaploGrep2 [32]. To identify heteroplasmy, sequencing data for each position of the mitochondrial genome was extracted by Samtools mpileup [33], and bases were further filtered by sequencing quality ($>= 20$). Heteroplasmy was then identified with the following criteria: 1) Sequencing coverage >200. 2) Minor allele frequency $>= 2\%$. 3) Minor allele must be observed at least twice from each strand.

Potential cross-sample contamination inspection

Potential cross-sample contamination was assessed in the UK10K project's original data processing by VerifyBamID [34] and "fraction skewed hets" [29]. Potential contaminated samples were already removed from the dataset. However, to be more conservative, we further tested contamination using mtDNA sequencing data, which has better sensitivity than nuclear DNA variants based methods. We evaluated potential contamination by 2 criteria 1) if a sample had extremely high heteroplasmy number (Q3 + 1.5IQR rule); by this criterion, samples having more than 8 heteroplasmies were suspected to be contaminated. 2) We constructed two consensus mtDNA sequences for each sample, one covering the major alleles at heteroplasmic sites, the other covering minor alleles. A sample was suspected to be contaminated if these two consensus sequences belonged to different haplogroups. If a sample met both criteria, we would recognize it as contamination and remove it from further analysis.

Annotation of mtDNA variants

Heteroplasmy and homoplasmy were annotated by customized scripts. Pathogenic potential of variants was predicted using Combined Annotation-Dependent Depletion (CADD) score (version 1.3) [35]. The CADD score integrated many diverse annotations of variants, including functionality, pathogenicity, experimentally measured effects etc., into a single score, which has been shown to have better performance than other predictive methods such as Grantham, SIFT and PolyPhen. As recommended, a scaled CADD score of 15 was used to define the pathogenic mutations. To avoid the bias of an arbitrary cutoff, a series of cutoffs from 12 to 22 were also applied to evaluate the variants' pathogenicity. The disease associated mtDNA mutations were obtained from the MITOMAP database [36].

mtDNA copy number estimation

Whole genome sequencing data of the study population were retrieved from the UK10K project [29]. To estimate mtDNA copy number, we further filtered mapped reads by the following criteria: 1) Mapping quality >20. 2) Reads were not PCR duplicates. 3) Mismatches <5. We proceeded with qualified reads for subsequent calculation. Sequencing coverage of each site in the reference genome was calculated by the Samtools mpileup function [33]. The average sequencing coverage was then calculated for each autosomal DNA and mtDNA locus. mtDNA copy number of each individual was further estimated based on Eq. 1 and Eq. 2. It has been shown that NUMTs have a negligible impact on mtDNA copy number estimation by this method [37].

$$\frac{mtDNA\ average\ coverage}{autosomal\ DNA\ average\ coverage} = \frac{mtDNA\ copies}{autosomal\ DNA\ copies} \quad (1)$$

$$mtDNA\ copy\ number = \frac{2*mtDNA\ coverage}{\frac{1}{22}\sum_{i=1}^{22} autosomal\ coverage} \quad (2)$$

Association testing for mtDNA copy number and heteroplasmy

Linear regression was carried out to test the association of mtDNA heteroplasmy with age, as well as copy number with age separately. To conveniently test the independence of the effects of age on mtDNA heteroplasmy and copy number, we performed a linear regression using age as response variable, and mtDNA copy number and heteroplasmy number as independent variables in the basic model (Eq. 3). WBC count and platelet count were reported to affect mtDNA copy number [24], thus these two factors were also included as covariates in the regression model for further analysis (Eq. 4). Down sampling was carried out by randomly sampling 0.06 million mtDNA reads from each individual and identifying heteroplasmy following the same criteria. The association of mtDNA copy number with mtDNA heteroplasmy number was assessed by linear regression, with age and mean nuclear coverage included as covariates. The influence of mtDNA variants (heteroplasmy, homoplasmy, haplogroup) on mtDNA copy number was tested by linear regression, with age and mean nuclear coverage included as covariates. For homoplasmy, variants present in more than 1% individuals were tested. The significance level was adjusted for multiple testing. Homoplasmic variants with P value <2.69e-4 were considered to be significant.

$$Age \sim mtDNA\ heteroplasmy\ number + mtDNA\ copy\ number \tag{3}$$

$$Age \sim mtDNA\ heteroplasmy\ number + mtDNA\ copy\ number \\ + WBC\ count + Platelet\ count \tag{4}$$

Phenotypic associations of mtDNA copy number and heteroplasmy

We assessed the associations of 32 phenotypes directly measured from blood samples with mtDNA copy number and heteroplasmy number. The details of phenotypic data measurements can be found from the UK10K project [29]. Linear regression models were applied to test for associations, and age was included as a covariate to adjust for its effects on these phenotypes. Significance levels were adjusted by Bonferroni correction. Effects were considered as significant if P value <0.0016.

Mitochondrial heteroplasmy load and SKAT test

The Sequence Kernel Association Test (SKAT) has been shown to have high statistical power under a variety of conditions [38]. We used it to test the association of mtDNA heteroplasmic mutation load with different phenotypes. To do the association test, we first constructed a genotype matrix containing mtDNA heteroplasmy information. Assuming that we have mtDNA sequences for n individuals, and there are, in total, m unique heteroplasmic variants among those individuals, then the genotype matrix could be constructed an n x m matrix X, where the entry a_{ij} in X represents the minor allele frequency of individual i at heteroplasmic site j. To calculate a_{ij}, the number of all possible bases (A, T, G, C) with sequencing quality >20 were counted for individual i at site j. If the minor allele count exceeded 5, minor allele frequency would be calculated by dividing the minor allele count by total coverage at the given site, otherwise a_{ij} would be set to 0. We also constructed another genotype matrix X', whose entries were either 0 or 1. This matrix will only consider whether a given site is heteroplasmic (as 1) or not (as 0), regardless of the minor allele frequency. Notably, the genotype matrices only contained the non-polymorphic heteroplasmies. After constructing the genotype matrix, a linear regression model can be considered (Eq. 5):

$$Y = \alpha_0 + C\alpha + X\beta \tag{5}$$

Where C is the covariates matrix, in which we included age, mtDNA copy number, and the top 2 PCs from principle component analysis of population structure. Phenotypes were log transformed to achieve normal distributed residuals. The CADD score of each heteroplasmic variant was used as the weights. This weighting scheme could upweight heteroplasmic variates which are predicted to be more deleterious. The tests were performed using the R package SKAT [38].

Results

Mitochondrial heteroplasmy is prevalent in UK10K TwinsUK cohort

The UK10K-cohorts arm [29] provided WGS data for healthy individuals from two British cohorts of European ancestry, namely the Avon Longitudinal Study of Parents and Children (ALSPAC) [39] and TwinsUK [40]. However, the genomic DNA used in these two cohorts was different. In ALSPAC, DNA was extracted from lymphoblastoid cell lines established in vitro, while DNA in TwinsUK was extracted from isolated PBMCs. We compared the mtDNA copy number distribution between the two cohorts and observed a dramatically higher mtDNA copy number in cell line DNA (Additional file 1: Figure S1). The higher mtDNA copy number in the cell line compared to PBMCs is likely attributable to the difference in biological material, rather than a genuine cohort difference. Therefore, in the current study, we only focused on individuals from the TwinsUK cohort for which PBMC DNA was available. In the UK10K project's original study design, in order to increase the genetic diversity and decrease the sequencing costs, only one individual from each of the twin pairs recruited was sequenced at the whole-genome level. After excluding 73 individuals with potential cross-sample contamination, we retained 1511 individuals for further analysis. The average age of these studied individuals was 55.5 years (SD = 12.8 years), ranging from 17.3 years to 84.5 years.

The average sequencing coverage of the mitochondrial genome in these individuals was ~568X, allowing us to reliably identify mtDNA heteroplasmy at 2% minor allele frequency (MAF) cutoff. After applying a series of criteria to filter out low-quality heteroplasmies, we identified 1348 mtDNA heteroplasmies in 1511 individuals; the detailed heteroplasmy information is summarized in Additional file 2: Table S1. 794 (52.5%) individuals harbored at least one heteroplasmy in their mitochondrial genome (Fig. 1a). Most heteroplasmies presented at low-to-medium frequency (62.7% of heteroplasmies have MAF < 5%, Fig. 1b). The gene-length normalized distribution of heteroplasmy frequency among mtDNA loci is shown in Fig. 1c. The distribution of homoplasmy frequency was also plotted. Heteroplasmies were observed over the entire mitochondrial genome. One exception was the control region, also known as the "hypervariable region" [41], which harbored the highest occurrence (normalized by region length) of both homoplasmic and heteroplasmic variants. Other regions were relatively homogenous with a few

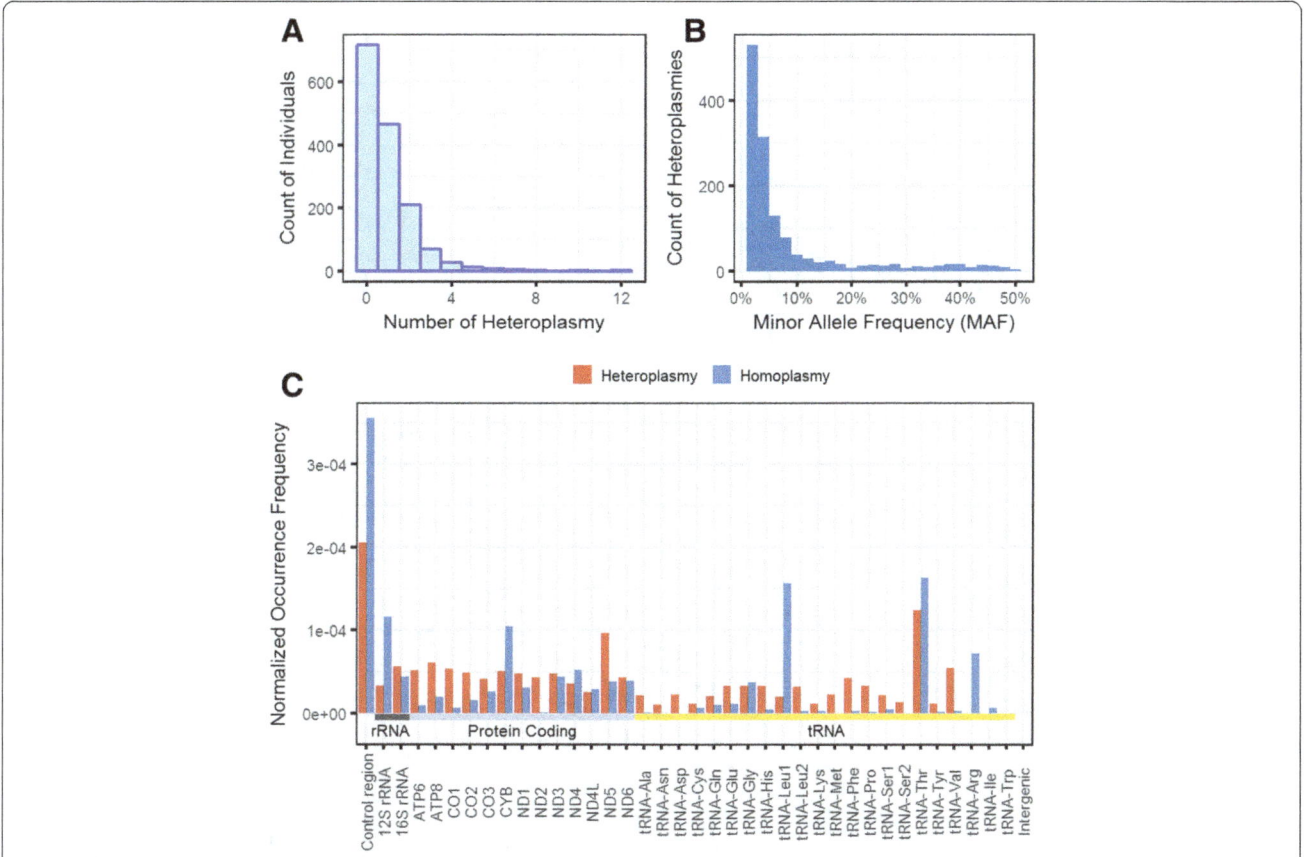

Fig. 1 Distribution of heteroplasmy in UK10K TwinsUK cohort. **a** Counts of individuals harboring a specific number of heteroplasmies (0–12 heteroplasmies at MAF > 2% cutoff). More than half of individuals (52.5%) carried at least one heteroplasmy in their genome. **b** Histogram for MAF of all heteroplasmies. 62.7% of heteroplasmies had MAF < 5% and 20.1% had MAF > 10%. **c** Normalized occurrence frequency distribution of heteroplasmies and homoplasmies. The frequency was normalized by the length of the mitochondrial loci. Dark gray, light blue and yellow bars indicate the genes in three different functional categories: rRNA, Protein coding and tRNA, respectively. The distribution of variants was relatively homogeneous among coding regions, except for some regions, such as higher frequency in ND5 (heteroplasmy) and tRNA Thr (heteroplasmy and homoplasmy)

exceptions (Additional file 1: Table S2): tRNA-Thr, which is positioned immediately upstream from the control region, had significantly higher frequencies than other tRNA genes in both heteroplasmy and homoplasmy ($P = 0.00015$ and 0.00297, respectively, Chi-squared outlier test). Interestingly, we observed that ND5 had a moderate occurrence frequency in homoplasmy, but significant higher frequency in heteroplasmy than other protein coding genes ($P = 0.00494$, Chi-squared outlier test).

Mitochondrial heteroplasmy has high pathogenic potential

Among the 1348 heteroplasmies, 192 (14.2%) were previously reported to be associated with diseases. 11.4% of individuals harboring at least one of these diseases-associated heteroplasmies. To further investigate the pathogenicity of the heteroplasmies, combined annotation dependent depletion (CADD) scores [35] were used to predict the potential pathogenicity of nonsynonymous mutations in

heteroplasmy and homoplasmy. We also annotated CADD scores for disease-associated mutations (retrieved from MITOMAP [36]) as a comparison. Disease-causing nonsynonymous mutations had a mean pathogenicity score of 17.43; in comparison, the mean CADD score of the 294 unique nonsynonymous heteroplasmic mutations was 14.32, which was significant higher than the 359 unique nonsynonymous homoplasmic mutations (10.84. $P = 3.967e-7$, Welch two sample t-test, Fig. 2a, Additional file 1: Figure S2). As suggested by the CADD score database, CADD score > 15 can be used as a cutoff to define mutations with high possibility to be pathogenic [35]. With this cutoff, the proportion of high pathogenic potential mutations in heteroplasmy was significantly higher than that in homoplasmy ($P = 0.00825$, Chi-square test). With the same criterion, 51.4% of heteroplasmies were high pathogenic potential while only 34.8% of homoplasmic mutations were high pathogenic potential. To avoid the potential bias of the arbitrary cutoff, we also applied a

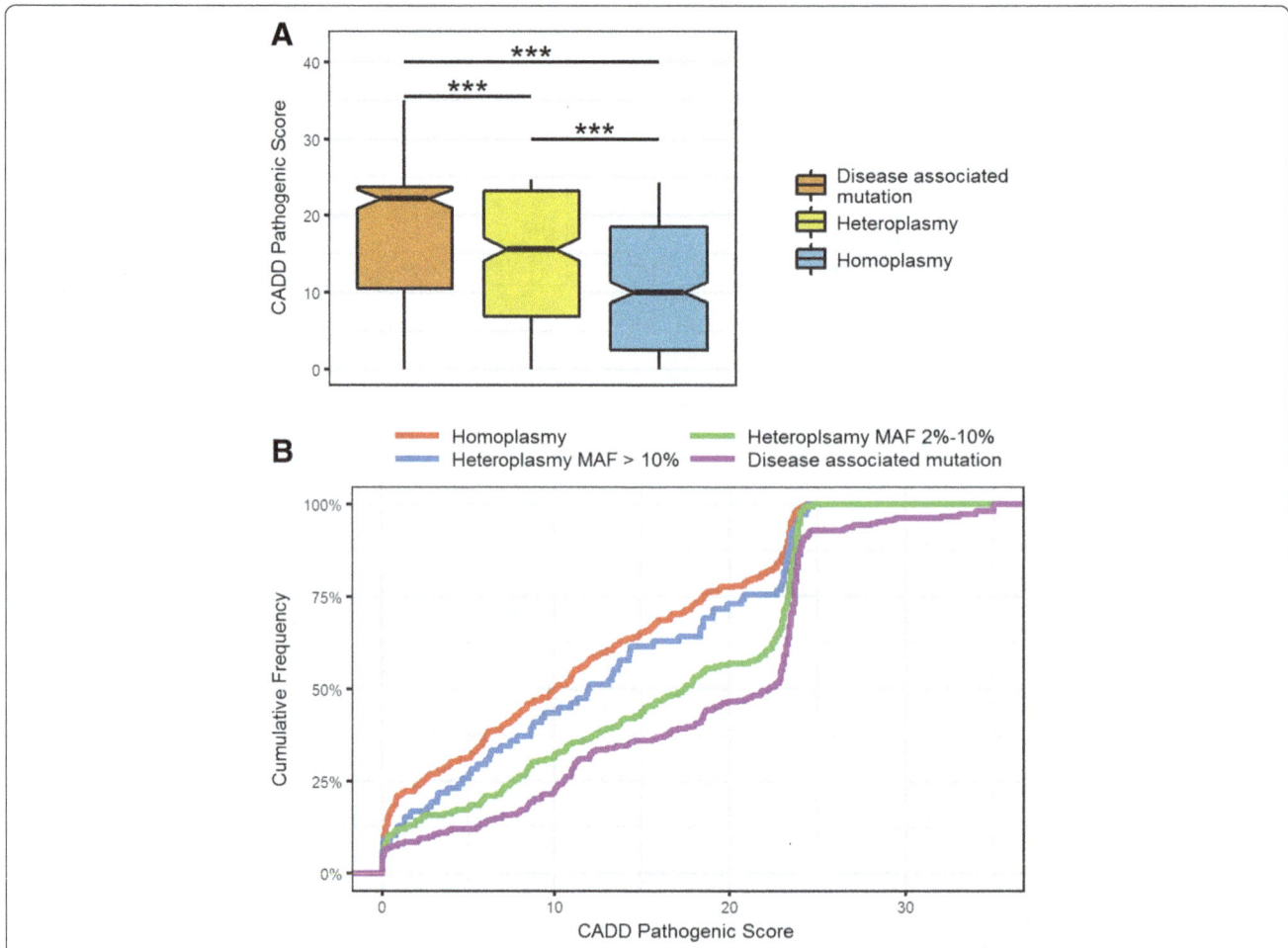

Fig. 2 Pathogenic potential for nonsynonymous heteroplasmies. **a** The box plot of CADD pathogenic score for disease associated mutations, nonsynonymous heteroplasmies and nonsynonymous homoplasmies (heteroplasmy and homoplasmy occurring in multiple individuals were counted only once). Heteroplasmies had significant higher pathogenic scores than homoplasmies ($P = 3.967e-7$) although still lower than disease associated mutations ($P < 2.2e-16$). **b** The cumulative distribution of CADD pathogenic scores of disease associated mutation, homoplasmy, low frequency heteroplasmy (MAF 2%–10%) and high frequency heteroplasmy (MAF > 10%). The distribution of low frequency heteroplasmy was close to disease associated mutations, indicating higher pathogenic potential

series of CADD score cutoffs from 12 to 22, and heteroplasmy was 1.42 to 1.94 times more likely to be high pathogenic potential than homoplasmy under different cutoffs (Additional file 1: Figure S3), consistent with the notion that more pathogenic mutations are more likely to be eliminated through purifying selection than less deleterious ones [8].

To further investigate this hypothesis, we separated nonsynonymous heteroplasmy into low frequency and high frequency groups using 10% MAF as a cutoff. The low frequency heteroplasmy group had significantly higher CADD scores than the high frequency group ($P = 0.019$, Welch two sample t test). Again, to avoid the potential bias of arbitrary cutoffs, we applied several MAF frequency cutoffs to separate low and high frequency groups (from 5% to 9%), and the difference remained significant until the cutoff was as

low as 6%. To visualize this difference, we plotted the cumulative distribution of CADD scores for each group. The distribution of CADD scores for low frequency heteroplasmies approached that of disease-associated mutations, whereas the distribution of high frequency heteroplasmies moved towards that of homoplasmic mutations (Fig. 2b, Additional file 1: Figure S4).

Mitochondrial heteroplasmy burden increases with age

The mtDNA haplogroup of each individual was identified using Haplogrep2 [32]. In this population, 48.5% (733) of the individuals belonged to H haplogroup, and there were 4 other haplogroups having more than 100 individuals: U (220), K (143), J (135) and T (125). This distribution is typical for a population of predominantly European descent. The haplogroups did not significantly contribute to the heteroplasmy variance ($P > 0.05$ for each haplogroup, Additional file 1: Figure S5), and thus were not considered in subsequent analysis.

mtDNA mutations have been thought to play an important role in aging. To investigate changes of heteroplasmies with age, we first applied linear regression and found that heteroplasmy number increased with age ($\beta = 0.011$, $P = 5.77e\text{-}6$, linear regression. Additional file 1: Figure S6). To better describe the changing heteroplasmy trend during aging, we further divided the 1511 individuals into five age groups. We observed a gradual and consistent increase of heteroplasmy number from the youngest group aged under 40-years to the oldest group aged over 70-years (Table 1). On average, individuals over 70-years old had 1.11 heteroplasmies, significantly higher than individuals under 40-years old (0.70 heteroplasmy, $P = 0.001593$, Welch two sample t test). We also separated heteroplasmy into low-to-medium MAF (2%–5%) and medium-to-high MAF (>5%) intervals, and found that this increasing trend was consistent for heteroplasmy in different MAF intervals. Individuals under 40-years old had 0.41 heteroplasmy with MAF 2%–5% and 0.29 heteroplasmy with MAF >5%, while individuals over 70-years old had 0.68 and 0.43, respectively.

Next, we evaluated the spectra of heteroplasmy in the five age groups. In all groups, heteroplasmy was predominantly present in protein coding regions, which was not surprising since protein coding sequences account for >67% of mtDNA. However, there was a tendency for the proportion of nonsynonymous heteroplasmies to increase with age. 25.9% of heteroplasmies were nonsynonymous in the under 40-years group, which increased to 28.6% in the over 70-years group, while the proportion of synonymous heteroplasmies did not significantly change (Table 2). Since nonsynonymous mutations are more likely to cause functional consequences than synonymous ones, this increased nonsynonymous proportion, together with the increased absolute heteroplasmy number in older individuals, could suggest that mtDNA integrity, or "quality" deteriorates with age.

Age has independent effects on mtDNA heteroplasmy and copy number

Because mtDNA heteroplasmy level can be affected by copies of mtDNA in blood, the age-related increase of heteroplasmy may reflect the consequences of decreased mtDNA copy number in older individuals [37]. We therefore investigated whether age acted on mtDNA heteroplasmy and copy number independently by incorporating data on heteroplasmy and copy number, simultaneously, in our analytical models with age. In normal human cells, there are two fixed copies of the nuclear genome, and therefore the ratio of average WGS sequencing coverage for mitochondrial and nuclear genomes can be used to estimate mtDNA copy number. Assuming that autosomal and mtDNA are processed and sequenced with no significant difference, average sequencing coverage should be proportional to DNA copy number for autosomal and mtDNA (Eq. 1), thus mtDNA copy number can be estimated using Eq. 2. By this method, we observed a broad range of mtDNA copy number among these individuals (Fig. 3a), from 65 to 573, with mean 169 and median 188. The distribution of mtDNA copy number was positively skewed ($P < 2.2e\text{-}16$, D'Agostino's test), with a coefficient of skewness of 1.55. Our results showed that mtDNA copy number and age were negatively correlated ($\beta = -0.395$, $P = 0.00972$, linear regression, Fig. 3b). For every 10 years, mtDNA copy number decreases about 4 copies. Similar to mtDNA heteroplasmy, mtDNA copy number was also not significantly affected by haplogroups (Additional file 1: Figure S7).

We next asked whether the effects of age on mtDNA copy number and mtDNA heteroplasmy were independent from one another. We tested this hypothesis by performing a linear regression between age and mtDNA copy number/heteroplasmy number (Eq. 3). In this regression model, both copy number and heteroplasmy number showed significant associations with age (Table 3), suggesting that age-related mtDNA copy number decrease and heteroplasmy increase are independent.

Table 1 Heteroplasmy number in different age groups

Age Group	< 40	40–50	50–60	60–70	> 70
Age Mean	30.12	45.24	55.27	64.28	74.07
Age SD	6.41	2.72	2.92	2.94	3.35
Individual count	166	267	464	447	167
Heteroplasmy count[a]	0.70; (44.6%)	0.72; (47.2%)	0.89; (55.0%)	0.98; (53.5%)	1.11; (60.0%)
Heteroplasmy count (MAF 2–5%)	0.41; (28.3%)	0.48; (32.2%)	0.56; (36.9%)	0.62; (38.7%)	0.68; (39.5%)
Heteroplasmy count (MAF > 5%)	0.29; (22.9%)	0.24; (20.6%)	0.33; (28.4%)	0.37;(27.7%)	0.43; (32.3%)

[a]numbers in parentheses indicate the proportion of individuals harboring heteroplasmy with specified MAF cutoffs. An individual can have heteroplasmies in both MAF groups

Table 2 Regional distribution of heteroplasmy in different age groups

	Control region	Intergenic region	rRNA	tRNA	Nonsynonymous	Synonymous
<40	18.1%	0.9%	12.9%	5.2%	25.9%	37.1%
40–50	24.9%	0.5%	13.0%	6.2%	25.9%	29.5%
50–60	27.3%	0.5%	9.9%	3.4%	26.8%	32.1%
60–70	22.0%	0.2%	12.0%	2.5%	28.2%	35.0%
>70	16.8%	0.0%	13.5%	4.9%	28.6%	36.2%

In addition, since DNA was extracted from blood cells, and WBC count and platelet count were reported to correlate with age and mtDNA copy number [24], we further included WBC count and platelet count as covariates (Eq. 4). These additional adjustments for possible confounding factors did not qualitatively alter the associations obtained in the basic model (Table 4). Additionally, because it was more likely to identify heteroplasmies in individuals with high mtDNA sequencing coverage, as a sanity check, we down sampled mtDNA sequencing reads in each individual to 0.06 million mtDNA reads (corresponding to mtDNA sequencing coverage of ~ 360X) and identified heteroplasmies at 2% MAF cutoff. The independent effects of age on mtDNA heteroplasmy and copy number were still significant in the down-sampled data (Additional file 1: Table S3, S4), confirming the robustness of this finding across the whole population.

Mitochondrial DNA copy number is associated with number of heteroplasmies

We next tested the correlation between mtDNA copy number and heteroplasmy. Since most heteroplasmies were unique to only one individual, especially those with high pathogenic potentials, instead of testing each single mtDNA heteroplasmy, our analysis was restricted to test the association between mtDNA copy number and the total number of heteroplasmies within an individual. With increasing heteroplasmy number, mtDNA copy number significantly decreased (Fig. 4. $\beta = -4.34$, $P = 0.007$, linear regression, adjusted for age and average nuclear DNA sequencing coverage).

We also tested whether single mtDNA homoplasmic variants were associated with copy number. We identified 186 unique homoplasmic single nucleotide variants, each presented in >1% of individuals in this study population. The associations between mtDNA copy number and these variants were tested using a linear model including age and mean nuclear DNA sequence coverage as covariates. After Bonferroni correction, none of these homoplasmic variants were significantly associated with mtDNA copy number (Additional file 1: Figure S8). Ridge et al. previously reported that 3 mtDNA variants (A9667G, T5277C and C6489A), belonging to haplogroups T2 and U5A1, were significantly associated with higher mtDNA copy number [42]. There were 26 individuals in our dataset harboring A9667G, but this variant was not associated with mtDNA copy number in our test ($P = 0.5669$). The other two variants were missing or found at a rare frequency (7 individuals) in our study, and thus were excluded from further association analysis.

Phenotypic associations of mtDNA copy number and heteroplasmy load

Age is the most significant risk factor for several diseases. It is possible that the effects of age on mtDNA

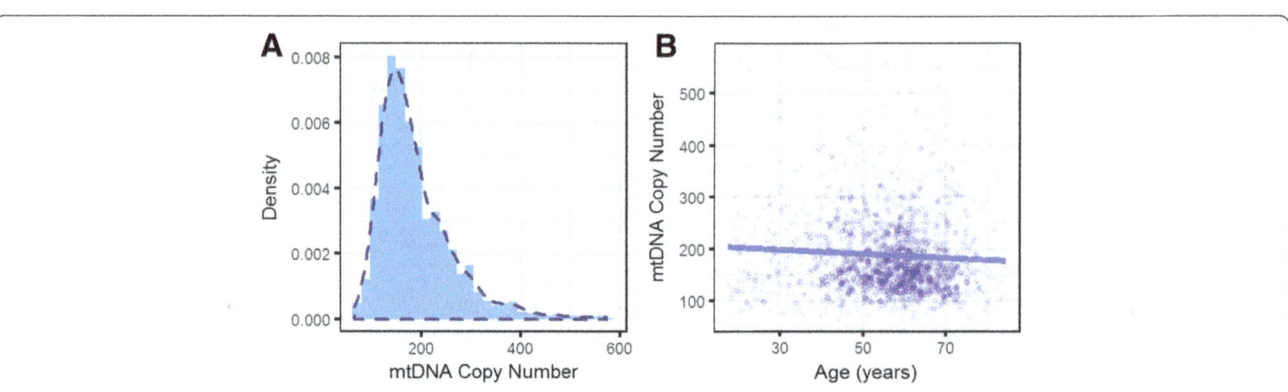

Fig. 3 Distribution of mtDNA copy number in the UK10K Twins cohort and its association with age. **a** mtDNA copy number was estimated using WGS data by comparing the mean sequencing coverage of mtDNA and nDNA. The distribution of mtDNA copy number was positively skewed, and most individuals had moderate numbers of mtDNA (mean 169 and median 188). **b** mtDNA copy number was negatively correlated with age ($\beta = -0.395$, $P = 0.00972$). Blue line represents the linear regression line. For every 10 years, mtDNA copy number decreases about 4 copies

Table 3 Correlation of age with mtDNA heteroplasmy number and copy number

Parameter	Parameter Estimate	SE	P Value
mtDNA heteroplasmy number	1.185	0.271	1.27e-5 ***
mtDNA copy number	−0.010	0.004	0.0228 *

Significance level (* $P < 0.05$, ** $P < 0.01$ and *** $P < 0.001$)

copy number and heteroplasmy could mediate these effects via their effects on physiological variables known to be perturbed in disease states and with aging. We examined the associations between 32 phenotypic traits provided by TwinsUK cohort and mtDNA copy number / heteroplasmy load (Additional file 3: Table S5). After correcting for multiple testing, mtDNA copy number was significantly associated with serum bicarbonate level ($P = 4.46e-5$, Fig. 5a) and WBC count ($P = 0.0006$, Fig. 5b).

Bicarbonate is an essential component of the pH buffering system and is indirectly related to mitochondrial oxidative reactions. In our analysis, there was a positive correlation between mtDNA copy number and serum bicarbonate level, such that, for each increase of 1 SD in mtDNA copy number (75.8 copies), serum bicarbonate level increased by 0.102 SD (0.27 mmol/L), indicating a potential interplay between the buffering system and mitochondrial activity. Conversely, WBC showed a significant negative correlation with mtDNA copy number. With each increase of 1 SD in mtDNA copy number, WBC count decreased by 0.116 SD ($0.2*10^9$ cell/L). WBC count is related to inflammation and immune senescence, so this observation indicated that mtDNA copy number could be associated with immune function.

Since the majority of the heteroplasmies were present in <1% of individuals in the samples, our ability to test their phenotypic associations were limited. Instead of performing analysis on each single heteroplasmy, we aggregated the heteroplasmic mutation information across the entire mitochondrial genome for each individual, and attempted to test the overall cumulative effects of heteroplasmy on different traits. We used the Sequence Kernel Association Test (SKAT) algorithm, which has been shown to have high statistical power under a variety of conditions [38]. Under SKAT default settings, the population frequencies of the variants were used as testing weights, since rare mutations were more susceptible to

Table 4 Correlation of age with mtDNA heteroplasmy number and copy number, adjusting for WBC and platelet counts

Parameter	Parameter Estimate	SE	P Value
mtDNA heteroplasmy number	0.901	0.252	0.00037 ***
mtDNA copy number	−0.013	0.004	0.00122 **
White blood cell count	−0.301	0.176	0.08787
Platelet count	0.009	0.005	0.08590

Significance level (* $P < 0.05$, ** $P < 0.01$ and *** $P < 0.001$)

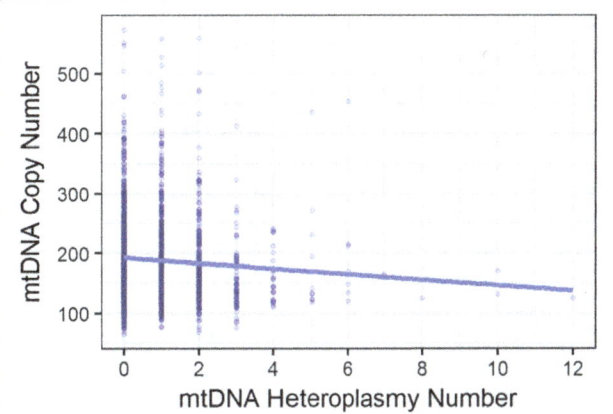

Fig. 4 Association between mtDNA heteroplasmy number and copy number. mtDNA copy number was significantly associated with the total heteroplasmy number within an individual, adjusting for age and mean nuclear sequencing coverage ($\beta = -4.34$, $P = 0.007$). Individuals harboring higher numbers of heteroplasmies were more likely to have low mtDNA copy number

being deleterious. Here, because most heteroplasmies were only found in one person, we used the predicted pathogenicity of heteroplasmy (CADD scores) as weights. We tried two different genotype matrices, one taking the heteroplasmy MAF into account, the other only considering whether a site was a heteroplasmy or not, regardless of the MAF. In both cases, after multiple test correction, we observed that mtDNA heteroplasmy load was significantly associated with blood apolipoprotein B (ApoB) level ($P = 1.33e-5$ and $3.73e-6$, respectively), but not with other phenotypes. Because ApoB is a component of the lipid transport system linked to cardiovascular disease risk [43–46], this suggested a potential link between mtDNA integrity and physiological lipid regulation.

Discussion

In this study, we first identified mtDNA heteroplasmies in 1511 generally healthy women using PBMC whole genome sequencing data from the UK10K project TwinsUK cohort, with an age range from 17 to 85 years of age. With 2% MAF cutoff, we demonstrated that more than half of the individuals (52.5%) harbor at least one heteroplasmy in their mitochondrial genome, and on average each individual had ~0.9 heteroplasmy. Both the proportion of individuals harboring heteroplasmy and the average heteroplasmy number per person (using the same MAF 2% cutoff) were lower compared to our previous heteroplasmy study using sequencing data from the 1000 genome project, which utilized lymphoblastoid cell lines as source of DNA [8]. This difference could be caused by the difference between the sources of biological material: PMBCs versus cell lines. In cell line transformation, only a small proportion of original cells are induced, hence the heteroplasmy identified in a cell line only represents the

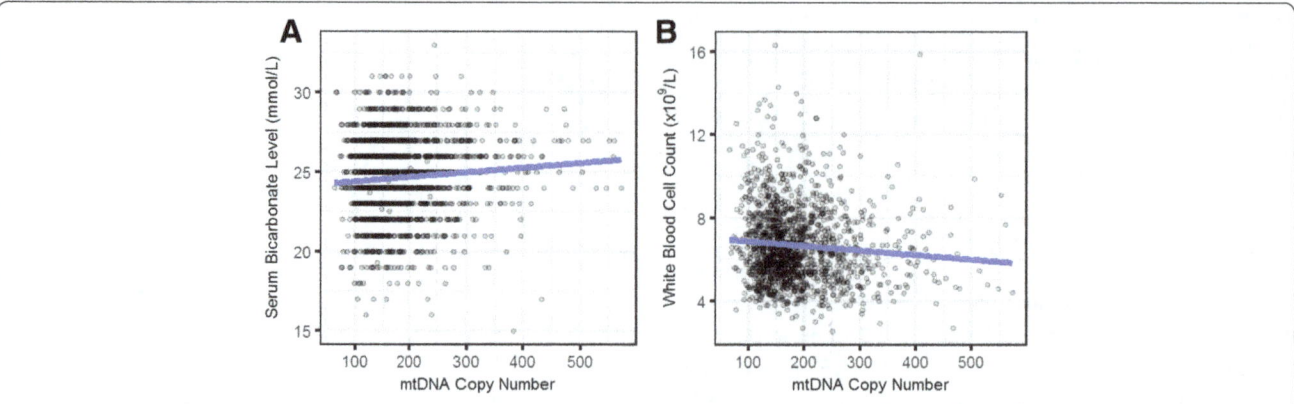

Fig. 5 mtDNA copy number association with phenotypic traits. **a** mtDNA copy number was positively associated with serum bicarbonate level ($P = 4.46e\text{-}5$). The reference range for bicarbonate level is 22–29 mmol/L (**b**) mtDNA copy number was negatively associated with WBC count ($P = 0.0006$). The reference range for WBC count is 4.0–11.0 ($\times 10^9$/L). The blue lines represent linear regression lines in each case

heteroplasmy pattern for a few cells instead of the whole cell populations [26]. Because the cell line and PBMC samples show big differences in mtDNA characteristics (Additional file 1: Figure S1), our study reinforces the notion that using DNA directly extracted from human samples is ideal for studying the impact of aging on mtDNA heteroplasmy and copy number. It is also important to point out that our analysis is limited to PBMCs, which are a mixture of different cell types. Further investigation of mtDNA characteristics in these different cell types and their relationship with aging will enable a better understanding on how mtDNA changes during aging.

We observed that heteroplasmy was not distributed uniformly across the mitochondrial genome, and several regions had enriched heteroplasmy: 1. The control region had the highest length normalized occurrence frequency; 2. Mutations located in tRNA-Thr had high occurrence frequency in both heteroplasmy and homoplasmy; 3. Notably, the normalized occurrence frequency of heteroplasmy in the ND5 gene was significantly higher than other protein coding genes, while the frequency of homoplasmy in ND5 was comparable to other genes. mtDNA mutations in the tRNA-Thr and ND5 regions have been reported to be implicated in diseases including Leigh syndrome, mitochondrial myopathy, Parkinson's disease and thyroid cancer [36]. The high occurrence of mtDNA heteroplasmy in those regions may be a potential source of future diseases, or could reflect an underlying prodromal disease state that independently promotes the accumulation of mtDNA defects.

Using CADD score as a measure of pathogenicity, we observed higher pathogenic potential of mtDNA heteroplasmy compared to homoplasmy, although still lower than disease-associated mutations. We further grouped heteroplasmies by their MAF, and found that heteroplasmies with lower MAF were more pathogenic than the ones with

higher MAF, implying that the selective pressure on highly pathogenic heteroplasmies could be stronger, which could occur during germline selection, and hence reduce those heteroplasmies to low frequency. Due to mitochondrial threshold effects [47], highly-pathogenic heteroplasmies can persist in healthy individuals at low frequency; however, once they reach high frequency, they could potentially contribute to mitochondrial dysfunction and further lead to the onset and/or progression of various age-related diseases, as previously suggested [48–50].

It has been proposed that patients with mitochondrial diseases experience a monoclonal expansion of a single deleterious mtDNA mutation (for example, 3243A > G), whereas aging is associated with a mosaic of multiple low-level mtDNA mutations accumulated during a lifetime [51]. mtDNA is replicated throughout the lifetime of an individual, independent of cell cycle. Both inherited and de novo mutations that emerged early in life could clonally expand to increase the heteroplasmy burden over time in a sub-population of cells. Several studies have reported that the amount of mtDNA mutation increases with age in several human tissues, including muscle, colon, putamen and heart [52–55]. Consistent with these reports, we observed that the heteroplasmy burden was elevated in older individuals. In the current sample, the absolute heteroplasmy number increased by 58.5% in individuals over 70-years (mean age 74.04) compared to individuals under 40-years (mean age 30.12). Meanwhile there was a trend for an increasing proportion of nonsynonymous heteroplasmy in older individuals. Given that the individuals involved in this study were generally healthy, it is possible that in aged individuals with diseases, more pronounced increases of heteroplasmy burden and pathogenicity would be observed. Future large-scale prospective studies should investigate the relationship between mtDNA heteroplasmy, disease status, and mortality.

Besides mtDNA quality, mtDNA quantity has also been suspected to be influenced by age. We estimated mtDNA copy number using WGS data, and found that mtDNA copy number was negatively correlated with age. These age-related mtDNA copy number changes were also reported in other studies. Wachsmuth et al. suggested that mtDNA copy number decreased with age in human muscle tissue [37] and Sahin et al. found a similar decrease in mice and rats in myocardial, hepatic, and hematopoietic cells [56, 57]. In measurements from whole blood, which are potentially confounded by several factors, Mengel-From et al. also reported a decline of 5.4 copies per decade of life in individuals above 48 years old [12].

However, no studies have investigated whether the effects of age on the two mitochondrial characteristics are independent, as it is possible that age can affect mtDNA copy number through age-related heteroplasmy changes or vice versa. In this study, we demonstrated that age was independently associated with mtDNA copy number and heteroplasmy. Furthermore, compared to previous studies, we also included WBC count and platelet count as covariates in the regression model to adjust for potential bias caused by blood cell contaminations. Mitochondrial biogenesis has been proposed as a marker of many age-related health outcomes or even the aging process itself [58]. Our results suggested that both mtDNA heteroplasmy and copy number should be included to establish this relationship. Mitochondrial mutations that occur early in life can clonally expand to cause mitochondrial dysfunction and further contribute to aging through a number of potential mechanisms including decreased oxidative capacity and energy production capacity, but also nuclear signaling and transcriptional dysregulation [59–63]. In addition, decreased mtDNA copy number may also lead to decreased energy production and/or decreased mitochondrial gene expression [57, 64]. Maintaining both mtDNA quality and quantity together may help to counteract or slow down the aging process.

Our data were also consistent with the idea that mtDNA copy number and heteroplasmy can influence each other. We observed a negative correlation between mtDNA copy number and total number of heteroplasmies in an individual. In mitochondrial diseases, a compensatory increase in mtDNA copy number via mitochondrial biogenesis may effectively compensate for heteroplasmic mtDNA mutations and mitochondrial dysfunction [65–67]. Thus, the observed age-related copy number decrease may result in a weaker copy number buffering effect during aging. In contrast, our results suggest that mtDNA haplogroups and homoplasmic variants were not strongly associated with mtDNA copy number. Although haplogroup T2 has been reported to be associated with higher mtDNA copy number [42], we did not observe this association in our dataset. This may be caused by different sample sizes for this specific haplogroup. Our data had 177 individuals belonging to T2 while the conclusions in the previous study [42] only included 12 individuals. Another study suggested that haplogroup J had higher copy number compared to haplogroup H [68]. However, neither Wachsmuth et al. [37] nor our data found this difference. It should be noted that our dataset only included UK females of European descent. To identify potential haplogroup-related effects on mtDNA copy number, further studies are needed to include a more ethnically-diverse range of populations, with both men and women, and larger sample sizes.

These age-related mitochondrial changes, combined with the fact that age is the main risk factor of many diseases in the population [69], further directed us to investigate mitochondrial associations with human physiological traits. After controlling for age, we found that serum bicarbonate level and WBC count were significantly associated with mtDNA copy number. The bicarbonate-carbon dioxide buffer system in blood can influence the pH gradient across the inner membrane of mitochondria, and thus may provide a link between systemic acid-base balance and regulation of mitochondrial metabolism [70, 71]. It has also been reported that reducing muscle hydrogen ion accumulation by sodium bicarbonate during running training was associated with greater improvements in both mitochondrial mass and mitochondrial respiration in rat models [72]. Our result was consistent with these reports and suggests a potential interplay between the bicarbonate buffer system and mitochondrial biogenesis with aging.

WBC count is a well-established marker for inflammation [73, 74]. Its negative correlation with mtDNA copy number indicated a potential change in mitochondrial biogenesis during the immune response. Decreased peripheral blood mtDNA copy number is observed in various diseases accompanied with inflammation, for example, COPD was associated with decreased leukocyte mtDNA copy number [75]. Decreased mtDNA copy number was also observed to be significantly associated with adverse clinical outcomes in peritoneal dialysis patients [76]. Mitochondria play an important role in inflammatory signaling; conversely, inflammation may also damage mtDNA, promoting a vicious inflammatory cycle [77]. However, because WBCs are a mixture of different immune cells, a change in the composition of different immune cells, or all immune cell types undergo similar age-related changes in mtDNA copy number, may both contribute to the decrease of mtDNA copy number detected here. Further studies are needed to elucidate this observation. The link to specific pro- and anti-inflammatory biomarkers will also be important to resolve.

Most heteroplasmic variants had very low frequency in the population, which limited our ability to test for associations. Inspired by studies on nuclear DNA rare variants [78, 79], instead of evaluating single variants, we aggregated heteroplasmic mutations across the entire mitochondrial genome as a "heteroplasmy mutation load", and tested the association between this mutation load and different healthy traits. By applying the SKAT algorithm, we found that mtDNA heteroplasmy load was significantly associated with blood ApoB level independent of age. Mitochondria play a critical role in fatty acid metabolism (eg, β-oxidation). Furthermore, ApoB is the main structural surface protein found on all beta-lipoproteins, which is important for lipid transportation. The ApoB level is predictive for atherosclerosis [80], and the onset of obesity is usually accompanied by overproduction of ApoB [81]. Our result suggests a potential interaction between mitochondrial function and ApoB metabolism. It has been reported that the suppression of the PPARα signaling pathway would result in disrupted mitochondrial integrity and upregulated hepatic *apoB* gene expression at both the transcriptional and translational level in liver [82], providing a potential mechanism for how mitochondrial dysfunction is connected with ApoB metabolism. Nonetheless, further studies are needed to elucidate this connection.

One limitation of our study is that all participates were female. Given sex differences in mtDNA copy number measured in whole blood [24, 25, 83], our findings may not be representative for both men and women. In a study of whole blood, mtDNA copy number was previously reported to be associated with waist circumference and waist-hip ratio, suggesting an association between mtDNA copy number and fat distribution and lipid metabolism [25]. In our study, we did not observe these associations, which could possibly be caused by sex differences, or by other confounding factors (platelets, cell-free DNA, or other) in previous studies compared to purified leukocytes in this study.

Conclusion

In conclusion, using WGS data from the UK10K project TwinsUK cohort, we conducted, to date, the first study addressing whether the age effects on mtDNA heteroplasmy and copy number are independent. Our analyses reveal that mtDNA copy number is inversely correlated with heteroplasmy number, and associated with serum bicarbonate level and WBC count. Moreover, heteroplasmy load is associated with blood ApoB level, suggesting future avenues for research aimed at understanding the role of mitochondrial dysfunction in human aging. Mitochondria play a central role in cellular energy metabolism and regulate a broad range of cellular activities, and alterations of mtDNA sequence integrity and copy number have been implicated in human disease. Therefore, it remains

promising to further investigate whether approaches to maintain mtDNA copy number and manage the expansion of mtDNA heteroplasmic mutations could help improve health status, especially in the elderly.

Abbreviations
ALSPAC: Avon longitudinal study of parents and children; ApoB: Apolipoprotein B; CADD Score: combined annotation-dependent depletion score; MAF: Minor allele frequency; mtDNA: Mitochondrial DNA; nDNA: Nuclear DNA; NUMT: Nuclear mitochondrial DNA; PBMC: Peripheral blood mononuclear cell; SKAT: Sequence kernel association test; WGS: Whole genome sequencing

Acknowledgments
We thank Mr. Yiping Wang, Drs. Xiaoxian Guo and Yudong Li for their discussion and comments on the manuscript. This study makes use of data generated by the UK10K Consortium, derived from samples from EGAD00001000741, EGAD00001000790, EGAD00001000740. A full list of the investigators who contributed to the generation of the data is available from http://www.uk10k.org. Funding for UK10K was provided by the Wellcome Trust under award WT091310.

Funding
This work was supported by various funds from Cornell University, National Science Foundation [MCB-1243588], and National Institute of Health [1R01AI085286], a research grant from ENN Science and Technology Development to Z. G.

Authors' contributions
RZ, YW and ZG designed the study. RZ, YW analyzed data. RZ, YW, KY, MP and ZG interpreted the data. RZ, YW, KY, MP and ZG wrote the manuscript. All authors have read and approved the manuscript.

Competing interests
The authors declare that they have no competing interests.

Author details
[1]Division of Nutritional Sciences, Cornell University, Ithaca, NY 14853, USA. [2]Department of Biological Statistics and Computational Biology, Cornell University, Ithaca, NY 14853, USA. [3]Department of Psychiatry, Division of Behavioral Medicine, Department of Neurology and Columbia Translational Neuroscience Initiative, Columbia Aging Center, Columbia University Medical Center, New York, NY 10032, USA.

References
1.   Schon EA, DiMauro S, Hirano M. Human mitochondrial DNA: roles of inherited and somatic mutations. Nat Rev Genet. 2012;13(12):878–90.
2.   Stewart JB, Chinnery PF. The dynamics of mitochondrial DNA heteroplasmy: implications for human health and disease. Nat Rev Genet. 2015;16(9):530–42.
3.   Picard M, Wallace DC, Burelle Y. The rise of mitochondria in medicine. Mitochondrion. 2016;30:105–16.

4. Calvo SE, Clauser KR, Mootha VK. MitoCarta2.0: an updated inventory of mammalian mitochondrial proteins. Nucleic Acids Res. 2016;44(D1):D1251–7.

5. Lane N, Martin W. The energetics of genome complexity. Nature. 2010; 467(7318):929–34.

6. Lightowlers RN, Chinnery PF, Turnbull DM, Howell N. Mammalian mitochondrial genetics: heredity, heteroplasmy and disease. Trends Genet. 1997;13(11):450–5.

7. Russell O, Turnbull D. Mitochondrial DNA disease—molecular insights and potential routes to a cure. Exp Cell Res. 2014;325(1):38–43.

8. Ye K, Lu J, Ma F, Keinan A, Gu Z. Extensive pathogenicity of mitochondrial heteroplasmy in healthy human individuals. Proc Natl Acad Sci. 2014; 111(29):10654–9.

9. Reznik E, Miller ML, Şenbabaoğlu Y, Riaz N, Sarungbam J, Tickoo SK, Al-Ahmadie HA, Lee W, Seshan VE, Hakimi AA, et al. Mitochondrial DNA copy number variation across human cancers. elife. 2016;5:e10769.

10. Schon EA, Manfredi G. Neuronal degeneration and mitochondrial dysfunction. J Clin Investig. 2003;111(3):303–12.

11. Kwak SH, Park KS, Lee K-U, Lee HK. Mitochondrial metabolism and diabetes. J Diab Investig. 2010;1(5):161–9.

12. Mengel-From J, Thinggaard M, Dalgård C, Kyvik KO, Christensen K, Christiansen L. Mitochondrial DNA copy number in peripheral blood cells declines with age and is associated with general health among elderly. Hum Genet. 2014;133(9):1149–59.

13. Lee JW, Park KD, Im JA, Kim MY, Lee DC. Mitochondrial DNA copy number in peripheral blood is associated with cognitive function in apparently healthy elderly women. Clin Chim Acta. 2010;411(7–8):592–6.

14. López-Otín C, Blasco MA, Partridge L, Serrano M, Kroemer G. The hallmarks of aging. Cell. 2013;153(6):1194–217.

15. Moskalev AA, Shaposhnikov MV, Plyusnina EN, Zhavoronkov A, Budovsky A, Yanai H, Fraifeld VE. The role of DNA damage and repair in aging through the prism of Koch-like criteria. Ageing Res Rev. 2013;12(2):661–84.

16. Itsara LS, Kennedy SR, Fox EJ, Yu S, Hewitt JJ, Sanchez-Contreras M, Cardozo-Pelaez F, Pallanck LJ. Oxidative stress is not a major contributor to somatic mitochondrial DNA mutations. PLoS Genet. 2014;10(2):e1003974.

17. Ballard JWO, Whitlock MC. The incomplete natural history of mitochondria. Mol Ecol. 2004;13(4):729–44.

18. Lynch M, Walsh B. The origins of genome architecture, vol. vol. 98. Sunderland: Sinauer Associates; 2007.

19. Ross JM, Coppotelli G, Hoffer BJ, Olson L. Maternally transmitted mitochondrial DNA mutations can reduce lifespan. Sci Rep. 2014;4:6569.

20. Ross JM, Stewart JB, Hagström E, Brené S, Mourier A, Coppotelli G, Freyer C, Lagouge M, Hoffer BJ, Olson L. Germline mitochondrial DNA mutations aggravate ageing and can impair brain development. Nature. 2013;501(7467):412–5.

21. Sondheimer N, Glatz CE, Tirone JE, Deardorff MA, Krieger AM, Hakonarson H. Neutral mitochondrial heteroplasmy and the influence of aging. Hum Mol Genet. 2011;20(8):1653–9.

22. Li M, Schroder R, Ni S, Madea B, Stoneking M. Extensive tissue-related and allele-related mtDNA heteroplasmy suggests positive selection for somatic mutations. Proc Natl Acad Sci U S A. 2015;112(8):2491–6.

23. Li M, Rothwell R, Vermaat M, Wachsmuth M, Schroder R, Laros JF, van Oven M, de Bakker PI, Bovenberg JA, van Duijn CM, et al. Transmission of human mtDNA heteroplasmy in the genome of the Netherlands families: support for a variable-size bottleneck. Genome Res. 2016;26(4):417–26.

24. Knez J, Winckelmans E, Plusquin M, Thijs L, Cauwenberghs N, Gu Y, Staessen JA, Nawrot TS, Kuznetsova T. Correlates of peripheral blood mitochondrial DNA content in a general population. Am J Epidemiol. 2016;183(2):138–46.

25. Ding J, Sidore C, Butler TJ, Wing MK, Qian Y, Meirelles O, Busonero F, Tsoi LC, Maschio A, Angius A, et al. Assessing mitochondrial DNA variation and copy number in lymphocytes of ~2,000 Sardinians using tailored sequencing analysis tools. PLoS Genet. 2015;11(7):e1005306.

26. Kang E, Wang X, Tippner-Hedges R, Ma H, Folmes CD, Gutierrez NM, Lee Y, Van Dyken C, Ahmed R, Li Y. Age-related accumulation of somatic mitochondrial DNA mutations in adult-derived human iPSCs. Cell Stem Cell. 2016;18(5):625–36.

27. Urata M, Koga-Wada Y, Kayamori Y, Kang D. Platelet contamination causes large variation as well as overestimation of mitochondrial DNA content of peripheral blood mononuclear cells. Ann Clin Biochem. 2008;45(5):513–4.

28. Hurtado-Roca Y, Ledesma M, Gonzalez-Lazaro M, Moreno-Loshuertos R, Fernandez-Silva P, Enriquez JA, Laclaustra M. Adjusting MtDNA quantification in whole blood for peripheral blood platelet and leukocyte counts. PLoS One. 2016;11(10):e0163770.

29. UKKC. The UK10K project identifies rare variants in health and disease. Nature. 2015;526(7571):82–90.

30. Langmead B, Salzberg SL. Fast gapped-read alignment with bowtie 2. Nat Meth. 2012;9(4):357–9.

31. Van der Auwera GA, Carneiro MO, Hartl C, Poplin R, Del Angel G, Levy-Moonshine A, Jordan T, Shakir K, Roazen D, Thibault J, et al. From FastQ data to high confidence variant calls: the genome analysis toolkit best practices pipeline. Curr Protoc Bioinformatics. 2013;43:11.10.11–33.

32. Weissensteiner H, Pacher D, Kloss-Brandstatter A, Forer L, Specht G, Bandelt HJ, Kronenberg F, Salas A, Schonherr S. HaploGrep 2: mitochondrial haplogroup classification in the era of high-throughput sequencing. Nucleic Acids Res. 2016;44(W1):W58–63.

33. Li H, Handsaker B, Wysoker A, Fennell T, Ruan J, Homer N, Marth G, Abecasis G, Durbin R. The sequence alignment/map format and SAMtools. Bioinformatics (Oxford, England). 2009;25(16):2078–9.

34. Jun G, Flickinger M, Hetrick Kurt N, Romm Jane M, Doheny Kimberly F, Abecasis Gonçalo R, Boehnke M, Kang Hyun M. Detecting and estimating contamination of human DNA samples in sequencing and Array-based genotype data. Am J Hum Genet. 2012;91(5):839–48.

35. Kircher M, Witten DM, Jain P, O'Roak BJ, Cooper GM. A general framework for estimating the relative pathogenicity of human genetic variants. Nat Genet. 2014;46(3):310–5.

36. Ruiz-Pesini E, Lott MT, Procaccio V, Poole JC, Brandon MC, Mishmar D, Yi C, Kreuziger J, Baldi P, Wallace DC. An enhanced MITOMAP with a global mtDNA mutational phylogeny. Nucleic Acids Res. 2007;35(suppl 1):D823–8.

37. Wachsmuth M, Hübner A, Li M, Madea B, Stoneking M. Age-related and Heteroplasmy-related variation in human mtDNA copy number. PLoS Genet. 2016;12(3):e1005939.

38. Ionita-Laza I, Lee S, Makarov V, Buxbaum Joseph D, Lin X. Sequence kernel association tests for the combined effect of rare and common variants. Am J Hum Genet. 2013;92(6):841–53.

39. Boyd A, Golding J, Macleod J, Lawlor DA, Fraser A, Henderson J, Molloy L, Ness A, Ring S, Davey Smith G. Cohort profile: the 'children of the 90s'—the index offspring of the Avon longitudinal study of parents and children. Int J Epidemiol. 2013;42(1):111–27.

40. Moayyeri A, Hammond CJ, Hart DJ, Spector TD. The UK adult twin registry (TwinsUK resource). Twin Res Hum Genet. 2013;16(1):144–9.

41. Stoneking M. Hypervariable sites in the mtDNA control region are mutational hotspots. Am J Hum Genet. 2000;67(4):1029–32.

42. Ridge PG, Maxwell TJ, Foutz SJ, Bailey MH, Corcoran CD, Tschanz JT, Norton MC, Munger RG, O'Brien E, Kerber RA, et al. Mitochondrial genomic variation associated with higher mitochondrial copy number: the Cache County study on memory health and aging. BMC Bioinformatics. 2014;15(7):S6.

43. Mahley RW, Innerarity TL, Rall SC, Weisgraber KH. Plasma lipoproteins: apolipoprotein structure and function. J Lipid Res. 1984;25(12):1277–94.

44. Andrikoula M, McDowell IFW. The contribution of ApoB and ApoA1 measurements to cardiovascular risk assessment. Diabetes Obes Metab. 2008;10(4):271–8.

45. Benn M. Apolipoprotein B levels, APOB alleles, and risk of ischemic cardiovascular disease in the general population, a review. Atherosclerosis. 2009;206(1):17–30.

46. Boekholdt SM, Arsenault BJ, Mora S, Pedersen TR, LaRosa JC, Nestel PJ, Simes RJ, Durrington P, Hitman GA, Welch K. Association of LDL cholesterol, non–HDL cholesterol, and apolipoprotein B levels with risk of cardiovascular events among patients treated with statins: a meta-analysis. JAMA. 2012;307(12):1302–9.

47. Rossignol R, Faustin B, Rocher C, Malgat M, Mazat JP, Letellier T. Mitochondrial threshold effects. Biochem J. 2003;370(Pt 3):751–62.

48. Bender A, Krishnan KJ, Morris CM, Taylor GA, Reeve AK, Perry RH, Jaros E, Hersheson JS, Betts J, Klopstock T. High levels of mitochondrial DNA deletions in substantia nigra neurons in aging and Parkinson disease. Nat Genet. 2006;38(5):515–7.

49. Corral-Debrinski M, Horton T, Lott MT, Shoffner JM, Beal MF, Wallace DC. Mitochondrial DNA deletions in human brain: regional variability and increase with advanced age. Nat Genet. 1992;2(4):324–9.

50. Wallace DC. A mitochondrial paradigm of metabolic and degenerative diseases, aging, and cancer: a dawn for evolutionary medicine. Annu Rev Genet. 2005;39:359–407.

51. Park CB, Larsson N-G. Mitochondrial DNA mutations in disease and aging. J Cell Biol. 2011;193(5):809–18.

52. Bua E, Johnson J, Herbst A, Delong B, McKenzie D, Salamat S, Aiken JM. Mitochondrial DNA–deletion mutations accumulate intracellularly to detrimental levels in aged human skeletal muscle fibers. Am J Hum Genet. 2006;79(3):469–80.

53. Greaves LC, Nooteboom M, Elson JL, Tuppen HAL, Taylor GA, Commane DM, Arasaradnam RP, Khrapko K, Taylor RW, Kirkwood TBL, et al. Clonal expansion of early to mid-life mitochondrial DNA point mutations drives mitochondrial dysfunction during human ageing. PLoS Genet. 2014;10(9):e1004620.

54. Williams SL, Mash DC, Züchner S, Moraes CT. Somatic mtDNA mutation spectra in the aging human Putamen. PLoS Genet. 2013;9(12):e1003990.

55. Cortopassi GA, Arnheim N. Detection of a specific mitochondrial DNA deletion in tissues of older humans. Nucleic Acids Res. 1990;18(23):6927–33.

56. Sahin E, Colla S, Liesa M, Moslehi J, Muller FL, Guo M, Cooper M, Kotton D, Fabian AJ, Walkey C, et al. Telomere dysfunction induces metabolic and mitochondrial compromise. Nature. 2011;470(7334):359–65.

57. Barazzoni R, Short KR, Nair KS. Effects of aging on mitochondrial DNA copy number and cytochromec oxidase gene expression in rat skeletal muscle, liver, and heart. J Biol Chem. 2000;275(5):3343–7.

58. Carré JE, Orban J-C, Re L, Felsmann K, Iffert W, Bauer M, Suliman HB, Piantadosi CA, Mayhew TM, Breen P. Survival in critical illness is associated with early activation of mitochondrial biogenesis. Am J Respir Crit Care Med. 2010;182(6):745–51.

59. Raffaello A, Rizzuto R. Mitochondrial longevity pathways. Biochim Biophys Acta. 2011;1813(1):260–8.

60. Kroemer G, Galluzzi L, Brenner C. Mitochondrial membrane permeabilization in cell death. Physiol Rev. 2007;87(1):99–163.

61. Green DR, Galluzzi L, Kroemer G. Mitochondria and the autophagy-inflammation-cell death axis in organismal aging. Science. 2011;333(6046):1109–12.

62. Bratic A, Larsson N-G. The role of mitochondria in aging. J Clin Invest. 2013;123(3):951–7.

63. Picard M, Zhang J, Hancock S, Derbeneva O, Golhar R, Golik P, O'Hearn S, Levy S, Potluri P, Lvova M. Progressive increase in mtDNA 3243A> G heteroplasmy causes abrupt transcriptional reprogramming. Proc Natl Acad Sci. 2014;111(38):E4033–42.

64. Clay Montier LL, Deng JJ, Bai Y. Number matters: control of mammalian mitochondrial DNA copy number. J Genet Genomics. 2009;36(3):125–31.

65. Kauppila TES, Kauppila JHK, Larsson N-G. Mammalian mitochondria and aging: an update. Cell Metab. 2017;25(1):57–71.

66. Giordano C, Iommarini L, Giordano L, Maresca A, Pisano A, Valentino ML, Caporali L, Liguori R, Deceglie S, Roberti M. Efficient mitochondrial biogenesis drives incomplete penetrance in Leber's hereditary optic neuropathy. Brain. 2014;137(2):335–53.

67. Yu-Wai-Man P, Sitarz KS, Samuels DC, Griffiths PG, Reeve AK, Bindoff LA, Horvath R, Chinnery PF. OPA1 mutations cause cytochrome c oxidase deficiency due to loss of wild-type mtDNA molecules. Hum Mol Genet. 2010;19(15):3043–52.

68. Suissa S, Wang Z, Poole J, Wittkopp S, Feder J, Shutt TE, Wallace DC, Shadel GS, Mishmar D. Ancient mtDNA genetic variants modulate mtDNA transcription and replication. PLoS Genet. 2009;5(5):e1000474.

69. Niccoli T, Partridge L. Ageing as a risk factor for disease. Curr Biol. 2012;22(17):R741–52.

70. Simpson DP, Hager SR. Bicarbonate-carbon dioxide buffer system: a determinant of the mitochondrial pH gradient. Am J Phys. 1984;247 (3 Pt 2):F440–6.

71. Durand T, Delmas-Beauvieux M-C, Canioni P, Gallis J-L. Role of intracellular buffering power on the mitochondria-Cytosol pH gradient in the rat liver Perfused at 4°C. Cryobiology. 1999;38(1):68–80.

72. Bishop DJ, Thomas C, Moore-Morris T, Tonkonogi M, Sahlin K, Mercier J. Sodium bicarbonate ingestion prior to training improves mitochondrial adaptations in rats. Am J Physiol Endocrinol Metab. 2010;299(2):E225–33.

73. Pearson TA, Mensah GA, Alexander RW, Anderson JL, Cannon RO, Criqui M, Fadl YY, Fortmann SP, Hong Y, Myers GL. Markers of inflammation and cardiovascular disease. Circulation. 2003;107(3):499–511.

74. Barati M, Alinejad F, Bahar MA, Tabrisi MS, Shamshiri AR, Bodouhi NO, Karimi H. Comparison of WBC, ESR, CRP and PCT serum levels in septic and non-septic burn cases. Burns. 2008;34(6):770–4.

75. Liu S-F, Kuo H-C, Tseng C-W, Huang H-T, Chen Y-C, Tseng C-C, Lin M-C. Leukocyte mitochondrial DNA copy number is associated with chronic obstructive pulmonary disease. PLoS One. 2015;10(9):e0138716.

76. Yoon C-Y, Park JT, Kee YK, Han SG, Han IM, Kwon YE, Park KS, Lee MJ, Han SH, Kang S-W, et al. Low mitochondrial DNA copy number is associated with adverse clinical outcomes in peritoneal dialysis patients. Medicine. 2016;95(7):e2717.

77. López-Armada MJ, Riveiro-Naveira RR, Vaamonde-García C, Valcárcel-Ares MN. Mitochondrial dysfunction and the inflammatory response. Mitochondrion. 2013;13(2):106–18.

78. Arnedo J, Svrakic DM, Del Val C, Romero-Zaliz R, Hernandez-Cuervo H, Fanous AH, Pato MT, Pato CN, de Erausquin GA, Cloninger CR, et al. Uncovering the hidden risk architecture of the schizophrenias: confirmation in three independent genome-wide association studies. Am J Psychiatry. 2015;172(2):139-53.

79. Lohmueller Kirk E, Sparsø T, Li Q, Andersson E, Korneliussen T, Albrechtsen A, Banasik K, Grarup N, Hallgrimsdottir I, Kiil K, et al. Whole-exome sequencing of 2,000 Danish individuals and the role of rare coding variants in type 2 diabetes. Am J Hum Genet. 2013;93(6):1072–86.

80. Olofsson SO, Boren J. Apolipoprotein B: a clinically important apolipoprotein which assembles atherogenic lipoproteins and promotes the development of atherosclerosis. J Intern Med. 2005;258(5):395–410.

81. Choi SH, Ginsberg HN. Increased very low density lipoprotein (VLDL) secretion, hepatic steatosis, and insulin resistance. Trends Endocrinol Metab. 2011;22(9):353–63.

82. Su Q, Baker C, Christian P, Naples M, Tong X, Zhang K, Santha M, Adeli K. Hepatic mitochondrial and ER stress induced by defective PPARα signaling in the pathogenesis of hepatic steatosis. Am J Physiol Endocrinol Metab. 2014;306(11):E1264–73.

83. Reiling E, Ling C, Uitterlinden AG, van't Riet E, Welschen LMC, Ladenvall C, Almgren P, Lyssenko V, Nijpels G, van Hove EC, et al. The Association of Mitochondrial Content with prevalent and incident type 2 diabetes. J Clin Endocrinoly Metab. 2010;95(4):1909–15.

4

Selection for female traits of high fertility affects male reproductive performance and alters the testicular transcriptional profile

Marten Michaelis[1,4*†], Alexander Sobczak[1†], Dirk Koczan[3], Martina Langhammer[2], Norbert Reinsch[2], Jennifer Schoen[1] and Joachim M. Weitzel[1,4*]

Abstract

Background: Many genes important for reproductive performance are shared by both sexes. However, fecundity indices are primarily based on female parameters such as litter size. We examined a fertility mouse line (FL2), which has a considerably increased number of offspring and a total litter weight of 180% compared to a randomly bred control line (Ctrl) after more than 170 generations of breeding. In the present study, we investigated whether there might be a parallel evolution in males after more than 40 years of breeding in this outbred mouse model.

Results: Males of the fertility mouse line FL2 showed reduced sperm motility performance in a 5 h thermal stress experiment and reduced birth rate in the outbred mouse line. Transcriptional analysis of the FL2 testis showed the differential expression of genes associated with steroid metabolic processes (Cyp1b1, Cyp19a1, Hsd3b6, and Cyp21a1) and female fecundity (Gdf9), accompanied by 150% elevated serum progesterone levels in the FL2 males. Cluster analysis revealed the downregulation of genes of the kallikrein-related peptidases (KLK) cluster located on chromosome 7 in addition to alterations in gene expression with serine peptidase activity, e.g., angiotensinogen (Agt), of the renin-angiotensin system essential for ovulation. Although a majority of functional annotations map to female reproduction and ovulation, these genes are differentially expressed in FL2 testis.

Conclusions: These data indicate that selection for primary female traits of increased litter size not only affects sperm characteristics but also manifests as transcriptional alterations of the male side likely with direct long-term consequences for the reproductive performance of the mouse line.

Keywords: Long-term selection mouse lines, Outbred mouse model, High-fertility, Fecundity, Testis, Reproductive fitness, Sperm motility, Casa

Background

Fertility is complex, involving numerous interactions of different pathways. Most knowledge concerning the interaction of genes has been obtained from exploratory models, such as transgenic or knockout mice. The database Mouse Genome Informatics (MGI - www.informatics.jax.org) harbors more than 2000 genotypes associated with reproductive phenotypes. Although there are cases of duplicated annotation, nearly all of these annotations refer to an infertile or at least subfertile phenotype. Only a

minority (<1%) of annotations have been associated with enhanced fertility or improved reproductive performance (see Table 1). This correlation is clear evidence for transgenic alterations resulting in a decreased fertility phenotype. Nevertheless, genetically interventional studies are unquestionably helpful in dissecting fertility-relevant pathways. The above comparison argues for a shortage of models of increased reproductive performance to understand this condition. In addition, a majority of transgenic models are monogenetic, despite the common view that (in)fertility is multi-causal, and a network of genes is essential for reproductive processes. As such, we expect not only a single gene but also a combination of genes to be reproductively responsive.

* Correspondence: michaelis@fbn-dummerstorf.de; weitzel@fbn-dummerstorf.de
†Equal contributors
[1]Institute of Reproductive Biology, University of Rostock, Rostock, Germany
Full list of author information is available at the end of the article

Table 1 Genotypes associated with a certain fertility phenotype, its amount and proportion [%] on total phenotype number. Data are extracted (May, 2016) from the database 'Mouse Genome Informatics' (http://www.informatics.jax.org/vocab/mp_ontology/MP:0002161)

Mammalian phenotype (number of matching genotypes)	
Abnormal fertility/fecundity (2479 genotypes)	total
Infertility (1462 genotypes)	59%
Reduced fertility (888 genotypes)	36%
Enhanced fertility (4 genotypes)	0.2%
Decreased litter size (441 genotypes)	18%
Increased litter size (25 genotypes)	1%

We report on a unique murine animal model selected for the 'high-fertility' phenotype over more than 170 generations for use as an alternative experimental strategy for overcoming the above-mentioned limitations. The additional advantage of this approach is its heterogeneity, which more closely mimics the phenotypic alterations in nature compared to single gene approaches to generate classical transgenic or knockout models.

In the 1970s, the Leibniz Institute for Farm Animal Biology (FBN) established two mouse models for high reproductive performance through long-term selective breeding: fertility lines 1 and 2 (FL1 and FL2). The improved fertility selection criteria included (i) number of offspring and (ii) total litter birth weight at first delivery. Both selection criteria have been combined in a breeding index = 1.6× litter size + litter birth weight. The selection for litter size and litter weight via breeding index avoided intrauterine growth retardation of the offspring. The factor 1.6 reflects the average birth weight of single pups. Animals of the largest and heaviest litters were recruited for breeding of the next generation. In parallel, an unselected control line (Ctrl) was developed based on the same initial founder population. All specifications regarding breeding and the proceedings of these mouse lines have been reviewed [1–3]. Until recently, these fertility lines have lacked a profound molecular biological description, and to fill this gap, we exclusively focused on FL2 characterization in the present study.

Based on previous studies, FL2 females ovulate more oocytes and harbor almost twice as many corpora lutea (CL) as Ctrl females [4]. The intrauterine growth and development of embryos was similar between both mouse lines [4]. Thus, the increased ovulation occurs in response to selection pressures. Moreover, FL2 females showed an increase in serum progesterone (P4) levels over the estrus cycle [4]. On the male side, FL2 bucks showed decreased lifetime expectancy and a more explorative behavior in an open field test [2, 5]. This result indicates that breeding focused on female high reproductive performance not only affects the females but also the males. Despite selection criteria exclusively visible on the dam side, we examined the physiological and genetic responses of the male side to selection for high female fecundity traits after more than 170 generations of breeding.

Many genes essential for reproductive performance are shared within the germinal organs of both sexes. Thus, we hypothesized that during the female-focused selection for high litter size, achieved through the increased ovulation rate in FL2 females, the male side is also affected, particularly in the male germinal organ.

The main task of the testis is the production of functional gametes; hence, we analyzed sperm motility as a physiological parameter for overall testicular function. Furthermore, we compared FL2 and Ctrl testicular transcriptional profiles on a global gene expression level.

Methods
Animal model & ethics statement
The animal experiments have been approved by the local authorities (Landesamt für Landwirtschaft, Lebensmittelsicherheit und Fischerei, Mecklenburg-Vorpommern, Germany). The mouse lines were maintained in a specific pathogen-free (since 2012) environment with a 12:12 h light-dark regime and ad libitum access to water and food (ssniff® M-Z, Soest, Germany) at the laboratory animal facility of the FBN.

All mouse lines were originally derived from the same genetic pool of a mixture of eight defined founder mouse lines. From this starting population, the high-fertility line FL2 was generated during a long-term selection experiment for more than 40 years referring to >170 generations. The selection was performed via breeding index, combining first litter size and total litter birth weight (Dummerstorf breeding index: 1.6× litter size + litter birth weight). Offspring of litters with the highest breeding indices were recruited to breed the next generation. Until the 23rd generation, the FL2 females were cycle synchronized using gestagen chlormadinone acetate. In parallel, an unselected Ctrl line was generated from the same starting population and has been maintained under identical housing conditions using a rotational mating scheme and avoiding full sib mating to decrease the average rate of inbreeding. To ensure the outbred character, the mouse lines were bred with a population size of 60–100 and 125–200 animals per generation for FL2 and Ctrl lines, respectively. However, reflecting the breeding process, the corresponding inbreeding values accounted for 0.175 for the Ctrl and 0.977 for the FL2 lines [6]. Details regarding the breeding procedures can be found in Dietl et al. and Schüler et al. [1–3]. The males of this investigation have been further described in a two-factorial breeding experiment to delineate the impact of males and females on fertility parameters [6]. Birth rates (accounted as: deliveries (living and nonliving litters) per mating) were extracted from standard breeding data acquired over a 4-year period and encompassing at least 840 pairings per line (14 generations with at least 60 breeding pairs in FL2 and 125 pairs in Ctrl). Statistical analysis was performed using Student's t-test.

Sperm motility analysis - sperm motility stress test

The sperm motility analysis was accomplished using CASA (computer-assisted sperm analysis) via SpermVision (Minitube, Germany). Prior to a 5 h thermal stress procedure and CASA measurements, 12-week-old males ($n = 10$ per line) were euthanized by CO_2 inhalation. To obtain sperm, the cauda epididymis was extracted, cleaned and minced (five cuts) in freshly prepared 300 µl of spermatozoa-suitable M199 media (M7528, Sigma-Aldrich, Germany) and incubated for 5 min at 37 °C for sperm release. Tissue remnants were subsequently filtered using 30-µm mesh.

To generate stress conditions, the spermatozoa suspension was continuously exposed to 37 °C for a 5-h thermal stress period. Every hour, the current sperm motility characteristics were determined by quickly applying a 3-µl aliquot to 37 °C tempered chamber slides (20 µm, Leja, Netherlands). Each sperm sample was considered as the average of 8 defined chamber partitions viewed and measured using the CASA system (Minitube, Germany).

RNA extraction

For microarray and qPCR studies, the testes of 12–13-week-old males were dissected and snap frozen in liquid N2 or preserved in RNAlater® (Ambion, Austin, USA). According to the manufacturer's instructions, total RNA was isolated using the RNeasy® Mini Kit (Qiagen, Germany) or the InviTrap Spin RNA Mini Kit (Stratec, Germany) with the simultaneous removal of genomic DNA traces. RNA integrity and quantity were assessed using capillary electrophoresis (Agilent 2100 Bioanalyzer, Santa Clara, USA).

Sample labeling and hybridization of microarray

For transcriptome profiling, we used the GeneChip® Mouse Transcriptome Array (MTA) 1.0 (released in May, 2015; currently termed Clariom™ D assay, Affymetrix Inc., Santa Clara, USA) with more than 66,100 coding and non-coding transcripts. We used two separate sets of microarray experiments (denoted as 1st and 2nd array set). The 1st array set was based on 8 testicular samples hybridized to individual gene chips. For validation, a 2nd array set was utilized, comprising 8 biologically independent replicates pooled in equivalent amounts and hybridized to a single 'pool-microarray' for each mouse line.

RNA labeling and hybridization were conducted at the Core Facility for Microarray Analysis, University of Rostock. Briefly, 200 ng of quality controlled total RNA was used for cDNA preparation and labeling with the GeneChip® WT PLUS Reagent Kit (Affymetrix, Santa Clara, USA). The fragmented (~100 bp) and biotinylated cDNA was hybridized for 16 h at 45 °C to Affymetrix Gene Chip® MTA 1.0, followed by washing and staining using the Affymetrix Fluidics Station 450 according to standard instructions. The chips were subsequently scanned at 0.7-µm resolution (GeneChip Scanner 3000 7G, Affymetrix). Furthermore, all hybridizations were assessed for quality requirements, and raw data were submitted to the Gene Expression Omnibus (GEO) database according to MIAME guideline (GSE86063).

Microarray data normalization and statistical analysis

Raw data cell intensity files were normalized by the Robust Multiarray Average (RMA) algorithm with a Signal Space Transformation (SST) employing the Affymetrix Expression Console software (EC, version 1.4.1.46). Within the same software package, data were further explored using principal component analysis (PCA). Additionally, unsupervised hierarchical cluster analysis was accomplished using Transcriptome Analysis Console software (TAC, Affymetrix, version 3.0.0.466). The same software package was used to calculate fold-changes (FC) and perform statistical analyses. For gene expression detection, an alternative splicing analysis algorithm in the Affymetrix software packages EC and TAC was applied. As recommended, a transcript was considered expressed when at least 50% of the eligible exons were detected above background (DABG, $p < 0.05$).

For data mining, we employed the web-based Database for Annotation, Visualization and Integrated Discovery (DAVID 6.8, david.ncifcrf.gov) [7, 8] and the PANTHER classification system (www.pantherdb.org) [9, 10].

Quantitative real-time PCR (RT-qPCR)

Microarray results were additionally confirmed for a selected group set of genes using quantitative real-time PCR (qPCR). To this end, 0.5 µg of total RNA was reverse transcribed using random hexamer primers and the iScript cDNA Synthesis kit (BioRad, Germany) in accordance to the manufacturer's instructions. Briefly, 2 µl of a 1:5 cDNA dilution were amplified with technical duplicates using a SYBR Green mix (BioRad, Germany) and 40 cycles of 30 s at 95 °C, 45 s at 56 °C, and 30 s at 72 °C, as previously described [5]. The primers were designed using the web-based software Primer-BLAST (www.ncbi.nlm.nih.gov/tools/primer-blast) [11]. When possible, special care was taken to select oligonucleotides binding to intron-spanning exons. In addition, specificity control was assessed using gel electrophoresis. The primers used in the present study are listed in the Additional file 1. The samples were normalized to a combination of reference genes (36B4, GAPDH, HPRT, and B2m) and statistically evaluated using the Relative Expression Software Tool (REST 2009) [12].

Hormone analysis

Serum progesterone was measured by ^{3}H–radioimmunoassay using a [1,2,6,7-3H] progesterone (Hartmann Analytic, Germany) tracer as previously described [13]. Briefly, 50 μl of serum in duplicate ($n = 14$ per group) was analyzed with an incubation step at 37 °C for 30 min and 4 °C for 2 h. The B/F separation was performed by a dextran-charcoal method. The radioactivity was quantified using a Liquid Scintillation Counter with an integrated RIA program (TriCarb 2900 TR; Perkin-Elmer, Waltham, USA). Intra- and interassay precision was 7.6% and 9.8%, respectively, for progesterone. The standard curve ranged from 6.25 to 1600 pg/ml, and the detection limit corresponded to 7 pg/ml for progesterone. Statistical analysis was performed using Student's t-test.

Results

Animal model

Studies in mice almost exclusively focus on a single gene approach using transgenic intervention in gain or loss-of-function analyses. We changed the perspective by investigating a mouse model for the long-term selection of the primary female trait of high-fertility visible in increased litter size. Thus, nature itself has selected the genetic alteration to match the selection criteria, in contrast to the classical transgenic approach.

After 172 generations, the randomly selected Ctrl line showed an average litter size of 11.4 ± 3.3 pups, while the total litter weight accounted for 20.8 ± 4.8 g. The extent of the delivery parameters has basically remained the same during the entire random selection process. In contrast, FL2 almost doubled in litter size (+92%) compared to the Ctrl line.

Similar results were obtained for the selection criterion of total litter birth weight (+94%), explaining why we detected no changes in individual newborn weights. An overview of the breeding status after 40 years, with respect to litter size and litter weight, is summarized in Table 2. The line differences in both traits were highly significant ($p < 0.0001$). The gain in reproductive

Table 2 Number of offspring per litter and total litter weight at birth for Ctrl and FL2 mouse lines after 172 generations of selection

Mouse line	Ctrl	FL2
Offspring per litter	11.4 ± 3.3	$21.9*** \pm 2.6$ (192%)
Total litter weight at birth [g]	20.8 ± 4.8	$40.4*** \pm 4.5$ (194%)

Proportions (%) to Ctrl are indicated. Data were tested for normal distribution and analyzed by Student's t-test (***, $p < .0001$)

performance was accompanied by no impairment of pub survival. Indeed, intrauterine growth and development were similar between lines, and the body weight of the individual newborns was not reduced in FL2 compared to the Ctrl line [4].

Previous studies have demonstrated that the selection process is associated with increased ovulation and alterations in germinal organs of FL2 females [4, 14]. However, molecular and functional information concerning whether FL2 male germinal organs are affected by the selection process was lacking. To address this issue, we examined sperm motility parameters using CASA and evaluated the overall birth rate of the mouse line acquired over a 4-year breeding period. In addition, we performed a gene expression analysis of the central male reproductive organs, the testis.

Sperm motility and birth rate

Computer-assisted sperm motility analysis (CASA) was conducted to characterize the males at physiological and functional levels, focusing on the quality of the sperm of FL2 and Ctrl males. To this end, spermatozoa of the cauda epididymis were released into media. The resulting sperm suspension was exposed to thermal stress at 37 °C for 5 h and analyzed hourly using CASA.

The data obtained from the sperm motility experiment are shown in Fig. 1a–e. For the starting point, no significant differences in any of the motility parameters between the FL2 and Ctrl lines were observed. At 0 h, sperm motility was $74.2 \pm 4.2\%$ (SD) and $74.8 \pm 6.3\%$ for FL2 and Ctrl bucks, respectively, whereas progressively motile sperms were observed at $61.3 \pm 4.9\%$ and $60.2 \pm 5.2\%$ in FL2 and Ctrl males, respectively. As expected, the percentage of motile and progressively motile sperms rapidly decreased with thermal stress duration (see Fig. 1a & b). Notably, the sperms of FL2 animals were considerably more sensitive to thermal stress than those of Ctrl animals. Quality characteristics, such as velocity parameters (e.g., VCL Fig. 1c; VAP and VSL data not shown), revealed a significantly higher reduction for FL2 compared to the Ctrl line during the entire incubation period, most substantially within the first 60 min. Consistent with these observations, the data for linearity characteristics (e.g., LIN = VSL/VCL) showed a significant increase over the incubation period (Fig. 1d). Presumably reflecting the enhanced loss of energy necessary to ensure suitable motility, the tendency towards straightforward but slow movement increased, whereas Ctrl sperm apparently can conserve more resources for extensive, more curvilinear, oscillatory movement. Coincidently, the beat-cross frequency (BCF) as an additional indicator of oscillation behavior was significantly more diminished over thermal stress duration for FL2 than that for Ctrl (Fig. 1e).

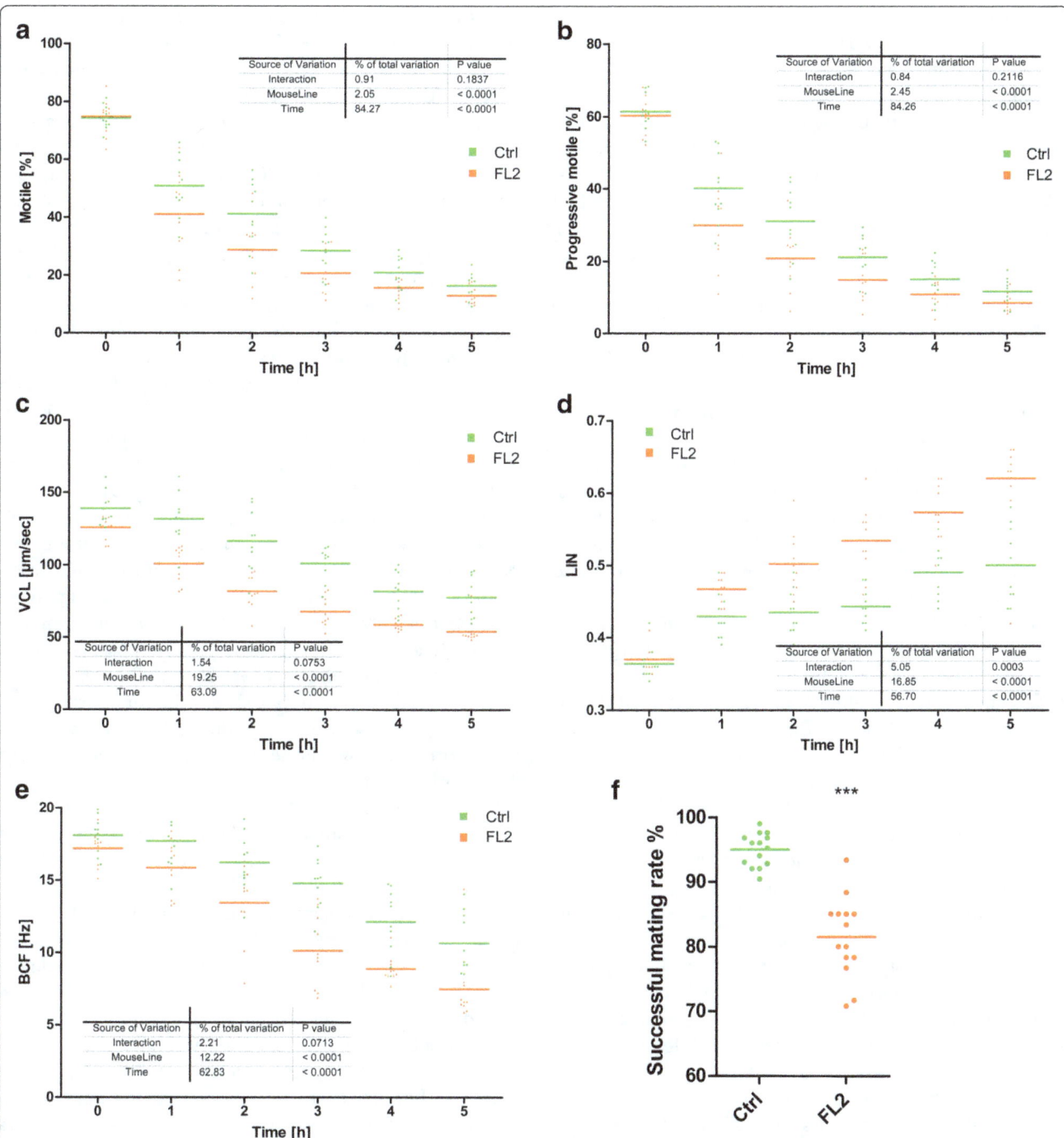

Fig. 1 Sperm motility and birth rate. Time-course changes during thermal stress response assay of the sperm motility parameters average motility (**a**), progressive motility (**b**), curvilinear velocity (VCL) (**c**), Beat Cross Frequency (BCF) (**d**) and Linearity (LIN) (**e**) after continuing incubation at 37 °C. For this analysis, cauda epididymis spermatozoa of FL2 and Ctrl mouse line (*n* = 10 per group) were extracted and incubated at 37 °C for 5 h. Each point represents the mean of 8 defined chamber partitions viewed and measured using the Computer-assisted sperm analysis (CASA) system. The experiment has been statistically evaluated for the two factors time and mouse line effect using two-way ANOVA (GraphPad Prism) as indicated in the graph. Birth rate (**f**) of 'high-fertility' mouse line FL2 (orange) and Ctrl line (green) is visualized providing information about portion of successful line-specific matings. Each of the 14 points represents the average birth rate per generation. These data were acquired over a 4-year period with generation population sizes ensuring at least 60 matings. The groups were tested for normal distribution and analyzed by two-tailed t-test (***, *p* < 0.001)

Birth rate showed a considerable difference between the FL2 and Ctrl lines based on the evaluation of breeding information acquired over a 4-year-period and comprising at least 840 matings per group. The mating period of general maintenance breeding lasted for two weeks, enabling the occurrence of at least one ovulation. The birth rate of the Ctrl mouse line accounted for 94.9 ± 2.6% per generation, whereas the birth rate of the FL2 mouse line were registered for only 81.5 ± 6.1% (Fig. 1f). Consequently, the breeding analysis revealed a significant decrease in deliveries for the FL2 line. However, we cannot exclude this effect as accomplished by not only male but also female animals.

Microarray analysis

Within the germinal organs, males and females share many genes essential for reproductive performance. We speculated that during selection for increased female reproductive performance, reflected in elevated ovulation rate, alterations were gradually established not only for the female but also for the male reproductive organs. To address this issue on a molecular level, we employed whole transcriptome analysis.

Testicular RNA was hybridized to individual or pooled MTA 1.0 microarrays (see microarray experimental design). The overall cell intensity distribution of each gene chip before and after normalization (SST-RMA) is depicted in the Additional file 2, illustrating an overall well-performed hybridization experiment.

The Principal component analysis (PCA) was used to assess overall transcriptional differences of both mouse lines in an unbiased manner based on the first microarray experiment using testicular individual hybridization. The PCA plot illustrates a clustering according to grouping with an overall PCA mapping of 57% (see Fig. 2a). Consistent with the inbreeding coefficient (see Methods), the PCA analysis illustrated an overall more homogenous clustering of FL2 mice and the heterogenic nature of the randomly selected Ctrl line. Hence, FL2 breeding effects are visible at the overall transcriptional level prior to supervised exploration.

Hierarchical cluster analysis was based on the 500 most differentially expressed genes over all samples. FL2 and Ctrl transcripts clustered corresponding to their groups (Fig. 2b). However, we detected more homogenous gene expression within FL2 animal than in Ctrl animals.

The microarray experimental design was based on two sets of microarray experiments. For the first microarray experiment, we used 8 animals per group, whose quality-controlled testicular RNA was hybridized to individual chips. To analyze differentially expressed genes (DEGs), we applied conventional filter criteria, such as fold-change (FC) and statistical significance. Instead of only selecting a few genes for validation (via qPCR), we employed a second microarray set of biologically independent testis replicates. We used the equivalent group size of 8 animals per group. To biologically validate the DEGs of the first hybridization setup, it was sufficient to analyze the second set of independent samples as a spot test. RNA in biological replicates was pooled in equivalent amounts and suspended onto one microarray per line regarded as a verification-serving Pool-Chip. Using this approach, we ensured the comprehensive verification of the entire gene expression results obtained from the first round of microarray hybridizations.

The nonparametric Spearman Correlation Analysis of these two independent microarray experiments distinctly illustrated the experimental benefit of this approach (Fig. 3). For all 65,770 PS, the correlation between FCs of the 1st and 2nd array set accounted for 0.3915 (Fig. 3a). When increasing the stringency blotting of only FCs of statistical significance in the first array set, the correlation coefficient increased from 0.6502 to 0.9505, depending on the statistical test applied (see Fig. 3b-d). Consistent with this finding, the number of probe sets passing these statistical criteria dramatically decreased from 18,098 using ANOVA $p < 0.05$ to 317 applying Bonferroni's correction. Thus, the PS in Quadrants II and IV (Fig. 3a-d) are regarded as 'false positive' DEGs, reflecting a lack of biological validation of the expression intensity and direction based on the first array sets. Hence, in the case of missing information in the second microarray set, it would be impossible to filter out these 'false positive' genes.

Gene expression detection was used to filter for genes expressed in the samples. To determine whether a PS is expressed, at least 50% of the given transcript isoforms have to be detected above background. To this end, we used the expression summary based on filtering for 'true' or 'false' provided by the TCA splicing analysis algorithm. This filter of expression detection was employed in both microarray sets (see experimental design). The abundance of PS detected as present among the entire 65,770 available genes is listed for each group in Fig. 4a.

For differentially expressed genes (DEGs), we used the following filtering criteria: i) PS denoted as upregulated was expressed at least in the FL2 group; transcripts assigned as downregulated were detected at least in the Ctrl group. ii) The PS abundance was considered differentially expressed when the FC of FL2 transcripts was at least higher than 1.5 or lower than −1.5 compared to the Ctrl group in the 1st and 2nd microarray settings to be regarded as verified. Iii) The ANOVA p-value of the 1st array set below 0.05 was used as statistical filter.

Based on these criteria, among the 65,770 PS represented on the microarray, we detected 92% and 87% PS as expressed in FL2 and Ctrl, respectively (Fig. 4a). Only 1103 PSs passed the FC filter criterion for differential

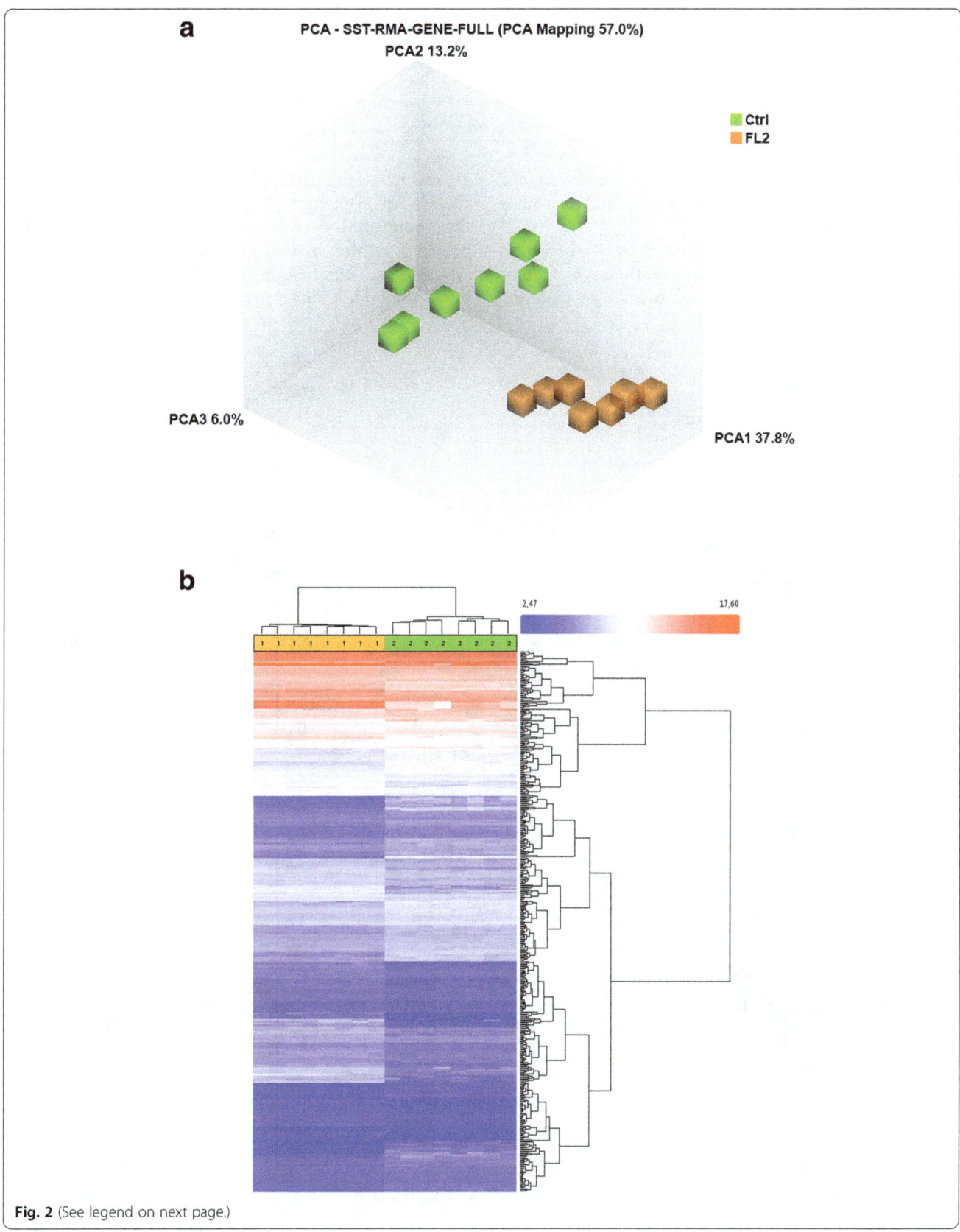

Fig. 2 (See legend on next page.)

(See figure on previous page.)
Fig. 2 Principal component analysis (PCA) plot (**a**) of overall mRNA expression data that characterize the entire testicular transcriptional profile of enrolled samples. Each cube represents one single microarray of FL2 (orange) and Ctrl (green) mice. The PCA plot was generated using Affymetrix Expression Console software. Unsupervised hierarchical cluster analysis (**b**) of the 500 most differentially expressed genes over all groups performed. Genes are indicated vertically, the individual testicular samples are designated horizontally (FL2 orange; Ctrl green). Genes displaying upregulation are visualized in red-, genes showing downregulation are specified in blue. Intensity of the red or blue color suggests signal intensity as annotated in the scale

expression, for which 1061 probe sets were determined as statistically significant according to ANOVA, with a p-value <0.05 (Fig. 3b).

When analyzing a high number of data, correcting for the amounts of tested null hypotheses is recommended. For microarray experiments, a common algorithm is the concept of the false discovery rate (FDR). An even more stringent test is constituted by the Bonferroni's correction. However, both methods do not reflect the biological diversity. Thus, we used ANOVA in combination with the experimental validation method. Using biologically independent sample

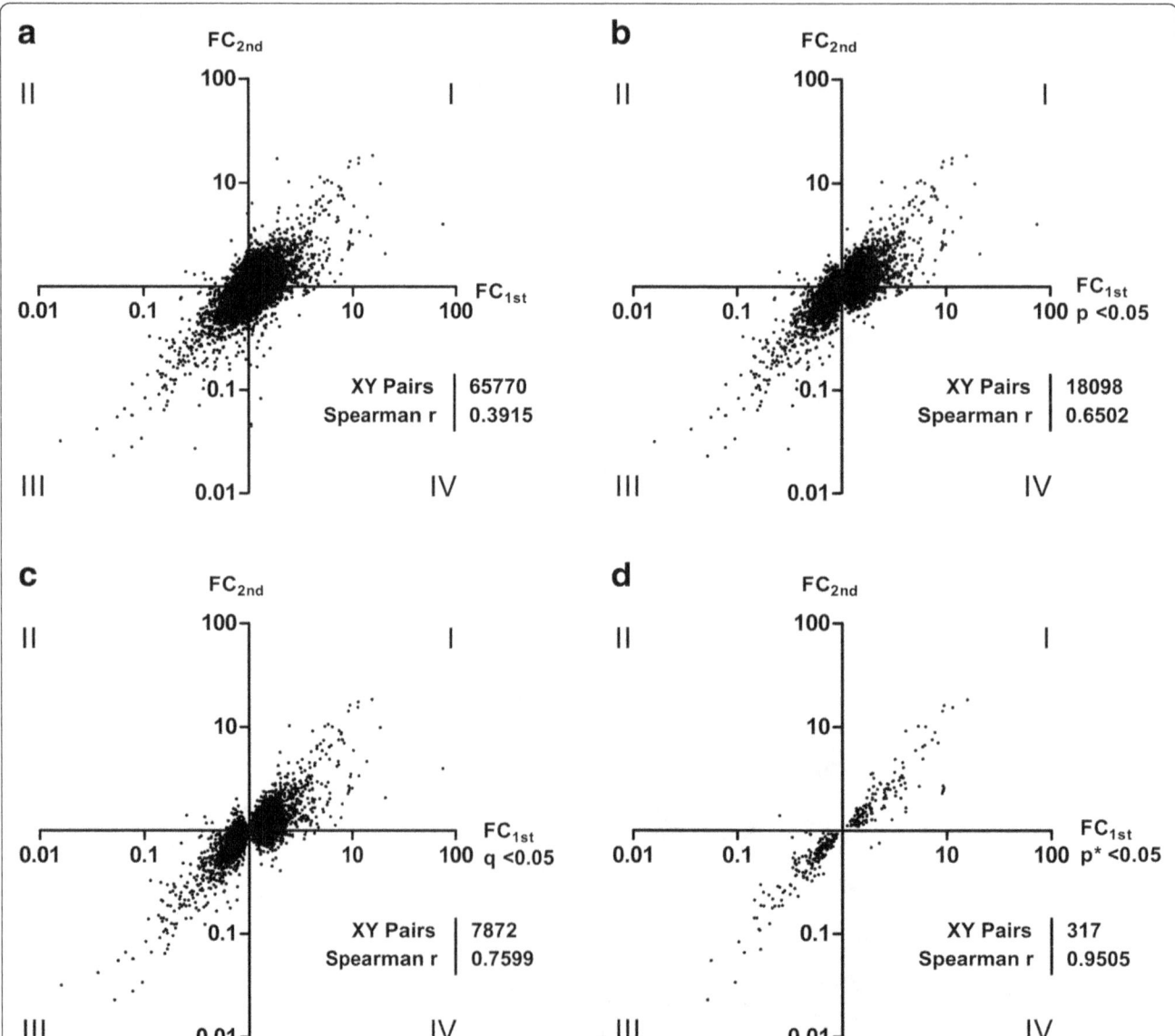

Fig. 3 Scatter blots showing the correlation of two sets of microarray experiments. Abscissae indicate FCs of the first microarray experiment filtered statistically significant as indicated (**a**, unfiltered; **b**, ANOVA $p < 0.05$; **c**, FDR q < 0.05; **d**, Bonferroni's correction $p^* < 0.05$). Ordinates represent FCs of the second array validation experiment. Correlation coefficient according to Spearman (r), significance level (p) and number of enrolled XY pairs (probe set pairs) are designated

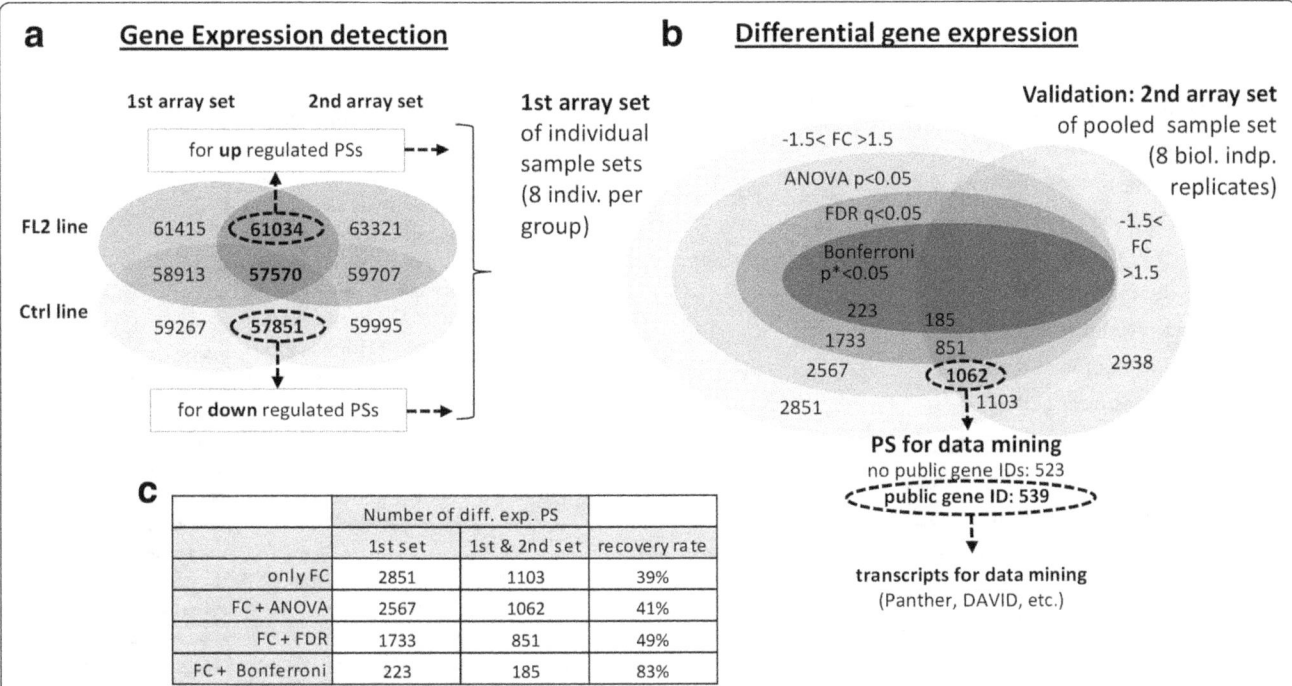

Fig. 4 Microarray filtering strategy of differentially expressed genes (DEGs): The microarray experiment is based on two biologically independent sample sets (1st, 2nd set) of two mouse lines (FL2, Ctrl). The 1st set analyzed for differential gene expression, while the 2nd set experimentally confirmed differential expression (see section microarray experimental design Material & Methods). The gene chip MTA 1.0 (Affymetrix) is equipped with 65,770 probe sets (PS). Using the gene expression detection the number of expressed genes within each sample set was evaluated and served as first filter criteria. (**a**) The Venn diagram illustrates the number of differentially expressed genes (FL2 vs. Ctrl) detected in 1st, 2nd and both sample sets independent of statistical filter (of 1st array set ANOVA, FDR, Bonferroni). (**b**) Experimentally verified DEGs were defined as: FC > 1.5; FC < −1.5 (of 1st and 2nd set); one-way ANOVA (of 1st set). The number and percentage of DEGs of the 1st set recovered in the 2nd set are given in table (**c**)

sets, we observed only moderate 'recovery' rates of 48%, 54% or 84% using ANOVA, FDR or Bonferroni's correction as a filter, respectively (Fig. 4c). Although a robust statistical filter ensures an increased prediction rate in unknown sample sets, as in the 2nd microarray set, the number of faithfully positive PS was only 48% to 84%, depending on the statistic filter used. Using a more stringent statistical filter would only reduce the number of DEGs, i.e., when applying Bonferroni's correction.

Gene ontology (GO) classification was performed for all differentially expressed transcripts filtered by our double-array-set validation. Among the 1061 differentially expressed PSs, almost 50% (523 PS) of these genes could not be annotated with official gene symbols or public gene IDs, reflecting the missing database knowledge of these mainly noncoding transcripts (Fig. 4b). Hence, we excluded these transcripts from functional interpretation processing. A list of the DEGs is provided in Additional file 3. GO classification identified hydrolase activity, protein binding and transferase activity as major molecular functional groups (geneontology.org) [9, 10]. However, within these three classifications, more than

twice as many transcripts were downregulated than upregulated, indicating reduced activity at the transcriptional level in the FL2 group (Fig. 5a). Similar results were obtained when classifying the biological processes (Fig. 5b). Among these broad biological annotations, the number of downregulated transcripts also exceeded the amount of upregulated DEGs. However, we observed that this GO alignment lacks almost 50% of the 539 applied transcripts, reflecting missing annotations within the database geneontology.org.

Cluster analysis was performed to investigate whether particular chromosomal regions are differentially expressed, potentially resulting from the breeding process. In general, particular genes of close relations can be organized into clusters within the eukaryotic genome. Such clusters are up to 300 kb in size [15–17]. Hence, we explored whether particular DEGs, showing similar expression, map to closely related genomic regions.

For example, seventeen transcripts of the immunoglobulin heavy chain (J558 family) cluster located on Chr12 62.59 cM are less active in FL2 than in Ctrl mice. However, it cannot be excluded that there might be an indirect effect resulting from a reduction in diversity and

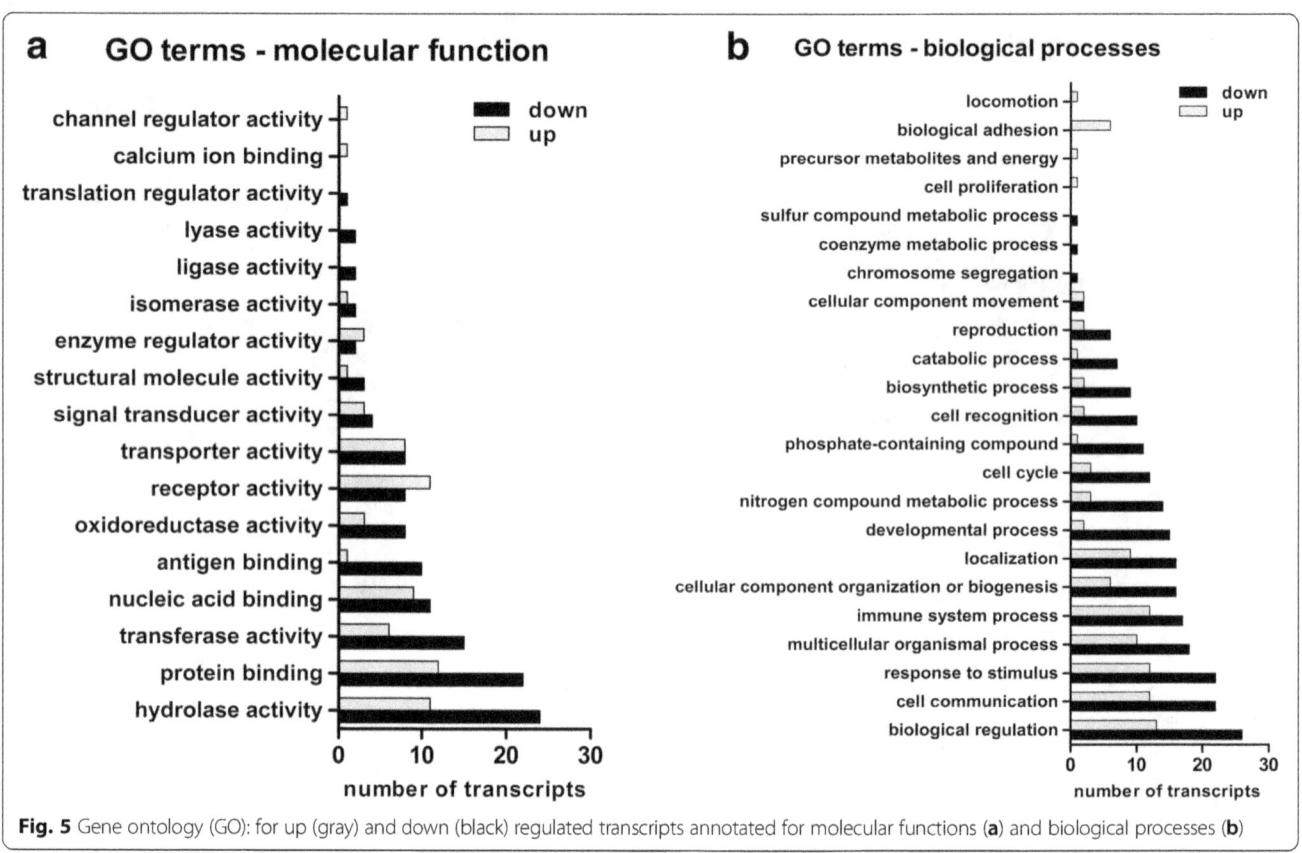

Fig. 5 Gene ontology (GO): for up (gray) and down (black) regulated transcripts annotated for molecular functions (**a**) and biological processes (**b**)

the breeding process and the genetic separation for more than 170 generations causing an undirected allelic drift rather than an active regulation. In addition, Chr9, 1.83–1.84 cM, harbors a region of 8 consecutive upregulated (FC 2.0–9.9) genes. These genes are encode transcripts of unknown function.

An additional locus of differential expression is the kallikrein-related peptidases (KLKs) cluster. This cluster is located on Chr7, 28.26–28.74 cM. Although the KLKs cluster comprises 26 KLKs, only 14 of these enzymes are downregulated Additional file 4. The subset of differentially expressed KLKs is expressed within the testis, whereas KLKs are generally unexpressed in the testis and are not actually detected as differentially expressed within the testis transcriptomics data. Thus, these findings suggest an apparent active regulation of these differentially expressed KLKs rather than decreased activity of the whole KLK cluster.

Gene set enrichment analysis (GSEA) was used to evaluate functional differences between the FL2 and Ctrl testicular transcriptomes. A total of 539 DEGs with official geneIDs were subjected to Database for Annotation, Visualization and Integrated Discovery (DAVID). Among the Gene ID transcript list, 477 genes matched to murine annotations within the GO database. However, when performing the GSEA, only 50% of these IDs could be

assigned to functional annotations in DAVID. A subset of the main enrichment functions and corresponding genes is presented in Table 4.

The bioinformatics GSEA disclosed a wide range of biological and molecular categories obviously affected in the testis of FL2 bucks during long-term breeding. The most prominently affected genes are those of the renin-angiotensin system, with a fold-enrichment of 27.5. Most of the DEGs within this category are constituted by serine proteases. Indeed, molecular functional classification on the one hand unveiled an overrepresentation of transcripts associated with serine-type endopeptidase activity (fold enrichment 6.7), most of which were located within the so-called KLK cluster. On the other hand, we detected genes with serine-type endopeptidase inhibitor activity (fold-enrichment 4.3). In addition to the peptides involved in proteolytic processes, the genes of the steroid hormone biosynthesis cascade were also observed (fold-enrichment 4.4), and most of these genes were associated with ovarian steroidogenesis (fold-enrichment 6.8). Moreover, Jak-STAT signaling pathway transcripts represented a third broad category of differentially expressed genes within the FL2 transcriptome (fold-enrichment 3.3), a majority of which were associated with cyclin-dependent protein serine/threonine kinase activity (fold-enrichment 9.8).

Mammalian phenotypes (MP) Based on the overall assumption to detect selection-induced alterations in testicular reproduction gene expression, we searched the DEG lists for genes associated with reproductive phenotypes within the database Mouse Genome Informatics (MGI - www.informatics.jax.org). We expected genes associated with male and female fecundity phenotypes. Among more than 2000 genotypes associated with reproductive phenotypes, we detected 17 differentially expressed genes in the FL2 mouse line. These genes are summarized in Table 3, with a selection of the most pronounced reproductive phenotypes. This list also included the genes Gdf9 and Agt, which are important for proper ovulation. Both genes were upregulated in FL2 testis. In contrast, Cyp19a1, a gene associated with asthenozoospermia, oligozoospermia and female fecundity, was less expressed in FL2 bucks.

Validation of microarray data using RT-qPCR
A select group of DEGs were re-analyzed using quantitative real-time PCR. The intention of this re-evaluation was to confirm the overall experimental pooling approach of the 2nd set of microarrays. Based only on equal amounts of pooled samples, this 2nd set missed individual variation. Hence, we used the same isolated RNA samples of the 2nd microarray set prior to pooling and analyzed the relative transcriptional abundance of each sample individually via qPCR. Both methods differed in the detection target. Microarray data are based on the binding of 20-bp probes distributed from 5′ to 3′. In contrast, the amplification of a 150–250-bp amplicons is the basis of real-time PCR. Although traces of genomic DNA were removed in the RNA extraction process, the primers were designed as oligonucleotides, particularly sequences binding to intron-overspanning exons, when possible. The overall result of the qPCR suitably demonstrated a correlation with the microarray data (Fig. 6b), thus also experimentally confirming both methods and evaluations as based on the same sample set of RNAs.

The analyzed transcripts were significantly expressed regarding Cyp21a1 (FC 8.1; $p < 0.01$), Sult1e1 (FC −1.8; $p = 0.011$), Ccnd2 (FC 4.2; $p < 0.01$), Cdkl4 (FC −2.9; $p < 0.01$) and Klk1b21 (FC −6.8; $p < 0.01$). The FCs and the p-value of the first microarray experiment exhibited significant similarity to the biologically independent replicates (2nd set) (see Fig. 6a, number in brackets). Consequently, the comparison convincingly verified the suitability of the applied microarray experimental design.

Progesterone concentration in blood
Among the most differentially expressed genes of the microarray results, Cyp21a1 was also detected (FC: 6.3***, 10.0; 1st and 2nd array set, respectively). This

gene encodes the enzyme steroid 21-hydroxylase, which is essential for converting progesterone (P4) to 11-deoxycorticosterone. Although this enzyme is generally associated with adrenal gland activity, there have also been reports of peripheral expression. Indeed, the average signal intensity of the microarray experiment was clearly above background (Ctrl: Avg Signal (log2): 5.84, 5.41 (1st, 2nd set); (FL2: Avg Signal (log2): 8.48, 8.73 (1st, 2nd set)).

To analyze whether this differential expression of Cyp21a1 results from hormonal alterations, the serum progesterone concentration of the males was examined. To minimize the influence of circadian fluctuation, all serum samples were collected at the same time of day.

In Ctrl males, we measured 2.9 ± 0.8 ng/ml total P4 from blood plasma samples (Fig. 7). This concentration was slightly higher compared with C57BL/6 male mice [18]. In contrast, bucks of the FL2 showed significantly increased P4 values of 4.2 ± 0.7 ng/ml ($p < 0.001$). We previously determined serum corticosterone values for FL2 by GC/MS [6]. Corticosterone acts downstream of P4 in the corticosteroid-aldosterone metabolic pathway and hence could be induced by enhanced P4. However, these corticosterone differences did not pass statistical significance (FL2: $30.8 \pm$ ng/ml; Ctrl: $26.7 \pm$ ng/ml).

Discussion
Research on reproductive phenotypes primarily focuses on animal models of reduced fertility. The majority of these models are based on transgenic or knockout approaches. Hence, currently, little is known about increased fertility and the gender specific molecular alteration of this phenotype. In the present study, we investigated whether and to what extent male animals are affected by long-term selected breeding for 'high' reproductive performance of exclusively female focused fecundity traits.

The study object was a worldwide unique mouse fertility line (FL2) generally distinguished by almost doubling number of pups per litter and total litter weight (Tab. 2), obviously reflecting increased ovulation rate, as previously described [4]. To further previously accomplished initial characterization studies founded on notable changes in endocrinology and behavior, the aim of the present study was to shed light on physiological traits, such as sperm motility and transcriptome alterations, which might have been generated on the male side in response to more than 170 generations of female focused selected breeding.

Sperm motility and overall functional sperms are the souls of male reproduction. Therefore, we examined sperm motility as a physiological characteristic. To this end, FL2 and Ctrl spermatozoa were released out of the cauda epididymis and subsequently exposed to 37 °C

Table 3 List of genes associated with reproductive phenotypes of the database Mouse Genome Informatics (MGI - www.informatics.jax.org) mapping to DEGs. Genes and the corresponding mammalian phenotypes (MP) are indicated

Gene Symbol	FC 1st set	FC 2nd set	Term	MP ID
Abca1	−3.2***	−2.8	Male infertility	MP:0001925
			Female infertility	MP:0001926
			Abnormal ovary morphology	MP:0001126
			Abnormal testis morphology	MP:0001146
Ctsh	−3.1***	−3.4	Reproductive system phenotype	MP:0005389
Gdf9	−3.0***	−2.3	Female infertility	MP:0001926
			Abnormal oogenesis	MP:0001931
			Abnormal ovarian folliculogenesis	MP:0001130
			Increased circulating fsh	MP:0001750
			Increased circulating lh	MP:0001751
Agt	−2.8***	−2.1	Decreased ovulation rate	MP:0003355
Gpr133/Adgrd1	−2.8**	−3.0	Female infertility	MP:0001926
Eno4	−2.8***	−2.4	Male infertility	MP:0001925
			Asthenozoospermia	MP:0002675
			Oligozoospermia	MP:0002687
Vcam1	−2.0***	−2.2	Reproductive system phenotype	MP:0005389
Pttg1	−1.7***	−1.8	Reduced fertility	MP:0001921
			Decreased litter size	MP:0001935
			Decreased testis weight	MP:0004852
Hoxa5	−1.6*	−1.6	Abnormal estrous cycle	MP:0001927
Nr2c2	−1.5**	−1.5	Reduced male fertility	MP:0001922
			Oligozoospermia	MP:0002687
Cyp19a1	1.7***	1.7	Male infertility	MP:0001925
			Female infertility	MP:0001926
			Impaired ovarian folliculogenesis	MP:0001129
			Asthenozoospermia	MP:0002675
			Oligozoospermia	MP:0002687
			Leydig cell hyperplasia	MP:0001152
			Increased circulating fsh	MP:0001750
			Increased circulating lh	MP:0001751
Adamts5	1.8***	1.7	Reproductive system phenotype	MP:0005389
Atxn7	1.8***	1.8	Male infertility	MP:0001925
			Female infertility	MP:0001926
Clec4a2	2.0***	1.6	Seminiferous tubule degeneration	MP:0001154
Antxr2	2.6**	2.6	Female infertility	MP:0001926
Ccnd2	3.1***	3.0	Abnormal ovulation	MP:0001928
			Abnormal ovarian folliculogenesis	MP:0001130
			Oligozoospermia	MP:0002687

FC of 1st (*, $p < 0.05$; **, $p < 0.01$; ***, $p < 0.001$) and 2nd microarray set are provided in the table

thermal stress for 5 h and hourly evaluated for several motility characteristics based on CASA system measurements. Initial assessment immediately after preparation (0 h measurement) showed no substantial difference among FL2 and Ctrl bucks with acquired data equivalent to the literature [19]. However, surprisingly, thermal exposure apparently prompted a considerable reduction in overall motility performance for FL2 spermatozoa. The amount of motile and progressively motile sperm was significantly decreased during 5 h thermal stress and

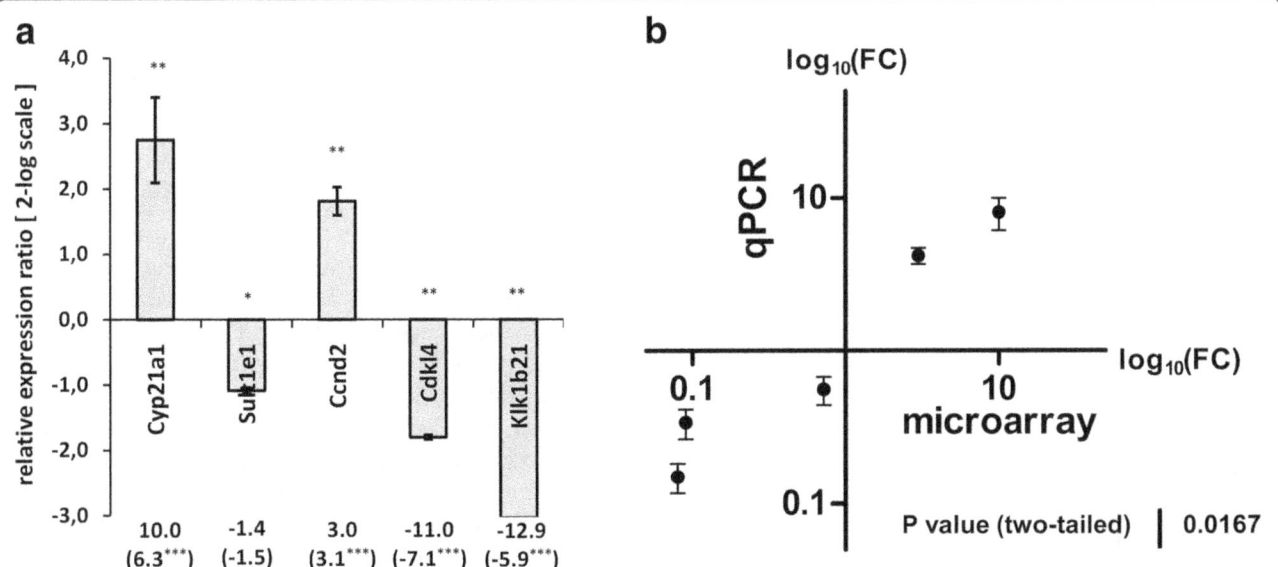

Fig. 6 Real-time qPCR. Validation of the 2nd set of microarrays to verify the pooling approach using quantitative real time PCR as independent method. RNA of individual FL2 and Ctrl mice was used prior of pooling for the microarray. Samples were normalized to a combination of reference genes (36B4, GAPDH, HPRT, B2m) and statistically evaluated employing the Relative Expression Software Tool (REST 2009) [12]. Relative expression ratio are presented as log2 bars (**a**) (* $p < 0.05$; *** $p < 0.001$). Numbers below bars indicate fold change (FC) values calculated for the 2nd set of microarrays comprising the same animals as pooled samples. Numbers in brackets are FCs of the 1st set of microarrays as biologically independent replicates. Scatter blot illustrates the FCs acquired by both microarray (abscissae) and qPCR (ordinates), based on the same set of animals (**b**)

substantial diminishing of entire quality parameters, such as speed of movement (e.g., VCL), Linearity (e.g., LIN) and oscillation (e.g., BCF). Hence, these findings suggest that FL2 sperm substantially suffer from temperature increases of approximately 3 °C higher than the physiological environment (34–35 °C). Thus, initially after exposure, FL2 spermatozoa begin to extensively lose motility capability continuing in minimized motility fractions over the whole experimental period. Spermatozoa able to persist viability are distinguished by speed impairments, which lead to short distances passed compared to Ctrl. Furthermore, consistent with prolonged thermal stress, FL2 spermatozoa showed to increasing loss in the maintenance of curvilinear and oscillatory paths during progression. Apparently under stress, the suitable metabolic activity required for vital motility associated with progressive, fast and wriggled, oscillatory movement can no longer be provided. Loss of granted energy leads to essentially slower and more straightforward FL2 spermatozoa movement. Consequently, the detection of enhanced linearity and oscillation parameters for FL2, such as LIN and BCF, during the thermal stress period is presumably a reflection of comprehensive spermatozoa damage. Furthermore, P4 triggers the hyperactivation of spermatozoa. FL2 females and males exhibit an increase in P4 level [4], see Fig. 7. These elevated P4 levels could reflect the selection for increased litter size to support larger pregnancies. Thus, FL2

spermatozoa might be adapted to high-progesterone intrauterine environments. FL2 sperms are more susceptible to P4 loss, as observed in M199 media, which was used for the CASA experiments in the present study. Hence, the observed decrease of FL2 sperm motility performance alleviated curvilinear velocity and increased linearity.

Thus, the causes of FL2 sperm motility reduction with respect to thermal noxa remain unknown, and clarification of the potential essential restrains in FL2 spermatozoa performance across the entire fertilization process remain elusive. For example, not only metabolic activity but also sperm capacitation and oocyte fertilization performance in single and competitive conditions should be assessed. These future assays together with the findings of the present sperm motility studies will hopefully lead to profound FL2 sperm characterization necessary to generalize FL2 sperm as substantially worse in performance compared to unselected Ctrl lines.

Birth rate breeding data were obtained during a 4-year period. However, these physiological data show a considerable reduction of birth rate for the FL2 line. These data are based on at least 60 breeding pairs per generation, reflecting the overall reduced delivery rate over 14 generations. In addition to these data, we collected data from two-factorial breeding experiments crossing males and females of the Ctrl and the FL2 line in all combinations. Within this setting, we registered no significant

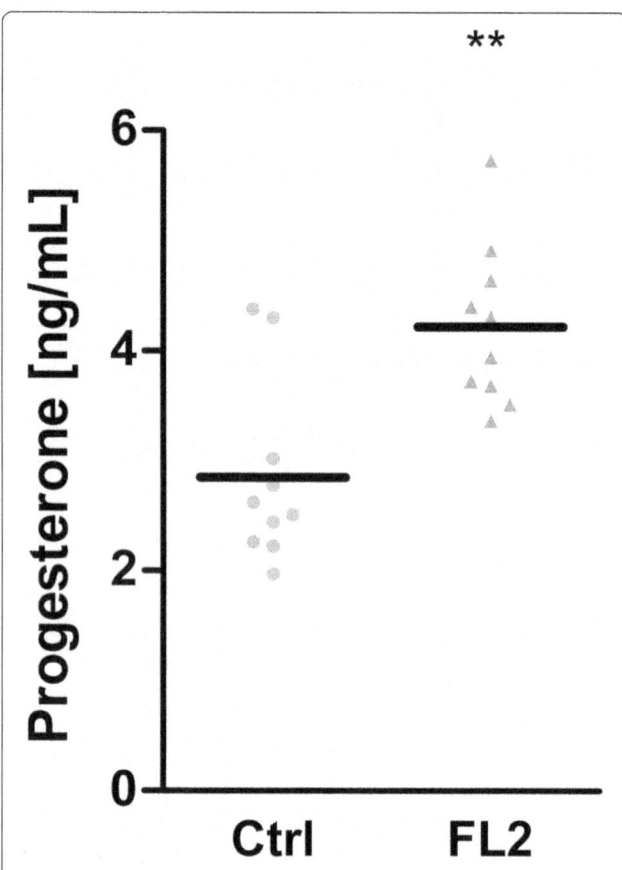

Fig. 7 Progesterone. Plasma progesterone concentrations of FL2 and Ctrl males were analyzed by RIA. Individual and mean levels are indicated. Groups were tested for normal distribution and analyzed by two-tailed t-test (**, $p < 0.01$)

differences in any of the groupings [6]. However, admittedly, the group size of this experiment was considerably less than the obtained breeding values over 4 years.

Furthermore, these findings do not exclusively imply that FL2 males contribute to alleviating mating success. However, previous data indicate no reduced female fecundity performance. For example, FL2 females ovulate approximately 24 oocytes. Moreover, based on in vivo studies, almost every FL2 oocyte was fertilized and developed to an embryo [4]. Taken together, these results may explain the weaker FL2 birth rate resulting from the male gender. Nevertheless, additional in vitro fertilization studies are needed in the future.

Transcriptional alterations (genes, phenotype, cluster)
The comparative whole transcriptome expression profile was used to identify testicular genes whose expression pattern has been potentially affected by the long-term selection for increased female reproductive performance. Thus, we expected some of detected differentially expressed genes (DEGs) to be directly causative of the

female focused breeding process. Other genes might be regulated as 'secondary' effect within the testis. However, in both cases, transcriptional alterations might actually contribute to the FL2 testicular phenotype. Hence, we discussed DEGs in light of male and female reproduction.

Functional annotation clustering revealed that most enrichment scores for the renin-angiotensinogen system (RAS) and for a set of RAS genes encoding serine-type endoprotease activity. RAS is a hormonal system for blood pressure regulation [20]. In addition, there is increasing evidence of the local synthesis of RAS components in various tissues [21]. A majority of differentially expressed RAS-associated genes in FL2 males mapped to tissue kallikrein-related peptidases (KLK). These KLKs are organized in tandem within a cluster on Chr7 [22]. In FL2, most of these differentially expressed KLK transcripts mapped to genes expressed in the murine testis. Other KLK cluster genes not expressed in the testis were indeed not altered within this dataset compared to the Ctrl line. This tissue-specific effect suggests the active regulation rather than decreased activity of the entire KLK cluster locus (Additional file 4).

The physiological function of the kallikrein system within the testis has been only partly elucidated. Using the term KLK in combination with testis, eight publications in the PubMed database were detected. Particularly, KLKs are discussed as prognostic tumor markers [23]. However, the beneficial effect on asthenozoospermia and oligozoospermia of systemic kallikrein administration has been recognized in clinical interventional studies [24, 25]. Consequently, the downregulation of KLKs might potentially explain the decreased sperm motility parameters observed in FL2 bucks.

In addition to the local function of KLKs in the testis [26, 27], there have been reports of kallikrein – kinin systems (KKS) interacting in RAS [28–30]. In fact, the differential expression of KLK genes was accompanied by alterations in the angiotensinogen (Agt) transcriptional level. Agt is one of the major players in the RAS, acting as a precursor of the angiotensin cascade. There is evidence for the direct interaction of kallikrein on the generation of angiotensin (Ang) II in male reproductive organs rather than the 'canonical' enzyme cascade via renin [31]. Nevertheless, the role for angiotensin effectors in the testis is only partially understood.

In addition to its implication in male reproduction, RAS and KLKs are of vital importance for ovarian function, including the systemic action of RAS on blood pressure control and vascular function. There is also evidence for local tissue events of RAS components [32, 33]. Thus, the dysregulation of the Agt gene in transgenic females decreases the ovulation rate [34]. In addition, antagonizing Ang II type 1 receptor (AGTR1) in dominant follicles blocks follicle growth and rupture and decreases Cyp19a1 and Ccnd2 mRNA expression. Surprisingly, the opposite pattern was observed in FL2 testis. The Agt transcript level was decreased, while Cyp19a1 and Ccnd2 transcripts were elevated. Furthermore, the perfusion of Ang II in rat ovaries induces progesterone (P4) synthesis. Consistently, we detected enhanced serum P4 concentrations in FL2 males.

Ccnd2 is important in the testicular and ovarian cell cycles [35, 36]. Genetic intervention leads to oligozoospermia and abnormal ovarian folliculogenesis [37]. The Ccnd2 gene is induced by FSH and regulates cell proliferation in the gonads [37]. Downregulation by dihydrotestosterone (DHT) induces cell cycle arrest in granulosa cells [38]. Moreover, D-type cyclins can be regulated via the intracellular Jak/STAT signaling pathway [39–41]. In the present study, functional annotation clustering indicated the enrichment of Jak/STAT pathway-associated genes in FL2 testis (see Table 4 for dedicated genes). This pathway has been implicated in the regulation of cell cycle progression mediating extracellular signaling. Within FL2 testis Jak/Stat dependent receptors (LIFR, CNTFR, Il3ra) and SOS1 are repressed, whereas intracellular Ccnd2 is upregulated. LIF and CNTF signaling acts on testis and ovarian functions [42–46]. Furthermore, we detected the decreased expression of cyclin-dependent kinase (Cdk) Cdkl4, Cdk14 and Cdk15 in FL2 testis compared to Ctrl. Although the action of cyclin/cdk complexes on cell cycle progression is widely known [47, 48], the interaction of these particular genes within male or female germinal organs remains unknown [49]. However, studies have shown that Jak/STAT is associated with homeostasis of stem cell niche within Drosophila testis and ovaries [50].

Moreover, a group of differentially expressed genes mapped to the steroid hormone synthesis and ovarian steroidogenesis. Steroids are of vital importance for male and female reproduction regulation. The Abca1 transcript is involved in cholesterol trafficking. Consequently, transgenic intervention leads to impaired cholesterol levels [51, 52], resulting in male and female infertility paired with abnormal reproductive organ morphology in mice [53–55]. This rate-limiting gene was decreased in FL2 testis. Further, Cyp19a1, the gene encoding aromatase, which is essential for converting testosterone to estradiol (E2), was upregulated in FL2. This gene has been targeted by several transgenic interventions leading to phenotypic alterations, such as impaired spermatogenesis and ovarian folliculogenesis, manifested in the reduced fertility of both sexes [56–59]. Interestingly, the expression level of Sult1e1 was also diminished in FL2 testis compared to Ctrl. E2 is inactivated by the steroid sulfatase Sult1e1, commonly regarded as the only relevant sulfotransferase for estrogens [60]. The enzyme is present in the ovary, whereas expression and testicular function have not been reported [61]. However, sulfated steroid inactivation and transport and local storage of steroids have been postulated [60]. Moreover, the overexpression of Sult1e1 counteracts the estrogen-mediated proliferation and decreased expression of D-cyclin [62]. Indeed, the level of Ccnd2 was elevated, while Sult1e1 mRNA was decreased in FL2 testis compared to Ctrl.

Cyp1b1 expression was decreased in FL2 males. This enzyme catalyzes several reactions among the hydroxylation of estradiol for the elimination process [63]. Although Cyp1b1 has been genetically targeted, information for its effects on the reproductive system is still lacking [64, 65]. However, polymorphisms in Cyp1b1 may augment the risk of abnormal sperm parameters [66].

A key enzyme in the biosynthesis of steroids is 3β-hydroxysteroid dehydrogenase, which is essential for the

Table 4 Gene Ontology (GO) term enrichment analysis for testis of FL2 versus Ctrl line using the platform Database for Annotation, Visualization and Integrated Discovery (DAVID). Functional Annotation Clustering of differentially expressed genes in FL2 versus Ctrl testis. The analysis for the top overrepresented GO terms with fold-enrichment scores and Benjamin's correction are indicated

Term	Genes	Fold Enrichment	Benjamini
Renin-angiotensin system	AGT, KLK1B21, KLK1B8, KLK1B22, KLK1B9, KLK1B11, KLK1B26, KLK1B24, KLK1B1, MAS1	27.5	6×10^{-09}
Serine-type endopeptidase activity	KLK1B21, KLK1B8, KLK1B22, KLK1B9, KLK1B11, KLK1B26, KLK1B4, KLK1B24, KLK1B1, EGFBP2, KLK1B16, KLK1B27, CTSH	6.7	9×10^{-05}
Serine-type endopeptidase inhibitor activity	SERPINE3, SERPINA3G, AGT, SERPINA3C, SERPINA3A	4.3	8×10^{-01}
Ovarian steroidogenesis	CYP1B1, HSD3B6, GDF9, CYP19A1	6.76	6×10^{-01}
Steroid hormone biosynthesis	CYP1B1, HSD3B6, CYP21A1, CYP19A1	4.43	7×10^{-01}
Fatty acid biosynthetic process	ELOVL3, AGMO, TECR, OXSM, NDUFS6	5.77	9×10^{-01}
Jak-STAT signaling pathway	CCND2, SOS1, LIFR, CNTFR, IL3RA	3.32	7×10^{-01}
Cyclin-dependent protein serine/threonine kinase activity	CDK15, CDK14, CDKL4	9.81	9×10^{-01}

production of progesterone (P4) and functions as a precursor for the synthesis of androgens, estrogens, glucocorticoids and mineralocorticoids [67]. In this experimental setting, the Hsd3b6 transcript was downregulated in FL2 testis compared to Ctrl. In mice, the isoform type VI (Hsd3b6) is expressed in the testis from puberty to adulthood in interstitial Leydig cells as and placental giant trophoblast cells, potentially suggesting embryonic-placental implantation [67, 68]. Decreased Hsd3b6 levels, as observed in FL2, have been associated with adverse male reproductive function in several studies [69–71], however we currently lack genetic interventional studies clearly dissecting its functional role in reproduction.

Furthermore, Cyp21a1 was considerably upregulated. This enzyme acts downstream of P4, converting the steroid to 11-deoxycorticosterone. To examine the differential mRNA expression of both enzymes, we analyzed the level of P4 in serum using RIA. P4 concentrations in Ctrl males were the same magnitude as previously published [18, 72]. Interestingly, the P4 levels of FL2 bucks were 50% elevated. Increased P4 levels have previously been observed in FL2 females compared to the Ctrl line [4]. In addition, FL2 females have more CLs [4]. However, no correlation between CLs and P4 levels has been reported [4]. For males, we currently have no indication for elevated steroidogenic cell numbers in FL2 testis (data not shown), suggesting the 'active' regulation of high P4 levels.

In terms of considering an integrating picture, some of the effects observed in FL2 males were consistent with the work of Harini et al., showing the effects of prenatal exposure to progesterone on male reproductive parameters at adulthood [73]. After birth, the females received three injections of P4 over pregnancy (7 mg P4/kg), adopting human interventional studies to mouse species. Mice of the resulting F1 generation were paired with normal cycling females for recording of reproduction traits. Intriguingly fetal exposure to high P4 levels apparently resulted in several adverse reproductive effects, e.g., decreased motile and viable sperms, low levels of Hsd3b activity and reduced fertility index, consistent with the observed effects on birth rate.

Although P4 is a 'female hormone', serum P4 levels in males do not differ from female levels out of luteal phase [74, 75] or in post-menopausal women [76, 77]. In men, P4 has been associated with processes of spermatogenesis and hence functional loss will lead to infertility [78, 79]. P4 triggers hyperactivation in spermatozoa, leading to a rapid increase in Ca^{2+}, which is potentially mediated by the $GABA_A$ receptor [80]. This hormone can also serve as a precursor for corticosterone to feed the corticosteroid/aldosterone pathway. We previously analyzed serum corticosterone levels in males by GC/MS [6]. Although corticosterone levels are slightly elevated in FL2 compared to Ctrl males,

these differences were not statistically significant. In males, this hormone is either testicular or of adrenal origin. However, P4 is known as a hormone for female reproduction and ovulation. Other approaches to obtain an integrative perspective of these FL2 transcriptomics data should consider the observed differential KLK expression: It is known that progesterins and other steroids control the expression of multiple kallikreins [81, 82]. For example, epostane, representing an antagonist of 3β-hydroxysteroid dehydrogenase (HSD3), can inhibit ovulation while exogenous P4 induces kallikrein and reverses the anovulatory effect of the antagonist [83]. Nevertheless, steroids induce KLK expression, whereas we detected the decreased expression of several KLKs.

Conclusion

To our knowledge, the present study is the first report depicting an in-depth transcriptomics analysis of high-fertility male mice with female-focused breeding on augmented reproductive performance compared to a still existing 'founder' population of an unselected control line.

In combination with additional physiological phenotype studies, we showed that long-term selection accompanied by an increased ovulation rate [4] in females also affects the testicular gene expression of transcripts associated with female fecundity and ovulation. In addition, decreased sperm motility parameters for FL2 bucks could be revealed when semen is exposed to stress response analyses. Furthermore, the FL2 line displayed a markedly alleviated birth rate in long-term breeding assessment studies. However, based on the birth rate data obtained from standard breeding values in the present study, we cannot differentiate to what extent the FL2 gender is responsible for the decreased mating rate. Hence, the results of the present study should be considered as a starting point to shed light on the definite transcriptomics and genotypic alterations distinctly manifested on male side during 40 years of murine outbred breeding towards high-fertility phenotypes.

Additional files

Additional file 1: List of RT-qPCR primers used. (DOCX 52 kb)

Additional file 2: Distribution of probe cell intensity of unprocessed raw data (.CEL data) (a) Distribution of signal intensity values (.CHP data) after normalization by the Robust Multiarray Average with Signal Space Transformation algorithm (SST-RMA). (b) Consideration of both plots implies an overall successful hybridization experiment for all processed MTA 1.0 microarrays. (ZIP 63 kb)

Additional file 3: List (Excel file) of differential expressed transcripts (FL2 vs. Ctrl). DEGs were defined as: FC > 1.5; FC < −1.5 (of 1st and 2nd set); one-way ANOVA (of 1st set). (XLS 988 kb)

Additional file 4: Kallikrein (KLK)-Cluster within the murine genome is located on Chromosome 7 and spans from 28.26 to 28.74 cM. The orientation and organization within the cluster is illustrated. Color intensity of up (red) and down (green) regulated gene expression in FL2 testis in relation to the Ctrl group is depicted. (TIFF 766 kb)

Abbreviations
BCF: Beat cross frequency; CASA: Computer- assisted sperm analysis; CL: Corpus luteum; Ctrl: Control mouse line; DABG: Detection above background; DEG: Differential expressed genes; E2: Estradiol; EC: Expression console software; FC: Fold-change; FDR: False discovery rate; FL2: Fertility mouse line 2; GO: Gene ontology; KKS: Kallikrein – kinin systems; KLK: Kallikrein-related peptidases; LIN: Linearity; P4: Progesterone; PCA: Principal component analysis; PS: Probe set; RAS: Renin-angiotensinogen system; RMA: Robust multiarray average (algorithm); SST: Signal space transformation; TAC: Transcriptome analysis console software; VCL: Curvilinear velocity

Acknowledgements
The authors would like to thank Ursula Antkewitz, Swanhild Rodewald, Hannelore Klückmann, and Petra Reckling for excellent technical assistance. In addition, the authors would also like to thank the staff of the Mouse Laboratory, Leibniz Institute for Farm Animal Biology, for conducting the selection experiments, animal care and sample collection.

Funding
This work was supported by a grant from the German Research Foundation (DFG) WE2458/10-1.

Authors' contributions
MM, AS, JS and JMW conceived the experimental design. ML and NR conducted the animal breeding and line selection and provided substantial section on the manuscript. AS and MM conducted the sperm motility assay. DK organized and performed the RNA preparation, RNA labeling and RNA microarray hybridization and worked on evaluation of the transcriptional data and provided substantial sections on the manuscript. MM conducted the statistical evaluation of microarray data and the qPCR re-analysis, interpreted the results and drafted the manuscript. AS, JS and JMW provided substantial suggestions on interpretation of the data sets and critically read and helped finalize the manuscript. All authors approved the final manuscript.

Competing interests
The authors declare that they have no competing interests.

Author details
[1]Institute of Reproductive Biology, University of Rostock, Rostock, Germany. [2]Institute of Genetics and Biometry, Leibniz Institute for Farm Animal Biology (FBN), Dummerstorf, Germany. [3]Institute of Immunology, University of Rostock, Rostock, Germany. [4]Leibniz Institute for Farm Animal Biology (FBN), Institute of Reproductive Biology, FBN Dummerstorf, Wilhelm-Stahl-Allee 2, 18196, Dummerstorf, Germany.

References
1. Dietl G, Langhammer M, Renne U. Model simulations for genetic random drift in the outbred strain Fzt: DU. Arch Tierzucht. 2004;47(6):595–604.
2. Langhammer M, Michaelis M, Hoeflich A, Sobczak A, Schoen J, Weitzel JM. High-fertility phenotypes: two outbred mouse models exhibit substantially different molecular and physiological strategies warranting improved fertility. Reproduction. 2014;147(4):427–33.
3. L SLB. Die reproduktive Lebensleistung auf Fruchtbarkeit selektierter Labormauslinien. Arch Tierzucht. 1982;25:275–81.
4. Spitschak M, Langhammer M, Schneider F, Renne U, Vanselow J. Two high-fertility mouse lines show differences in component fertility traits after long-term selection. Reprod Fertil Dev. 2007;19(7):815–21.
5. Michaelis M, Langhammer M, Hoeflich A, Reinsch N, Schoen J, Weitzel JM. Initial characterization of an outbreed mouse model for male factor (in)fertility. Andrology. 2013;1(5):772–8.
6. Langhammer M, Michaelis M, Hartmann MF, Wudy SA, Sobczak A, Nurnberg G, Reinsch N, Schon J, Weitzel JM. Reproductive performance primarily depends on the female genotype in a two-factorial breeding experiment using high-fertility mouse lines. Reproduction. 2017;153(3):361–8.
7. Huang d W, Sherman BT, Lempicki RA. Systematic and integrative analysis of large gene lists using DAVID bioinformatics resources. Nat Protoc. 2009; 4(1):44–57.
8. Huang d W, Sherman BT, Lempicki RA. Bioinformatics enrichment tools: paths toward the comprehensive functional analysis of large gene lists. Nucleic Acids Res. 2009;37(1):1–13.
9. Harris MA, Clark J, Ireland A, Lomax J, Ashburner M, Foulger R, Eilbeck K, Lewis S, Marshall B, Mungall C, et al. The gene ontology (GO) database and informatics resource. Nucleic Acids Res. 2004;32(Database issue):D258–61.
10. Mi H, Lazareva-Ulitsky B, Loo R, Kejariwal A, Vandergriff J, Rabkin S, Guo N, Muruganujan A, Doremieux O, Campbell MJ, et al. The PANTHER database of protein families, subfamilies, functions and pathways. Nucleic Acids Res. 2005;33(Database issue):D284–8.
11. Ye J, Coulouris G, Zaretskaya I, Cutcutache I, Rozen S, Madden TL. Primer-BLAST: a tool to design target-specific primers for polymerase chain reaction. BMC bioinformatics. 2012;13:134.
12. Pfaffl MW, Horgan GW, Dempfle L. Relative expression software tool (REST) for group-wise comparison and statistical analysis of relative expression results in real-time PCR. Nucleic Acids Res. 2002;30(9):e36.
13. Brussow KP, Schneider F, Wollenhaupt K, Tuchscherer A. Endocrine effects of GnRH agonist application to early pregnant gilts. The Journal of reproduction and development. 2011;57(2):242–8.
14. Alm H, Kuhlmann S, Langhammer M, Tuchscherer A, Torner H, Reinsch N. Occurrence of polyovular follicles in mouse lines selected for high fecundity. The Journal of reproduction and development. 2010;56(4):449–53.
15. Hurst LD, Pal C, Lercher MJ. The evolutionary dynamics of eukaryotic gene order. Nat Rev Genet. 2004;5(4):299–310.
16. Lercher MJ, Urrutia AO, Hurst LD. Clustering of housekeeping genes provides a unified model of gene order in the human genome. Nat Genet. 2002;31(2):180–3.
17. Raghupathy N, Durand D. Gene cluster statistics with gene families. Mol Biol Evol. 2009;26(5):957–68.
18. Schneider JS, Stone MK, Wynne-Edwards KE, Horton TH, Lydon J, O'Malley B, Levine JE. Progesterone receptors mediate male aggression toward infants. Proc Natl Acad Sci U S A. 2003;100(5):2951–6.
19. Carroll M, Luu T, Robaire B. Null mutation of the transcription factor inhibitor of DNA binding 3 (id3) affects spermatozoal motility parameters and epididymal gene expression in mice. Biol Reprod. 2011;84(4):765–74.
20. Williams B. Drug discovery in renin-angiotensin system intervention: past and future. Ther Adv Cardiovasc Dis. 2016;10(3):118–25.
21. Bader M, Ganten D. Update on tissue renin-angiotensin systems. J Mol Med. 2008;86(6):615–21.
22. Lundwall A. Old genes and new genes: the evolution of the kallikrein locus. Thromb Haemost. 2013;110(3):469–75.
23. Schmitt M, Magdolen V, Yang F, Kiechle M, Bayani J, Yousef GM, Scorilas A, Diamandis EP, Dorn J. Emerging clinical importance of the cancer biomarkers kallikrein-related peptidases (KLK) in female and male reproductive organ malignancies. Radiol Oncol. 2013;47(4):319–29.
24. Izzo PL, Canale D, Bianchi B, Meschini P, Esposito G, Menchini Fabris GF, Fasani R, Ohnmeiss H. The treatment of male subfertility with kallikrein. Andrologia. 1984;16(2):156–61.
25. Kamidono S, Hazama M, Matsumoto O, Takada KI, Tomioka O, Ishigami J. Kallikrein and male subfertility. Usefulness of high-unit kallikrein tablets. Andrologia. 1981;13(2):108–20.
26. Fink E, Schill WB, Fiedler F, Shimamoto K, Krassnigg F, Frick J: The tissue kallikrein-kinin system in human seminal plasma–biochemical and functional aspects. Advances in experimental medicine and biology 1986, 198 Pt A:291–297.
27. Schill WB, Miska W. Possible effects of the kallikrein-kinin system on male reproductive functions. Andrologia. 1992;24(2):69–75.
28. Su JB. Different cross-talk sites between the renin-angiotensin and the kallikrein-kinin systems. Journal of the renin-angiotensin-aldosterone system: JRAAS. 2014;15(4):319–28.

29. Campbell DJ. The renin-angiotensin and the kallikrein-kinin systems. Int J Biochem Cell Biol. 2003;35(6):784–91.

30. Shen B, El-Dahr SS. Cross-talk of the renin-angiotensin and kallikrein-kinin systems. Biol Chem. 2006;387(2):145–50.

31. Maruta H, Arakawa K. Confirmation of direct angiotensin formation by kallikrein. The Biochemical journal. 1983;213(1):193–200.

32. Goncalves PB, Ferreira R, Gasperin B, Oliveira JF. Role of angiotensin in ovarian follicular development and ovulation in mammals: a review of recent advances. Reproduction. 2012;143(1):11–20.

33. Herr D, Bekes I, Wulff C. Local renin-angiotensin system in the reproductive system. Front Endocrinol. 2013;4:150.

34. Hefler LA, Gregg AR. Influence of the angiotensinogen gene on the ovulatory capacity of mice. Fertil Steril. 2001;75(6):1206–11.

35. Lee J, Kanatsu-Shinohara M, Morimoto H, Kazuki Y, Takashima S, Oshimura M, Toyokuni S, Shinohara T. Genetic reconstruction of mouse spermatogonial stem cell self-renewal in vitro by Ras-cyclin D2 activation. Cell Stem Cell. 2009;5(1):76–86.

36. Shimizu T, Hirai Y, Miyamoto A. Expression of cyclins and cyclin-dependent kinase inhibitors in granulosa cells from bovine ovary. Reproduction in domestic animals = Zuchthygiene. 2013;48(5):e65–9.

37. Sicinski P, Donaher JL, Geng Y, Parker SB, Gardner H, Park MY, Robker RL, Richards JS, McGinnis LK, Biggers JD, et al. Cyclin D2 is an FSH-responsive gene involved in gonadal cell proliferation and oncogenesis. Nature. 1996; 384(6608):470–4.

38. Pradeep PK, Li X, Peegel H, Menon KM. Dihydrotestosterone inhibits granulosa cell proliferation by decreasing the cyclin D2 mRNA expression and cell cycle arrest at G1 phase. Endocrinology. 2002;143(8):2930–5.

39. Singh SR, Chen X, Hou SX. JAK/STAT signaling regulates tissue outgrowth and male germline stem cell fate in drosophila. Cell Res. 2005;15(1):1–5.

40. Arbouzova NI, Zeidler MP. JAK/STAT signalling in drosophila: insights into conserved regulatory and cellular functions. Development. 2006;133(14): 2605–16.

41. Li Z, Shen J, WK W, Yu X, Liang J, Qiu G, Liu J. Leptin induces cyclin D1 expression and proliferation of human nucleus pulposus cells via JAK/STAT, PI3K/Akt and MEK/ERK pathways. PLoS One. 2012;7(12):e53176.

42. Murphy MJ, Halow NG, Royer PA, Hennebold JD. Leukemia Inhibitory Factor is Necessary for Ovulation in Female Rhesus Macaques. Endocrinology. 2016:en20161283.

43. Ilha GF, Rovani MT, Gasperin BG, Antoniazzi AQ, Goncalves PB, Bordignon V, Duggavathi R. Lack of FSH support enhances LIF-STAT3 signaling in granulosa cells of atretic follicles in cattle. Reproduction. 2015;150(4):395–403.

44. Watanobe H, Habu S. Ciliary neurotrophic factor, a gp130 cytokine, regulates preovulatory surges of luteinizing hormone and prolactin in the rat. Neuroendocrinology. 2001;74(5):281–7.

45. Jenab S, Morris PL. Testicular leukemia inhibitory factor (LIF) and LIF receptor mediate phosphorylation of signal transducers and activators of transcription (STAT)-3 and STAT-1 and induce c-fos transcription and activator protein-1 activation in rat Sertoli but not germ cells. Endocrinology. 1998;139(4):1883–90.

46. Bornstein SR, Rutkowski H, Vrezas I. Cytokines and steroidogenesis. Mol Cell Endocrinol. 2004;215(1–2):135–41.

47. Gerard C, Goldbeter A. Dynamics of the mammalian cell cycle in physiological and pathological conditions. Wiley interdisciplinary reviews Systems biology and Medicine. 2016;8(2):140–56.

48. Duronio RJ, Xiong Y. Signaling pathways that control cell proliferation. Cold Spring Harb Perspect Biol. 2013;5(3):a008904.

49. Sobinoff AP, Sutherland JM, McLaughlin EA. Intracellular signalling during female gametogenesis. Mol Hum Reprod. 2013;19(5):265–78.

50. Zhang Z, Pan C, Zhao Y. Hedgehog in the drosophila testis niche: what does it do there? Protein & cell. 2013;4(9):650–5.

51. McNeish J, Aiello RJ, Guyot D, Turi T, Gabel C, Aldinger C, Hoppe KL, Roach ML, Royer LJ, de Wet J, et al. High density lipoprotein deficiency and foam cell accumulation in mice with targeted disruption of ATP-binding cassette transporter-1. Proc Natl Acad Sci U S A. 2000;97(8):4245–50.

52. Juan T, Veniant MM, Helmering J, Babij P, Baker DM, Damore MA, Bass MB, Gyuris T, Chhoa M, Li CM, et al. Identification of three loci affecting HDL-cholesterol levels in a screen for chemically induced recessive mutations in mice. J Lipid Res. 2009;50(3):534–45.

53. Christiansen-Weber TA, Voland JR, Wu Y, Ngo K, Roland BL, Nguyen S, Peterson PA, Fung-Leung WP. Functional loss of ABCA1 in mice causes severe placental malformation, aberrant lipid distribution, and kidney glomerulonephritis as well as high-density lipoprotein cholesterol deficiency. Am J Pathol. 2000;157(3):1017–29.

54. Wahrle SE, Jiang H, Parsadanian M, Kim J, Li A, Knoten A, Jain S, Hirsch-Reinshagen V, Wellington CL, Bales KR, et al. Overexpression of ABCA1 reduces amyloid deposition in the PDAPP mouse model of Alzheimer disease. J Clin Invest. 2008;118(2):671–82.

55. Hamon Y, Broccardo C, Chambenoit O, Luciani MF, Toti F, Chaslin S, Freyssinet JM, Devaux PF, McNeish J, Marguet D, et al. ABC1 promotes engulfment of apoptotic cells and transbilayer redistribution of phosphatidylserine. Nat Cell Biol. 2000;2(7):399–406.

56. Robertson KM, Simpson ER, Lacham-Kaplan O, Jones ME. Characterization of the fertility of male aromatase knockout mice. J Androl. 2001;22(5):825–30.

57. Robertson KM, O'Donnell L, Jones ME, Meachem SJ, Boon WC, Fisher CR, Graves KH, McLachlan RI, Simpson ER. Impairment of spermatogenesis in mice lacking a functional aromatase (cyp 19) gene. Proc Natl Acad Sci U S A. 1999;96(14):7986–91.

58. Fisher CR, Graves KH, Parlow AF, Simpson ER. Characterization of mice deficient in aromatase (ArKO) because of targeted disruption of the cyp19 gene. Proc Natl Acad Sci U S A. 1998;95(12):6965–70.

59. Britt KL, Drummond AE, Cox VA, Dyson M, Wreford NG, Jones ME, Simpson ER, Findlay JK. An age-related ovarian phenotype in mice with targeted disruption of the Cyp 19 (aromatase) gene. Endocrinology. 2000;141(7): 2614–23.

60. Geyer J, Bakhaus K, Bernhardt R, Blaschka C, Dezhkam Y, Fietz D, Grosser G, Hartmann K, Hartmann MF, Neunzig J, et al. The role of sulfated steroid hormones in reproductive processes. J Steroid Biochem Mol Biol. 2016;

61. Brown KA, Dore M, Lussier JG, Sirois J. Human chorionic gonadotropin-dependent up-regulation of genes responsible for estrogen sulfoconjugation and export in granulosa cells of luteinizing preovulatory follicles. Endocrinology. 2006;147(9):4222–33.

62. Xu Y, Liu X, Guo F, Ning Y, Zhi X, Wang X, Chen S, Yin L, Li X. Effect of estrogen sulfation by SULT1E1 and PAPSS on the development of estrogen-dependent cancers. Cancer Sci. 2012;103(6):1000–9.

63. Li F, Zhu W, Gonzalez FJ. Potential role of CYP1B1 in the development and treatment of metabolic diseases. Pharmacol Ther. 2017;

64. Buters JT, Doehmer J, Gonzalez FJ. Cytochrome P450-null mice. Drug Metab Rev. 1999;31(2):437–47.

65. Libby RT, Smith RS, Savinova OV, Zabaleta A, Martin JE, Gonzalez FJ, John SW. Modification of ocular defects in mouse developmental glaucoma models by tyrosinase. Science. 2003;299(5612):1578–81.

66. Hu W, Yang H, Sun J, Zhang Q, Wang J, Lu L, Zhang J, Qin Y, Xia Y, Wang X. Polymorphisms in CYP1B1 modify the risk of idiopathic male infertility with abnormal semen quality. Clin Chim Acta. 2011;412(19–20):1778–82.

67. Yamamura K, Doi M, Hayashi H, Ota T, Murai I, Hotta Y, Komatsu R, Okamura H. Immunolocalization of murine type VI 3beta-hydroxysteroid dehydrogenase in the adrenal gland, testis, skin, and placenta. Mol Cell Endocrinol. 2014;382(1):131–8.

68. Baker PJ, Sha JA, McBride MW, Peng L, Payne AH, O'Shaughnessy PJ. Expression of 3beta-hydroxysteroid dehydrogenase type I and type VI isoforms in the mouse testis during development. Eur J Biochem. 1999;260(3):911–7.

69. Abarikwu SO, Iserhienrhien BO, Badejo TA. Rutin- and selenium-attenuated cadmium-induced testicular pathophysiology in rats. Hum Exp Toxicol. 2013;32(4):395–406.

70. Pasha HF, Rezk NA, Selim SA, Abd El Motteleb DM. Therapeutic effect of spermatogonial stem cell on testicular damage caused by lead in rats. Gene. 2016;592(1):148–53.

71. Swathy SS, Panicker S, Indira M. Effect of exogenous selenium on the testicular toxicity induced by ethanol in rats. Indian J Physiol Pharmacol. 2006;50(3):215–24.

72. Gonzalez Deniselle MC, Liere P, Pianos A, Meyer M, Aprahamian F, Cambourg A, Di Giorgio NP, Schumacher M, De Nicola AF, Guennoun R. Steroid Profiling in Male Wobbler Mouse, a Model of Amyotrophic Lateral Sclerosis. Endocrinology. 2016;157(11):4446–4460.

73. Harini C, Sainath SB, Reddy PS. Recovery of suppressed male reproduction in mice exposed to progesterone during embryonic development by testosterone. Reproduction. 2009;137(3):439–48.

74. Zumoff B, Miller L, Levin J, Levit CD, Miller EH, Heinz U, Kalin M, Denman H, Jandorek R, Rosenfeld RS. Follicular-phase serum progesterone levels of nonsmoking women do not differ from the levels of nonsmoking men. Steroids. 1990;55(12):557–9.

75. Zumoff B, Miller L, Poretsky L, Levit CD, Miller EH, Heinz U, Denman H, Jandorek R, Rosenfeld RS. Subnormal follicular-phase serum progesterone levels and elevated follicular-phase serum estradiol levels in young women with insulin-dependent diabetes. Steroids. 1990;55(12):560–4.

76. Trabert B, Falk RT, Stanczyk FZ, McGlynn KA, Brinton LA, Xu X. Reproducibility of an assay to measure serum progesterone metabolites that may be related to breast cancer risk using liquid chromatography-tandem mass spectrometry. Horm Mol Biol Clin Invest. 2015;23(3):79–84.

77. Oettel M, Mukhopadhyay AK. Progesterone: the forgotten hormone in men? The aging male: the official journal of the International Society for the Study of the Aging Male. 2004;7(3):236–57.

78. Aquila S, De Amicis F. Steroid receptors and their ligands: effects on male gamete functions. Exp Cell Res. 2014;328(2):303–13.

79. Abid S, Gokral J, Maitra A, Meherji P, Kadam S, Pires E, Modi D. Altered expression of progesterone receptors in testis of infertile men. Reprod BioMed Online. 2008;17(2):175–84.

80. Kon H, Takei GL, Fujinoki M, Shinoda M. Suppression of progesterone-enhanced hyperactivation in hamster spermatozoa by gamma-aminobutyric acid. The Journal of reproduction and development. 2014;60(3):202–9.

81. Lawrence MG, Lai J, Clements JA. Kallikreins on steroids: structure, function, and hormonal regulation of prostate-specific antigen and the extended kallikrein locus. Endocr Rev. 2010;31(4):407–46.

82. Yousef GM, Diamandis EP. An overview of the kallikrein gene families in humans and other species: emerging candidate tumour markers. Clin Biochem. 2003;36(6):443–52.

83. Tanaka N, Espey LL, Stacy S, Okamura H. Epostane and indomethacin actions on ovarian kallikrein and plasminogen activator activities during ovulation in the gonadotropin-primed immature rat. Biol Reprod. 1992;46(4):665–70.

5

Association of variation in the sugarcane transcriptome with sugar content

Prathima P. Thirugnanasambandam[1,2†], Nam V. Hoang[1,3†], Agnelo Furtado[1], Frederick C. Botha[4] and Robert J. Henry[1,5*] (iD)

Abstract

Background: Sugarcane is a major crop of the tropics cultivated mainly for its high sucrose content. The crop is genetically less explored due to its complex polyploid genome. Sucrose synthesis and accumulation are complex processes influenced by physiological, biochemical and genetic factors, and the growth environment. The recent focus on the crop for fibre and biofuel has led to a renewed interest on understanding the molecular basis of sucrose and biomass traits. This transcriptome study aimed to identify genes that are associated with and differentially regulated during sucrose synthesis and accumulation in the mature stage of sugarcane. Patterns of gene expression in high and low sugar genotypes as well as mature and immature culm tissues were studied using RNA-Seq of culm transcriptomes.

Results: In this study, 28 RNA-Seq libraries from 14 genotypes of sugarcane differing in their sucrose content were used for studying the transcriptional basis of sucrose accumulation. Differential gene expression studies were performed using SoGI (*Saccharum officinarum* Gene Index, 3.0), SAS (sugarcane assembled sequences) of sugarcane EST database (SUCEST) and SUGIT, a sugarcane Iso-Seq transcriptome database. In total, about 34,476 genes were found to be differentially expressed between high and low sugar genotypes with the SoGI database, 20,487 genes with the SAS database and 18,543 genes with the SUGIT database at FDR < 0.01, using the Baggerley's test. Further, differential gene expression analyses were conducted between immature (top) and mature (bottom) tissues of the culm. The DEGs were functionally annotated using GO classification and the genes consistently associated with sucrose accumulation were identified.

Conclusions: The large number of DEGs may be due to the large number of genes that influence sucrose content or are regulated by sucrose content. These results indicate that apart from being a primary metabolite and storage and transport sugar, sucrose may serve as a signalling molecule that regulates many aspects of growth and development in sugarcane. Further studies are needed to confirm if sucrose regulates the expression of the identified DEGs or vice versa. The DEGs identified in this study may lead to identification of genes/pathways regulating sucrose accumulation and/or regulated by sucrose levels in sugarcane. We propose identifying the master regulators of sucrose if any in the future.

Keywords: Sucrose, Transcriptome, High and low sugar genotypes, Sucrose genes, Sugarcane transcriptome

* Correspondence: robert.henry@uq.edu.au
†Equal contributors
[1]Queensland Alliance for Agriculture and Food Innovation, The University of Queensland, St. Lucia, QLD 4072, Australia
[5]The University of Queensland, Room 2.245, Level 2, The John Hay Building, Queensland Biosciences Precinct [#80], 306 Carmody Road, St Lucia, QLD 4072, Australia
Full list of author information is available at the end of the article

Background

Among the domesticated grasses, sugarcane and sweet sorghum have undergone extensive selection for high accumulation of sucrose that serves as the primary sources of sugars for human and animal consumption, as well as ethanol production for fuel [1].The maturing sugarcane culm represents both an economically important and physiologically interesting experimental system to study the dynamics of carbohydrate partitioning and metabolism associated with the accumulation of high concentrations of sucrose. A distinctive feature of sugarcane is that high levels of sucrose storage occurs only in the culm parenchyma cells as against in other plants where storage of sugar or other storage molecule/s occurs in terminal sink organs such as tubers, grains, or fleshy fruits. Sucrose concentration that peaks in the sugarcane culm during the end of the vegetative cycle (called ripening) is utilized for the sexual reproductive phase and the remaining reserve is re-mobilized to produce new vegetative structures unlike the pattern in monocarpic annuals where there is a single cycle of storage and utilization for the reproductive phase [2]. In addition, sucrose is the only major form in which reduced carbon is exported from the source and hence all cellular processes outside the source are dependent on the mobilisation and utilisation of sucrose. Sucrose is the dominant storage reserve in sugarcane in contrast to most other plant stems that store polysaccharides such as starch or fructans with a low concentration of sucrose. As sugarcane matures, there is a shift in carbon partitioning from that of insoluble and respiratory components towards the osmotically active sucrose [3].

Although sugarcane stores the highest concentration (reaching about 0.7 M) of sucrose in the plant kingdom, studies on the physiological, biochemical and genetic basis of sucrose synthesis and accumulation have been limited compared to those in model plants like Arabidopsis or rice that do not accumulate high levels of sucrose. There are very few studies of sucrose accumulation primarily focusing on the sugarcane culm. Often these studies in sugarcane have reported a network of genes related to cell wall metabolism, carbohydrate metabolism, stress responses and regulatory processes [4–11]. Microarray analysis of sugarcane genotypes that varied in sucrose content revealed that many of the genes associated with high sucrose content showed overlap with drought data sets, but appeared to be mostly independent from abscisic acid signalling [12]. A large expressed sequence tag (EST) study of the sugarcane transcriptome and physiological, developmental and tissue-specific gene regulation was initiated in Brazil [13]. Sugarcane cultivars differing in both maximum sucrose accumulation (in Brix) capacity and accumulation dynamics during growth and culm maturation were studied cDNA microarrays and developmentally

regulated genes related to hormone signalling, stress response, sugar transport, lignin biosynthesis and fibre content were identified [12]. An expression profiling of a set of genes associated with sucrose accumulation was studied using quantitative real time reverse transcription PCR (qRT-PCR) in 13 genotypes of sugarcane and its progenitor species including *S. officinarum*, *S. spontaneum* and related genera *Erianthus arundinaceus* [14]. High brix genotypes exhibited increased expression of sucrose nonfermenting related kinases and cellulose synthases in an expression study comparing high and low brix genotypes of sugarcane using qRT-PCR [15]. In another transcriptome study using next generation sequencing (NGS) [16] enrichment of transcripts involved in a network of sucrose synthesis, accumulation, storage and retention in relation to the agronomic characteristics of the genotypes contrasting for rust resistance was observed. Casu et al. [9] proposed that sucrose accumulation may be regulated by a network of genes induced during culm maturation which included clusters of genes with roles that contribute to key physiological processes including sugar translocation and transport, fibre synthesis, membrane transport, vacuole development and function, and abiotic stress tolerance. These studies show that the sugarcane culm is a composite organ associated with numerous diverse functions other than sucrose storage. A gene networking pattern involving genes associated with culm maturation and sucrose accumulation, sugar transport, vacuole development, lignification, suberisation and abiotic-stress tolerance can be inferred from these studies. The present study aimed to identifying transcripts that were associated with sucrose accumulation using a set of seven high sugar and seven low sugar genotypes by expression profiling of mature and immature culm tissues and bioinformatic analyses of culm transcriptomes. The upregulation of several thousands of transcripts associated with sucrose biosynthesis was demonstrated in the high sugar, and maturing culm of sugarcane. This is the first transcriptome study showing the association of expression of a large number of genes with sucrose synthesis and accumulation in the sugarcane culm tissue.

Methods

Plant material and phenotypic data collection

Fully grown, disease free 12 months old plants grown in the field in a randomized complete block design were selected for analysis. The genotypes were derived from a sugarcane population provided by Sugar Research Australia (SRA), Brisbane, Australia, previously described in [17]. Sugar content measured as Brix (a measure of the soluble solids in sugarcane juice) was used for classifying the genotypes as high and low sugar genotypes (Table 1). The low sugar genotypes had a Brix range of 17–18.4 while the high sugar genotypes had a Brix range of 19.4–21.4. The Brix at

Table 1 Sugar, Brix, fibre and pedigree information for each genotype

Genotype	Code	Brix (degree)	Fibre (%)	Pedigree	Group
QC02–402	G01	18.3	31.39	Commercial hybrid	Low sugar
QA02–1009	G02	18.3	43.36	Commercial hybrid	Low sugar
QN05–803	G10	17.8	47.74	Commercial hybrid	Low sugar
KQB07–24739	G16	18.4	48.2	Introg. BC1 (*S. spont*)	Low sugar
KQB09–23137	G18	17.7	33.53	BC1 (*S. spont*)	Low sugar
KQB09–20620	G19	17.8	39.88	Introg. BC1 (*S. spont*)	Low sugar
KQB09–20432	G20	18.3	49.79	Introg. BC1 (*S. spont*)	Low sugar
QN05–1743	G04	21.4	34.62	Commercial hybrid	High sugar
QN05–1509	G05	20.1	40.43	Commercial hybrid	High sugar
QS99–2014	G06	20.9	31.21	Commercial hybrid	High sugar
QA96–1749	G07	19.4	46.33	Commercial hybrid	High sugar
Q200	G09	19.7	37.6	Commercial hybrid	High sugar
KQB07–23990	G13	20	36.25	Introg. BC1 (*S. spont*)	High sugar
KQ08–2850	G14	20.3	43.84	Introg. BC3 (*Erianthus* sp)	High sugar

Introg-introgression; BC-back cross; *S. spont- Saccharum spontaneum*

the point of collection was used for defining the high and low sugar groupings. These genotypes may have high or low sugar content in other environments. A wide variation in sugar content was not obtained as these genotypes were commercial cultivars and introgression lines in the breeding pipeline with a sugar content above 16 and fibre content below 15 on a fresh weight basis. Culm samples (from both top and bottom tissues, the 4th internode from top and 3rd internode from the bottom of the cane) were collected from four representative stalks and pooled for each internode sample. All samples were collected between 10 am to 2 pm to limit the diurnal fluctuations in the transcriptome. After collection, the samples were immediately flash frozen in liquid nitrogen and stored at −80 °C until RNA extraction. In addition, HPLC (high performance liquid chromatography) and NIR (near infrared spectroscopy) was used to measure the sugar composition and fibre content on a fresh weight basis. A sub-sample of each genotype was processed through a mechanical grinder, a component of the SpectraCane system (Biolab, Australia) and scanned by NIR for fibre content, Brix and sugar content (commercial cane sugar - CCS). For details see Additional file 1: Tables S1a, S2, S3; Figure S1.

Sample collection and preparation for RNA-Seq

The frozen sugarcane samples were pulverized using a Retsch TissueLyser (Retsch, Haan, Germany) at a frequency of 30/S for 1 min 30 s and about 1 g of ground sample powder was used for RNA extraction. RNA extractions were conducted as described by Furtado et al. (2014) [18] employing a Trizol kit (Invitrogen) and a Qiagen RNeasy Plant minikit (#74134, Qiagen, Valencia, CA, USA). For RNA quality and quantity assessment, a NanoDrop8000 spectrophotometer (ThermoFisher Scientific,

Wilmington, DE, USA), and an Agilent Bioanalyser 2100 with the Agilent RNA 6000 Nano kit (Agilent Technologies, Santa Clara, CA, USA) were used. Only RNA samples with a RIN value of >7.5 were chosen for library preparations. About 3 μg each of 28 internodal RNA samples was used for indexed-library preparation (average insert size of 200 bp) with a TruSeq stranded with Ribo-Zero Plant Library Prep Kit for preparing total RNA library (Illumina Inc.) as described in [19]. The library was subjected to sequencing in two lanes (equimolar) using an Illumina HiSeq4000 instrument to obtain paired-end (PE) read of 150 bp. The library preparation and sequencing was conducted at the Translational Research Institute, The University of Queensland, Australia.

RNA-Seq data processing

Read adapter and quality trimming were performed in CLC Genomics Workbench v9.0 (CLC-GWB, CLC Bio-Qiagen, Aarhus, Denmark) with a quality score limit of <0.01 (equivalent to Phred Q score ≥ 20), and allowing a maximum of two ambiguous nucleotides. Only PE reads with a length ≥ 35 bp were kept for further analyses. Further information on the RNA Bioanalyser profiles, raw RNA-Seq reads, trimming, quality parameters including size distribution and GC content is described in detail in [19] and in Additional file 1; Table S1b. Table 2 gives the details of reads from each genotype (top and bottom internode tissues) after quality trimming.

Differential gene expression (DGE) analyses

Using the CLC-GWB v9.5.1 software, RNA-Seq experiments were performed with a minimum length fraction of 0.9 and a minimum similarity fraction of 0.8. The

Table 2 RNA sequence data obtained for each genotype

Low sugar genotypes			High sugar genotypes		
Code	Number of paired end reads		Code	Number of paired end reads	
	Top	Bottom		Top	Bottom
G01	19,224,245	41,184,328	G04	24,837,412	22,869,941
G02	42,714,286	79,760,012	G05	8,578,357	47,973,820
G10	15,278,340	80,811,427	G06	12,971,379	24,967,893
G16	28,875,842	46,037,690	G07	25,204,740	23,694,629
G18	64,789,858	66,628,598	G09	36,810,635	36,366,607
G19	24,882,334	30,954,783	G13	19,443,531	24,196,385
G20	6,529,582	45,500,054	G14	37,845,062	19,172,922

number of reads per kilobase per million mapped reads (RPKM) was used for normalization [20]. The CLC-GWB provides a comprehensive RNA-Seq tool for differential gene expression accompanied by statistical analyses. The Baggerley's test that is used in this case [21] is the proportion-based statistical analysis that uses raw count data (un-transformed, not-previously-normalized) as input for setting up the experiment and uses total or unique gene/exon reads for calculating the differentially expressed genes. This test compares counts by considering the proportions that the counts for each gene make-up of the total sum of counts in each group. That is, it takes into account the proportion of every genotype in a group for a gene to be considered as differentially expressed. When Edge test [22] in CLC-GWB (an equivalent tool to EdgeR available in R Package v3.4.0) was used, this consistency (of differential expression across all genotypes in a group) was not observed. Similarly, the Differential expression for RNA-Seq tool available in the recent version of CLC GWB 10.1.1 gave a different set of results for the DGE experiments (data not shown here) and hence was not included for further analyses. As sugarcane genotypes differ genetically among and between each other, the criterion was to select only those genes that were differentially expressed despite the genetic differences inherent to the genotypes. For example, a gene was considered differentially expressed only when it is consistently differentially expressed in all the seven genotypes in one group in comparison with all the seven genotypes in the other group. Further the Baggerley test also corrects for the differences in the sample sizes (within and between library variations) by comparing the expression levels at the level of proportions rather than raw counts [21]; CLC manual).

The reads for each genotype in the high and low sugar groups were separately mapped against reference databases, the *Saccharum officinarum* gene indices (SoGI), the sugarcane Iso-Seq transcriptome database (SUGIT, TSA accession number GFHJ01000000) and the sugarcane assembled sequences (SAS) from the sugarcane

expressed sequence tags database (SUCEST). The SoGI database was downloaded from the DFCI gene indices [23] which had adequate gene or protein function descriptions. In the present case, the SoGI dataset represented 282,683 ESTs that resulted in 121,342 unique sequences after clustering. A collection of ~240,000 ESTs generated by the SUCEST project from 26 cDNA libraries from different sugarcane tissues sampled at various developmental stages [24] were assembled into 43,141 distinct contigs using CAP3 [25]. This set of 43,141 contigs make up the SAS database. The SAS database was not annotated and the annotation was performed using the BLASTX against the nr protein database with an e value of 10^{-5}, for 100 hits using the high-performance computing facility (HPC), at The University of Queensland, Australia. In addition, we used a newly constructed SUGIT, sugarcane long reads database described in [26]. In brief, the database was derived from a pooled RNA sample collection including those genotypes used in this study, plus leaf and root tissue samples of 22 commercial and introgressed sugarcane genotypes. The basic descriptions of the databases are given in Table 3 and the methodology is summarised in Fig. 1.

Identification of differentially expressed transcripts

For all the RNA-Seq experiments, involving high and low sugar groups, low sugar samples were used as the references, for comparing top and bottom (immature and mature culm tissues respectively), bottom internode sample was used as the reference for identifying DEGs that were upregulated or down regulated. This means, if one transcript was up-regulated in the reference group, it was down-regulated in the group being compared, and vice versa. Proportion based statistical analysis (Baggerley's Test) and a Volcano plot were used to compare gene expression levels in the two groups that were considered for differential gene expression (high and low sugar, top and bottom internode samples) in terms of the $\log_2$ fold change (at FDR 0.01). The DEGs were further sorted and selected at three different fold change levels, i.e., above and equal to 2, above and equal to 10 fold and below 2

Table 3 Databases used for RNA Seq experiments

Features	SoGI	SUGIT	SAS
Total number of contigs	121,342	107,598	43,141
Total number of bases	88,397,709	166,929,028	35,730,322
Longest contig (bp)	4854	18,858	6193
Shortest contig (bp)	100	307	56
N50 (bp)	729	1994	827
N75 (bp)	642	1269	641

SoGI-*Saccharum officinarum* gene indices; SUGIT-Sugarcane Iso-Seq transcriptome database; SAS-sugarcane assembled sequences; bp- base pairs

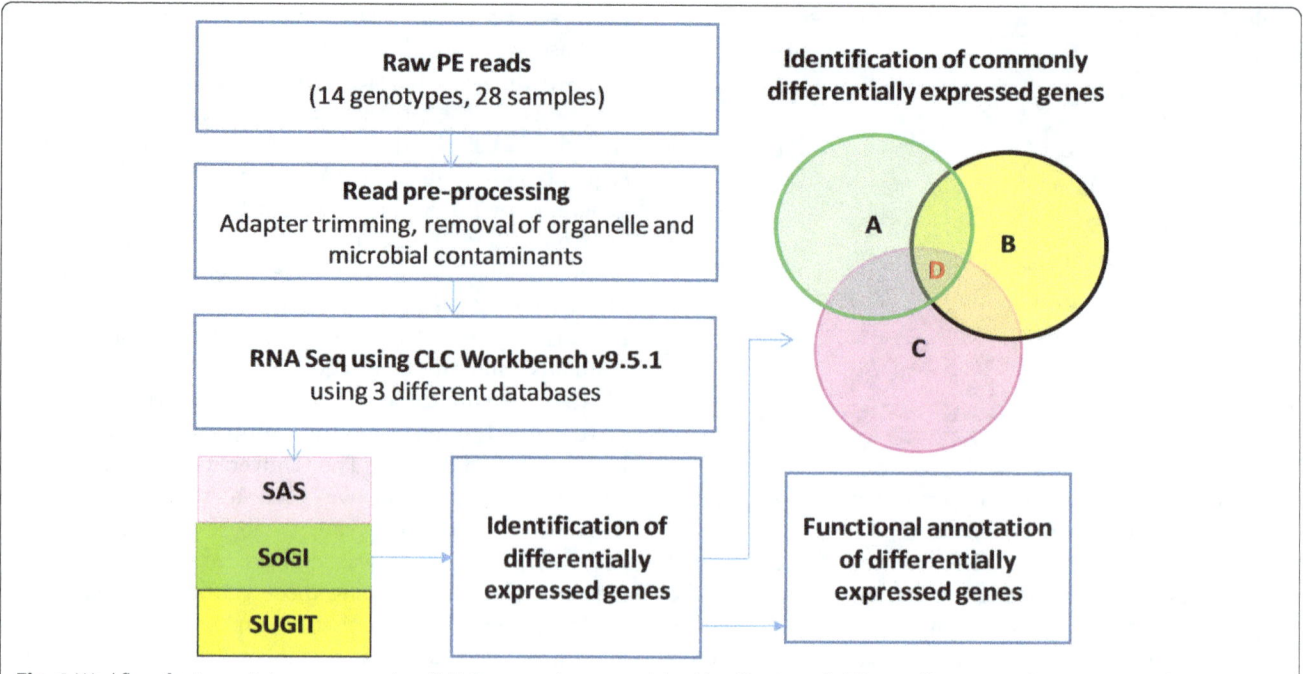

Fig. 1 Workflow for transcriptome sequencing, RNA-Seq experiments and the identification of differentially expressed transcripts. High and low sugar genotypes as well as mature and immature culm samples of sugarcane genotypes were compared in the study

fold change to identify highly expressed and those expressed at low levels.

Functional annotation of identified differentially expressed transcripts

Functional annotation of the transcripts was performed using MapMan categories [27] using BlastX (e-value $\leq 10^{-5}$, with a cut off value of 80% similarity) against *Arabidopsis thaliana* and *Oryza* sp. and SwissProt/UniProt Plant Proteins. In addition Blast2GO [28] followed by KEGG pathway mapping analyses were performed for the DEGs.

Validation of gene expression using quantitative real-time PCR (qPCR)

In addition, a correlation analysis was performed to validate the expression levels of eight selected transcripts from the RNA-Seq analyses in this study using qPCR expression values of the same transcripts extracted from a separate study [29]. The RPKM values obtained for four tissue samples (two top and two bottom internodes) of two genotypes (QC02–402 and QN05–803) were correlated against the respective qPCR expression values (Cq qPCR normalised gene expression), using Microsoft Excel 2013.

Results

RNA-Seq analyses and identification of differentially expressed genes

The mapping of reads to each reference database is shown in the Table 4. The results of the different RNA-

Seq experiments (hereafter SoGI-DGE, SUGIT-DGE and SAS-DGE) are listed in Table 5 and the differential gene expression patterns are depicted as Volcano plots in the Fig. 2. For all DEGs, FDR 0.01 and a fold change of ≥ 2 were used as cut off values. In SoGI-DGE, with high and low sugar bottom internode samples (HSB vs LSB), out of the total 121,342 transcripts, 34,375 showed upregulation and 101 transcripts showed down regulation in high sugar genotypes when compared to low sugar genotypes. When low sugar top and low sugar bottom internode samples (LST vs LSB) were compared, 30,723 transcripts were differentially expressed, upregulated in the low sugar top internode sample, while 86 transcripts were down regulated. When high sugar top and high sugar bottom intermodal samples (HST vs HSB) were considered, 31 transcripts were found to be upregulated in high sugar bottom internode sample compared to the corresponding top. In SUGIT-DGE, out of 107,598 transcripts, 18,411 transcripts were upregulated while 132 transcripts were down regulated in high sugar bottom internode sample compared to low sugar bottom internode sample (HSB vs LSB). 11,713 transcripts were differentially expressed between low sugar top and low sugar bottom intermodal samples (LST vs LSB), wherein 11,599 transcripts were upregulated and 114 transcripts were down regulated. In the SAS-DGE, 19,808 transcripts showed differential expression (19,782 upregulated, 26 down regulated) out of 43,141 transcripts of the SAS reference database in the HSB vs LSB comparison and 20,487 transcripts were differentially expressed

Table 4 Reads from each genotype with details of mapping to three different databases SoGI, SUGIT and SAS

Code	Top internode				Bottom internode			
	Trimmed reads	Reads mapped (%)			Trimmed reads	Reads mapped (%)		
		SoGI	SUGIT	SAS		SoGI	SUGIT	SAS
High sugar genotypes								
G04	24,837,412	52.12	70.65	54.86	22,869,941	46.03	69.38	52.64
G05	8,578,357	48.63	69.34	57.65	47,973,820	49.63	69.35	52.56
G06	12,971,379	52.41	74.20	56.09	24,967,893	47.72	71.51	54.77
G07	25,204,740	51.85	70.79	55.45	23,694,629	50.42	70.45	55.07
G09	36,810,635	52.11	73.64	56.87	36,366,607	54.16	80.44	56.56
G13	19,443,531	52.45	69.10	51.22	24,196,385	50.35	68.45	54.57
G14	37,845,062	52.38	67.85	56.98	19,172,922	51.38	68.90	55.90
Low sugar genotypes								
G01	19,224,245	47.95	70.76	54.87	41,184,328	52.45	78.06	55.75
G02	42,714,286	45.45	66.43	52.42	79,760,012	54.89	70.48	57.26
G10	15,278,340	51.09	72.10	54.05	80,811,427	55.42	81.09	56.65
G16	28,875,842	52.19	75.47	56.07	46,037,690	55.19	80.31	56.07
G18	64,789,858	54.13	71.65	56.69	66,628,598	54.47	72.14	53.71
G19	24,882,334	46.73	67.67	54.39	30,954,783	56.50	72.57	56.97
G20	6,529,582	55.51	73.44	55.82	45,500,054	55.51	80.15	55.82

SoGI-*Saccharum officinarum* gene indices; SUGIT-Sugarcane Iso-Seq transcriptome database; SAS-sugarcane assembled sequences

(20,449 up regulated, 38 down regulated) in the LST vs LSB comparison (see Table 5 for details). However, in the SAS-DGE, there were more DEGs in the HST vs HSB comparison, with 2826 DEGs. This comparison resulted in only 21 and 31 DEGs with the SUGIT and SoGI DGEs respectively (Fig. 3). In addition, the common and unique transcripts among the three comparisons in three different

Table 5 Details of differentially expressed genes obtained from three different RNA-Seq experiments at FDR 0.05 and 0.01

Experiment	FDR 0.01			FDR 0.05			
	Up	Down	Total	Up	Down	Total	Reference
SoGI-DGE							
HST vs HSB	–	31	31	1	58	59	HSB
HSB vs LSB	101	34,375	34,476	111	43,109	43,220	LSB
LST vs LSB	86	30,723	30,809	102	42,149	42,251	LSB
SUGIT-DGE							
HST vs HSB	–	21	21	–	38	38	HSB
HSB vs LSB	132	18,411	18,543	140	30,129	30,269	LSB
LST vs LSB	114	11,599	11,713	142	26,626	26,768	LSB
SAS-DGE							
HST vs HSB	2591	235	2826	4752	383	5135	HSB
HSB vs LSB	38	20,449	20,487	41	24,233	24,274	LSB
LST vs LSB	26	19,782	19,808	36	24,438	24,474	LSB

HSB, high sugar bottom internode; HST, high sugar top internode; LSB-low sugar bottom internode; LST-low sugar top internode; SoGI-*Saccharum officinarum* gene indices; SUGIT-Sugarcane Iso-Seq transcriptome database; SAS-sugarcane assembled sequences

DGEs were found (Figs. 4 and 5). For additional information on experimental set up and statistical analyses, see Additional files 2: Table S4-S6 (complete list of DEGs in the SOGI-DGE), Additional files 3: Table S7-S9 (complete list of DEGs in the SUGIT-DGE), and Additional files 4: Table S10-S12 (complete list of DEGs in the SAS-DGE). The results of the qPCR expression values were found to be significantly correlated with the RPKM values for selected genes ($r = 0.629$, $p < 0.001$, $n = 32$, $df = 30$). The details of qPCR validation analysis are provided in Additional file 5: Table S13 and Figure S2.

Identification of consistently differentially expressed transcripts between high and low sugar genotypes

The results of the DGE analyses are given in the Table 5. The DEGs at different fold change cut off values, i.e. ≥ 2, ≥ 10 and < 2 fold changes were identified. This resulted in the identification of DEGs that are expressed at high levels (10 fold and above), low levels (< 2) apart from the cut off of 2 and above (Table 6). In addition, to check for specific sucrose/sugar related transcripts, filtering was done for "sucrose" and "sugar" as key words in the DGE experiment files as the DEGs were in large numbers. Although some transcripts related to sucrose/sugar may have been missed, this approach helped screening the large number of DEGs. At the fold change value of 2 and above, the sucrose and sugar related genes were 63, 68 and 49 in HSB vs LSB and 75, 74, and 60 in in LST vs LSB using SoGI-DGE, SUGIT-

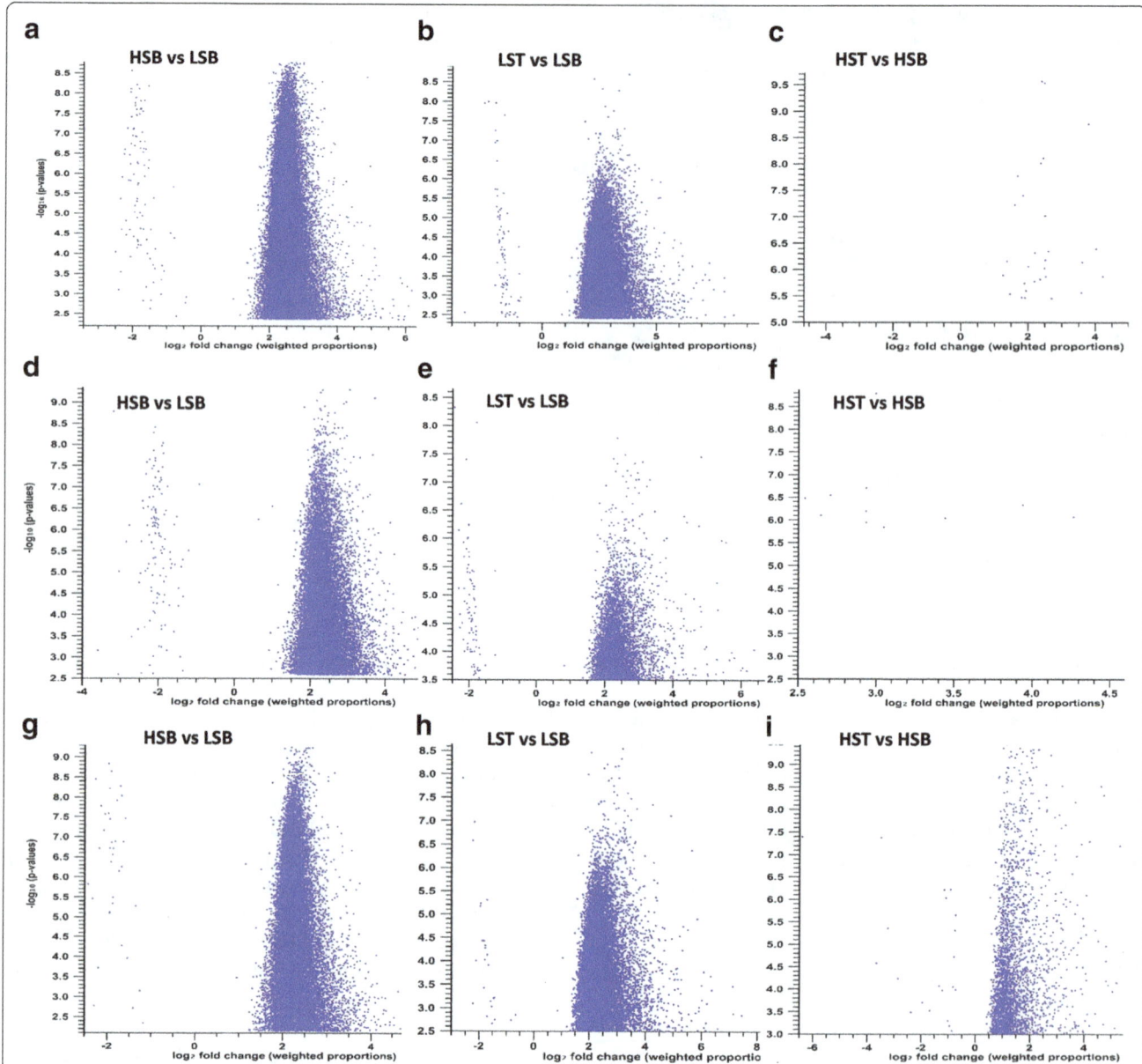

Fig. 2 Volcano plot depiction of the DGEs in different groups using **a, b, c)** *Saccharum officinarum* gene indices, SoGI; **d, e, f)** Sugarcane Iso-Seq transcriptome database, SUGIT; **g, h, i)** Sugarcane assembled sequences, SAS. LST, low sugar top internode; LSB, low sugar bottom internode; HST, high sugar top internode; HSB, high sugar bottom internode; In HSB vs LSB, HSB shows upregulation of transcripts, whereas in LST vs LSB, LST shows upregulation of transcripts; in HST vs HSB, HSB shows a clear upregulation of transcripts using SAS database, though very few transcripts showed differential expression in other two databases. Please note that there was no DGE detected in HST vs LST

DGE and SAS-DGE respectively. These transcripts are listed in the Additional file 6: Tables S14–22 and some are listed in the Tables 7, 8 and 9. At the fold change value of 10 and above, the sucrose/sugar related transcripts were very few in number and included sucrose synthase (SuSy), sucrose transporter (SuT), sucrose phosphate synthase (SPS) and a SWEET transporter (Table 6). Further, SuSy2 and SuT3 were consistently present in all three sets of DEGs for LST vs LSB, at the maximum fold change

value of 10 and above, showing upregulation in LST. In HSB vs LSB, SuSy 2 and SuT 2 were observed in SoGI-DGE and SUGIT-DGE, upregulated in HSB, whereas no sucrose/sugar related transcripts were present in SAS-DGE at this fold change. At the fold change value of below 2, there were no sucrose/sugar related transcripts for these two comparisons in any of the DGEs. In HST vs HSB, sucrose/sugar related transcripts were not found in SoGI- and SUGIT-DGEs, however, at the fold change cut off value of

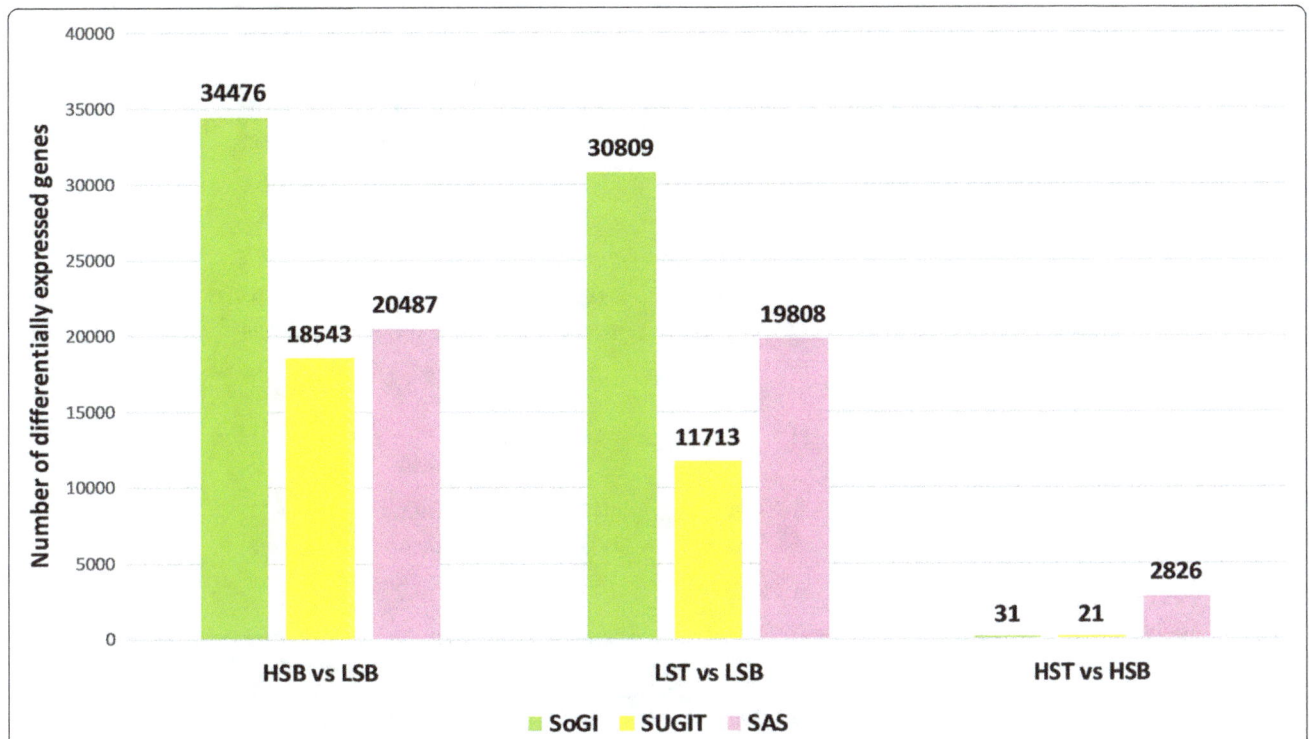

Fig. 3 Graphical representation of DGE experiments with three different databases, *Saccharum officinarum* gene indices, SoGI; Sugarcane Iso-Seq transcriptome database, SUGIT; Sugarcane assembled sequences, SAS. LST, low sugar top internode; LSB, low sugar bottom internode; HST, high sugar top internode; HSB, high sugar bottom internode

below <2, sucrose phosphate phosphatase (SPP) 2, SuSy, SWEET 16 like transporter and a sugar phosphate phosphate translocator were found in SAS-DGE, showing upregulation in HSB. Interestingly, the DEGs at fold change ≥10 in HST vs HSB were related to phenyl propanoid pathway genes like terpene cyclase (TC), phenyl ammonia lyase (PAL), chalcone synthase (CHS), cinnamoyl CoA reductase (CCoAR), ferruloyl esterase (FE), laccase 7-like (LAC), β-expansin (BE) 1a and ethylene responsive transcripts etc., in SoGI, SUGIT and SAS-DGEs (Table 6). Overall, genes specific to sucrose

synthesis and accumulation were enriched in the HSB vs LSB and LST vs LSB experiments, while genes for secondary metabolites, were found to be enriched in the case of the HST vs HSB comparison. There were no DEGs in HST vs LST experiment in the three DGE analyses.

Gene ontology annotation

The gene ontology annotation using MapMan resulted in grouping and classification of the DEGs into different functional categories. The DGE analysis between LST and LSB was almost similar in number and composition to the

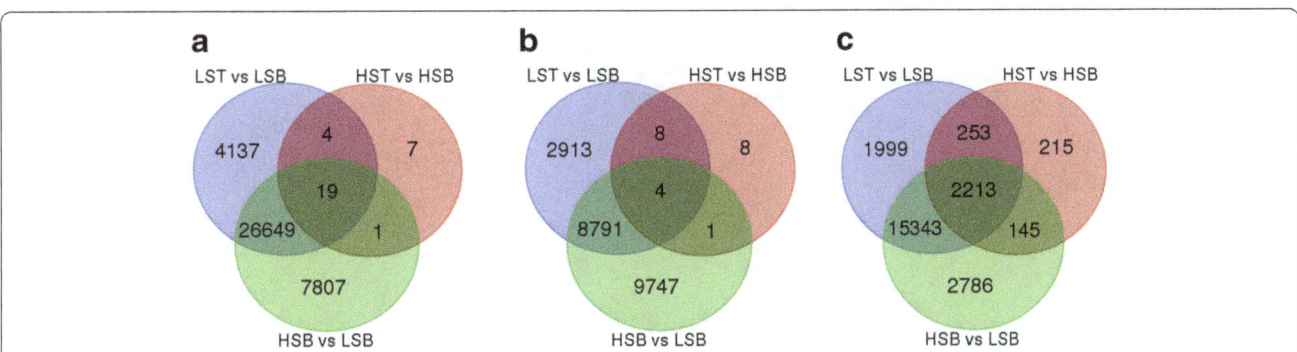

Fig. 4 Venn diagrams depicting the common and unique DEGs obtained from three different comparisons between mature and immature culm tissues of high and low sugar genotypes. LST, low sugar top internode; LSB, low sugar bottom internode; HST, high sugar top internode; HSB, high sugar bottom internode; **a**, **b** and **c**), RNA-Seq using *Saccharum officinarum* gene indices, SoGI; Sugarcane Iso-Seq transcriptome database, SUGIT; Sugarcane assembled sequences, SAS, respectively

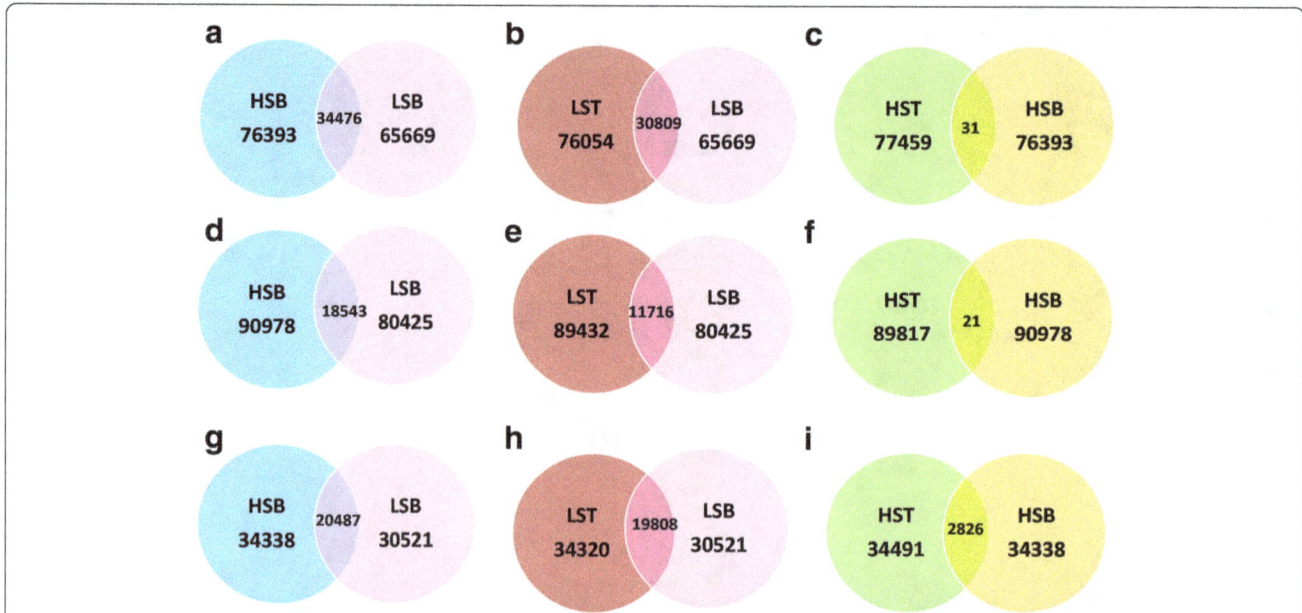

Fig. 5 A schematic representation of global and differential gene expression between and within the groups using three databases. **a, b, c)** *Saccharum officinarum* gene indices, SoGI (121,342 ESTs); **d, e, f)** SUGIT-Sugarcane Iso-Seq transcriptome database (107,598 contigs) and **f, g, h)** Sugarcane assembled sequences, SAS (43,141) contigs. LST, low sugar top internode; LSB, low sugar bottom internode; HST, high sugar top internode; HSB, high sugar bottom internode. The numbers within the intersection are the number of DEGs between the groups compared. The numbers in the circle gives the number of transcripts expressed against the reference database with a RPKM cut off of >0

DEGs obtained between HSB vs LSB (Additional file 7: Figure S3, Table S23).

Upregulated transcripts in high sugar genotypes when compared with low sugar genotypes

The DGE analyses of HSB vs LSB and LST vs LSB were showing a similar trend and the two sets of DEGs had an extensive overlap (see Additional file 7: Figure S3). About 89.3% (with SoGI) of the transcripts differentially expressed were similar in both the comparisons (63% in SUGIT and 96.8% in case of SAS). Hence only the DEGs of HSB vs LSB is considered for further discussion. Only a few transcripts are discussed here. For a complete list DEGs of all the three DGE analyses, refer Additional files 2, 3 and 4: Tables S4-S12. In addition, a list of unique and commonly expressed transcripts in each group was prepared (Additional File 8: Tables S24–28) for

Table 6 Details of differentially expressed genes obtained from three different RNA seq experiments at FDR 0.01 at three different fold change settings

	HSB vs LSB			LST vs LSB			HST vs HSB		
	Fold change			Fold change			Fold change		
	≥2	≥10	<2	≥2	≥10	<2	≥2	≥10	<2
SoGI- DGE	(34476) *SuSy 2* **SuT2**	(1723) **SuSy 2** **SuT2**	(7)	(30809) **SuSy 2** **SuT2**	(3279) **SuSy 2** SuT	(5)	(31)	(5) PAL CHS CCoAR	–
SUGIT- DGE	(18543) **SuSy 2** SuT2a	(952) **SuSy 2** SuT2a	(2)	(11716) **SuSy 2** SuT2	(872) **SuSy 2** SuT3	(2)	(21)	(3) TC	–
SAS-DGE	(20487) **SuSy 2** SuT3	(575)	(4)	(19808) **SuSy 2** SuT2	(1706) **SuSy 1** SuT3 SWEET3 SPS1	(2)	(2826)	(74+) (3-) FE, LAC, BE	(794+) (172-) SPP2 SWEET3 SWEET16 SPT

SuSy-sucrose synthase; *SuT*-sucrose transporter; *SWEET*-bidirectional sugar transporter sweet; *FL*-Ferruloyl esterase; *LAC*-laccase; *BE*-beta expansin; *SPP*-sucrose phosphate phosphatase; *SPT*-sugar phosphate phosphate translocator; HSB, high sugar bottom internode; HST, high sugar top internode; LSB-low sugar bottom internode; LST-low sugar top internode; SoGI-*Saccharum officinarum* gene indices; SUGIT-Sugarcane Iso-Seq transcriptome database; SAS-sugarcane assembled sequences. The numbers in brackets indicate the number of DEGs obtained at that fold change setting, while the sucrose and sugar related genes within the DEGs are indicated below them. (+) and (−) denote upregulation and down regulation respectively. The genes in bold letters are present in all the three DEGs

Table 7 DEGs obtained between high sugar bottom (HSB) vs low sugar bottom (LSB) internode samples using three databases SoGI, SUGIT and SAS. Shown here are some of the sucrose/sugar related transcripts

Feature ID	Description	Fold change (original values)	FDR < 0.01
SoGI-DGE			
CA255667	Sucrose synthase 2	−28.85	0.01
CA207180	Sucrose transporter 2	−18.14	5.53E-03
CA267680	Sugar-phosphate isomerase-like protein	−17.31	4.84E-03
TC112923	Sugar-starvation induced protein	−15.38	7.76E-03
TC121981	Sucrose synthase 3	−8.42	1.88E-04
TC146639	UDP-sugar pyrophosphorylase	−8.39	3.65E-04
TC113610	Sucrose phosphate phosphatase	−8.29	1.66E-03
CA072415	Sucrose non-fermenting related protein kinase	−7.5	3.33E-05
CA258700	Possible sugar transferase	−6.69	1.04E-04
TC140141	Impaired sucrose induction 1-like protein	−6.53	4.84E-04
CA233504	Sugar phosphate exchanger 2	−6.15	1.68E-03
TC131675	Sucrose phosphate synthase III	−5.74	2.64E-06
TC136732	Sucrose transporter SUT4	−5.49	8.38E-06
TC146044	Sugar transporter ERD6-like 5	−5.21	3.08E-06
TC129039	Sugar efflux transporter	−4.92	1.24E-04
SUGIT-DGE			
c94324f1p42760	sugar transporter type 2a	−12.62	9.96E-03
c98328f1p02743	Sucrose synthase 2	−10.46	3.32E-03
c98146f1p0774	sugar transporter (ERD6)	−9.43	1.00E-02
c88771f1p01741	Bidirectional sugar transporter SWEET	−9.2	3.59E-03
c32435f3p21876	Sucrose non-fermenting related kinase 1b	−8.47	8.64E-03
c41415f1p01118	Sucrose transporter 1	−6.16	6.80E-03
c29857f1p01086	UDP-sugar pyrophosphorylase	−5.72	2.93E-04
c106308f1p04384	Sucrose phosphate synthase A	−5.33	2.54E-04
SAS-DGE			
SCUTFL1058E04.g	sugar phosphatase -like	−7.8	6.02E-04
SCEQAM1036A06.g	sucrose-phosphate synthase 3	−7.47	4.37E-06
SCEZAM2031D12.g	UDP-sugar pyrophosphorylase	−7.29	2.74E-04
SCEQRT1031C11.g	bidirectional sugar transporter SWEET14-like	−7.17	2.06E-03
SCEPAM2014B12.g	sucrose transport SUC3	−7.08	2.03E-05
SCEPCL6023F02.g	sucrose synthase 2	−6.89	2.30E-04
SCBGSD2049G08.g	sugar transport 7	−6.72	8.51E-05
SCAGLR1021A01.g	sugar phosphate phosphate translocator	−6.34	9.12E-05
SCCCRT2001F10.g	sucrose non-fermenting 4	−6.08	1.81E-06
SCCCLR1C06G07.g	sucrose-phosphate synthase 1	−5.63	4.97E-06
SCCCRZ1004G04.g	impaired sucrose induction 1	−5.21	7.47E-05
SCEPLR1008A12.g	sucrose transport SUC4-like	−4.81	4.82E-06
SCSBST3096E12.g	sucrose-phosphatase 2-like	−5.21	3.55E-05

SoGI-*Saccharum officinarum* gene indices; SUGIT-Sugarcane Iso-Seq transcriptome database; SAS-sugarcane assembled sequences

all the DGEs. The description below gives an overview of the DEGs obtained in the three DGE analyses at FDR 0.01 without any filtering.

Sucrose, starch and other sugar derivatives

In the SoGI-DGE, there were 71 sucrose related transcripts consisting of sucrose synthases 2 and 3, sucrose phosphate

Table 8 DGEs obtained between low sugar top (LST) and low sugar bottom (LSB) internode samples with three databases SoGI, SUGIT and SAS. Shown here are some of the sucrose/sugar related transcripts

Feature ID	Description	Fold change (original values)	FDR < 0.01
SoGI-DGE			
CA267680	Sugar-phosphate isomerase-like protein	−24.39	2.84E-03
TC123316	Sucrose synthase 2	−19.2	0.01
TC150523	Glycosyltransferase sugar-binding region	−11.03	0.01
TC153302	ADP-sugar diphosphatase	−8.41	7.81E-04
CA240368	Sucrose non-fermenting related protein kinase	−8.31	2.30E-03
TC141576	Sucrose phosphate phosphatase	−7.69	7.81E-04
CA136361	UDP-sugar pyrophosphorylase	−7.43	4.96E-03
TC140141	Impaired sucrose induction 1-like protein	−7.14	2.07E-03
TC136732	Sucrose transporter SUT4	−7.08	1.02E-03
TC113476	Sucrose phosphate synthase II	−6.44	9.73E-04
TC146044	Sugar transporter ERD6-like 5	−5.84	9.77E-04
CA233504	Sugar phosphate exchanger 2	−5.74	5.88E-03
TC117267	Sucrose phosphate synthase III	−5.61	9.44E-04
CA291037	Sucrose synthase 3	−5.4	1.86E-03
SUGIT-DGE			
c26397f1p01230	Sucrose synthase	−12.81	0.01
c96752f1p02674	sugar transporter type 2a	−9.5	0.01
c10824f1p0909	SUT2-h1	−7.52	0.01
c29857f1p01086	UDP-sugar pyrophosphorylase	−6.95	0.00131
c65976f2p01948	SUT4-h1	−6.6	0.0081
c1589f4p31134	Bidirectional sugar transporter SWEET	−5.14	0.00455
c42187f1p11882	Sucrose non-fermenting related kinase 1b	−5.44	0.01
c106308f1p04384	Sucrose phosphate synthase A	−5.12	0.00358
SAS-DGE			
SCJLHR1025D07.g	bidirectional sugar transporter SWEET3	−14.92	9.46E-03
SCCCRZ1002G07.g	sucrose-phosphate synthase 1	−13.59	9.84E-03
SCEQRT2090C11.g	Sucrose transport SUC3	−12.97	5.05E-04
SCQGST3153F06.g	sugar transport 5-like	−9.65	1.12E-03
SCCCLR2C03H09.g	sugar transporter ERD6-like 6	−9.36	1.34E-03
SCEZAM2031D12.g	UDP-sugar pyrophosphorylase	−9	3.23E-04
SCAGLR1021A01.g	sugar phosphate phosphate translocator	−8.78	3.46E-04
SCBGSD2049G08.g	sugar transport 7	−6.25	6.77E-03
SCCCRT2001F10.g	sucrose non-fermenting 4	−6.15	3.42E-04
SCEZRZ1013G04.g	Galactinol-sucrose galactosyltransferase 2	−6.09	1.07E-03
SCEPLR1008A12.g	sucrose transport SUC4-like	−5.94	2.87E-04
SCCCRZ1004G04.g	impaired sucrose induction 1	−5.76	5.49E-04
SCEZLR1031D07.g	sucrose-phosphatase 2	−5.59	2.62E-04
SCSGHR1068D07.g	UDP-sugar transporter	−4.91	5.61E-04
SCEPCL6023F02.g	sucrose synthase 2	−4.89	9.76E-04
SCEQAM1036A06.g	sucrose-phosphate synthase 3	−4.41	0.01

SoGI-*Saccharum officinarum* gene indices; SUGIT-Sugarcane Iso-Seq transcriptome database; SAS-sugarcane assembled sequences

Table 9 DEGs obtained between high sugar top (HST) vs high sugar bottom (HSB) internode samples with three different databases SoGI, SUGIT and SAS. Shown here are some of the sucrose/sugar related transcripts

Feature ID	Description	Fold change (original values)	FDR < 0.01
SoGI-DGE			
TC125737	Phenylalanine ammonia-lyase	−16.27	4.08E-03
TC131133	Chalcone synthase 5	−13.79	4.24E-05
CA207335	Cinnamoyl-CoA reductase	−12.11	5.19E-03
CA212197	Beta-amyrin synthase	−11.85	0.01
CA113829	LIM transcription factor homolog	−5.75	7.51E-03
CA065092	Universal stress protein family protein ERD65	−5.43	1.48E-09
TC124516	4-coumarate coenzyme A ligase	−4.59	4.16E-03
TC137240	Serine/threonine-protein kinase Nek5	−3.5	0.01
SUGIT-DGE			
c98442f1p02354	Terpene cyclase mutase family	−18.71	0.00608
c61441f1p11782	Phenylalanine ammonia lyase	−6.44	0.00382
SAS-DGE			
SCJFRT1010B12.g	Sugar phosphate phosphate translocator	2.43	3.49E-08
SCEZLR1031D07.g	Sucrose phosphate phosphatase 2	2.17	6.51E-03
SCACSB1117F03.g	Sucrose synthase	1.24	4.07E-03
SCEZSD1079C10.g	Bidirectional sugar transporter SWEET16-like	1.12	1.87E-04

SoGI-*Saccharum officinarum* gene indices; SUGIT-Sugarcane Iso-Seq transcriptome database; SAS-sugarcane assembled sequences

synthase (SPS) 2 and 3, sucrose phosphate phosphatase (SPP), *sucrose non-fermenting related* protein kinases; *impaired sucrose induction 1*-like protein and sucrose transporters (SuT) 2 and 4. About 22 transcripts were sugar related including transport, efflux, and glycosyltransferases. Ten transcripts were related to alkaline/neutral invertases and three transcripts with homology to sucrase from *Oryza sativa* were found. There were ten *high-glucose regulated protein 8-like* transcripts. Forty six transcripts, were related to intermediary metabolism of fructose phosphates, the most expressed being fructose-bisphosphate aldolase cytoplasmic isozyme. Sixteen transcripts were related to xylose metabolism and β-glucosidase related transcripts were observed. Fifteen hexose related transcripts were transporters, while 18 transcripts were related to triose phosphates metabolism. Fifty three UDP-related transcripts were found, out of which six were UDP-glucose-dehydrogenases. There were also UDP-sugar, arabinose, xylose, galactose transporters, –epimerases and -pyrophosphorylase related transcripts. Fucoses are hexose sugars and nine transcripts associated with them include fucosidases and fucosyltransferases. Thirteen mannose, trehalose and sorbitol related transcripts were found. Glucans metabolizing genes were another prominent group found to be highly expressed with 45 transcripts including β-1, 4 glucan synthases and endoglucanases. Nine transcripts related to alpha amylases were also upregulated in high sugar genotypes. In addition, 85 transcripts were found to be related to kinases including hexokinases, fructokinases (1, 2 and 3), phosphofructokinases,

carbohydrate kinases and galactokinases. In the SUGIT-DGE, there were 208 sucrose related transcripts. In addition to the transcripts observed in the SoGI-DGE, sugar transport 5 and 7, sugar transporter ERD6 like, bidirectional sugar transporter SWEET1 and 4 like, and an abundance of ABC transporters B, C, D, E, G, F, and I for sugar were found in SUGIT-DGE. In the SAS-DGE, 75 transcripts were related to sucrose consisting of galactinol-sucrose galactosyltransferase 1,2 and 6, sucrose transporters SUC3 and its isoform X2, SUC4, SPP 2 and SPS 1, 3 and 4, bidirectional sugar transporter2a, 4, 14 and 16, sugar transporter ERD6-like 5, 6 and 16, sugar transport 5 and its isoform X1, 7 and 9 transcripts for sugar phosphate phosphate translocator. Interestingly one transcript for invertase inhibitor and one transcript for sulfofructose kinase like transcript which were not detected in the other two DGEs were found. Starch synthases II b and c, III, IV and starch branching and debranching (pullulanase and isoamylase) enzymes were found to be upregulated. The KEGG pathway map for starch and sucrose related DEGs are shown in the Additional file 9: Figure S4.

Vacuole and transporters

Transcripts related to transporters comprising of sucrose, sugar, sugar efflux, sugar phosphate exchanger, hexose, nitrate, GDP-mannose, aquaporins, vacuolar ATP synthase subunit C, vacuolar H^+ ATP synthase subunit C, vacuolar H^+ pyrophosphatase, vacuolar proton pumps, vacuolar targeting receptors, vacuolar protein sorting proteins (1, 13, 13A, 22, 25, 33, 36, 41, 55, DUF1162) and vacuolar H^+- inorganic pyrophosphatase were found to be upregulated. A transcript was found to match the bacterial sugar transport system probably due to contaminating sequences. An abundance of ABC transporters could be observed in all DGEs.

Hormones

Auxin related transcripts were *auxin response factors* 1, 3, 4, 5, 7, 9, 13, 15, 16, 17, 22, 23, 26, 27 and 31, *auxin responsive* proteins, auxin influx/efflux carriers, auxin transporters 1, 2, and *auxin binding protein* 4 were found. With respect to ethylene, 43 transcripts including *ethylene over-producer* like proteins, *ethylene responsive* transcription factors, elongation factors, calmodulin binding factors, element binding factors, small GTP binding proteins, ethylene receptors, and *ethylene insensitive* 2, and 3 proteins were found. Transcripts related to abscisic acid (ABA) and gibberellic acid (GA) and very few jasmonate and brassinosteroid related transcripts were found in the DEGs.

Organellar

Transcripts related to the chloroplast, notably chloroplastic group IIB intron splicing facilitator CRS2, alpha-glucan water dikinase, rubisco large subunit alpha binding, chloroplast post-illumination chlorophyll fluorescence increase protein, starch synthases II b and c, III, IV and starch branching and debranching (pullulanase and isoamylase) enzymes to name a few from the three DGEs. The ribosomal proteins were one of the most upregulated transcripts in all the DGEs comprising of nuclear, cytoplasmic, chloroplast and mitochondrial ribosome related functions especially of 30S, 40S, 50S, and 60S and acidic ribosomal transcripts.

Senescence/ripening/stress

Transcripts related to senescence including senescence-inducible chloroplast *stay-green* protein and *leaf senescence* proteins, *senescence-inducible chloroplast stay-green* protein, heat shock related transcripts of DNA and chloroplast, *wound inducible* protein, ripening ABA induced, autophagy, programmed cell death, cell death related protein, and *defender against cell death*, vascular death associated transcript were found. Transcripts were related to stress (light, water, heat, salt, ozone-responsive, bio-stress) and pathogenesis related transcripts, hypersensitive induced response proteins, 22 kDa drought inducible proteins, dehydrins and transcripts related to proline were found to be upregulated.

Flowering

Flowering related transcripts including pistil, pollen, immature pollen, flowering-time protein isoforms, phytochrome and *flowering time, flowering locus, GIGANTEA,* OVA4 ovule abortion 4, and *fertilization independent* were upregulated in high sugar genotypes. Proteins related to the egg

apparatus, seed maturation, *shrunken seed* and seed starch branching enzyme related transcripts were upregulated. *HASTY 1* flower development, *agamous*-like MADS box AGL12, *photoperiod-independent early flowering* 1, *early flowering 3, flowering time control FY, luminidependens* are some of the flowering related transcripts found across the DEGs.

Signalling

Transcripts of signalling related to DNA damage, signal recognition, pollen, and integral membrane, *14–3-3* like proteins. Out of a large number of kinases, serine/threonine phosphatases, appeared to have a dominant role during sucrose accumulation. Also, it was observed that several signalling events can be inter related with others from the pattern of gene expression observed to be upregulated in high sugar genotypes (Additional file 9: Figure S5).

Fibre/cellulose

In the SoGI-DGE, transcripts matching with *fibre proteins* 11, 12, 15, 19 and 34 of cotton and *Hyacinthus* sp. were found. There was a transcript weakly similar to *cement protein* 3b from the marine worm *Phragmatopoma californica*. Vegetative and secondary cell wall proteins, cell wall hydrolases, cell envelope and cell shape, cell wall beta 1,3, endoglucanase cellulose synthases, bundle sheath cell specific proteins, 50 transcripts for cellulose synthases 2, 3, 4, 5, 6, E6, D3, A and 7, cellulose 1,4, beta-cellobiosidase were upregulated in high sugar genotypes. Also, transcripts of phenyl ammonia lyase (PAL), 6 caffeic acid-o-methyl transferase (COMT), caffeoyl CoA 3-O-methyl transferases (CCoAOMT), glutathione S-transferase, 6 dihydroxyacetone kinase, and transcripts related to chorismate, succinyl, cinnamoyl alcohol of shikimate pathway, caffeoylshikimate esterase, expansins A2 and A13, transcripts for vegetative cell wall, and secondary cell wall related transcripts were found.

Light/photosynthesis

Transcripts related to light/photosystem including light induced, light responsive proteins. *De-etiolated* 1, phytochrome, *rubisco* sub unit binding proteins, chloroplast post-illumination chlorophyll fluorescence increase protein, cryptochrome, photosystems I 700 and II 680 chlorophyll A apoprotein, photosystem reaction centre subunits II, III, VIII, IX, XI, 23 are few to mention. Interestingly, there were eight non-photosynthetic NADP-malic enzymes transcripts from *Zea mays* in SoGI-DGE. In SUGIT-DGE, transcripts of *CIRCADIAN TIMEKEEPER*, blue light photoreceptor PHR2, *negatively light regulated*, light-stress responsive one helix like, light inducible CPRF2 and *WEAK CHLOROPLAST MOVEMENTUNDER BLUE LIGHT* 1 like. Transcripts related to photosynthetic NDH subunit of subcomplex B chloroplastic, *light dependant short hypocotyls* 4 like, *high light induced chloroplastic* like, blue light photoreceptor PHR2 etc. were found in SAS-DGE. Nitrogen (N) related transcripts comprising of nitrogen utilization substrate protein, nitrogenase, nitrilase, *nitrate extrusion* proteins and nitrate reductase, bifunctional nitrilase nitrile hydratase NIT4A were up regulated in high sugar genotypes.

Uncharacterized
Interestingly, in SoGI-DGE, about 6552 transcripts were found to match the chromosomal regions of *Vitis vinifera* (SoGI annotation) which are whole genome shotgun sequences. In SUGIT-DGE, 243 transcripts were uncharacterized and in SAS-DGE, 320 transcripts were found to be uncharacterized.

Down regulated transcripts in high sugar genotypes
The transcripts down regulated in high sugar genotypes included 17S, 18S, 26S, ribosomal RNA genes, cytochrome P450, and photosystem I 700, a stem specific transcript and leaf specific transcript from *Saccharum* hybrid cultivar, rRNA intron encoded homing endonuclease, zinc finger protein and uncharacterized transcripts in the three DGEs.

Discussion
Two groups of genotypes, high sugar and low sugar, were formed based on the sugar content in terms of Brix as in sugarcane most of the soluble solids in the juice (70–91%) correspond to sucrose [12, 30]. Differential expression of genes was studied between the two groups and between top and bottom internodal samples (immature and mature) of the two groups. Therefore, gene expression changes were studied among high sugar top internode (HST), high sugar bottom internode (HSB), low sugar top internode (LST) and low sugar bottom internode (LSB)samples in various comparisons. Thus, the HST vs LST and HSB vs LSB were comparisons between the high and low sugar genotypes, whereas HST vs HSB and LST vs LSB were comparisons between top and bottom intermodal samples. For the DGE analyses, three databases were used as references individually wherein a large number of DEGs were identified from each. The databases were chosen to be specific for sugarcane. SoGI and SAS are derived from 26 different cDNA libraries [24] as a result, a large number of DEGs where obtained. The SUCEST database which encompasses SoGI and SAS is reported to cover >90% of the sugarcane genes [31]. The SUGIT database is essentially a long reads database sequenced using the latest Iso-Seq technology [17] which can further be used for refining the DEGs for isoform/allelic information. This database covers approximately 71% of the total predicted genes in sugarcane [17]. The common and unique transcripts

from each database are not discussed further as the main objective of this paper was to find the DGE for sugar content. A subset of sucrose /sugar related DEGs were derived, which is interesting as several other studies on sucrose accumulation in sugarcane reported that sucrose related genes were less abundant or not expressed during the maturation stage [6, 12, 32]. There were approximately 70 transcripts related to sugar/sucrose in each DGE. Sucrose synthase (SuSy) and sucrose transporters (SuTs) were consistently found to be highly expressed in high sugar genotypes. Similar association was reported in [6, 8, 14]. The identity of the exact isoform of these two genes could not be found due to the varying annotations of the three databases used, which needs further studies. SuSy is reported to contribute to increasing the sink capacity, building cell wall materials and starch while sucrose transporters facilitate transportation of sucrose that leads to steady increase in sucrose content [30]. Further work on the isoforms/allelic expression of these genes would certainly be useful for understanding the finer details of their regulatory roles. The functioning of the two sucrose synthesis enzymes, SuSy and SPS and their regulation, has not yet been well demonstrated in sugarcane. SPS, sucrose non-fermenting related kinases, bidirectional sugar transporter SWEET, UDP-sugar pyrophosphorylase, *impaired sucrose induction* 1 -like proteins were the other genes that were consistently present at lower fold changes. Interestingly, an *invertase inhibitor* gene was found to be highly expressed in LST (13 folds) in LST vs LSB in SAS-DGE. Invertase inhibitors have been previously reported to be highly expressed during the sucrose accumulation stages in sugarcane [33].

In addition to the above genes, the gene expression pattern in our study reveals a clear association between different gene networks during sucrose accumulation similar to earlier reports [9, 12]. It is possible to make a direct parallel between sucrose content and gene expression levels for almost all the DEGs though the difference in the sugar content between the two groups is very narrow. Sucrose is a carbohydrate compound and was originally recognized only as an energy source for metabolism in plants but was recently shown to also function as a signalling molecule involved in regulation of various physiological processes in plants such as root growth, fruit development and ripening, and hypocotyl elongation [34]. Sugars serve as key components reflecting the plant's energy status and, therefore, the ability to continuously sense sugar levels and control energy status is a key to survival and therefore transcript levels of thousands of genes respond to changing sugar levels [34]. Further, different sugars can have different regulatory roles in physiological processes, and the developmental stage of the plant further determines the response to sugars [35–37]. Recently, it was observed that glucose facilitates the juvenile to adult phase change in Arabidopsis by repressing microRNA (miRNA) 156 expression [38–40]. Consequently, mutants in sugar signalling or starch metabolism display an altered juvenile phase [38]. At high concentrations, sugars can induce meristem quiescence as observed in the arrest of development of seedlings germinated on high sugar levels [34]. Sugar induced quiescence of the stem can be seen through the expression of several transcripts for *no apical meristem* and *indeterminate spikelet* transcripts in addition to senescence related transcripts. Transcripts related to less abundant and lignocellulosic sugars identified in this study included xylose, trehalose, galactose, arabinose, fucose, mannose, taurine and the sugar alcohols inositol and sorbitol. The constant synthesis and breakdown of sucrose into its hexose components helps regulating various physiological events associated with these less abundant sugars and maintain a reserve for tackling any stimuli including the accumulation of sugar in the form of sucrose. It could be possible that breeding programmes for high sucrose genotypes have resulted in selection for these sugars (total sugars, in addition to sucrose) and gene expression changes of certain regulatory genes [15]. Therefore, diverse phenotypes may stem from multiple effects of sucrose and other sugars as signal and storage compounds when accumulated in various developmental and compartmental patterns resulting from differential gene expression and regulation.

The vacuole occupies as much as 90% of most mature cells and can accumulate and store sucrose, glucose and fructose and serves as a primary pool of free calcium ions in plant cells. Furthermore, the space-filling function of the vacuole is essential for cell growth, as the cell enlargement is mainly through the expansion of the vacuole rather than of the cytoplasm [41]. A vast majority of the differentially expressed gene transcripts were vacuole related including aquaporins, glucans related, aspartic proteinases, endopeptidases, ABC transporters, TIPs, V-ATP synthases, vacuolar protein sorting proteins, proton pumps, Ca^{2+} ATPases, calmodulins, that showed higher expression levels in high sugar genotypes. Further molecular characterization of vacuolar and tonoplast sugar transporters should advance our understanding of vacuole function, sugar transport and sugar accumulation in sugarcane.

Several transcripts related to plant defense, wounding, and disease were upregulated in high sugar genotypes together with the ripening and senescence related transcripts. Further, water stress and dehydration related gene transcripts were upregulated in the high sugar genotypes. Apart from the ripening and senescence related transcripts which indicate the physiological state of the stem, the up regulation of transcripts encoding plant disease resistance proteins suggests that the defense system

of sugarcane was activated by high sugar levels which might contribute to protecting from the extreme stress caused by the high sucrose levels during the maturity stages. It may also create steep osmotic gradients between compartments with varying sucrose concentrations (more negative than –2.0 MPa during sucrose accumulation). The increased commitment to fibre synthesis in the maturing stem is evident in the upregulation of several fibre and cellulose related transcripts in the high sugar genotypes highlights the need to maintain the structure of the stem in conjunction with sucrose accumulation. These may act to restrict apoplastic movement of solutes between the vascular bundles and the sucrose-storage parenchyma cells [42]. Transcripts related to proline and glyoxalase were highly expressed in high sugar genotypes. The differential expression of genes related to fibre, cellulose and lignin synthesis shows that the osmotic regulation and structural maintenance as directed by the sugar levels. Though sucrose content in the sugarcane culm ranges from 14 to 42% of the culm dry weight [3], the majority of carbohydrates in sugarcane is lignocellulose, a major component in the cell wall. As cell elongation and sucrose accumulation ceases in the maturing sugarcane internodes, there is a major increase in cell wall thickening and lignification [43]. Cellulose accounts for about 42–43% in sugarcane and energy cane cultivars [44] and can be a prominent competing sink for carbon in sugarcane. Cellulose synthases 1, 2, 3, 4, 5, 6, 7, and 9 along with a novel transcript matching for a *cement protein* like gene that is upregulated in high sugar genotypes indicates that there are several aspects of sugarcane cell wall composition remain to be explored [45]. S-adenosylmethionine (SAM) produced by SAM-synthase is required as the methylation donor in lignin and suberin biosynthesis and secondary metabolism. It is also required as a precursor for SAM-decarboxylase, which is also up-regulated and important in polyamine synthesis, a response to osmotic stress. Elevated SPS activity is consistently correlated with high rates of cellulose synthesis and secondary wall deposition [46]. UDP-Glucose, apart from being the precursor for sucrose synthesis, is a nucleotide sugar central to diverse pathways of polysaccharide biosynthesis, leading to starch and cellulose, hemicellulose and callose synthesis. About 10 major monosaccharides in cell wall polymers are converted from glucose through UDP-Glucose related interconversion pathways. All UDP-Glucose related transcripts including UDP-Glucose dehydrogenases, pyrophosphorylases were upregulated in high sugar genotypes indicating a high correlation with sugar contents. Ethylene is often related to the lignification of plant tissues by increasing the expression of genes involved in the phenylpropanoid pathway [47]. This explains the parallel upregulation of cellulose

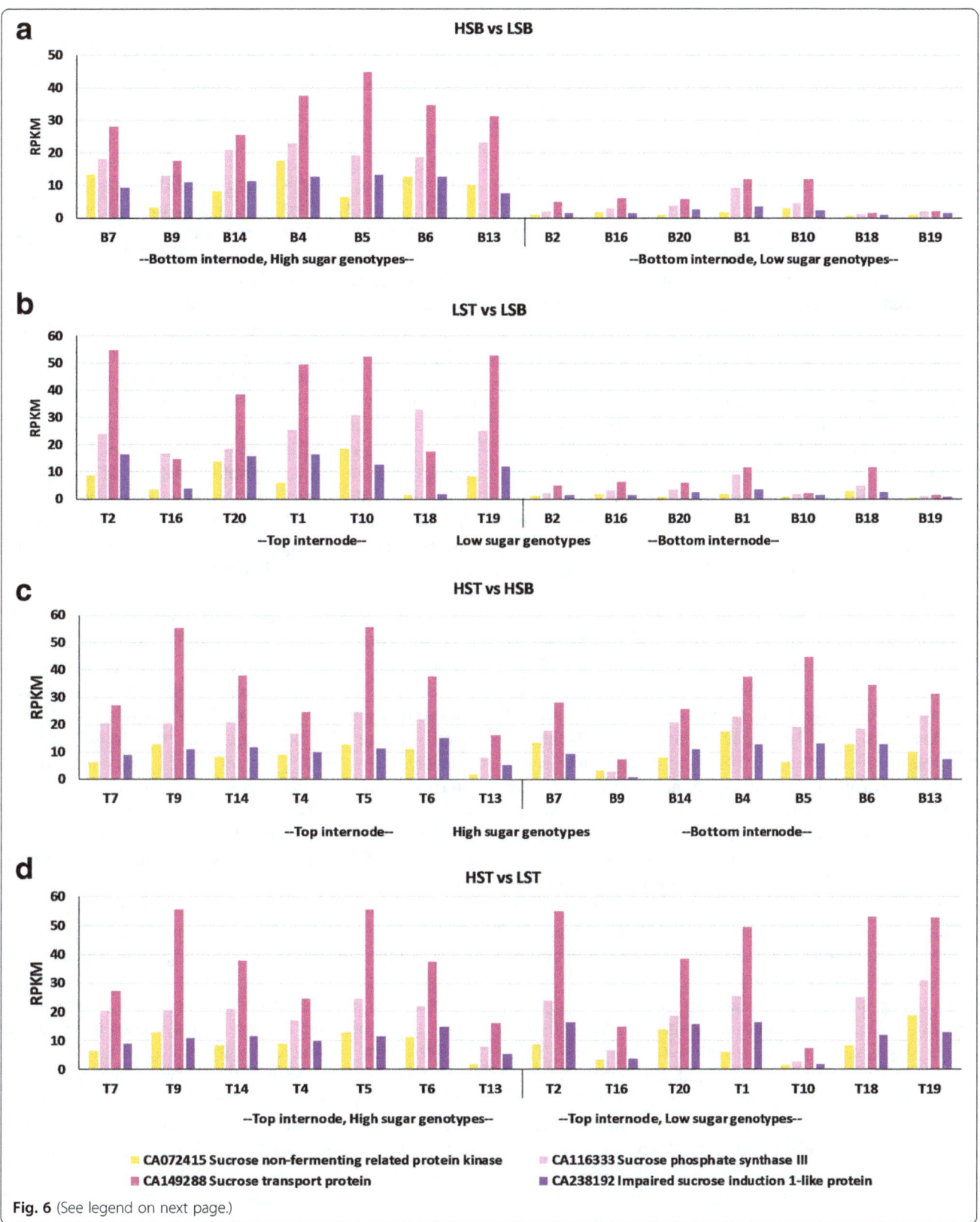

Fig. 6 (See legend on next page.)

synthases, ethylene related transcripts, as well as SPS in the DGEs. The mechanisms regulating cell wall biosynthesis and source-sink relations in sugarcane will be crucial constituents of any efforts to alter carbon partitioning between fibre and sugar in the culm. In addition, the alteration of cell wall biosynthesis genes in association with sucrose (Brix) content is an interesting indication of a correlation between these processes. Silencing or overexpression of some of these genes may lead to altered cell wall or increased sucrose content. Interestingly, when comparing two genotypes contrasting for lignin content, Vicentini et al. [45] found that a simple correlation between lignin content and differential expression of lignin genes is not always straightforward and most of the lignin biosynthetic genes did not show increased transcript levels in the high lignin genotype.

Sugar signals and the circadian clock are part of a complex network that controls floral transition. In sugarcane, sugar levels peak just before flowering induction. The signalling for senescence, arrest of apical growth, high sucrose levels and flowering induction are well coordinated. The upregulation of several flowering related genes like *flowering locus D*, pollen and pistil related transcripts in the high sugar genotypes clearly shows that the crop has attained its maximum sugar levels and was in a transition state to flowering though many are commercial cultivars that do not or flower rarely. Sugarcane has been selected for higher sugar content that involved strategies for delayed flowering and seed set, due to which a majority of sugarcane cultivars now are either sterile or the reproductive cycle has been delayed, or dormant for years [48]. Trehalose and its phosphate derivative trehalose-6- phosphate have recently gained importance as signalling molecules involved in carbon partitioning and also linking sugar status and diurnal rhythm to floral transition, in plants [49, 50]. For example, high sucrose and trehalose-6-phosphate (T6P) levels signal a cellular sugar abundance status [37, 51]. In addition to other sugar forms, the role of trehalose in sugarcane sucrose metabolism needs further studies as corroborated by the upregulation of several transcripts for trehalose phosphate synthase and trehalases.

Light interception and the *stay-green* trait are considered as major factors influencing the level of carbohydrates in the internodes [51]. Leaf angle is a genetic trait and higher sucrose yield in sweet sorghum can be achieved by genetic adjustment of leaf angles to optimum light interception. In addition, *stay-green* varieties of sweet sorghum were found to have higher stem sugar concentrations than senescing lines [52]. This may be due to the reduced need for re-mobilizing stem sucrose in addition to prolonged photosynthetic capacity [53]. Similarly, the upregulation of *stay-green* gene transcripts in the high sugar genotypes indicates an association between high sugar levels and higher photosynthetic capacity as the C4 enzymes are mainly localized in the chloroplasts. Further, high expression levels of photosynthetic, light harvesting, etiolation, starch, chlorophyll, gene transcripts were observed in high sugar genotypes. In addition, transcripts related to non- photosynthetic NADP malic enzymes [54] were upregulated in high sugar genotypes for which the functional significance is unknown in sugarcane. The rapid cycling of sugars in non-photosynthetic cells has been referred to as a 'futile cycle' [55] because of the continuous and simultaneous synthesis and degradation of sucrose. However, it is recognised that these cycles allow cells to respond in a highly sensitive manner to small changes in the balance between the supply of sucrose and the demand for carbon for respiration and biosynthesis and thus resulting in a strong sink [30]. This remobilisation of stored sucrose as a food supply results in rapid regrowth following stress or in germination of axillary buds of the internode [56]. Photosynthesis, growth and yield are strongly linked to N availability especially in C4 crops [57].The upregulation of N related transcripts in high sugar genotypes indicate that this is an ongoing process even if the crop has reached maturity.

The general cell related functions and growth, organellar and nuclear functions, biosynthetic pathways of pigments, amino acids, metabolites, hormonal signalling, transcription factors, various other transporters, proteins of transposons, root/stem/leaf related transcripts, were upregulated in the high sugar genotypes. The functions enriched in genes that are differentially expressed between different tissues in each comparison are consistent with the physiological changes associated with the development of that tissue, mainly sucrose content (Figs. 6 and 7). The absence of DEGs in HST vs LST suggests that the top internodes are metabolically active irrespective of their sucrose contents (i.e. high or low sugar genotype). The absence of sucrose related DEGs in HST vs HSB, where the top and bottom internodes of high sugar genotypes show almost similar expression patterns, indicates homogeneity for sucrose content throughout the culm. Further, the high sugar genotypes seem to invest in

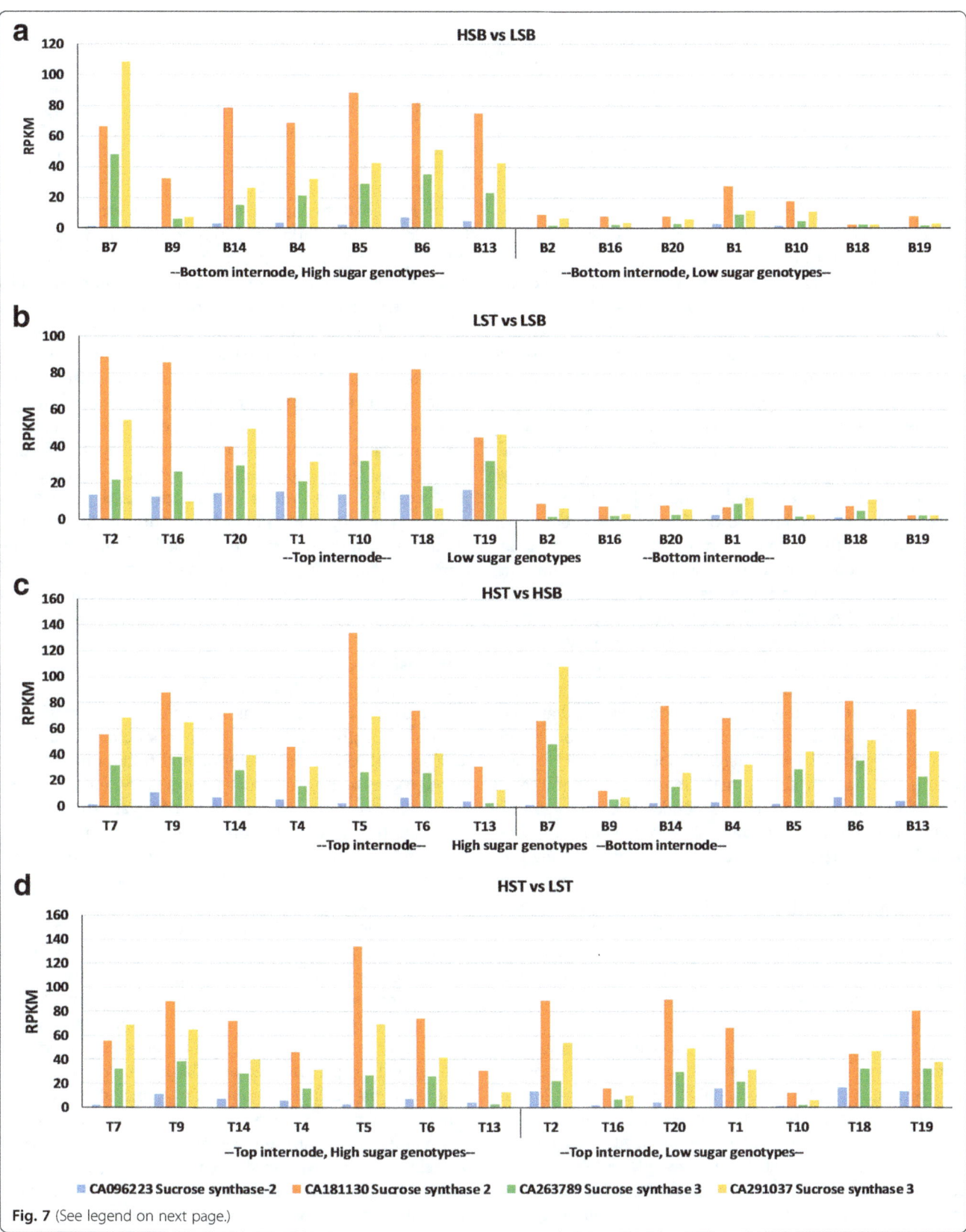

Fig. 7 (See legend on next page.)

(See figure on previous page.)

Fig. 7 Graphical representation of the expression pattern of some of SuSy transcripts in various comparisons. LST, low sugar top internode sample; LSB, low sugar bottom internode sample; HST, high sugar top internode sample; HSB, high sugar bottom internode sample; T-top tissue; B-bottom tissue. Shown here are the sucrose phosphate synthase III, sucrose non-fermented related protein kinase, sucrose transport protein, impaired sucrose induction 1- like protein transcripts from *Saccharum officinarum* gene indices, SoGI database, showing differential expression in top two comparisons (**a**) HSB vs LSB; (**b**) LST vs LSB while there is no differential expression in lower two comparisons (**c**) HST vs HSB; (**d**) HST vs LST at FDR < 0.01. X-axis shows the genotypes while Y-axis represents RPKM values

more fibre and cellulose as revealed by the nature of the transcripts that are differentially expressed between top and bottom internodes (Table 6). Meanwhile, a large number sucrose related DEGs in LST vs LSB shows that a gradient for sucrose exists in the low sugar genotypes. This observation can be inferred in two possible ways. One is that the low sugar genotypes have an active top internode compared to bottom leading to sucrose futile cycling, resulting in less accumulation or the other way could be that the bottom internodes have slowed down metabolically over time, reaching their physiological threshold levels of sucrose. The former is unlikely as there are no acid invertases (cell wall/vacuolar) expression observed in the DEGs which are involved in the sucrose breakdown. The latter is likely to be the reason and the bottom internodes play a major role in the sucrose content of the genotypes. Also, the bottom internode of high sugar genotypes shows high expression of sucrose related genes. Feedback inhibition or post translational regulation could possibly be involved in the low sugar genotypes having higher expression of the sucrose related genes in the top internode and in turn having a low sugar content. In addition, the low sugar genotypes could also be late maturing genotypes, as some of them are introgression lines (other than the commercial hybrids) not having an established sugar profile or maturity indices yet (for e.g., fibre: sugar ratio). Many factors besides Brix, like ratoonability, vigour, softness, several resistance mechanisms, secondary metabolites, starch, etc., may also differ among the genotypes taken that remain to be evaluated. There were 7814 transcripts unique to HSB vs LSB, and 3667 transcripts unique to LST vs LSB (of the 34,476 DEGs in HSB vs LSB and 30,809 DEGs in LST vs LSB). These transcripts may indicate tissue specificity of the genes or their isoforms which is to be explored. When these unique transcripts were filtered for sucrose/sugar genes, SPS, SPP, SuSy, and sugar transporter genes were more specific to HSB vs LSB whereas, only some of the sugar transporter genes were specific to LST vs LSB (Additional file 8: Tables S24–28).

It was proposed that sucrose accumulation may be regulated by a network of genes induced during culm maturation that contribute to key physiological processes including sugar translocation/transport, fibre synthesis, membrane transport, vacuole development and function, and abiotic stress tolerance [9, 12]. We found a similar trend in this profiling study, in addition to a very high number of differentially expressed sucrose and sugar related transcripts that might help bridge missing links in the interlinking of biosynthetic pathways and their regulatory factors. It is to be studied if sucrose regulates the large number of genes or large number genes is required for controlling this trait. As sucrose emerges as a signalling molecule as seen in the recent studies [34, 50], the all-pervasive nature of this sugar is likely to regulate the growth and developmental processes of the plant. It can be speculated for the presence of master switches or the major regulatory genes of this trait as further genomic information is obtained in the future. Many novel genes, like caffeoyl shikimate esterase that was recently discovered in Arabidopsis and reported to be absent in sugarcane [44] where found in the upregulated transcripts in our study. Further mining of the transcriptomes would certainly lead to new targets and new aspects for sucrose synthesis and accumulation in sugarcane.

Conclusion

The data reported here provide a comprehensive resource for sucrose related as well as culm maturation related studies in sugarcane. Further studies on a large data set with different developmental time points for genotypes contrasting for sugar content or energy canes that do not accumulate high levels of sugars should indicate targets for further biotechnological approaches. A dedicated analysis of transcription factors, and regulatory elements will further help understanding the complexity of the sugar network. Sucrose accumulation is very dynamic and unlike fruiting organs, the sugarcane culm is continuously exposed to every possible stimulus in the crop, soil and water continuum which results in a plethora of genes that are expressed at any point of time (approximately about 33,000). Although the present study identified more than 30,000 genes regulated and differentially expressed between high and low sugar genotypes, it is hard to pinpoint any particular group of genes or a gene to be responsible for the sucrose content and maintenance. Further, it is not possible for a gene to be lacking or not expressed in either of the groups as sucrose is a primary metabolite and principal transport sugar in sugarcane, which shows that the trait is quantitative and it is under transcriptional control. The machinery for sucrose synthesis is conserved across species and it is supposed that the complexity of sugarcane genome must play an important role in the sucrose levels that are observed in sugarcane. With multiple forms of each enzyme, with their own isoforms, various localizations, compartmentalized processes, the availability of large vacuoles and a unique stem morphology together contributes to the sugarcane stem sucrose content. Further, the availability of multiple isoforms or alleles gives the crop the advantage of buffering against any functional disruption which is the main reason for the instability of transformation events in sugarcane [58]. With these challenges in the sugarcane crop, a multitude of strategies are required for any genetic manipulation or for identification of regulatory genes for important traits particularly sucrose.

Additional files

Additional file 1: Table S1a. Sugar profile of the genotypes taken for the study. **Table S1b** Quality Report of RNA Seq reads of samples used in the study. **Table S2** Genotypes and their transcriptome samples taken based on sugar content (Brix). **Table S3** Additional information with regard to genotypes selected. **Figure S1** Graphical representation of the sugar profiles of the genotypes selected for the study. (XLSX 2248 kb)

Additional file 2: Table S4. Complete list of DEGs in the SOGI-DGE. **Table S5** List of DEGs of SOGI HSB VS LSB. **Table S6** List of DEGs of SOGI HST VS HSB. (XLSX 8800 kb)

Additional file 3: Table S7. List of DEGs in the SUGIT, HST VS HSB. **Table S8** List of DEGs in the SUGIT, LST VS LSB. **Table S9** List of DEGs in the SUGIT, HST VS HSB. (XLSX 3817 kb)

Additional file 4: Table S10. List of DEGs in the SAS-DGE. **Table S11** List of DEGs in the SAS-DGE. **Table S12** List of DEGs in SAS-DGE (Excel workbook). (XLSX 5447 kb)

Additional file 5: Table S13. qPCR for selected transcripts and correlation analysis with RNA-Seq expression values. **Figure S2** Correlation analysis of RNA- Seq and qRT-PCR expression values for selected transcripts. (XLSX 28 kb)

Additional file 6: Table S14. DEGs of the experiment high sugar bottom vs low sugar bottom with SoGI database. **Table S15** DEGs in the experiment low sugar top vs low sugar bottom with SoGI database. **Table S16** DEGs in the experiment high sugar top vs high sugar bottom SoGI database. **Table S17** DEGs in high sugar bottom vs low sugar bottom with SUGIT database. **Table S18** DEGs obtained in High sugar top vs high sugar bottom with SUGIT database. **Table S19** DEGs obtained in low sugar top vs low sugar bottom with SUGIT database. **Table S20** DEGs in high sugar top vs high sugar bottom experiment with SAS database. **Table S21** DEGs in low sugar top vs low sugar bottom experiment with SAS database. **Table S22** DEGs in high sugar bottom vs low sugar bottom experiment with SAS database. (DOCX 45 kb)

Additional file 7: Figure S3. Functional classification of the DEGs obtained in three different DGEs using Mapman annotation. 1. High sugar top vs high sugar bottom internode samples (HST vs HSB), 2. Low sugar top vs low sugar bottom internode samples (LST vs LSB), 3. High sugar bottom vs low sugar bottom internode samples; a, d, g) *Saccharum officinarum* gene indices, SoGI; b, e, h) Sugarcane long read database, SLRD; c, f, i) Sugarcane assembled sequences, SAS. **Table S23** Functional classification of the DEGs obtained in three different DGEs using Mapman annotation. 1. High sugar top vs high sugar bottom internode samples (HST vs HSB), 2. Low sugar top vs low sugar bottom internode samples (LST vs LSB), 3. High sugar bottom vs low sugar bottom internode samples; a, d, g) *Saccharum officinarum* gene indices, SoGI; b, e, h) Sugarcane long read database, SLRD; c, f, i) Sugarcane assembled sequences, SAS. (XLSX 101 kb)

Additional file 8: Table S24. Common and unique transcripts between LST vs LSB and HSB vs LSB (SoGI-DGE). **Table S25** Common and unique transcripts between LST vs LSB and HSB vs LSB (SUGIT-DGE). **Table S26** Common and unique transcripts between LST vs LSB and HSB vs LSB (SAS-DGE). **Table S27** Unique transcripts (sucrose/sugar related) in HSB vs LSB (SoGI-DGE). **Table S28** Unique transcripts (sucrose/sugar related) in LST vs LSB (SoGI-DGE). (XLSX 3047 kb)

Additional file 9: Figure S4. Blast2GO and KEGG Mapping for DEGs in the SOGI-DGE with respect to starch and sucrose metabolism. **Figure S5** Sucrose emerges as a signalling molecule regulating most of the interlinked plant functions in sugarcane. The gene expression pattern in culm tissue during sucrose accumulation in the genotypes studied reveals several networks of genes regulated by sucrose, and correlating with the sucrose content of the genotypes studied. (XLSX 469 kb)

Abbreviations

ABA: Abscisic acid; BC: Back cross; BE: β-expansin; Bp: Basepair; CCoAOMT: Caffeoyl CoA 3-O-methyl transferases; CCoAR: Cinnamoyl CoA reductase; CCS: Commercial cane sugar; cDNA: Complementary DNA; CHS: Chalcone synthase; COMT: 6 caffeic acid-o-methyl transferase; DEGs: Differentially expressed genes; DGE: Differential gene expression; EST: Expressed sequence tag; FDR: False discovery rate; FE: Ferruloyl esterase; GA: Gibberellic acid; GDP: Guanosine diphosphate; GO: Gene ontology; GTP: Guanidine tri phosphate; HPC: High performance computing; HPLC: High performance liquid chromatography; HSB: High sugar bottom internode; HST: High sugar top internode; KEGG: Kyoto Encyclopedia of Genes and Genomes; LAC: Laccase; LSB: Low sugar bottom internode; LST: Low sugar top internode; MPa: Megapascal; mRNA: Messenger RNA; NADP: Nicotinamide adenosine diphosphate; NCBI: National Center for Biotechnology Information; NGS: Next generation sequencing; NIR: Near infrared; Nr database: Non-redundant database; ORF: Open reading frame; PAL: Phenyl ammonia lyase; PE: Paired end; qRT-PCR: Quantitative real-time polymerase chain reaction; RNA-seq: Ribonucleic acid sequencing; RPKM: Reads per kilobase per million mapped reads; SAM: S-adenosylmethionine; SAS: Sugarcane assembled sequences; SNP: Single nucleotide polymorphism; SoGI: *Saccharum officinarum* Gene Index; SPP: Sucrose phosphate phosphatase; SPS: Sucrose phosphate synthase; SPT: Sugar phosphate translocator; SRA: Sugar Research Australia; SSR: Simple sequence repeats or microsatellites; SUCEST: Sugarcane expressed sequence tag database; SUGIT: Sugarcane Iso-Seq transcriptome database; SuSy: Sucrose synthase; SuT: Sucrose transporter; TC: Terpene cyclase; TCA: Tricarboxylic acid; TSA: Transcriptome shot-gun assembly; UDP: Uridine diphosphate; V-ATP: Vacuolar adenosine triphosphate

Acknowledgements

We gratefully acknowledge the financial support to PPT from the Department of Biotechnology, Government of India, for the Indo-Australian Career Boosting Gold Fellowship. We are grateful to the Australian Agency for International Development (AusAID) for financial support through an Australian Awards Scholarship to NVH. We thank SRA staff in Brandon station, Burdekin, Queensland, Australia for helping with the sample collecting and processing; Ravi Nirmal for helping us in sample collection and transport.

Funding

This work was funded by the Queensland Government and Sugar Research Australia (SRA). The funders had no role in the design of the study, collection, analysis, and interpretation of data, nor in writing the manuscript.

Authors' contributions

RJH, AF, FCB and NVH conceived and designed the experiments. AF and NVH collected the samples. PPT, NVH and AF conducted analyses. PPT prepared the first draft. RJH, AF, FCB and NVH critically revised the manuscript. All authors read and approved the final manuscript. PPT and NVH contributed equally to the manuscript.

Competing interests

The authors declare that they have no competing interests.

Author details

[1]Queensland Alliance for Agriculture and Food Innovation, The University of Queensland, St. Lucia, QLD 4072, Australia. [2]ICAR - Sugarcane Breeding Institute, Coimbatore, Tamil Nadu, India. [3]College of Agriculture and Forestry, Hue University, Hue, Vietnam. [4]Sugar Research Australia, Indooroopilly, QLD 4068, Australia. [5]The University of Queensland, Room 2.245, Level 2, The John Hay Building, Queensland Biosciences Precinct [#80], 306 Carmody Road, St Lucia, QLD 4072, Australia.

References

1. Slewinski T, Baker R, Stubert A, Braun D. Tie-dyed2 encodes a callose synthase that functions in vein development and affects symplastic trafficking within the phloem of maize leaves. Plant Physiol. 2012;160:1540–50.
2. Slewinski TL. Non-structural carbohydrate partitioning in grass stems: a target to increase yield stability, stress tolerance, and biofuel production. J Exp Bot. 2012;63(13):4647–70.
3. Whittaker A, Botha F. Carbon partitioning during sucrose accumulation in sugarcane internodal tissue. Plant Physiol. 1997;115:1651–9.
4. Welbaum GE, Meinzer FC. Compartmentation of solutes and water in developing sugarcane stalk tissue. Plant Physiol. 1990;3:1147–53.
5. Casu R, Dimmock C, Thomas M, Bower N, Knight D, Grof C, McIntyre L, Jackson P, Jordan D, Whan V. Genetic and expression profiling in sugarcane. In: Proc Int soc sugar cane Technol: 2001; 2001. p. 542–6.
6. Casu R, Grof C, Rae A, McIntyre C, Dimmock C, Manners J. Identification of a novel sugar transporter homologue strongly expressed in maturing stem vascular tissues of sugarcane by expressed sequence tag and microarray analysis. An Int J on Mol Biol, Mol Genet and Biochem. 2003;52(2):371–86.
7. Casu R, Dimmock C, Chapman S, Grof C, McIntyre C, Bonnett G, Manners J. Identification of differentially expressed transcripts from maturing stem of sugarcane by in silico analysis of stem expressed sequence tags and gene expression profiling. An Int J on Mol Biol, Mol Genet and Biochem. 2004; 54(4):503–17.
8. Casu R, Rae A, Nielsen J, Perroux J, Bonnett G, Manners J. Tissue-specific transcriptome analysis within the maturing sugarcane stalk reveals spatial regulation in the expression of cellulose synthase and sucrose transporter gene families. An Int J on Mol Biol, Mol Genet and Biochem. 2015;89(6):607–28.
9. Casu RE, Manners JM, Bonnett GD, Jackson PA, McIntyre CL, Dunne R, Chapman SC, Rae AL, Grof CPL. Genomics approaches for the identification of genes determining important traits in sugarcane. Field Crops Res. 2005; 92(2–3):137–47.
10. da Silva JA, Bressiani JA. Sucrose synthase molecular marker associated with sugar content in elite sugarcane progeny. Genet Mol Biol. 2005;28(2):294–8.
11. Watt DA, McCormick AJ, Govender C, Carson DL, Cramer MD, Huckett BI, Botha FC. Increasing the utility of genomics in unravelling sucrose accumulation. Field Crops Res. 2005;92(2–3):149–58.
12. Papini-Terzi FS, Rocha FR, Vêncio RZN, Felix JM, Branco DS, Waclawovsky AJ, Del Bem LEV, Lembke CG, Costa MDL, Nishiyama MY, et al. Sugarcane genes associated with sucrose content. BMC Genomics. 2009;10:120.
13. Arruda P. Sugarcane transcriptome. A landmark in plant genomics in the tropics. Genet Mol Biol. 2001;24(1–4):36.
14. Iskandar HM, Casu RE, Fletcher AT, Schmidt S, Xu J, Maclean DJ, Manners JM, Bonnett GD. Identification of drought-response genes and a study of their expression during sucrose accumulation and water deficit in sugarcane culms. BMC Plant Biol. 2011;11:12.
15. Waclawovsky AJ, Sato PM, Lembke CG, Moore PH, Souza GM. Sugarcane for bioenergy production: an assessment of yield and regulation of sucrose content.(report). Plant Biotechnol J. 2010;8:263.
16. Cardoso-Silva CB, Costa EA, Mancini MC, Balsalobre TWA, Canesin LEC, Pinto LR, Carneiro MS, Garcia AAF, de Souza AP, Vicentini R. De novo assembly and Transcriptome analysis of contrasting sugarcane varieties. PLoS One. 2014;9(2):e88462.
17. Hoang NV, Furtado A, Donnan L, Keeffe EC, Botha FC, Henry RJ. High-throughput profiling of the fiber and sugar composition of sugarcane biomass. Bioenerg Res. 2016;10(2):400–16.
18. Furtado A. RNA extraction from developing or mature wheat seeds. Methods Mol Biol. 2014;1099:23–8.
19. Hoang NV. Analysis of genes controlling biomass traits in the genome of sugarcane (Saccharum spp. hybrids) PhD Thesis, Queensland Alliance for Agriculture and Food Innovation, The University of Queensland. 2017. doi: 10.14264/uql.2017.602.
20. Ali M, Brian AW, Kenneth M, Lorian S, Barbara W. Mapping and quantifying mammalian transcriptomes by RNA-Seq. Nat Methods. 2008;5(7):621.
21. Baggerly KA, Deng L, Morris JS, Aldaz CM. Differential expression in SAGE: accounting for normal between-library variation. Bioinformatics. 2003;19(12): 1477–83.
22. Robinson MD, McCarthy DJ, Smyth GK. edgeR: a bioconductor package for differential expression analysis of digital gene expression data. Bioinformatics. 2010;26(1):139–40.
23. DFCI. 2016. ftp://occams.dfci.harvard.edu/pub/bio/tgi/data/. Accessed 27 Feb 2016. In.
24. Vettore AL, Da Silva FR, Kemper EL, Arruda P. The libraries that made SUCEST. Genet Mol Biol. 2001;24(1–4):1–7.
25. Huang X, Madan A. CAP3: a DNA sequence assembly program. Genome Res. 1999;9(9):868.
26. Hoang NV, Furtado A, Mason PJ, Marquardt A, Kasirajan L, Thirugnanasambandam PP, Botha FC, Henry RJ. A survey of the complex transcriptome from the highly polyploid sugarcane genome using full-length isoform sequencing and de novo assembly from short read sequencing. BMC Genomics. 2017;18(1):395.
27. Thimm O, Bläsing O, Gibon Y, Nagel A, Meyer S, Krüger P, Selbig J, Müller LA, Rhee SY, Stitt M. MAPMAN: a user-driven tool to display genomics data sets onto diagrams of metabolic pathways and other biological processes. Plant J. 2004;37(6):914.
28. Conesa A, Götz S. Blast2GO: a comprehensive suite for functional analysis in plant genomics. Int J Plant Genomics. 2008;2008:619832.
29. Hoang NV, Furtado A, O'Keeffe AJ, Botha FC, Henry RJ. Association of gene expression with biomass content and composition in sugarcane. PLoS One. 2017;12(8):e0183417.
30. Moore PH. Temporal and spatial regulation of sucrose accumulation in the sugarcane stem. Aust J Plant Physiol. 1995;22(4):661–79.
31. Vettore AL, Da Silva FR, Kemper EL, Souza GM, Da Silva AM, Ferro MIT, Henrique-Silva F, Giglioti EA, Lemos MVF, Coutinho LL, et al. Analysis and functional annotation of an expressed sequence tag collection for tropical crop sugarcane. Genome Res. 2003;13(12):2725.
32. Ferreira S, Hotta C, Poelking V, Leite D, Buckeridge M, Loureiro M, Barbosa M, Carneiro M, Souza G. Co-expression network analysis reveals transcription factors associated to cell wall biosynthesis in sugarcane. An Int J on Mol Biol, Mol Genet and Biochem. 2016;91(1):15–35.
33. Prathima P, Suparna T, Anishma S, Punnya R, Ramalashmi K. Cloning and characterization of a differentially regulated invertase inhibitor gene during sucrose accumulation in sugarcane. J Sugarcane Res. 2014;4:21–8.
34. Lastdrager J, Hanson J, Smeekens S. Sugar signals and the control of plant growth and development. J Exp Bot. 2014;65(3):799–807.
35. Eveland AL, Jackson DP. Sugars, signalling, and plant development. J Exp Bot. 2012;63(9):3367–77.
36. Rolland F, Baena-Gonzalez E, Sheen J. SUGAR SENSING AND SIGNALING IN PLANTS: conserved and novel mechanisms. Annu Rev Plant Biol. 2006;57: 675–709.
37. Tognetti JA, Pontis HG, Martínez-Noël GMA. Sucrose signaling in plants: a world yet to be explored. Plant Signal Behav. 2013;8(3):e23316.
38. Matsoukas IG, Massiah AJ, Thomas B. Starch metabolism and antiflorigenic signals modulate the juvenile-to-adult phase transition in arabidopsis. Plant Cell Environ. 2013;36(10):1802–11.

39. Yang L, Xu M, Koo Y, He J, Poethig RS. Sugar promotes vegetative phase change in Arabidopsis Thaliana by repressing the expression of MIR156A and MIR156C. elife. 2013;2:e00260.

40. Yu S, Cao L, Zhou C-M, Zhang T-Q, Lian H, Sun Y, Wu J, Huang J, Wang G, Wang J-W. Sugar is an endogenous cue for juvenile-to-adult phase transition in plants. elife. 2013;2:e00269.

41. Maeshima M. TONOPLAST TRANSPORTERS: organization and function. Annu Rev Plant Physiol Plant Mol Biol. 2001;52:469–97.

42. Jacobsen KR, Fisher DG, Maretzki A, Moore PH. Developmental changes in the anatomy of the sugarcane stem in relation to phloem unloading and sucrose storage. Botanica Acta. 1992;105(1):70–80.

43. Botha FC and Black KG. Sucrose phosphate synthase and sucrose synthase activity during maturation of internodal tissue in sugarcane. Funct Plant Biol. 2000;27(1):81–5.

44. Kim M, Day D. Composition of sugar cane, energy cane, and sweet sorghum suitable for ethanol production at Louisiana sugar mills. Official J Soc Ind Microbiol. 2011;38(7):803–7.

45. Vicentini R, Bottcher A, Brito M, Santos A, Creste S, Landell G, Cesarino I, Mazzafera P. Large-scale Transcriptome analysis of two sugarcane genotypes contrasting for lignin content. PLoS One. 2015;10(8):e0134909.

46. Babb VM, Haigler CH. Sucrose phosphate synthase activity rises in correlation with high-rate cellulose synthesis in three heterotrophic systems (1). Plant Physiol. 2001;127(3):1234.

47. PDC S, Palhares AC, Taniguti LM, Peters LP, Creste S, Aitken KS, Van Sluys M-A, Kitajima JP, MLC V, Monteiro-Vitorello CB. RNAseq transcriptional profiling following whip development in sugarcane smut disease.(research article)(report). PLoS One. 2016;11(9):e0162237.

48. Rae A, Perroux J, Grof C. Sucrose partitioning between vascular bundles and storage parenchyma in the sugarcane stem: a potential role for the ShSUT1 sucrose transporter. An Int J Plant Biol. 2005;220(6):817–25.

49. Eastmond PJ, van Dijken AJH, Spielman M, Kerr A, Tissier AF, Dickinson HG, Jones JDG, Smeekens SC, Graham IA. Trehalose-6-phosphate synthase 1, which catalyses the first step in trehalose synthesis, is essential for Arabidopsis embryo maturation. Plant J. 2002;29(2):225.

50. Griffiths CA, Sagar R, Geng Y, Primavesi LF, Patel MK, Passarelli MK, Gilmore IS, Steven RT, Bunch J, Paul MJ, et al. Chemical intervention in plant sugar signalling increases yield and resilience. Nature. 2016;540(7634):574–8.

51. Lunn JE, Feil R, Hendriks JHM, Gibon Y, Morcuende R, Osuna D, Scheible W-R, Carillo P, Hajirezaei M-R, Stitt M. Sugar-induced increases in trehalose 6-phosphate are correlated with redox activation of ADPglucose pyrophosphorylase and higher rates of starch synthesis in Arabidopsis Thaliana. The Biochem J. 2006;397(1):139.

52. Cook MG and Evans LT. The roles of sink size and location in the partitioning of assimilates in wheat ears. Func Plant Biol. 1983;10(3):313–27.

53. Borrell AK, Hammer GL. Nitrogen dynamics and the physiological basis of stay-green in sorghum. Crop Sci. 2000;40(5):1295.

54. Duncan RR, Bockholt AJ, Miller FR. Descriptive comparison of senescent and nonsenescent sorghum genotypes. Agronomy Journal. 1981;73(5):849–53.

55. Maurino VG, Saigo M, Andreo CS, Drincovich MF. Non-photosynthetic 'malic enzyme' from maize: a constituvely expressed enzyme that responds to plant defence inducers. Plant Mol Biol. 2001;45(4):409.

56. Dancer J, Hatzfeld W-D, Stitt M. Cytosolic cycles regulate the turnover of sucrose in heterotrophic cell-suspension cultures of Chenopodium Rubrum L. An Int J Plant Biol. 1990;182(2):223–31.

57. Bull TA, Glasziou K. The evolutionary significance of sugar accumulation in Saccharum. Aust J Biol Sci. 1963;16(4):737.

58. Birch R, Bower R, Elliott A. Highly efficient, 5'-sequence-specific Transgene silencing in a complex Polyploid. An International Journal devoted to original research in tropical plants. 2010;3(2):88–97.

6

Global gene expression in two potato cultivars in response to 'Candidatus Liberibacter solanacearum' infection

Julien G. Levy[1][*], Azucena Mendoza[2], J. Creighton Miller Jr.[1], Cecilia Tamborindeguy[2] and Elizabeth A. Pierson[1]

Abstract

Background: Transcriptomic analyses were performed to compare the molecular responses of two potato varieties previously shown to differ in the severity of disease symptoms due to infection by *"Candidatus* Liberibacter solanacearum" (Lso), the causative agent of Zebra Chip in potato. A factorial design utilizing the two varieties and psyllids either harboring Lso or without bacteria was used to discriminate varietal responses to pathogen infection versus psyllid feeding. Plant response was determined from leaf samples 3 weeks after infection.

Results: In response to Lso infection, 397 genes were differentially expressed in the variety Atlantic (most susceptible) as compared to 1027 genes in Waneta. Over 80% of the transcriptionally-changed genes were down-regulated in both varieties, including genes involved in photosynthesis or primary and secondary metabolism. Many of the Lso-responsive genes involved in stress responses or hormonal pathways were regulated differently in the two potato varieties.

Conclusions: This study focused on the time point just prior to the onset of symptom development and provides valuable insight into the mechanisms of *Liberibacter* pathogenicity, especially the widespread suppression of plant gene expression, including genes involved in plant defenses.

Keywords: Transcriptome, Potato, Zebra Chip, *Candidatus* Liberibacter solanacearum, Psyllids

Background

'*Candidatus* Liberibacter solanacearum' (Lso) are Gram-negative, phloem-limited, nonculturable bacteria. Lso is the causative agent of the Zebra Chip in potato [28, 48]. This pathogen is vectored to potato and other solanaceous hosts by the potato psyllid, *Bactericera cockerelli* (Hemiptera: Triozidae) [19, 34]. Lso has impacted potato production in Mexico since 1994, in Texas since 2000, and more recently in the Pacific Northwest of the US [9, 10]. Zebra Chip (ZC) is also a threat in Central America and New Zealand [1, 28, 47]. Lso affects other crops such as carrot, celery and parsnip in several countries in Europe and Africa [18], where it is vectored by several species of carrot psyllids. Lso can be seed transmitted in carrot, celery [6, 51] and chili pepper [7, 43]. Because of the emergence of Lso as a worldwide threat to different crops, studies such as this one aimed at elucidating plant

defense mechanisms targeting Ca. Liberibacter species have global importance.

Potato (*Solanum tuberosum* L.) is the most important vegetable crop in the US and the third most important worldwide based on human consumption. Potato yields, on average per acre, more food and protein than cereals [13, 37]. Zebra Chip affects all cultivated potatoes, resulting in increased production costs and revenue losses in the US, Mexico, Central America and New Zealand [33]. Zebra Chip symptoms in potato include curling, purple coloration, and chlorosis on the youngest leaves. As the disease progresses, plants develop shortened and swollen internodes, aerial tubers and wilting; ultimately, plants may die prematurely [28, 34]. Fresh tubers from infected plants develop a characteristic brown discoloration when cut [10]; however, fresh tuber symptoms do not appear in all potato varieties. The trademark of the disease is the development of dark stripes when tubers are fried, resulting in tubers that are unsuitable for the production of potato chips or French fries [47]. Previous studies showed that all potato market

* Correspondence: Julienlevy@tamu.edu
[1]Department of Horticultural Sciences, Texas A&M University, College Station, TX 77843, USA
Full list of author information is available at the end of the article

classes tested, including fresh market, chip, fry and processing varieties, were susceptible to ZC [27, 32]. Although all commercial potato varieties are susceptible to ZC, understanding the host molecular responses associated with Lso infection could facilitate the identification of genes and pathways involved in pathogen virulence and host recognition and/or disease development. Knowledge of these molecular interactions is important for the development of disease management strategies such as directed breeding or gene editing for resistance.

Transcriptomic analysis is a useful tool for investigating the effects of biotic and abiotic stresses on genome-wide gene expression patterns. However, like many other crops, cultivated potato varieties are polyploids; they are typically autotetraploids (4n = 48), so there are four interchangeable genes at each locus. Despite this complexity, transcriptomic analyses have been used successfully in other potato studies [2, 31, 58]. In the current study, we used a transcriptomic approach to analyze plant responses to Lso. A two-by-two factorial design was used that included two potato varieties: Atlantic, a susceptible variety, and Waneta (formerly known as NY138), a variety that develops milder and delayed ZC symptoms [26], and two infection treatments: infestation of potato by psyllids infected with Lso (Lso+) or not (Lso-). The plants were sampled 3 weeks after infestation, a time point that corresponds with the detection of Lso by PCR and qPCR in both varieties, but the onset of the first aerial symptoms only in Atlantic [26]. We focused on this time point because it was likely to show the greatest difference between varieties in their response to Lso infection. The aims of this study were to identify plant responses to Lso infection that were either conserved or different among two potato varieties with different degrees of susceptibility to Zebra Chip, in terms of the rate of symptom progression. The results of the present study provide novel insights into the genome-wide responses of the two varieties to Lso infection, including the expression of genes involved in photosynthesis, cell wall synthesis and metabolism and abiotic/biotic stress signaling.

Methods

Plant material and plant growth
Two potato varieties were used in the study: Atlantic and Waneta. The Atlantic seed pieces were produced in Dalhart, TX and the Waneta seed pieces were obtained from Childstock farms, NY. For each variety, eight tuber seeds were planted individually in one-gallon pots containing autoclaved (1 h 121 °C) potting mix (Metro-Mix 300). The plants were maintained in a growth chamber at 24 °C with 50% humidity and a 16:8 h day:-night light cycle. Plants were watered three times a week to field capacity.

Insect source
Potato psyllids (*Bactericera cockerelli*) were obtained from a colony harboring both LsoA and LsoB haplotypes (Lso+) or from a colony that does not harbor Lso (Lso-) [56]. The colonies were tested every 6 weeks prior their use for the presence (Lso+) or absence (Lso-) of Lso by PCR [25]. All insects were from the Northwestern haplotype.

Experimental procedure and tissue sampling
Four weeks after sprouting, plants from each cultivar were randomly assigned to each treatment (Lso + or Lso-). In total, there were eight replicate plants for each cultivar (Waneta and Atlantic); four plants of each cultivar received one or the other of the two insect treatments (Lso + or Lso-) (Fig. 1). Each plant was infested with three adult insects: either Lso + or Lso-. Insects were maintained in an organza bag placed on a single leaf located in the middle tier of plant leaves. After a 7-day inoculation access period, the leaf with the organza bag containing the insects was removed from the plant so that there would be no opportunity for insect escape and cross contamination.

Samples for RNA extraction were collected from the upper-most leaf of each plant at 3 weeks post infestation (wpi). All leaf samples were immediately flash frozen in liquid nitrogen and stored at –80 °C until RNA extraction. After sampling, the plants were maintained in the same conditions until the end of the experiment, which was 7 weeks post infestation. At the end of the experiment, leaf samples were collected and tested for Lso infection via PCR analysis of DNA samples extracted from the leaf mid-vein. Tuber samples were chipped and fried to detect ZC symptoms [27]. Only three replicates were used in the transcriptomic analysis.

DNA extraction and Lso detection using PCR
DNA extraction and Lso detection by PCR were performed following methods regularly used in our laboratory. For DNA extraction from plant tissue, the protocol previously published [40] was followed, and for insects, the fast protocol [25] was used. The PCR primers used were LsoTX [36, 40].

RNA extraction and cDNA library preparation
Total RNA from 100 mg of leaf tissue (including the central vein) from twelve plants was isolated: e.g., three biological replicates from each cultivar (Atlantic and Waneta) under two Lso treatments (Lso + and Lso-). Total RNA was isolated using the RNeasy Plant kit (Qiagen) according to the manufacturer's protocol. Samples were treated with Turbo DNase (Ambion) following manufacturer recommendations to remove genomic DNA contamination. Absence of DNA contamination in

Fig. 1 Leaf and chip symptoms 3 and 7 weeks post infestation. The two potato varieties Atlantic and Waneta developed ZC symptoms in aerial tissues and potato tubers. As reported previously, Waneta plants developed modest symptoms between weeks 3–4 compared to Atlantic plants, which typically develop obvious symptoms during that time, whereas both varieties generally have prominent symptoms by week 7 [26, 27]. Atlantic typically experiences a rapid decline immediately after week 6, leading to death between weeks7–8, whereas the decline is slower in Waneta . Thus, the two photo dates show minimal differences between the Lso + plants of the two varieties at 3 wpi, concurrent with the onset of symptom development, and at 7 wpi, where both have prominent symptoms

the RNA samples was verified by PCR. RNA quantification and quality assurance were performed using a Bioanalyzer (Agilent Technologies). The construction of the cDNA libraries (TruSeq RNA, Illumina) and sequencing were performed by the Texas A&M AgriLife Genomics & Bioinformatics Services. One lane of the Illumina HiSeq 2000 was used for the 12 samples with the sequencing format of 130 bp single read.

Transcriptomic analyses

Raw reads were trimmed to remove left over adapters using the program Cutadapt v1.1 [http://dx.doi.org/10.14806/ej.17.1.200], and then remaining bases with a minimum quality score of 15 were trimmed and filtered to a minimum length of 30 bp using the FASTX-Toolkit (http://hannonlab.cshl.edu/fastx_toolkit/). Low quality reads were discarded. The Texas A&M Brazos Cluster was used for file manipulation. All the bioinformatics analyses were performed using the cloud computing in CyVerse within the Discovery Environment [17]. The RNA-seq reads that passed the quality filters (FASTQC tools, [4]) were mapped to the potato reference genome (PGSC_DM_v3.4 gene models ensemblv19-preinstalled in CyVerse) using the Tuxedo RNA-seq pipeline [53]. The Tuxedo pipeline was comprised of TopHat2 (mapping), Cufflinks2 (transcript assembly), Cuffmerge2 (transcript merging), and Cuffdiff2 (differential gene and transcript expression). For TopHat2 the standard (per default) parameters for an Illumina 1.9 (PHRED33) were used. Determination of differential gene expression in response to Lso infection was made with Cuffdiff2 using the defaults parameters with multiple hit correction and upperquartile normalization (Minimum per-locus counts

for significance testing: 10). Analyses to identify Lso-responsive genes were performed separately for each potato variety by comparing the transcriptomic responses of plants treated with Lso- insects to those treated with Lso + insects. We chose the significantly DEGs based on q value below 0.05.

For Gene Ontology (GO) enrichment analysis, the web based tool gprofiler (http://biit.cs.ut.ee/gprofiler/) was used with the default settings and Benjamini–Hochberg False Discovery Rate for significance threshold with a value of 0.05 [42]. The Open Source MapMan software was used to generate diagrams of the biologic and metabolic pathways that were differentially expressed in Atlantic and Waneta in response to Lso infection [52]. The data discussed in this publication have been deposited in the NCBI Gene Expression Omnibus [12] and are accessible through GEO Series accession number GSE92312 (https://www.ncbi.nlm.nih.gov/geo/query/acc.cgi?acc= GSE92312).

Real-time PCR (RT-qPCR) quantitative analysis

Real-time quantitative PCR analyses were conducted to confirm the results obtained by RNA-seq analysis. Synthesis of cDNA was performed on eight different samples obtained from two separate plants from each treatment, e.g., plants infested with Lso + or Lso- psyllids from each potato variety (Atlantic, Waneta). Reverse transcription of RNA samples was performed using the Verso cDNA Synthesis kit (Thermo, Waltham, MA) according to the manufacturer's instructions. Elongation factor 1a (Ef1a) was chosen as the reference gene [38].

A total of 19 genes were chosen for qPCR analysis: ten differentially expressed genes (DEGs) identified in both varieties as well as five genes that were transcriptionally-changed only in Atlantic and four genes transcriptionally changed only in Waneta. Primers were obtained from idtDNA (www.idtDNA.com) using the idtDNA primer design software, and all primer sequences are listed in Additional file 1: Table S1.

For qPCR amplification, each reaction contained 5 ng of cDNA, 250 nM of each primer and 1X of SensiFAST SYBR Hi-ROX Master Mix (Bioline, Taunton, MA); the volume was adjusted with nuclease-free water to 10 μL. The real-time PCR program was 95 °C for 2 min followed by 40 cycles at 95 °C for 5 s and 60 °C for 30 s. Real-time PCR assays were performed using an Applied Biosystems ABI 7300 real-time PCR Thermocycler (Applied Biosystems) according to the manufacturer's recommendations. For RT-qPCR, two technical replicates for each of the eight synthetized cDNAs were performed, with negative controls in each run. The qPCR results were analyzed using the Pfaffl equation [39] of the comparative Ct method ($2^{-\Delta\Delta Ct}$). A Pearson product-moment correlation test was performed to evaluate the correlation between the RNA-seq and the RT-qPCR results.

Results

All infected plants showed aerial symptoms such as chlorosis and wilting by the end of the experiment (7 wpi), whereas these symptoms did not develop in non-infected plants. All plants were tested at 7 wpi for the presence of Lso in leaves by PCR and for ZC symptoms in tubers by frying. DNA was extracted from the mid-vein of a leaf from each plant and all tubers from each plant were chipped and fried [27]. All plants infested with Lso + insects tested positive for Lso by PCR and showed ZC symptoms in the frying test, whereas plants infested with Lso- insects tested negative for Lso in leaves and tubers had no ZC defects when fried (Fig. 1).

Transcriptome data

To understand how Lso infection affects plant gene expression for each potato variety, we compared the transcriptomes of plants 3 weeks after infestation with potato psyllids harboring or not harboring 'Candidatus Liberibacter solanacearum', (Lso + and Lso-, respectively). This time point corresponds to the onset of symptoms in Atlantic, e.g., when the Lso + plants showed slight chlorosis at the base of the upper leaf. These symptoms were not yet visible on most Waneta plants. Plant samples were taken from leaf tissues.

We obtained approximately 20 million reads per sample (Table 1). After quality filtering, reads were mapped to the double haploid *Solanum tuberosum* reference genome DMI3.4 ensembl 19; using Tophat2 in CyVerse. Three biological replicates were sequenced for each of the four treatments: e.g., cultivar (Atlantic and Waneta) by bacterial infection status (Lso + and Lso-). However,

Table 1 Summary of potato transcriptomic data

Variety	Treatment	Reads	Mapped reads	Mapped reads (%)
Atlantic	Lso+	28,874,601	18,865,812	65.3
Atlantic	Lso+	16,489,980	9,972,190	60.5
Atlantic	Lso-	14,819,808	10,060,020	67.9
Atlantic	Lso-	16,106,431	9,567,021	59.4
Atlantic	Lso-	18,219,259	10,890,012	59.8
Waneta	Lso-	26,727,692	18,435,375	69
Waneta	Lso-	35,950,390	24,050,379	66.9
Waneta	Lso-	29,583,541	21,154,857	71.5
Waneta	Lso+	15,547,755	10,354,955	66.6
Waneta	Lso+	15,214,633	10,649,743	70
Waneta	Lso+	25,997,451	17,814,346	68.5

one of the Atlantic samples in the Lso + treatment was removed after discovering that it was also infected with *Potato virus S* (PVS). In this study, between 59 to 71% of the reads in each library mapped to the potato genome; at least 9 million reads per sample mapped to the potato genome (Table 1).

Differentially expressed genes

Using Cuffdiff2, the following comparisons were performed: Waneta Lso- versus Waneta Lso+, and Atlantic Lso- versus Atlantic Lso+. The objective was to identify genes that were differentially expressed in plants in response to Lso infection, by variety. A total of 397 differentially expressed genes (DEGs) were identified in the Atlantic comparison (Additional file 1: Table S2 and Figure S1), whereas in the Waneta comparison there were 1027 DEGs (Additional file 1: Table S3 and Figure S1). In order to characterize the plant processes potentially affected by Lso infection, DEGs were classified into MapMan functional plant categories [52] and Gene Onotology (GO) enrichment analyses (http://www.geneontology.org/page/go-enrichment-analysis) were performed on sets of DEGs to identify which GO terms (biological process, molecular function, or cellular component) were over- or under-represented, based on gene set annotations. In both Atlantic and Waneta the DEGs were associated predominantly with photosynthesis, primary and secondary metabolism and biotic and abiotic signal recognition and stress responses. Results are discussed under the headers transcriptomic overview, plant metabolism (including photosynthesis), and plant stress response.

Transcriptomic overview

In both varieties, most of the DEGs were down-regulated in the Lso + samples (Atlantic =323 genes, Waneta =852 genes, which corresponds to approximately 82% of the DEGs in each comparison), indicating that the transcriptomic response to Lso infection by both varieties was primarily down-regulation in gene expression. Of the total DEGs, 111 were identified in both the Waneta and Atlantic comparisons: 61 of these conserved DEGs were down-regulated in both comparisons, nine were up-regulated in both comparisons, and 41 were oppositely regulated in the two comparisons (Fig. 2, Additional file 1: Tables S4A-D).

Of the 70 similarly down- or up-regulated DEGs identified in both the Waneta and Atlantic comparisons, GO enrichment analysis showed that GO terms with Molecular Function types associated with photosynthesis, such as GO 0046906 and GO 0016168, were significantly enriched (Additional file 1: Table S5A). Among the 41 Lso-responsive DEGs common to both comparisons but oppositely regulated in Atlantic and Waneta, two GO terms with Molecular Function types were significantly enriched, both associated with transcription factor activity: GO:0001071 and GO:0003700 (Additional file 1: Table S5B).

Of the total DEGs in Atlantic and Waneta, 13% and 21%, respectively, were annotated as conserved genes of unknown function. Since these genes have not yet been characterized, it is possible that they could be associated with specialized functions and may be worthy of future consideration.

Plant metabolism

In response to Lso infection, a similar number of DEGs involved in plant metabolism were identified in Atlantic and Waneta, although these accounted for a greater percentage of the total number of DEGs identified in Atlantic (110 DEGs = 29% of the total DEGs vs 103 DEGs =

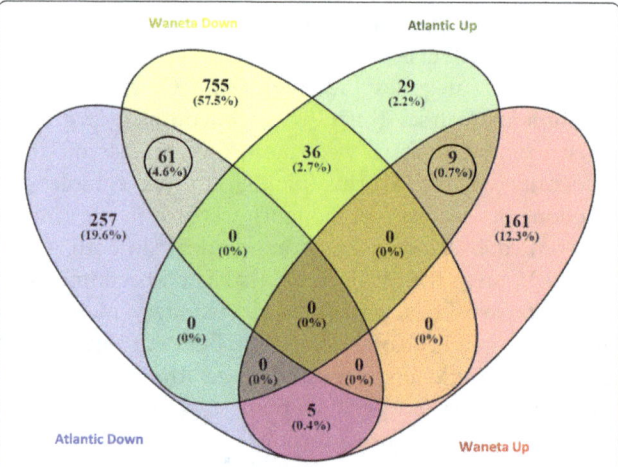

Fig. 2 Venn diagram summarizing DEGs in Atlantic (AT) and Waneta(W) in response to Lso infection. Up and Down refer to genes with significantly higher or lower expression in Lso + versus Lso- plants (e.g., up-regulated vs. down-regulated DEGs in response to Lso infection). The number of DEGs and the percentage of the total DEGs are given. DEGs similarly up- or down-regulated in both potato varieties in response to Lso infection are circled

10% of the total DEGs). Furthermore, in Atlantic the DEGs involved in plant metabolism had some of the highest fold changes: 90 of the 110 DEGs were down-regulated and had over a 10-fold decrease in expression in response to Lso infection. GO enrichment analysis indicated that these DEGs were significantly enriched for molecular function terms (e.g., GO:0016798 hydrolase activity, acting on glycosyl bonds; GO:0004553 hydrolase activity, hydrolyzing O-glycosyl compounds; and GO:0003824 catalytic activity). For example, the two DEGs with the highest expression in Atlantic were annotated as the steroidal glycoalkaloid (SGA) pathway enzyme UDP-glucosyl transferase and a peroxidase. The highly expressed DEGS in Atlantic also were significantly enriched for biological process terms associated with photosynthesis and cell wall degradation. In Waneta, the GO enrichment analysis of down-regulated DEGs in response to Lso infection revealed a much wider array of functions (Additional file 1: Table S10).

In both the Atlantic and Waneta comparisons, photosynthesis-related DEGs were primarily down-regulated in response to Lso infection. These included genes annotated as functioning in light reactions, tetrapyrrole synthesis, and the Calvin cycle. In particular, several chlorophyll-binding proteins were down regulated in both Waneta and Atlantic. Only one gene in Waneta was up-regulated in those pathways; this gene is annotated as an ATP synthase gamma chain (Fig. 3 and Additional file 1: Table S6).

Similarly, DEGs involved in carbohydrate metabolism and transport were primarily down-regulated in both varieties in response to Lso infection. Interestingly, among them were several genes that typically function in controlling the balance between starch and sucrose (Fig. 3, Additional file 1: Table S7). For example, one gene annotated as UDP glycosyl transferase was down-regulated in Atlantic, whereas an invertase was down-regulated and a transferase was up- regulated in Waneta. Similarly, DEGs annotated as functioning in glycolysis were down-regulated in both varieties. Down-regulated genes in Atlantic included an aldehyde dehydrogenase, a malate synthase and an ATP citrate synthase, and in Waneta included a phosphoglycerate/biphosphoglycerate mutase and a ribose-5-phosphate isomerase. Other DEGs potentially involved in minor carbohydrate metabolism also were primarily down-regulated in both varieties. The one exception was a galactinol synthase that was up-regulated in Atlantic (Additional file 1: Table S7).

In both varieties, Lso infection induced changes in the expression of genes involved in cell wall synthesis and modification, as well as of genes encoding cell wall proteins (Table 2, Fig. 3). Among the 35 DEGs in Atlantic, most DEGs in these categories were down-regulated in response to Lso infection; only one was up-regulated and it encoded a cell wall protein (RGP3). Similarly in

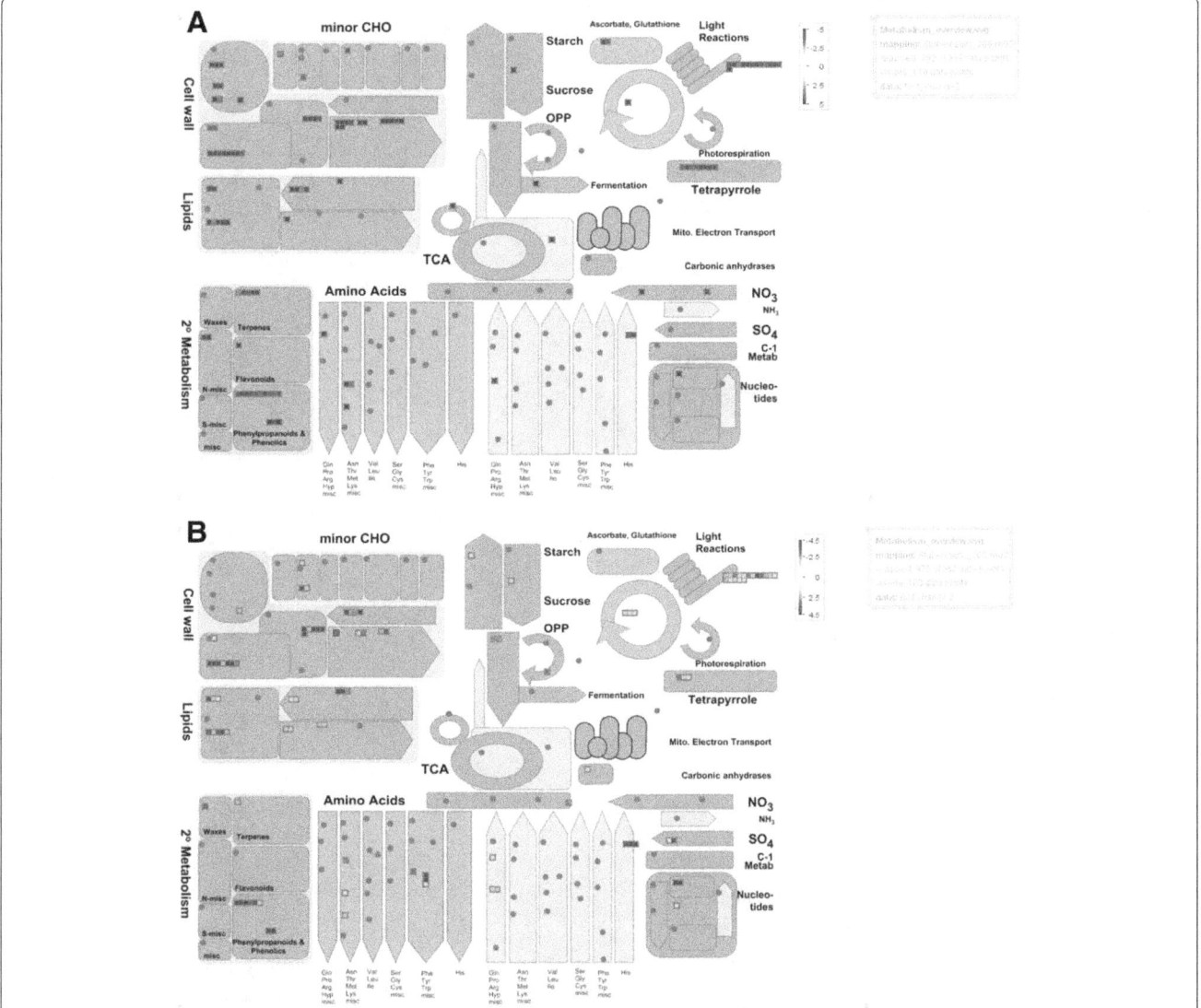

Fig. 3 MapMan overview of DEGs annotated as contributing to metabolic pathways in Atlantic (**a**) and Waneta (**b**). Up-regulated DEGs (transcript abundance higher in Lso + samples than Lso- samples) appear in red and down-regulated genes appear in blue. Fold changes are expressed on a logarithmic scale. DEGs in the photosynthesis pathway, carbohydrate metabolism and glycolysis were primarily down-regulated in response to Lso infection in both Atlantic and Waneta

Waneta, there were 25 cell wall-related DEGs, the majority of which were down-regulated; only four were up-regulated in response to Lso infection (Table 2).

DEGs annotated as functioning in lipid metabolism also were identified: in Atlantic 13 of these DEGs were identified and all were down-regulated in response to Lso, whereas in Waneta, 15 DEGs were identified and all but two were down-regulated (Fig. 3, Additional file 1: Figure S2 and Table S8). Additionally, 11 Lso-responsive DEGs annotated as being involved in amino acid metabolism were identified in both comparisons, all of which were down-regulated (Fig. 3).

Most of the DEGs involved in secondary metabolism identified in both potato varieties were down-regulated, although some differences in the way genes were regulated were notable. For instance, in Atlantic there were five terpene-related DEGs and all were down-regulated, whereas in Waneta only one terpene-related DEG was identified and it was down-regulated. One flavonoid-related gene, annotated as an isoflavone reductase, was down-regulated in Atlantic whereas none were differentially expressed in Waneta. Nine (out of ten) phenylpropanoid related DEGs were down-regulated in Atlantic, whereas in Waneta all six phenylpropanoid related DEGs were down-regulated. Additionally three DEGs in the phenolic pathway were identified in Atlantic (two were up-regulated) and two in Waneta (both down-regulated) (Fig. 3, Additional file 1: Figure S2 and Table S9).

Table 2 Mapman analysis of DEGs related to cell wall synthesis in Atlantic and Waneta in response to Lso infection. Negative Fold Change values denote the fold decrease in transcript abundance (FPKM) in Lso + samples compared to Lso- samples, e.g., down-regulated genes in response to Lso infection; positive Fold Change values denote up-regulated genes. Bin name and descriptors are provided

Bin Name	Transcript id	Fold Change Atlantic	Fold Change Waneta	Gene Description
cell wall.cellulose synthesis	PGSC0003DMT400007050	−28.6	–	Cellulose synthase-like A1
cell wall.cellulose synthesis	PGSC0003DMT400026046	–	−5.78	Transferase, transferring glycosyl groups
cell wall.cellulose synthesis	PGSC0003DMT400068221	−10.36	–	UPA15
cell wall.modification	PGSC0003DMT400003608	−4.43	−2.07	Expansin
cell wall.modification	PGSC0003DMT400022769	−6.02	–	Expansin
cell wall.modification	PGSC0003DMT400025776	−5.35	–	Expansin11
cell wall.modification	PGSC0003DMT400038406	–	−10.23	Xyloglucan endotransglucosylase-hydrolase XTH3
cell wall.modification	PGSC0003DMT400038411	–	−16.06	Xyloglucan endotransglucosylase-hydrolase XTH3
cell wall.modification	PGSC0003DMT400038412	–	−8.31	Xyloglucan endotransglucosylase-hydrolase XTH3
cell wall.modification	PGSC0003DMT400044562	–	−34.42	Xyloglucan endotransglycosylase
cell wall.modification	PGSC0003DMT400047948	–	−5.33	Expansin
cell wall.modification	PGSC0003DMT400055138	−13.58	–	Xyloglucan endotransglucosylase-hydrolase XTH7
cell wall.modification	PGSC0003DMT400061841	−6.14	–	Xyloglucan endotransglucosylase/hydrolase protein A
cell wall.modification	PGSC0003DMT400063689	−49.43	2.82	Xyloglucan endotransglucosylase/hydrolase 1
cell wall.modification	PGSC0003DMT400067358	−3.78	–	Xyloglucan endotransglycosylase/hydrolase 16 protein
cell wall.modification	PGSC0003DMT400078182	−14.74	–	Expansin18
cell wall.precursor synthesis.AXS	PGSC0003DMT400018431	–	−4.09	UDP-apiose/xylose synthase
cell wall.precursor synthesis.UGD	PGSC0003DMT400003666	–	−2.46	UDP-glucose dehydrogenase 2
cell wall.precursor synthesis.GAE	PGSC0003DMT400020223	–	−19.16	UDP-glucuronate 5-epimerase
cell wall.precursor synthesis.GAE	PGSC0003DMT400029181	–	−2.58	UDP-glucuronate 5-epimerase
cell wall.cellulose synthesis.cellulose synthase	PGSC0003DMT400008183	–	−5.39	Glycosyltransferase, CAZy family GT2
cell wall.cellulose synthesis.cellulose synthase	PGSC0003DMT400008184	–	−14.79	Glycosyltransferase, CAZy family GT2
cell wall.cellulose synthesis.cellulose synthase	PGSC0003DMT400009764	–	−2.2	Cellulose synthase
cell wall.cellulose synthesis.cellulose synthase	PGSC0003DMT400030678	−30.63	−5.39	Cellulose synthase
cell wall.cellulose synthesis.cellulose synthase	PGSC0003DMT400073085	−4.76	–	Cellulose synthase catalytic subunit
cell wall.cellulose synthesis.COBRA	PGSC0003DMT400065840	–	−13.12	Protein COBRA
cell wall.cell wall proteins.AGPs.AGP	PGSC0003DMT400033202	−29.85	–	Fasciclin-like arabinogalactan protein 13
cell wall.cell wall proteins.AGPs.AGP	PGSC0003DMT400044769	−15.53	–	Fasciclin-like arabinogalactan protein 19
cell wall.cell wall proteins.AGPs.AGP	PGSC0003DMT400076660	−30.56	–	Fasciclin-like arabinogalactan protein 10
cell wall.cell wall proteins. Proline rich proteins	PGSC0003DMT400004126	−11.75	−2.09	Proline-rich protein

Table 2 Mapman analysis of DEGs related to cell wall synthesis in Atlantic and Waneta in response to Lso infection. Negative Fold Change values denote the fold decrease in transcript abundance (FPKM) in Lso + samples compared to Lso- samples, e.g., down-regulated genes in response to Lso infection; positive Fold Change values denote up-regulated genes. Bin name and descriptors are provided *(Continued)*

Bin Name	Transcript id	Fold Change Atlantic	Fold Change Waneta	Gene Description
cell wall.cell wall proteins.LRR	PGSC0003DMT400015439	−7.07	–	Leucine-rich repeat/extensin
cell wall.cell wall proteins.LRR	PGSC0003DMT400043527	−6.26	–	Leucine-rich repeat family protein / extensin family protein
cell wall.cell wall proteins.RGP	PGSC0003DMT400039494	−5.13	–	GRP 2
cell wall.cell wall proteins.RGP	PGSC0003DMT400063316	3.38	–	GRP 2
cell wall.degradation.cellulases and beta −1,4-glucanases	PGSC0003DMT400008105	−8.23	–	Endo-1,4-beta-glucanase
cell wall.degradation.cellulases and beta −1,4-glucanases	PGSC0003DMT400009667	−11.21	–	Endo-1,4-beta-glucanase
cell wall.degradation.cellulases and beta −1,4-glucanases	PGSC0003DMT400012809	−9.03	3.45	Endo-beta-1,4-glucanase
cell wall.degradation.cellulases and beta −1,4-glucanases	PGSC0003DMT400015233	−10.73	–	Hydrolase, hydrolyzing O-glycosyl compounds
cell wall.degradation.cellulases and beta −1,4-glucanases	PGSC0003DMT400023082	−4.55	–	Endo-1,4-beta-glucanase
cell wall.degradation.cellulases and beta −1,4-glucanases	PGSC0003DMT400032028	−8.27	–	Endo-beta-1,4-D-glucanase
cell wall.degradation. mannan-xylose-arabinose-fucose	PGSC0003DMT400009055	−19.24	3.03	Endo-beta-mannanase
cell wall.degradation. mannan-xylose-arabinose-fucose	PGSC0003DMT400012716	−6.72	–	Xylanase Xyn2
cell wall.degradation. mannan-xylose-arabinose-fucose	PGSC0003DMT400076470	–	−1.92	LEXYL2 protein
cell wall.degradation.pectate lyases and polygalacturonases	PGSC0003DMT400001638	−4.7	–	Polygalacturonase
cell wall.degradation.pectate lyases and polygalacturonases	PGSC0003DMT400027955	−17.69	–	Pectate lyase
cell wall.degradation.pectate lyases and polygalacturonases	PGSC0003DMT400064881	−24.45	–	Dehydration-responsive protein RD22
cell wall.degradation.pectate lyases and polygalacturonases	PGSC0003DMT400076251	−8.1	−2.32	Pectase lyase
cell wall.degradation.pectate lyases and polygalacturonases	PGSC0003DMT400079602	−32.82	−3.43	Polygalacturonase-1 non-catalytic subunit beta
cell wall.pectin*esterases.PME	PGSC0003DMT400023725	–	−2.94	Pectinesterase 3
cell wall.pectin*esterases.PME	PGSC0003DMT400035577	–	1.71	Pectinesterase
cell wall.pectin*esterases.PME	PGSC0003DMT400016326	−3.55	–	Glutamyl-tRNA reductase
cell wall.pectin*esterases. Acetyl esterase	PGSC0003DMT400062026	−3.61	–	PAE

Interestingly, in Waneta, among the highly up-regulated genes were two genes annotated as encoding patatins (PGSC0003DMG400014104 and PGSC0003DMG4020170 90); however, the gene encoding patatin 3 (PGSC0003D MG400010022) was down-regulated. Patatins are glyco-proteins that serve as the major storage protein in tubers. These proteins also are reported to have esterase/lipase activity [44]. Moreover, genes encoding other tuber-specific proteins such as PGSC0003DMG400007994 and PGSC 0003DMG400010296 (Patellin-4) also were differentially regulated in Waneta, but these were down-regulated. The detection of tuber-specific DEGs in response to Lso infection in leaf samples was somewhat unexpected, because the functions of these genes in leaves are unknown.

Plant stress response

As compared to DEGs involved in metabolism where a similar number of DEGs were identified, in both the Atlantic and Waneta comparisons in response to Lso (110 and 103, respectively) almost three times more DEGs involved in stress response were identified in the Waneta compared to Atlantic (340 and 131, respectively). Still, this amounted to about one third of the total DEGs identified in both the Waneta and Atlantic comparisons having a role in biotic/abiotic stress response. In this category, most of the DEGs were annotated as being involved in signaling, hormone pathways, transcription, defense against pathogens, or abiotic stress (Figs. 4 and 5).

In Waneta, 115 signaling-related DEGs were identified, of which only nine were up-regulated in response to Lso infection, whereas in Atlantic there were 22 DEGs, of which 12 were up-regulated (Additional file 1: Table S11). In both varieties, a large number of DEGs were annotated as encoding receptor-like kinases and genes involved in calcium signaling. GO enrichment analysis of Waneta down-regulated genes with high fold change (73 DEGs with more than 10× fold change) showed enrichment in GO terms associated with calcium ion transmembrane transporter activity (GO:0015085); calcium-transporting ATPase activity (GO:0005388); and ATPase activity, coupled to transmembrane movement of ions, phosphorylative mechanism (GO:0015662). MapMan identified a total of 35 DEGs involved in calcium signaling in Waneta and all were down-regulated, whereas the four DEGs involved in calcium signaling in Atlantic were up-regulated (Fig. 4, Additional file 1: Figure S3 and Table S11). For instance, in Waneta, 17 DEGs related to calcium signaling and13 DEGs annotated as calmodulins were identified, whereas in Atlantic only two DEGs involved in calcium binding and two DEGs annotated calmodulin genes were identified.

Similar receptor-like protein kinases were among the DEGs identified in Atlantic and Waneta in response to Lso infection (Fig. 5a, b). Among the Atlantic DEGs, three were annotated as DUF26, three as LRR III and 11 as LRR XI genes. The DEG with the highest difference in expression was the LRR gene, which was down-regulated 12 fold in Lso-infected Atlantic (PGSC0003DMG400013898, Hcr2-0B). This gene is a homolog of the tomato Cf-5 disease resistance gene [11], and LRR is an important gene family involved in plant microbe interactions. Among the DEGs in Waneta, 27 were annotated as DEGLRR XI and 33 as DUF26. Of these seven were up-regulated (Fig. 5b) in response to Lso infection. Among the DEGs in Waneta, one WAK and two S-locus glycoprotein-like genes were down-regulated in response to Lso. No DEGs in these categories were identified in Atlantic. Also in Waneta, 31 DEGS annotated as receptor-like cytoplasmic kinases were down-regulated, whereas in Atlantic, only five DEGs in this category were identified.

Lso-responsive DEGs involved in several hormone pathways were identified in both varieties. Overall, in Waneta, 41 DEGs related to auxin, gibberellic acid, brassinosteroids, cytokinin, JA, and SA synthesis, degradation, and/or signaling were identified. Thirty-two of these were down-regulated, whereas nine were up-regulated in Lso infected plants. The ethylene and auxin pathways were the hormone pathways with the highest number of DEGs (19 for ethylene, only two of which were up-regulated, and 15 for auxin, of which four were up-regulated). Additionally in Waneta, six DEGs involved in brassinosteroid signaling and two DEGs in jasmonic acid signaling were identified, whereas in Atlantic, there were 12 DEGs in these categories (Additional file 1: Figure S3 and Table S12).

Interestingly, although many hormonal pathways were affected by Lso infection in both potato varieties, the responses often were opposite. For example, in Waneta, Lso infection resulted in down-regulation in the expression of genes involved in abscisic acid (ABA) signaling, such as genes annotated as encoding GRAM domain-containing protein / ABA-responsive protein-related, and in genes involved in ethylene signaling such as *ERF-1,2,3,4,5,9*. However, in Atlantic, the expression of genes involved in these hormone signaling pathways were up-regulated (Fig. 4). Similarly, in Atlantic, a DEG involved in the salicylic acid (SA) synthesis (S-adenosyl-L-methionine:benzoic acid carboxyl methyltransferase) was down-regulated in response to Lso, whereas in Waneta, two LOX genes, involved in jasmonic acid (JA) synthesis were down-regulated. In addition to the genes recognized by MapMan in each hormonal signaling pathway, 13 down-regulated and five up-regulated DEGs in gibberellic acid signaling and five DEGs in cytokinin signaling in Waneta were identified. The gibberellic acid signaling pathway has been implicated in plant development and plant-microbe interactions [15, 16].

The majority of DEGs annotated as transcription factors in Waneta (125) were down-regulated, whereas there were only 25 such DEGs in Atlantic (Fig. 4, Additional file 1: Table S13). The most represented family among the differentially expressed transcription factors was the WRKY family, with 22 down-regulated genes in Waneta and two up-regulated genes in Atlantic. In Waneta, 11 DEGS related to ethylene signaling in the AP2 EREBP family were identified, eight were down-regulated, whereas in Atlantic only one DEG from this family was identified. Similarly, 14 DEGs annotated as GRAS genes were found in Waneta (11 were down-regulated), whereas no such DEGs found in Atlantic. GRAS genes are gibberellic acid-dependent transcriptional regulators.

GO enrichment analysis of the up-regulated DEGs in Waneta in response to Lso infection indicated they were enriched in terms related to nucleic acid binding transcription factor activity (GO:0001071), transcription factor activity, and sequence-specific DNA binding (GO:0003700). The same GO terms were enriched among the 74 up-regulated DEGs in Atlantic: however, it was interesting that these were not the same genes as those that were differentially expressed in Waneta.

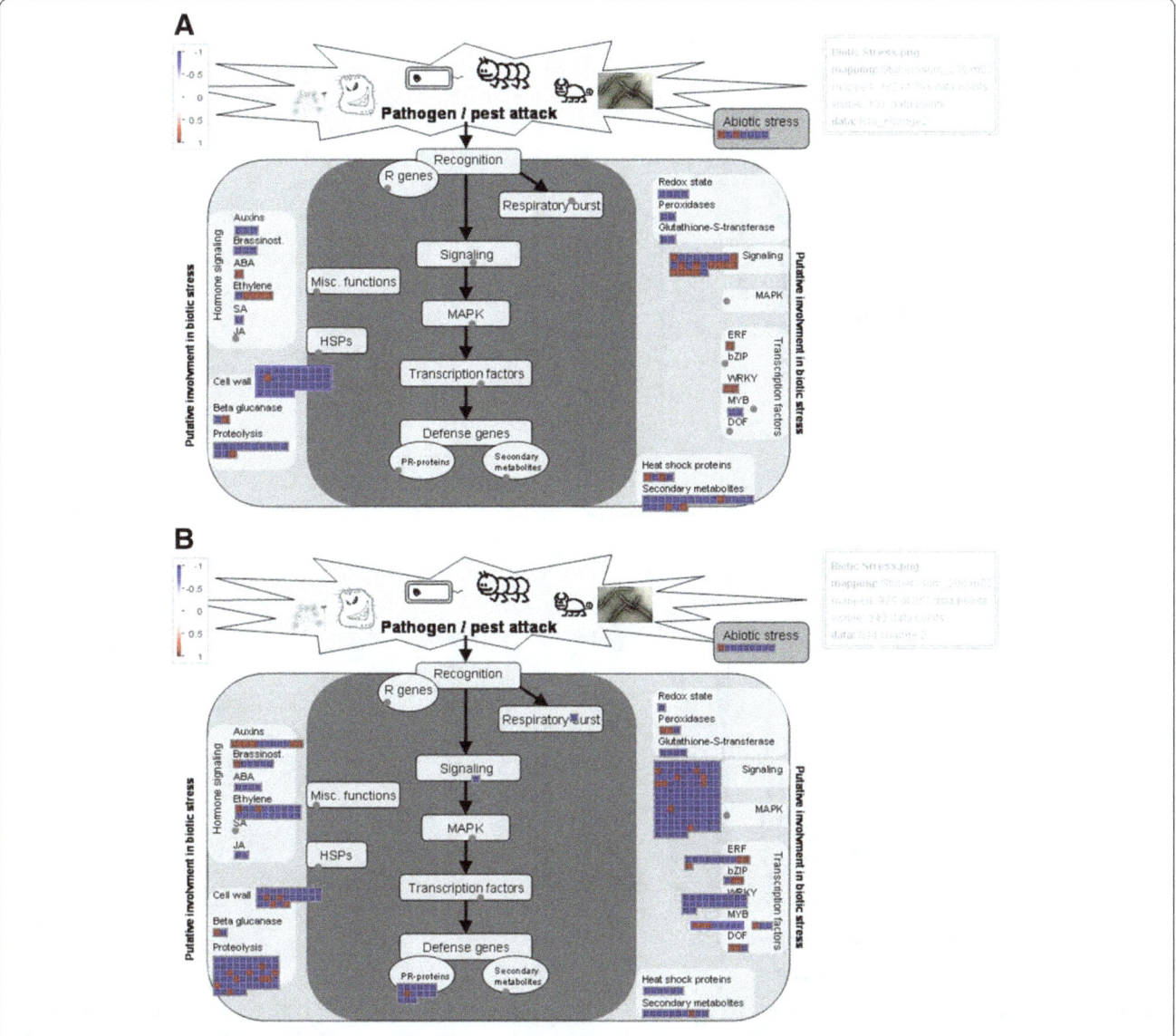

Fig. 4 MapMan overview of DEGs annotated as contributing to biotic and abiotic stress-related pathways in Atlantic (**a**) and Waneta (**b**). Up-regulated DEGs (transcript abundance higher in Lso + samples than Lso- samples) appear in red and down-regulated genes appear in blue. Fold changes are expressed on a logarithmic scale

Additionally, in Waneta, over 100 DEGs annotated as being associated with plant defense mechanisms were identified. Among these, 18 genes annotated as Avr9/Crf9 elicited proteins were down-regulated in Lso-infected plants, but one was up-regulated; five of these genes were also differentially expressed in Atlantic. Several Lso-responsive DEGs involved in proteolysis were identified in both Waneta and Atlantic: 55 vs 13 genes, respectively. Some of these DEGs were annotated as encoding F-box family proteins and ubiquitin-protein ligases. Such proteins have been shown to be vital to plant stress response. Similarly, 14 pathogenesis-related (PR) genes (most of them belonging to the TIR-NBS-LRR class) were down-regulated in Waneta in response

to Lso infection, whereas only one PR gene was up-regulated. None of the genes encoding these putative disease resistance proteins was differentially expressed in Atlantic.

Several stress-associated Lso-responsive DEGs were identified. In Waneta, several genes encoding salt-responsive proteins were down-regulated, including the genes with the two highest fold changes (126× and 86×, respectively, encoding salt responsive protein 2 genes PGSC0003DMG400000332 and PGSC0003DMG400010 713, as well as a stress associated protein 11 gene PGSC0003DMG400000512. A salt responsive protein 1 gene (PGSC0003DMG400022888) was down-regulated in both Waneta and Atlantic in response to Lso infection

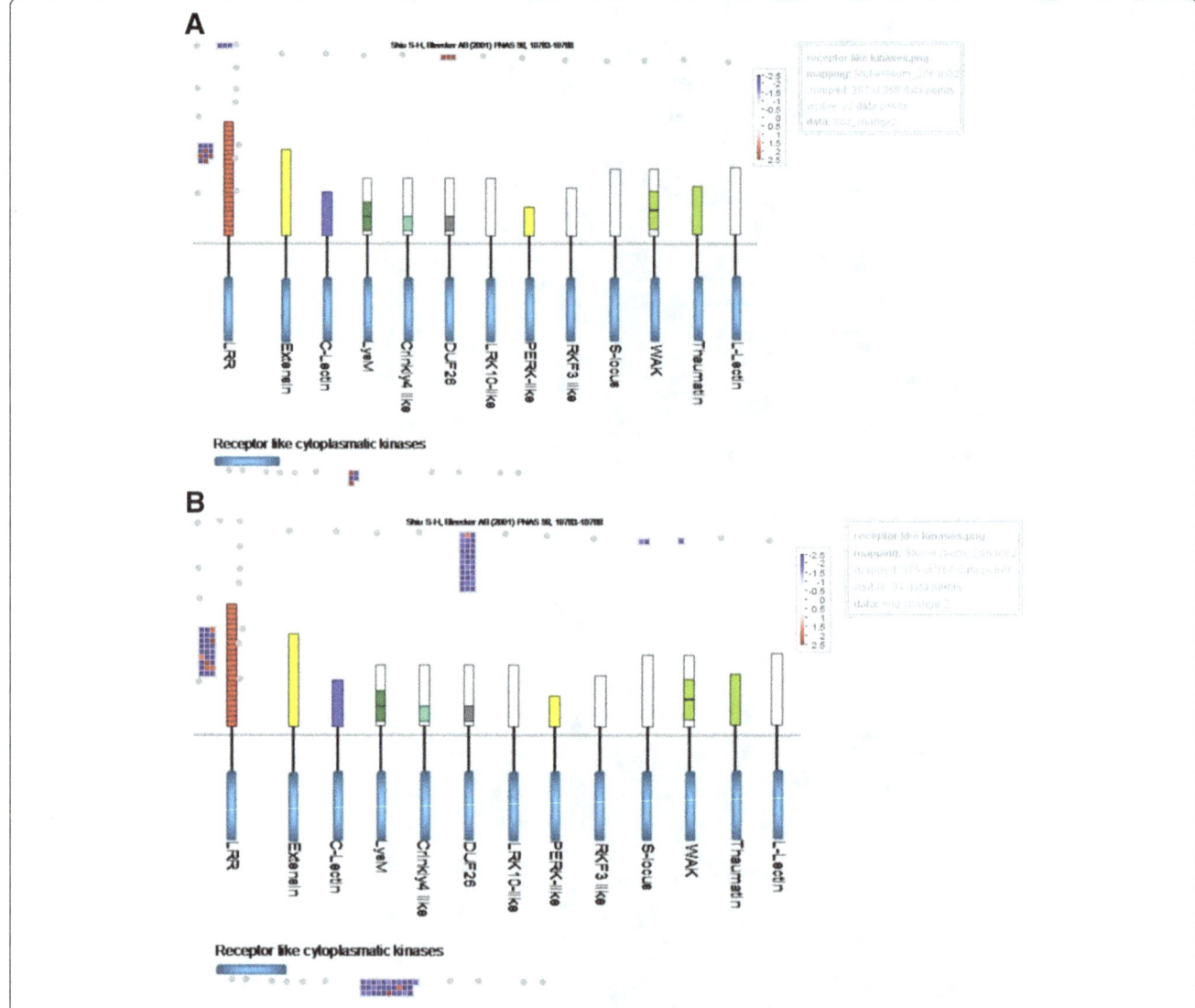

Fig. 5 MapMan overview of DEGs annotated as receptor kinases in Atlantic (**a**) and Waneta (**b**). Up-regulated DEGs (transcript abundance higher in Lso + samples than Lso- samples) appear in red and down-regulated genes appear in blue. Fold changes are expressed on a logarithmic scale

(2.85× and 2.97×, respectively). Moreover, a gene annotated as encoding a salt overly sensitive protein (PGSC0003DMG400010630) was highly induced (5.42×) in response to Lso infection in Atlantic.

Quantitative RT-PCR validation

The expression of 15 genes in Atlantic and 14 genes in Waneta was verified by RT-qPCR, using Ef1α as the reference gene. Ten of these genes were differentially expressed in response to Lso infection in both Atlantic and Waneta, and the remainder of the genes were differentially expressed in only one of the cultivars. In each case, similar gene expression patterns were observed using both the RT-qPCR and the transcriptomic analyses (Table 3). The relative gene expression levels obtained by the RNA-seq and RT-qPCR methods were compared using RPKM and

ΔCt values for each variety. The Pearson product-moment correlation test showed a correlation between the result from the RNA-seq and the RT-qPCR methods in Atlantic ($r = 0.659365899$, $n = 30$, $P = 6.41221E-05$) and in Waneta ($r = 0.724606906$, $n = 28$, $P = 1.03377E-05$) with a significant P value. Therefore, the RT-qPCR data support the transcriptomic analyses presented above.

Discussion

Zebra Chip disease is spreading quickly throughout the potato growing regions of the Americas and the world, without the availability of disease resistance in market varieties. In this study, next-generation transcriptomic sequencing was used to identify potato genes that are differentially expressed in response to infection by Lso. The design of the experiment compared Lso-infected

Table 3 Relative gene expression determined by real time RT-qPCR ($2^{-\Delta\Delta Ct}$). Gene expression was normalized to the expression of Ef1α and is presented as mean value of the Lso + treated plants compared to the Lso- treated plants for each variety

GENE ID	Gene function	Mean $2^{-\Delta\Delta Ct}$ value in Atlantic	Mean $2^{-\Delta\Delta Ct}$ value in Waneta
Genes down-regulated in Atlantic and Waneta after Lso treatment			
PGSC0003DMG400012822	Stem-specific protein TSJT1	0.0291977	0.1709327
PGSC0003DMG400004301	Chlorophyll a,b binding protein type I	0.0536232	0.0844505
PGSC0003DMG400033084	Chlorophyll a/b-binding protein (cab-12)	0.1027244	0.0373299
PGSC0003DMG400009869	DNAJ	0.2373127	0.0143763
PGSC0003DMG400011751	2-oxoglutarate-dependent dioxygenase	0.02778	0.134926
PGSC0003DMG400012763	C-4 sterol methyl oxidase	0.0117969	0.0495508
Genes up-regulated in Atlantic and down-regulated in Waneta after Lso treatment			
PGSC0003DMG400006179	Nodulin family protein	9.564544	0.1504254
PGSC0003DMG400036566	Ethylene response factor 5	2.2506224	0.1328974
PGSC0003DMG400031457	Phenylalanine ammonia-lyase 1	11.298391	0.036989
PGSC0003DMG402007970	Conserved gene unknown function	30.013692	2.2496471
Genes regulated in Atlantic			
PGSC0003DMG400011740	SGA	0.0212225	
PGSC0003DMG400027453	Ribonuclease t2	0.1067257	
PGSC0003DMG400009513	Aspartic protease inhibitor 5	0.1006475	
PGSC0003DMG400032792	Calmodulin-binding protein	13.664482	
PGSC0003DMG400011633	WRKY-type transcription factor	4.8602585	
Genes regulated in Waneta			
PGSC0003DMG400000332	Salt responsive protein 2		0.0393988
PGSC0003DMG400010713	Salt responsive protein 2		0.0777253
PGSC0003DMG400019964	Conserved gene unknown function		0.3503861
PGSC0003DMG400025721	Conserved gene f unknown function		7.2729239

and uninfected plants of two different potato varieties, the ZC sensitive Atlantic and the more ZC tolerant Waneta. Transcriptome-wide expression of potato genes in response to Lso infection at 3 wpi was determined separately for each variety, and differences between Atlantic and Waneta in transcriptomic responses are discussed. A total of 397 genes was differentially expressed in response to Lso infection in Atlantic as compared to 1027 genes in Waneta. Significantly, over 80% of the DEGs were down-regulated in both potato varieties. Although more DEGs were identified in Waneta than in Atlantic, it must be noted that one of the Atlantic Lso + samples was discarded and only two biological replicates were used in the Atlantic comparison. Thus, it is possible that the smaller number of replicates could have contributed to the identification of fewer DEGs.

In both varieties, primary and secondary metabolism were strongly altered by Lso infection: e.g., 1/3 of the DEGs identified in Atlantic and 1/10 of the DEGs in Waneta. In both varieties, many of the down-regulated DEGs were related to photosynthetic functions. The down-regulation of DEGs associated with photosynthesis in Atlantic was not surprising since yellow chlorosis on

some of the upper leaves of the plants was observed by 3 wpi. Down-regulation of DEGs related to photosynthesis also were observed in Waneta although fewer changes in the expression of genes involved in light reaction and tetrapyrrole metabolism were detected, as compared to Atlantic. This probably reflects the 1–2 week delay in symptom development we observed in this study and reported previously [26]. These results suggest that although chlorosis was just beginning in Atlantic and not yet observed in Waneta plants at sampling 3 wpi, transcriptomic changes that would lead to the development of this symptom already were occurring.

Because carbohydrate metabolism and transport in plants are closely linked with photosynthesis regulation, reductions in these functions also were expected in Lso-infected plants. Changes in sugar transport and sucrose and starch metabolism following Lso infection have been reported previously in both the stems and tubers of potato plants [3, 54]. In these studies, the changes were linked to the development of ZC symptoms such as chlorosis in the leaves, the formation of aerial tubers, and the darkened medullary rays in the tubers particularly visible upon frying. Previously, it was suggested that

Lso infected stems of potato plants may be reprogrammed to exhibit tuber-like physiological properties, including the accumulation of tuber storage proteins such as patatins [3]. In *Arabidopsis thaliana* members of the patatin-related family were found to be expressed in several different tissues [21] and potentially function in plant signal transduction as phospholipase A in response to auxin and pathogens [50]. The present study revealed that significant up-regulation in the expression of two of the patatin genes in the aerial tissues occurred by 3 wpi in Waneta, although the reason for this differential gene expression has yet to be determined.

Other metabolic pathways altered in response to Lso were cell wall synthesis and metabolism. Changes in cell wall metabolism associated with pathogen infection are common because the plant cell wall is the first barrier pathogens encounter following dispersal to plant surfaces. In contrast, Lso is inoculated into the sieve elements by psyllids and appears to be limited by the plant to phloem tissues. The plant cell wall is composed of carbohydrates, complex phenolic polymers, and structural proteins including receptors, which can modify the cell wall structure [49]. Changes in cell wall structure and composition in response to Lso infection could affect tissue structure and nutrient exchange between sink and source tissues. Therefore, the down-regulation of genes involved in cell wall metabolism or encoding cell wall structural proteins or other components could be an important element in ZC symptom development, including changes in cell growth, stunting of plant growth, and potentially development of aerial tubers. Previously, increased levels of total phenolic compounds or activity of polyphenol oxidase enzymes were reported in Lso-infected stems [3] and tubers [54]. It is interesting that we find these pathways up-regulated at the transcription level by 3 wpi, coincident with the onset of symptoms.

Genes involved in pathogen recognition, signaling, and defenses were differentially expressed in response to Lso infection in both varieties. Although more DEGs involved in stress responses were identified in Waneta than in Atlantic, in both varieties this category comprised ~1/3 of the DEGs identified. Receptor-like protein kinases are involved in many signaling pathways, but some of them, such as the LRR family, are involved in early steps of plant recognition of pathogenic signatures such as flagellin or lipopolysaccharides and defense. Although the transient expression of a conserved 22 AA-peptide of Lso flagellin did not induce cell death or Reactive Oxygen Species (ROS) in potato plants [20], it is possible that LRR mediates the interaction between Lso and plants through other bacterial signals. The expression levels of several LRR genes were either up- or down-regulated in each of the varieties. Other DEGs belonged to the DUF26 receptor kinase family, which is involved in oxidative stress and hormone and plant defenses signaling. Although the specific function of each protein in these families has not been determined in potato, the regulation of plant signaling mechanisms induced in recognition of and defense against Lso may be useful for controlling Lso infection in the future.

Effector Triggered Immunity (ETI) is a defense mechanism that is activated in response to specific pathogen effectors. In response to Lso infection, the expression of Avr9/Cf-9 genes were down-regulated. Avr9/Cf-9 function in the initial development of the defense response upon perception of pathogen molecules and are known to be up-regulated 15 to 30 min after infection [45]. The activation of ETI triggers a cascade of responses including calcium, MAPK and oxidative burst. Three percent of the DEGs in Waneta were related to calcium signaling. Calcium is involved in plant-microbe signaling in both symbiotic [24] and pathogenic interactions [41]. The down-regulation of signaling genes, ROS related enzymes, and Avr9/Cf-9 genes may indicate that Lso can efficiently repress plant defenses as suggested previously [8].

Following the perception of the stressor and signaling, plants regulate the expression of defenses. Many of the same phytohormone pathways were differentially regulated in Atlantic and Waneta, but not always in the same direction. The auxin and ethylene were the main phytohormone-signaling pathways similarly altered in response to Lso infection. Auxins are involved in regulation of the legume-rhizoba interactions [23], which is interesting given Lso as a member of the Rhizobiaceae is somewhat taxonomically related to rhizobia. The main differences between varieties were in the differential expression of genes related to the cytokinin and jasmonate pathways in Waneta, but not Atlantic; and the differential expression of genes in the salicylic acid pathway in Atlantic, but not Waneta. Hormones and the crosstalk between hormone signaling pathways play a critical role in plant perception and response to pathogen infection. Our results suggest that differences in susceptibility (apparent as rate of symptom development) between Atlantic and Waneta may be related to the observed differences in the regulation of some of the phytohormone pathways.

Overall, there were 41 DEGs that were regulated in the opposite direction in the two potato varieties. In addition to differences in phytohormone pathways, GO enrichment analysis identified enrichment in transcription factor terms among these genes (Additional file 1: Table S5B). The enrichment in DEGs that may be transcription factors in both Waneta and Atlantic is consistent with the extensive reprogramming of gene expression that was observed in plants following Lso infection.

In this study, plant responses to Lso infection were identified in two potato varieties with different degrees of susceptibility to ZC, and key differences in phytohormone signaling between both varieties were uncovered. However, it is important to note that in spite of the differences between Atlantic and Waneta in their responses to pathogen infection observed, both varieties are susceptible to Lso infection and ZC disease development. In both varieties, Lso populations grow slowly, but can be reliably detected by PCR 3 weeks after infection, with no difference in bacterial titer [26]. The main difference between the varieties is the rate of symptom development, which typically results in a one to several week delay in the onset of symptoms and the time to plant death in Waneta relative to Atlantic [26]. The present study was performed at three wpi in order to capture the transcriptomic differences between varieties that may be contributing to the rate of disease symptom development. Although earlier time points may reveal the initial pathogen recognition responses, 3 weeks after infection is a key time point in the development of Zebra Chip disease symptoms and thus in identifying the underlying causes.

Even though the specific Ca. Liberibacter virulence mechanisms triggering plant disease development are still somewhat unclear, some progress has been made in identifying pathogenicity and virulence traits (reviewed in [55]. Currently, it is unclear which factors play the most important role in the development of disease symptoms, i.e., whether symptoms are due primarily to the metabolic activity or virulence factors produced by the pathogen, the host plant responses to these, or some combination. Interestingly, recent findings suggest that disease symptoms in specific organs such as tubers may develop independently from the physiological changes in photosynthesis, nutrient transport, or metabolism occurring in the aerial plant organs. For example, one study revealed that plants infested with psyllids only 48 h before vine kill still developed symptoms in tubers left to mature on the ground for 1 month [46]. Moreover, tubers of plants infected just 4 days before harvest became infected despite the plants and tubers being asymptomatic and testing negative for Lso at harvest. Collectively, these findings demonstrate that despite the importance of understanding how whole plant physiology affects disease development, there is still much to be learned about the organ specific response to Lso infection.

This study is the first to observe the transcriptomic consequences of Lso infection in potato, and was focused on the time of symptom development and the responses of potato varieties differing in the rate of symptom development. The results of our study complement the few previous transcriptomic analyses of *Liberibacter*-citrus interactions which focused on citrus tree response to infection resulting in huanglongbing (citrus greening) disease [5, 29, 30, 57] or Lso-psyllid vector interactions which centered on the insect's transcriptomic response to infection and/or the bacterial transcriptome in the insect [14, 22, 35, 56].

Conclusion

This study provides insights into the complex network of changes that occur in potato plants following Lso infection at a timepoint coincident with the onset of symptom development. Those changes were characterized by a preponderance of genes being down-regulated. We identified multiple pathways that are responsive to Lso infection. Our analysis suggests that prior to disease symptom development, dramatic changes in transcriptome-wide gene expression have already occurred in the plant host that underly the physiological changes leading to symptom development. Substantial reprogramming of both primary and secondary plant metabolism were revealed, especially the down-regulation in the expression of genes related to photosynthesis and in the expression of genes involved in phytohormone regulation. The results support the hypothesis that Lso can repress plant physiology and metabolism as well as signaling and defense mechanisms leading to disease development.

Additional file

Additional file 1: Table S1. Primers for qPCR. **Table S2**. DEGs in Atlantic. **Table S3**. DEGs in Waneta. **Table S4**. Common DEGs in Waneta and Atlantic. (A) 61 genes down regulated after Lso treatment in both varieties. (B) 9 genes up regulated after Lso treatment. (C) 5 genes up regulated in Waneta and down regulated in Atlantic. (D) 36 genes down regulated in Waneta and up regulated in Atlantic. **Table S5**. Gprofiler results for Go enrichment in the genes commonly regulated. (A) Go Term enrichment for the 70 genes regulated in the same direction. T-type BP: Biological Process, CC. (B) Go Term enrichment for the 41 genes regulated in opposite direction. **Table S6**. Photosynthesis-related DEGs in Waneta and Atlantic (mapman analysis). **Table S7**. List of carbohydrates metabolism DEGs in Waneta and Atlantic (mapman analysis). **Table S8**. List of lipid metabolism DEGs in Waneta and Atlantic (mapman analysis). **Table S9**. List of secondary metabolism DEGs in Waneta and Atlantic (mapman analysis). **Table S10**. Gprofiler analysis of genes down regulated in AT (A) and in Waneta (B). **Table S11**. Signaling-related DEGs in Waneta and Atlantic (mapman analysis). **Table S12**. Hormone-related DEGs in Waneta and Atlantic (mapman analysis). **Table S13**. DEGs Transcription factors in Waneta and Atlantic (mapman analysis). **Figure S1**. Volcano plot of differentially expressed genes between control (Lso-) and Lso-infected (Lso+) plants for A) Atlantic and B) Waneta. Volcano plots were obtained from cuffdiff2 output. **Figure S2**. Heat map showing the expression pattern of DEGs between the biological treatment Lso- and Lso + for the DEGs annotated as contributing to metabolic pathways in Mapman in Atlantic (left panel) and Waneta (right panel). **Figure S3**. MapMan overview of DEGs annotated as contributing to Regulation overview in Atlantic (A) and Waneta (B). (DOCX 588 kb)

Abbreviations

DEGs: Differentially expressed genes; Lso: Candidatus Liberibacter solanacearum; ZC: Zebra Chip

Acknowledgements

We thank the Texas Potato Breeding Program and Douglas Scheuring for support in obtaining potato tubers and advice in potato cultivation. We thank Dr. Freddy Ibanez and Dr. Ordom Huot for support in maintenance of insect colonies and plants and for discussion. We thank Dr. Luciano Cosme for discussion and support for data analysis. We thank Dr. Denis Gross and Dr. Aravind Ravindran for discussion on Zebra Chip. The open access publishing fees for this article have been covered by the Texas A&M

University Open Access to Knowledge Fund (OAKFund), supported by the University Libraries and the Office of the Vice President for Research.

Funding
This project was supported by the Texas A&M AgriLife Research Zebra Chip Management Program grant number 06–407135 and the Texas A&M AgriLife Research Insect Vector Disease grant number 06-L701774 for JGL. salary.

Authors' contributions
JL, AM carried out the plant and molecular experiments, JL CT analyzed the data, performed the sequence alignment and the identification of DEGs. JL, AM, CT, JCMJr, EA participated in the design of the study. JL, CT, EA conceived the study. JL, CT drafted the manuscript with editorial input from EA. All authors read, edited and approved the final manuscript.

Competing interests
The authors declare that they have no competing interests.

Author details
[1]Department of Horticultural Sciences, Texas A&M University, College Station, TX 77843, USA. [2]Department of Entomology, Texas A&M University, College Station, TX 77843, USA.

References
1. Aguilar E, Sengoda VG, Bextine B, McCue KF, Munyaneza JE. First report of "Candidatus Liberibacter solanacearum" on tomato in Honduras. Plant Dis. 2013;97:1375.
2. Ali A, Alexandersson E, Sandin M, Resjö S, Lenman M, Hedley P, Levander F, Andreasson E. Quantitative proteomics and transcriptomics of potato in response to Phytophthora Infestans in compatible and incompatible interactions. BMC Genomics. 2014;15:1–18.
3. Alvarado VY, Odokonyero D, Duncan O, Mirkov TE, Scholthof HB. Molecular and physiological properties associated with zebra complex disease in potatoes and its relation with Candidatus Liberibacter contents in psyllid vectors. PLoS One. 2012;7:e37345.
4. Andrews, S. 2010. FastQC: a quality control tool for high throughput sequence data. [online]. Available: http://www.bioinformatics.babraham. ac.uk/projects/fastqc.
5. Aritua V, Achor D, Gmitter FG, Albrigo G, Wang N. Transcriptional and microscopic analyses of citrus stem and root responses to Candidatus Liberibacter asiaticus infection. PLoS One. 2013;8:e73742.
6. Bertolini E, Teresani GR, Loiseau M, Tanaka FAO, Barbé S, Martínez C, Gentit P, López MM, Cambra M. Transmission of 'Candidatus Liberibacter solanacearum' in carrot seeds. Plant Pathol. 2015;64:276–85.
7. Camacho-Tapia M, Rojas-Martinez RI, Zavaleta-Mejia E, Hernandez-Deheza MG, Carrillo-Salazar JA, Rebollar-Alviter A, Ochoa-Martinez DL. Aetiology of chili pepper variegation from Yurecuaro, Mexico. J Plant Pathol. 2011;93:331–5.
8. Casteel CL, Hansen AK, Walling LL, Paine TD. Manipulation of plant defense responses by the tomato psyllid (Bactericerca cockerelli) and its associated endosymbiont Candidatus Liberibacter psyllaurous. PLoS One. 2012;7(4): e35191. doi:10.1371/journal.pone.0035191
9. Crosslin JM, Hamm PB, Eggers JE, Rondon SI, Sengoda VG, Munyaneza JE. First report of zebra chip dDisease and "Candidatus Liberibacter solanacearum" on potatoes in Oregon and Washington state. Plant Dis. 2011;96:452.
10. Crosslin JM, Munyaneza JE, Brown JK, Liefting LW. Potato zebra chip disease: a phytopathological tale. Plant Health Prog. 2010; https://doi.org/10.1094/PHP-2010-0317-01-RV.
11. Dixon MS, Hatzixanthis K, Jones DA, Harrison K, Jones JD. The tomato Cf-5 disease resistance gene and six homologs show pronounced allelic variation in leucine-rich repeat copy number. Plant Cell. 1998;10:1915–25.
12. Edgar R, Domrachev M, Lash AE. Gene expression omnibus: NCBI gene expression and hybridization array data repository. Nucleic Acids Res. 2002;30:207–10.
13. FAO's Land and Water Division. 2008. International Year of the Potato [Online]. Available: http://www.fao.org/potato-2008/en/world/.
14. Fisher T, Vyas M, He R, Nelson W, Cicero J, Willer M, Kim R, Kramer R, May G, Crow J, Soderlund C, Gang D, Brown J. Comparison of potato and Asian citrus psyllid adult and nymph transcriptomes identified vector transcripts with potential involvement in circulative, propagative Liberibacter transmission. Pathogens. 2014;3:875–907.
15. Floss DS, Levy JG, Levesque-Tremblay V, Pumplin N, Harrison MJ. DELLA proteins regulate arbuscule formation in arbuscular mycorrhizal symbiosis. Proc Natl Acad Sci U S A. 2013;110:E5025–34.
16. Fonouni-Farde C, Tan S, Baudin M, Brault M, Wen J, Mysore KS, Niebel A, Frugier F, Diet A. DELLA-mediated gibberellin signalling regulates nod factor signalling and rhizobial infection. Nat Commun. 2016;7:12636.
17. Goff SA, Vaughn M, McKay S, Lyons E, Stapleton AE, Gessler D, Matasci N, Wang L, Hanlon M, Lenards A, Muir A, Merchant N, Lowry S, Mock S, Helmke M, Kubach A, Narro M, Hopkins N, Micklos D, Hilgert U, Gonzales M, Jordan C, Skidmore E, Dooley R, Cazes J, McLay R, Lu Z, Pasternak S, Koesterke L, Piel WH, Grene R, Noutsos C, Gendler K, Feng X, Tang C, Lent M, Kim S-J, Kvilekval K, Manjunath BS, Tannen V, Stamatakis A, Sanderson M, Welch SM, Cranston KA, Soltis P, Soltis D, O'Meara B, Ane C, Brutnell T, Kleibenstein DJ, White JW, Leebens-Mack J, Donoghue MJ, Spalding EP, Vision TJ, Myers CR, Lowenthal D, Enquist BJ, Boyle B, Akoglu A, Andrews G, Ram S, Ware D, Stein L, Stanzione D. The iPlant collaborative: Cyberinfrastructure for plant biology. Front Plant Sci. 2011;2:34.
18. Haapalainen M. Biology and epidemics of Candidatus Liberibacter species, psyllid-transmitted plant-pathogenic bacteria. Ann Appl Biol. 2014;165:172–98.
19. Hansen AK, Trumble JT, Stouthamer R, Paine TD. A new Huanglongbing species, "Candidatus Liberibacter psyllaurous," found to infect tomato and potato, is vectored by the psyllid Bactericera cockerelli (Sulc). Appl Environ Microb. 2008;74:5862–5.
20. Hao GX, Pitino M, Ding F, Lin H, Stover E, Duan YP. Induction of innate immune responses by flagellin from the intracellular bacterium, 'Candidatus Liberibacter solanacearum'. BMC Plant Biol. 2014;14. https://doi.org/10.1186/s12870-014-0211-9
21. Holk A, Rietz S, Zahn M, Quader H, Scherer GFE. Molecular identification of cytosolic, patatin-related phospholipases a from Arabidopsis with potential functions in plant signal transduction. Plant Physiol. 2002;130:90–101.
22. Ibanez F, Levy J, Tamborindeguy C. Transcriptome analysis of "Candidatus Liberibacter solanacearum" in its psyllid vector, Bactericera cockerelli. PLoS One. 2014;9:e100955.
23. Laplaze L, Lucas M, Champion A. Rhizobial root hair infection requires auxin signaling. Trends Plant Sci. 2015;20:332–4.
24. Levy J, Bres C, Geurts R, Chalhoub B, Kulikova O, Duc G, Journet EP, Ane JM, Lauber E, Bisseling T, Denarie J, Rosenberg C, Debelle F. A putative Ca2+ and calmodulin-dependent protein kinase required for bacterial and fungal symbioses. Science. 2004;303:1361–4.
25. Levy J, Hancock J, Ravindran A, Gross D, Tamborindeguy C, Pierson E. Methods for rapid and effective PCR-based detection of 'Candidatus Liberibacter solanacearum' from the insect vector Bactericera cockerelli: streamlining the DNA extraction/purification process. J Econ Entomol. 2013;106:1440–5.
26. Levy J, Ravindran A, Gross D, Tamborindeguy C, Pierson E. Translocation of "Candidatus Liberibacter solanacearum", the zebra chip pathogen, in potato and tomato. Phytopathology. 2011;101:1285–91.
27. Lévy J, Scheuring D, Koym J, Henne D, Tamborindeguy C, Pierson E, Miller JC Jr. Investigations on putative zebra Chip tolerant potato selections. Am J Potato Res. 2015;92(3):417–25.
28. Liefting LW, Perez-Egusquiza ZC, Clover GR, Anderson JAD. A new 'Candidatus Liberibacter' species in Solanum tuberosum in New Zealand. Plant Dis. 2008;92:1474.
29. Mafra V, Martins PK, Francisco CS, Ribeiro-Alves M, Freitas-Astua J, Machado MA. Candidatus Liberibacter americanus induces significant reprogramming of the transcriptome of the susceptible citrus genotype. BMC Genomics. 2013;14: 247. doi:10.1186/1471-2164-14-247.

30. Martinelli F, Uratsu SL, Albrecht U, Reagan RL, Phu ML, Britton M, Buffalo V, Fass J, Leicht E, Zhao WX, Lin DW, D'Souza R, Davis CE, Bowman KD, Dandekar AM. Transcriptome profiling of citrus fruit response to Huanglongbing disease. PLoS One. 2012;7

31. Massa A, Childs K, Lin H, Bryan G, Giuliano G, Buell C. The Transcriptome of the reference potato genome Solanum Tuberosum group Phureja clone DM1-3 516R44. PLoS One. 2011;6(10):e26801. doi:10.1371/journal.pone. 0026801

32. Munyaneza J, Buchman J, Sengoda V, Fisher T, Pearson C. Susceptibility of selected potato varieties to zebra Chip potato disease. Am J Potato Res. 2011;88:435–40.

33. Munyaneza JE. Zebra chip disease of potato: biology, epidemiology, and management. Am J Potato Res. 2012;89:329–50.

34. Munyaneza JE, Crosslin JM, Upton JE. Association of Bactericera cockerelli (Homoptera: Psyllidae) with "zebra chip," a new potato disease in southwestern United States and Mexico. J Econ Entomol. 2007;100:656–63.

35. Nachappa P, Levy J, Tamborindeguy C. Transcriptome analyses of Bactericera cockerelli adults in response to "Candidatus Liberibacter solanacearum" infection. Mol Gen Genomics. 2012a;287:803–17.

36. Nachappa P, Shapiro AA, Tamborindeguy C. Effect of 'Candidatus Liberibacter solanacearum' on fitness of its vector, Bactericera cockerelli (Hemiptera: Triozidae) on tomato. Phytopathology. 2012b;102:41–6.

37. NASS 2015. Crop production 2014 summary.

38. Nicot N, Hausman JF, Hoffmann L, Evers D. Housekeeping gene selection for real-time RT-PCR normalization in potato during biotic and abiotic stress. J Exp Bot. 2005;56:2907–14.

39. Pfaffl MW. A new mathematical model for relative quantification in real-time RT-PCR. Nucleic Acids Res. 2001;29(9):e45.

40. Ravindran A, Levy J, Pierson E, Gross DC. Development of primers for improved PCR detection of the potato zebra Chip pathogen, 'Candidatus Liberibacter solanacearum'. Plant Dis. 2011;95:1542–6.

41. Reddy ASN, Ali GS, Celesnik H, Day IS. Coping with stresses: roles of calcium- and calcium/Calmodulin-regulated gene expression. Plant Cell. 2011;23:2010–32.

42. Reimand J, Kull M, Peterson H, Hansen J, Vilo J. g:Profiler—a web-based toolset for functional profiling of gene lists from large-scale experiments. Nucleic Acids Res. 2007;35:W193–200.

43. Rojas-Martinez RI, Camacho-Tapia M, Zavaleta-Mejia E, Levy J. First report of the presence of haplotypes a and B of 'Candidatus Liberibacter solanacearum' in the central region of Mexico. J Plant Pathol. 2016; In press

44. Rosahl S, Schell J, Willmitzer L. Expression of a tuber-specific storage protein in transgenic tobacco plants - demonstration of an esterase-activity. EMBO J. 1987;6:1155–9.

45. Rowland O, Ludwig AA, Merrick CJ, Baillieul F, Tracy FE, Durrant WE, Fritz-Laylin L, Nekrasov V, Sjolander K, Yoshioka H, Jones JDG. Functional analysis of Avr9/Cf-9 rapidly elicited genes identifies a protein kinase, ACIK1, that is essential for full Cf-9-dependent disease resistance in tomato. Plant Cell. 2005;17:295–310.

46. Rush CM, Workneh F, Rashed A. Significance and epidemiological aspects of late-season infections in the Management of Potato Zebra Chip. Phytopathology. 2015;105:929–36.

47. Secor GA, Rivera-Varas VV. Emerging diseases of cultivated potato and their impact on Latin America. Rev Latinoam Papa (Suppl). 2004;1:1–8.

48. Secor GA, Rivera VV, Abad A, Lee IM, Clover GR, Liefting LW, Li X, De Boer H. Association of 'Candidatus Liberibacter solanacearum' with zebra Chip disease of potato established by graft and Psyllid transmission, electron microscopy, and PCR. Plant Dis. 2009;93:574–83.

49. Seifert GJ, Blaukopf C. Irritable walls: the plant extracellular matrix and signaling. Plant Physiol. 2010;153:467–78.

50. Senda K, Yoshioka H, Doke N, Kawakita K. A Cytosolic Phospholipase A2 from potato tissues appears to be Patatin. Plant Cell Physiol. 1996;37:347–53.

51. Teresani GR, Bertolini E, Alfaro-Fernández A, Martínez C, Tanaka FAO, Kitajima EW, Roselló M, Sanjuán S, Ferrándiz JC, López MM, Cambra M, Font MI. Association of 'Candidatus Liberibacter solanacearum' with a vegetative disorder of celery in Spain and development of a real-time PCR method for its detection. Phytopathology. 2014;104:804–11.

52. Thimm O, Bläsing O, Gibon Y, Nagel A, Meyer S, Krüger P, Selbig J, Müller LA, Rhee SY, Stitt M. MAPMAN: a user-driven tool to display genomics data sets onto diagrams of metabolic pathways and other biological processes. Plant J. 2004;37:914–39.

53. Trapnell C, Hendrickson DG, Sauvageau M, Goff L, Rinn JL, Pachter L. Differential analysis of gene regulation at transcript resolution with RNA-seq. Nat Biotech. 2013;31:46–53.

54. Wallis CM, Chen JC, Civerolo EL. Zebra chip-diseased potato tubers are characterized by increased levels of host phenolics, amino acids, and defense-related proteins. Physiol Mol Plant Pathol. 2012;78:66–72.

55. Wang N, Pierson EA, Setubal JC, Xu J, Levy JG, Zhang Y, Li J, Rangel LT, Martins J Jr. The Candidatus Liberibacter–Host Interface: Insights into Pathogenesis Mechanisms and Disease Control. Annu Rev Phytopathol. 2017;55:451–82.

56. Yao J, Saenkham P, Levy J, Ibanez F, Noroy C, Mendoza A, Huot O, Meyer DF, Tamborindeguy C. Interactions 'Candidatus Liberibacter solanacearum' – Bactericera cockerelli: haplotype effect on vector fitness and gene expression analyses. Front Cell Infect Microbiol. 2016;6:62. doi:10.3389/fcimb. 2016.00062

57. Zheng ZL, Zhao YH. Transcriptome comparison and gene coexpression network analysis provide a systems view of citrus response to 'Candidatus Liberibacter asiaticus' infection. BMC Genomics. 2013;14:27. doi:10.1186/ 1471-2164-14-27

58. Zuluaga AP, Solé M, Lu H, Góngora-Castillo E, Vaillancourt B, Coll N, Buell CR, Valls M. Transcriptome responses to Ralstonia solanacearum infection in the roots of the wild potato Solanum Commersonii. BMC Genomics. 2015;16:1–16.

7

Comparative genomic analysis of *Brevibacterium* strains: insights into key genetic determinants involved in adaptation to the cheese habitat

Nguyen-Phuong Pham[1], Séverine Layec[1], Eric Dugat-Bony[1], Marie Vidal[2], Françoise Irlinger[1] and Christophe Monnet[1]* (iD)

Abstract

Background: *Brevibacterium* strains are widely used for the manufacturing of surface-ripened cheeses, contributing to the breakdown of lipids and proteins and producing volatile sulfur compounds and red-orange pigments. The objective of the present study was to perform comparative genomic analyses in order to better understand the mechanisms involved in their ability to grow on the cheese surface and the differences between the strains.

Results: The genomes of 23 *Brevibacterium* strains, including twelve strains isolated from cheeses, were compared for their gene repertoire involved in salt tolerance, iron acquisition, bacteriocin production and the ability to use the energy compounds present in cheeses. All or almost all the genomes encode the enzymes involved in ethanol, acetate, lactate, 4-aminobutyrate and glycerol catabolism, and in the synthesis of the osmoprotectants ectoine, glycine-betaine and trehalose. Most of the genomes contain two contiguous genes encoding extracellular proteases, one of which was previously characterized for its activity on caseins. Genes encoding a secreted triacylglycerol lipase or involved in the catabolism of galactose and D-galactonate or in the synthesis of a hydroxamate-type siderophore are present in part of the genomes. Numerous Fe^{3+}/siderophore ABC transport components are present, part of them resulting from horizontal gene transfers. Two cheese-associated strains have also acquired catecholate-type siderophore biosynthesis gene clusters by horizontal gene transfer. Predicted bacteriocin biosynthesis genes are present in most of the strains, and one of the corresponding gene clusters is located in a probable conjugative transposon that was only found in cheese-associated strains.

Conclusions: *Brevibacterium* strains show differences in their gene repertoire potentially involved in the ability to grow on the cheese surface. Part of these differences can be explained by different phylogenetic positions or by horizontal gene transfer events. Some of the distinguishing features concern biotic interactions with other strains such as the secretion of proteases and triacylglycerol lipases, and competition for iron or bacteriocin production. In the future, it would be interesting to take the properties deduced from genomic analyses into account in order to improve the screening and selection of *Brevibacterium* strains, and their association with other ripening culture components.

Keywords: *Brevibacterium*, Horizontal gene transfer, Comparative genomics, Cheese rind, Cheese ripening, Iron acquisition, Bacteriocin, Lanthipeptide, Lantibiotic, BreLI

* Correspondence: christophe.monnet@inra.fr
[1]UMR GMPA, AgroParisTech, INRA, Université Paris-Saclay, 78850 Thiverval-Grignon, France
Full list of author information is available at the end of the article

Background

Microbial communities from rinds of surface-ripened cheeses are composed of various bacteria, yeasts and molds, which contribute to the flavor, texture and appearance of the final products. These microorganisms may come from the milk, the ripening environment or from ripening cultures that are widely used in the cheese industry. The function of the ripening cultures is to provide specific organoleptic properties, to ensure a better regularity of manufacturing, and to outcompete pathogens or spoilage microorganisms [1]. However, strains from ripening cultures frequently do not establish themselves in cheeses [2]. Even if they are massively inoculated, these strains are sometimes outcompeted by the resident "house flora" due to insufficient fitness in the cheese surface habitat. The ability to grow on the cheese surface depends on various properties such as efficient salt tolerance and iron acquisition systems, or on the ability to use the energy compounds present in the cheese [3]. In addition, growth is influenced by the other microorganisms present at the cheese surface, with which they may have positive or negative interactions.

One example of a ripening culture component that may have problematic growth in cheeses is *Brevibacterium* [4–9]. This microorganism contributes to the breakdown of lipids and proteins, and produces volatile sulfur compounds that are key aroma impact compounds, as well as red-orange pigments [10–12]. For a long time, *Brevibacterium linens* was considered to be the major *Brevibacterium* species in cheeses. In 2004, it was broken down into three species: *B. linens*, *B. antiquum* and *B. aurantiacum* [13]. These three species, together with *B. casei* and other not-yet described *Brevibacterium* species, have been isolated from cheeses [14].

In order to improve the strategies for selecting *Brevibacterium* strains for ripening cultures, it is important to better understand the mechanisms involved in their ability to grow on the cheese surface. This can be investigated by genomic analyses. For example, the study of the genomes of the cheese strains *Glutamicibacter arilaitensis* Re117 (formerly *Arthrobacter arilaitensis* Re117) and *Corynebacterium variabile* DSM 44702 revealed several metabolic capabilities that were considered to play roles in growth on cheese [15, 16]. In addition, a recent study provided evidence of extensive horizontal gene transfer (HGT) in cheese-associated bacteria, including *Brevibacterium* strains, for which genes involved in iron acquisition were particularly abundant in the transferred islands [17].

The aim of the present study was to investigate, in *Brevibacterium* strains, key genetic determinants known to be important for growth in cheese: the catabolism of energy compounds present in cheeses, iron acquisition, salt tolerance and bacteriocin production. For that

purpose, we sequenced the genome of 13 *Brevibacterium* strains, including eleven strains isolated from cheeses. We performed comparative analyses of these genomes and of ten other *Brevibacterium* genomes from strains isolated from diverse environments and already present in the Integrated Microbial Genomes (IMG) database [18].

Methods

Growth conditions and DNA extraction

The *Brevibacterium* strains were grown under aerobic conditions (rotary shaker at 150 rpm) for three days at 25 °C in 50-ml conical flasks containing 10 ml of brain heart infusion broth (Biokar Diagnostics, Beauvais, France). Bacterial cells were recovered by centrifugation at 4500 x g for 10 min from 5 ml of culture, washed once with 5 ml of TE buffer (Tris-HCl 10 mM, EDTA 1 mM, pH 8) and resuspended in 500 μl of the same buffer. Seventy-five μl of lysis solution containing lysozyme (40 mg/ml) and lyticase (1333 U/ml) were added and the suspensions were incubated for 30 min at 37 °C. After addition of 60 μl of 0.5 M EDTA pH 8.0, 20 μl of proteinase K (20 mg/ml) and 100 μl of 20% SDS, the samples were incubated for 1 h at 55 °C and subsequently transferred to 2-ml bead-beating tubes containing 100 mg of 0.1 mm-diameter zirconium beads and 100 mg of 0.5 mm-diameter zirconium beads. After cooling on ice, 500 μl of phenol-chloroform-isoamyl alcohol (25:24:1; saturated with 10 mM Tris, pH 8.0, and 1 mM EDTA) were added and the tubes were shaken in a bead beater (FastPrep-24, MP Biomedicals, Illkirch, France) using two 45-s mixing sequences at a speed of 6.0 m/s. The tubes were cooled on ice for 5 min after each mixing. The content of the tubes was transferred to Phase Lock Gel Heavy tubes (5 PRIME, Hilden, Germany), which were then centrifuged at 18,500 × g for 30 min at 20 °C. The aqueous phases were subsequently transferred to new Phase Lock Gel tubes. After adding 500 μl of phenol-chloroform-isoamyl alcohol and gentle mixing, centrifugation was performed at 18,500 × g for 20 min at 20 °C. Five hundred μl of chloroform were then added to the aqueous phase, and the tubes were centrifuged at 18,500 × g for 20 min at 20 °C after gently mixing. The aqueous phases (approximately 200 μl) were recovered, mixed with 1 μl of RNase A (20 mg/ml), and incubated for 30 min at 37 °C. DNA was then precipitated by adding 200 μl of cold isopropanol and the tubes were incubated for 10 min at 4 °C. The DNA was recovered by centrifugation for 15 min at 18,500 x g and 4 °C, and the pellets were subsequently washed two times with 500 μl of 70% (vol/vol) ethanol. The pellets were then dried for 15 min at 42 °C and dissolved in 100 μl of water. Finally, DNA cleanup was performed using the DNeasy Blood & Tissue Kit (Qiagen,

Courtaboeuf, France) according to the manufacturer's instructions. DNA yield and purity (absorbance ratio at 260/280 nm) were determined using a NanoDrop ND-1000 spectrophotometer (Labtech, Palaiseau, France). DNA integrity was verified by electrophoresis in a 1% agarose gel (Qbiogene, Illkirch, France) in 1X TAE buffer (40 mM Tris, 20 mM acetic acid, 1 mM EDTA, pH 8.3) stained with SYBR® Safe 1X (Invitrogen, Carlsbad, CA, USA).

Genome sequencing, assembly and annotation

Libraries were generated using the TruSeq DNA Sample Preparation Kit (Illumina, San Diego, CA, USA) according to the manufacturer's instructions. Sequencing was carried out on an Illumina MiSeq apparatus at the INRA GeT-PlaGE platform (http://get.genotoul.fr) in order to generate paired-end reads (250 bases in length). For each strain, the paired-end reads were merged using FLASH

[19]. De novo assembly was performed using SPAdes version 3.1.1 [20]. Only contigs with length > 1000 bp were considered for further study. Gene predictions and annotations were performed automatically using the Integrated Microbial Genomes (IMG) database and comparative analysis system [21], as described in the corresponding standard operating procedure [22]. PHASTER [23] was used to predict prophages. Sequences involved in the production of secondary metabolites and bacteriocins were searched using antiSMASH 3.0 [24], BAGEL3 [25] and BACTIBASE [26].

Comparative genomic analysis and phylogenetic classification

Comparative genomic analyses were performed considering the 13 genomes sequenced in the present study as well as the ten other *Brevibacterium* genomes present in the IMG database in September 2016 (Table 1). A

Table 1 Information about the *Brevibacterium* strains and genomes investigated in the present study

Species	Strain	Source	Bioproject	Sequence accession numbers	Status	Authors
B. antiquum	CNRZ 918	Beaufort cheese	PRJEB19830	FXZD01000001-FXZD01000049	Permanent draft	This study
B. antiquum	P10	Murol cheese	PRJEB19831	FXZE01000001-FXZE01000059	Permanent draft	This study
B. aurantiacum	ATCC 9175[T]	Camembert cheese	PRJEB19815	FXZB01000001-FXZB01000070	Permanent draft	This study
B. aurantiacum	CNRZ 920	Beaufort cheese	PRJEB19800	FXZG01000001-FXZG01000073	Permanent draft	This study
B. aurantiacum	6(3)	Langres cheese	PRJEB19867	FXYZ01000001-FXYZ01000091	Permanent draft	This study
B. aurantiacum	8(6)	Reblochon cheese	PRJEB19868	FXZI01000001-FXZI01000097	Permanent draft	This study
B. casei	CIP 102111[T]	Cheddar cheese	PRJEB19871	FXZC01000001-FXZC01000024	Permanent draft	This study
B. linens	ATCC 9172[T]	Harzer cheese	PRJEB19834	FXYY01000001-FXYY01000080	Permanent draft	This study
B. linens	Mu101	Munster cheese	PRJEB19836	FXZA01000001-FXZA01000081	Permanent draft	This study
B. sp.	239c	Camembert cheese	PRJEB19828	FXZH01000001-FXZH01000068	Permanent draft	This study
B. sp.	Mu109	Munster cheese	PRJEB19840	FXZF01000001-FXZF01000126	Permanent draft	This study
B. iodinum	ATCC 49514[T]	Cow milk	PRJEB19872	FXYX01000001-FXYX01000065	Permanent draft	This study
B. jeotgali	SJ5-8[T]	Seafood	PRJEB19841	FXZM01000001-FXZM01000047	Permanent draft	This study
B. aurantiacum	ATCC 9174[a]	Romadur cheese	PRJNA405	AAGP01000001-AAGP01000076	Permanent draft	DOE JGI, 2005 (direct submission)
B. album	DSM 18261[T]	Soil	PRJNA195785	AUFJ01000001-AUFJ01000016	Permanent draft	Kyrpides et al., 2013 (direct submission)
B. sandarakinum	DSM 22082[T]	Wall surface	PRJEB16423	LT629739	Complete	Varghese, 2016 (direct submission) Kämpfer et al., 2010 [91]
B. linens	AE038–8	Fresh water	PRJNA268212	JTJZ01000001-JTJZ01000029	Permanent draft	Maizel et al., 2015 [92]
B. siliguriense	DSM 23676[T]	Fresh water	PRJNA303729	LT629766	Complete	Varghese, 2016 (direct submission)
B. sp.	VCM10	Fresh water	PRJNA234061	JAJB01000001-JAJB01000141	Permanent draft	Muthukrishnan et al., 2014 (direct submission)
B. casei	S18	Human associated	PRJNA174308	AMSP01000001-AMSP01000043	Permanent draft	Sharma et al., 2012 (direct submission)
B. ravenspurgense	5401308[T]	Human associated	PRJNA159637	CAJD01000001-CAJD01000026	Permanent draft	Roux et al., 2012 [93]
B. mcbrellneri	ATCC 49030[T]	Human associated	PRJNA34583	ADNU01000001-ADNU01000096	Permanent draft	Qin et al., 2010 (direct submission)
B. senegalense	JC43[T]	Human associated	PRJEA82613	CAHK01000001-CAHK01000070	Permanent draft	Kokcha et al., 2012 [94]

[a]The "*Brevibacterium linens* BL2" genome in the JGI database is in fact the genome of strain ATCC 9174 [95]

phylogenetic analysis was performed for the 23 *Brevibacterium* genomes using the sequences of 40 marker genes, as described by Mende et al. [27]. The genomes of *Glutamicibacter arilaitensis* Re117 [16] (Project accession number PRJEA50353) and *Corynebacterium casei* LMG S-19264 [28] (Project accession number PRJNA186910) were used as outgroups. An in-house database of the 40 marker genes present in 388 bacterial strains, which included 117 strains isolated from dairy products [29], was used to detect the 40 marker genes in the *Brevibacterium* genomes using tBlastN [30]. The best hits were selected with at least 60% sequence identity and 80% coverage. Each marker gene was translated into an amino acid sequence using T-Coffee [31] and aligned using MUSCLE [32]. The 40 individual alignments were then concatenated to a single one, which was used to build the tree using FastTree 2 [33] with the following parameters: –gamma –pseudo –mlacc 3 –slownni and the default bootstrap procedure (1000 resamples). The tree was visualized and annotated using MEGA7 [34].

Homologous gene families were computed using the OrthoMCL procedure implemented in GET-HOMOLOGUES software [35]. Amino acid sequences of the CDSs from the 23 *Brevibacterium* genomes were grouped into clusters using 75% identity and 75% coverage thresholds with a BlastP cutoff E value <1e-05. The inflation index of OrthoMCL algorithm was set to 1.5, as recommended by Li et al. [36]. Strain ATCC 9175^T was set as the first reference genome, the subsequent ones were randomly chosen by GET-HOMOLOGUES. Functional category assignment to each cluster was done according to the Clusters of Orthologous Groups (COG) database [37]. The Average Nucleotide Identity (ANI), implemented in GET-HOMOLOGUES software, was computed on homologous genes for all possible pairs of genomes.

Results

General genomic features

Assembly and annotation metrics for the 13 newly sequenced *Brevibacterium* genomes and those concerning the ten additional genomes (which included one cheese isolate) for the comparative analysis are detailed in Additional file 1. *Brevibacterium* genomes show considerable size heterogeneity, ranging from about 2.3 to 4.5 Mbp. There is no clear relationship between genome size and the habitat from which the strains were isolated. However, genome size is quite similar among the 12 cheese isolates, 4 Mbp on average (min: 3.7; max: 4.5), and these genomes harbor an average of 3712 genes (min: 3395; max: 4154). The two smallest genomes corresponded to human-associated strains: *B. ravenspurgense* 5401308^T and *B. mcbrellneri* ATCC 49030^T (2.3 and 2.6 Mbp, respectively). The G + C content varied from 58.0% (*B. mcbrellneri* ATCC49030^T) to 70.9% (*B. album*

DSM 18261^T), and was between 62 and 65% for most of the strains. Clustered regularly interspaced short palindromic repeat (CRISPR) candidates were found in most genomes (19 out of 23). Complete CRISPR-Cas systems consist of an array of CRISPRs interspaced by spacers and an adjacent *cas* gene cluster. Such a complete structure was observed only in the genomes of *B. casei* S18 (scaffold ID: S272_Contig15.15; 17 spacers), *B. album* DSM 18261^T (scaffold ID: K318DRAFT_scaffold00001.1; 2 + 1 + 6 + 35 + 12 spacers) and *B. ravenspurgense* 5401308^T (scaffold ID: Y1ADRAFT_CAJD01000011_1.11; 36 spacers). They all belong to the Type I CRISPR-Cas system but their overall gene content is variable. The PHASTER tool identified only one complete prophage region, in strain *B. sp.* Mu109. This 11.4 kb region (scaffold ID: Ga0063700_1029; locus tag Ga0063700_02875 to Ga0063700_02886) contains attachment sites (*attL* and *attR*) and CDSs encoding putative transposase, recombinase, integrase and capsid scaffolding proteins. However, because most genomes are draft genomes, other prophage regions might be present as regions split on several contigs and, consequently, difficult to detect.

Phylogenomic analyses and orthology

The genomic data from the 23 *Brevibacterium* strains were used to assess their intra-genus phylogenetic relationships. The phylogenomic tree partitioned the strains into two major lineages (Fig. 1). Lineage 1 contains two *Brevibacterium* strains isolated from human-associated samples, i.e., *B. mcbrellneri* ATCC49030^T and *B. ravenspurgense* 5401308^T. Lineage 2, containing the 21 other strains, is composed of two branches corresponding to the groups 2.A and 2.B. Group 2.A contains four strains isolated from diverse habitats: *B. album* DSM 18261^T (saline soil), *B. senegalense* JC43^T (human stool), *B. sp.* Mu109 (cheese) and *B. jeotgali* SJ5-8^T (fermented seafood). Group 2.B contains all the *Brevibacterium* strains isolated from cheese except *B. sp.* Mu109. It can be divided into three clades: 2.B.1, 2.B.2 and 2.B.3. Clade 2.B.1 contains nine strains, including eight strains isolated from cheese, and *B. sandarakinum* DSM 22082^T, which was isolated from a house wall. Clade 2.B.2 consists of six strains, including two from cheeses and three from freshwater environments. Clade 2.B.3 contains two strains belonging to the *B. casei* species, i.e., CIP 102111^T, isolated from cheese and S18, isolated from human skin. In the following, clades 2.B.1, 2.B.2, 2.B.3 will be referred to as group "*aurantiacum/sandarakinum/antiquum*", group "*linens/siliguriense/iodinum*" and group "*casei*", respectively. The taxonomy of the *Brevibacterium* genus is still in process of reclassification [11]. According to Stackebrandt et al. [38], 16S rRNA (*rrs*) gene sequences and DNA-DNA hybridization (DDH) should be considered as molecular criteria for

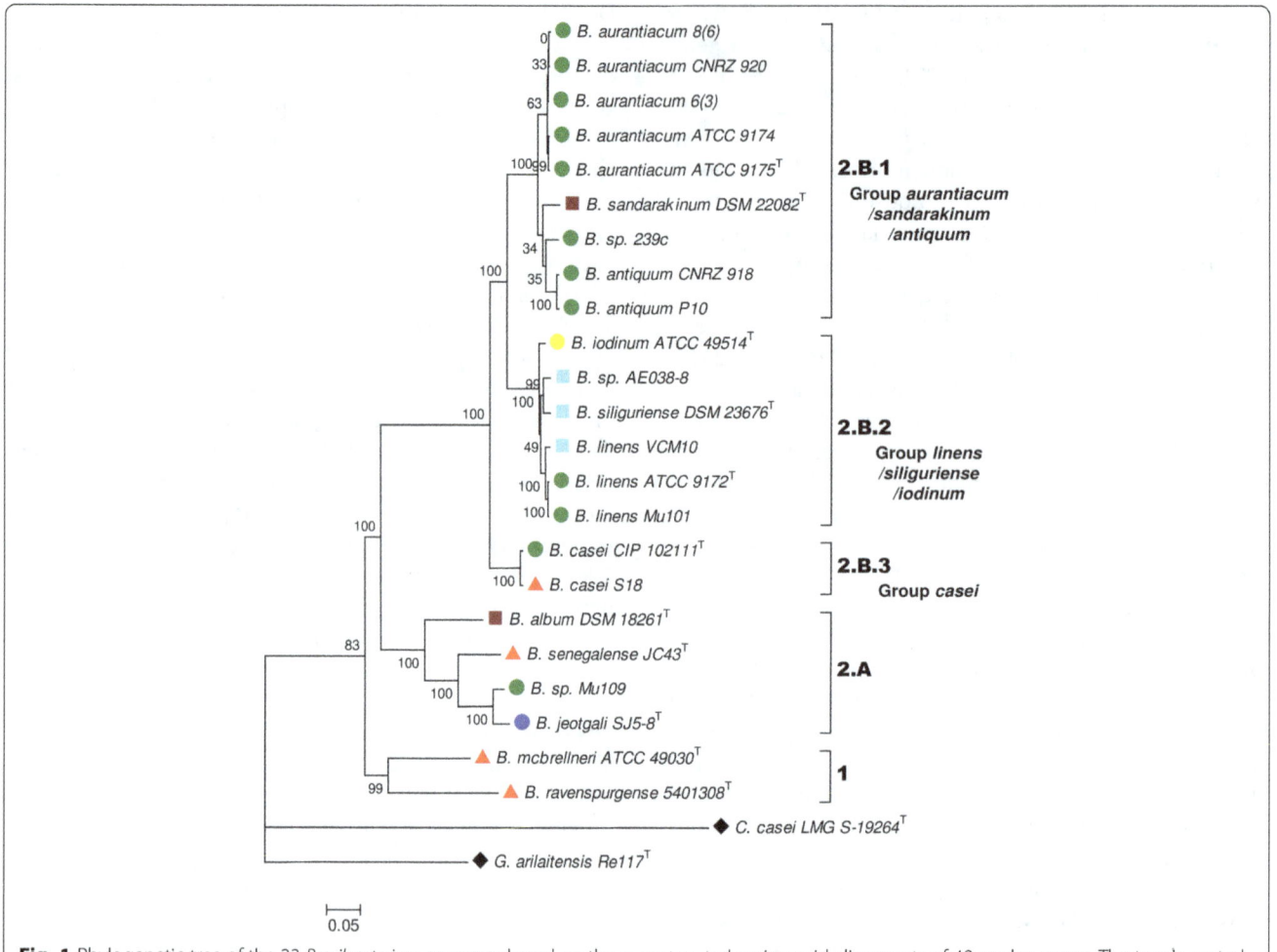

Fig. 1 Phylogenetic tree of the 23 *Brevibacterium* genomes based on the concatenated amino acid alignments of 40 marker genes. The tree is rooted using *Glutamicibacter arilaitensis* Re117 and *Corynebacterium casei* LMG S-19264 as outgroups (◆). *Brevibacterium* strains are labeled according to their habitats (●: cheese; ●: milk; ●: seafood; ■: fresh water; ■: soil or wall surface; ▲: human associated). Bootstrap support values are shown as a percentage before the respective nodes; the scale bar indicates the number of substitutions per site. Phylogenetic clusters are indicated on the right of the tree

species delineation. However, the Average Nucleotide Identity (ANI) has recently been proposed to replace DDH values [39–41]. In the present study, we used two criteria for delineating species: two strains were considered to belong to the same species if (i) their *rrs* gene sequence identity was ≥98% [42] and (ii) their ANI was ≥95% [40]. The results are presented in Additional file 2. Based on these criteria, two strains isolated from cheese products, i.e. 239c and Mu109 could not be linked to any sequenced species. In addition, *B. linens* AE038–8 and *B. linens* ATCC 9172^T have an ANI value of only 91.2%, suggesting that strain AE038–8 does not belong to the *B. linens* species.

Pan-genome analysis of the 23 *Brevibacterium* strains resulted in 25,376 orthologous gene clusters, containing a total of 78,702 protein-coding genes (Additional files 3 and 4). Of these clusters, only 263 (1%) are shared by all the strains (the core genome), reflecting a high

intragenus genomic variability. We further investigated the *B. aurantiacum* species, in which all five of the sequenced strains were isolated from cheese. The pan-genome of this group contains 5988 orthologous gene clusters, of which 2684 are core and 3304 are accessory, i.e., variable among strains (Fig. 2A). Functional prediction revealed an overrepresentation of some COG categories in the accessory genome relative to the core genome, especially the categories [V] (Defense mechanisms) and [X] (Mobilome: prophages, transposons), in which the ratio accessory/core was 4.4 and 45.3, respectively (Fig. 2B).

Catabolism of energy compounds present in cheeses
Catabolism of lactose, galactose and D-galactonate
During the manufacturing of cheeses, lactose is consumed by lactic acid bacteria, but some lactose may still be present at the beginning of ripening. When the lactic

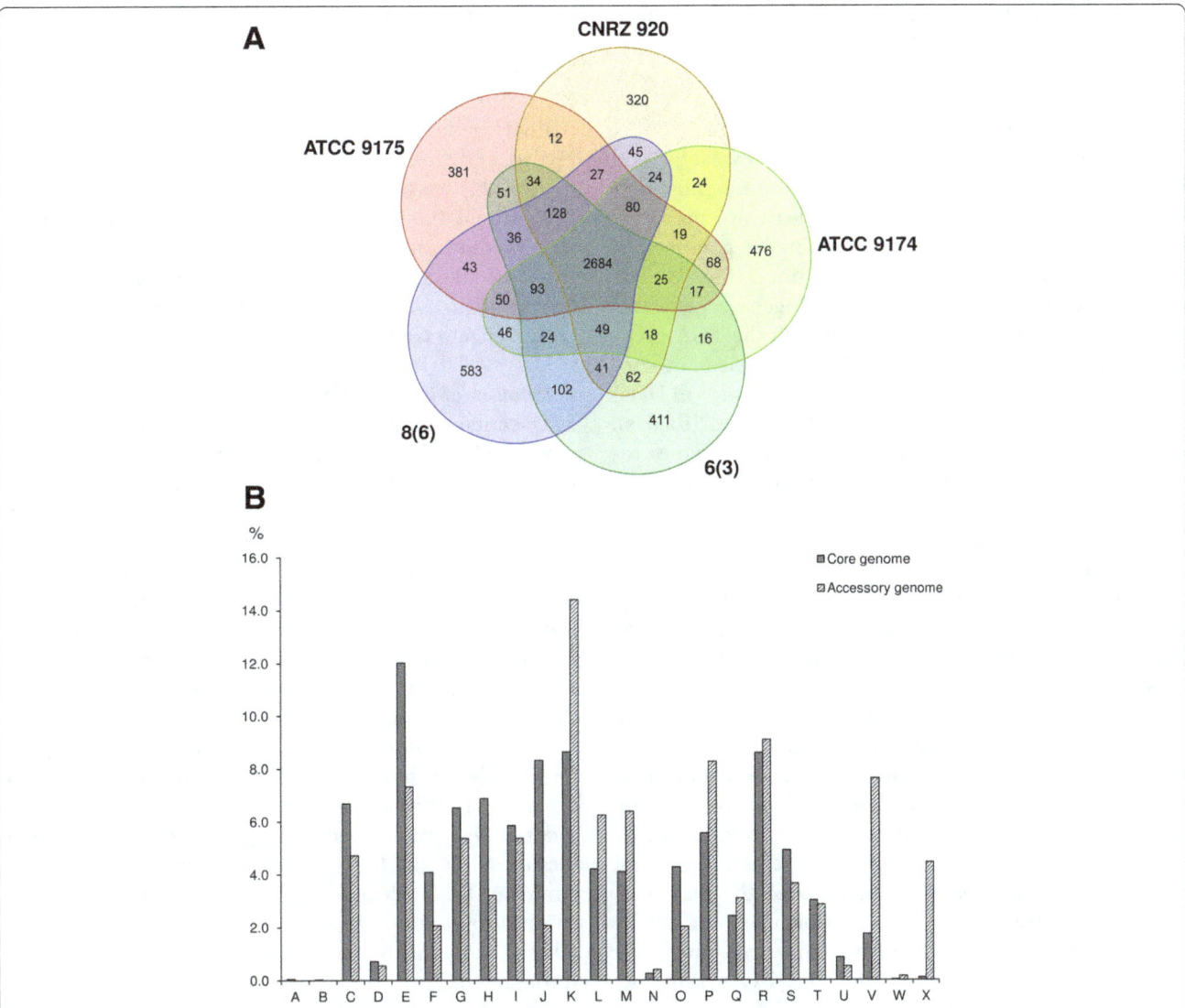

Fig. 2 Orthologous gene clusters in the five *B. aurantiacum* genomes. **a** Venn diagram of the distribution of orthologous gene clusters. **b** Functional categories of the core and the accessory genome. Functional assignments were performed using the Integrated Microbial Genomes (IMG) platform; functional categories were labeled according to the COG database (https://www.ncbi.nlm.nih.gov/COG). A: RNA processing and modification; B: Chromatin structure and dynamics; C: Energy production and conversion; D: Cell cycle control, cell division, chromosome partitioning; E: Amino acid transport and metabolism; F: Nucleotide transport and metabolism; G: Carbohydrate transport and metabolism; H: Coenzyme transport and metabolism; I: Lipid transport and metabolism; J: Translation, ribosomal structure and biogenesis; K: Transcription; L: Replication, recombination and repair; M: Cell wall/membrane/envelope biogenesis; N: Cell motility; O: Post-translational modification, protein turnover, chaperones; P: Inorganic ion transport and metabolism; Q: Secondary metabolites biosynthesis, transport and catabolism; R: General function prediction only; S: Function unknown; T: Signal transduction mechanisms; U: Intracellular trafficking, secretion and vesicular transport; V: Defense mechanisms; X: Mobilome: prophages, transposons; W: Extracellular structures

starter culture contains *Streptococcus thermophilus* strains, some galactose is produced from lactose, and this compound may be present in the cheese curd for several weeks [43]. No beta-galactosidase gene was identified in the 23 *Brevibacterium* genomes, which is consistent with the fact that most *Brevibacterium* strains are not able to consume lactose [11]. However, the genomes of twelve of the 23 *Brevibacterium* strains encode the four enzymes of the Leloir pathway for galactose

utilization (Additional file 5). The corresponding strains belong to the phylogenetic groups *aurantiacum/sandarakinum/antiquum* and *linens/siliguriense/iodinum*. A complete pathway for D-galactonate catabolism is present in strains from the phylogenetic groups *aurantiacum/sandarakinum/antiquum*, *linens/siliguriense/iodinum* and *casei*. The genes are organized in a cluster, that encodes a transcriptional regulator, a 2-dehydro-3-deoxyphosphogalactonate aldolase [EC 4.1.2.21], a 2-

dehydro-3-deoxygalactonokinase [EC 2.7.1.58], a galactonate dehydratase [EC 4.2.1.6] and one or two D-galactonate importers.

Catabolism of lactate, acetate, ethanol and citrate

Lactate, which is produced from lactose by lactic acid bacteria, is an important energy substrate for most aerobic cheese ripening bacteria [44]. At least two predicted lactate permease genes are present in the genomes of the 23 *Brevibacterium* strains (Additional file 5). The oxidation of lactate by lactate dehydrogenase generates pyruvate that is later catabolized through the TCA cycle. There are two types of lactate dehydrogenases in bacteria, NAD-dependant lactate dehydrogenases (nLDHs) and NAD-independent lactate dehydrogenases (iLDHs), the latter generally being considered as the enzymes responsible for lactate oxidation in bacteria [45]. Four types of iLDHs were identified in the genomes of the *Brevibacterium* strains: a quinone-dependent D-iLDH, a quinone-dependent L-iLDH, a Dld II-type quinone or cytochrome-dependent D-iLDH [46] and a three subunit quinone or cytochrome-dependent L-iLDH complex LldEFG [46]. Dld II-type quinone or cytochrome-dependent D-iLDHs were present only in group *aurantiacum/sandarakinum/antiquum*. A lactate permease gene is adjacent to the quinone-dependent L-iLDH gene in 14 strains. In cheese, acetate is produced from lactose by heterofermentative lactic acid bacteria, from citrate by citrate-utilizing lactic acid bacteria, or from lactate by *Pediococcus* and *Propionibacterium* strains. Twenty of the 23 *Brevibacterium* genomes encoded the monocarboxylic acid transporter MctC, which is an uptake system for acetate, propionate and pyruvate [47] (Additional file 5). Genes encoding acetyl-CoA synthase, which channels acetate toward the TCA cycle, are present in all the 23 strains. In cheese, ethanol is produced from lactose by yeasts such as *Kluyveromyces lactis* and *K. marxianus*, and by heterofermentative lactic acid bacteria. It is a potential energy substrate for aerobic microorganisms during cheese ripening but, to our knowledge, this has never been investigated. There are numerous candidate genes encoding enzymes involved in the catabolism of ethanol to acetate in the genomes of the *Brevibacterium* strains: between one and nine for the alcohol dehydrogenase and between nine and 31 genes for the acetaldehyde dehydrogenase (Additional file 5). In *Corynebacterium glutamicum*, the genes encoding the alcohol dehydrogenase (*adhA*) and the acetaldehyde dehydrogenase (*ald*) are responsible for ethanol catabolism [48]. All of the 23 *Brevibacterium* genomes encode an ortholog of *adhA* (65 to 69% identity at the amino acid level) and an ortholog of *ald* (67 to 82% identity), and in 21 strains, these genes form a cluster. Milk contains citrate (~8 mM in cow's milk), which

may persist during cheese ripening if the lactic starter culture does not contain citrate-utilizing lactic acid bacteria. Citrate importers belonging to the Citrate-Mg2+:H+ (CitM)/Citrate-Ca2+:H+ (CitH) Symporter (CitMHS) family (TC no. 2.A.11) and to the 2-HydroxyCarboxylate Transporter (2-HCT) family (TC no. 2.A.24) were detected in 16 *Brevibacterium* strains (Additional file 5). The genomes of the two *B. casei* strains did not encode any citrate transporter, and those of the six strains from the phylogenetic group *linens/siliguriense/iodinum* did not encode 2-HCT family citrate transporters.

Catabolism of lipids and glycerol

Milk contains large amounts of triglycerides (~35 g/l in cow's milk), which can be used by cheese microorganisms as an energy source. Lipid catabolism involves the release of free fatty acids and glycerol and the subsequent breakdown of these compounds. The *Brevibacterium* genomes encode between 8 and 37 proteins with putative lipase or esterase activity (Additional file 6). Two types of secreted lipases / esterases were identified. The first type is a triacylglycerol lipase [EC 3.1.1.3], that was identified in 15 genomes, corresponding to eight of the nine strains of the phylogenetic group *aurantiacum/sandarakinum/antiquum*, to all six strains of the group *linens/siliguriense/iodinum* and to only one of the eight other strains (the cheese-associated strain Mu109). The second type is a glycerophosphoryl diester phosphodiesterase [EC 3.1.4.46], which was identified only in the strains of the phylogenetic groups *aurantiacum/sandarakinum/antiquum* and *linens/siliguriense/iodinum*. Uptake of fatty acids can be done by passive diffusion through the membrane lipid bilayer or by protein-facilitated transfer [49]. Genomic analysis revealed the presence of short-chain fatty acid uptake (AtoE) family proteins (TC no. 2.A.73) in all of the 23 *Brevibacterium* strains and it is noteworthy that *Brevibacterium* strains have a greater number of transporters from this family than other *Actinobacteria* strains (1.7 vs. 0.1 genes per genome), based on assignations to the COG2031 for the 5726 *Actinobacteria* strains present in April 2017 in the IMG database. A complete beta-oxidation pathway for fatty acid degradation was identified in all of the investigated strains, except in strains ATCC 49030[T] and 5401308[T], for which genes encoding L-3-hydroxyacyl-CoA dehydrogenase or 3-ketoacyl-CoA thiolase seemed to be lacking. Beta-oxidation of odd-chain-length fatty acids yields propionyl-CoA in addition to acetyl-CoA. Propionyl-CoA may be catabolized via the methylcitrate cycle, which oxidizes it to pyruvate. Three enzymes are characteristic of this cycle: the methylcitrate synthase, the methylcitrate dehydratase and the 2-methylisocitrate lyase, encoded by *prpC*, *prpD* and *prpB*, respectively

[50]. In all of the investigated *Brevibacterium* genomes, these genes form a cluster (*prpDBC*). It is noteworthy that in all of the strains of the phylogenetic groups *aurantiacum/sandarakinum/antiquum*, *linens/siliguriense/iodinum* and *casei*, *prpDBC* is located upstream of the genes involved in the catabolism of glycerol (glycerol uptake protein, glycerol kinase and glycerol-3-phosphate dehydrogenase). This glycerol pathway occurs in aerobic conditions because glycerol-3-phosphate dehydrogenase reduces quinones of the respiratory chain [51]. The genome of the two *B. casei* strains also encodes a glycerol dehydrogenase and a dihydroxyacetone kinase, which constitute another glycerol utilization pathway.

Catabolism of proteins and amino acids

Cheese contains a large amount of proteins, mainly caseins, which can be degraded by various proteolytic microorganisms. The amino acids resulting from proteolysis can be used as an energy source by the cheese ripening microorganisms and they are also precursors of key flavor compounds. Genomic analysis of the 23 *Brevibacterium* genomes by the MEROPS peptidase BLAST search tool [52] and manual curation of the results revealed the presence of between four and 16 putative excreted enzymes with proteolytic activities, depending on the strain (Additional file 7). About 15% of these enzymes have a LPXTG motif (TIGR01167) at their C-terminus and are thus probably cell-wall-associated. Interestingly, in many cases the corresponding genes are contiguous. In addition, there was a complete identity between the sequence of one of the two predicted cell-wall-associated proteases of *B. aurantiacum* ATCC 9174 (locus tag BlinB01003410) and the first 20 N-terminal amino acid sequence of an extracellular protease purified from this strain [53]. The presence of amino acid degradation pathways was inferred from the annotation of the 23 genomes (Additional file 7). All of them encode a bifunctional proline dehydrogenase/L-glutamate gamma-semialdehyde dehydrogenase [EC 1.5.5.2 / EC 1.2.1.88], which catalyzes oxidation of proline to glutamate using a membrane-bound quinone and NAD as the electron acceptor. They also encode a NAD-specific glutamate dehydrogenase [EC 1.4.1.2] that produces NADH and alpha-ketoglutarate, an intermediate of the TCA cycle, and the enzymes involved in the degradation of threonine and serine. The histidine and alanine catabolic pathways seem to be present in all of the strains except *B. senegalense* JC43[T] (histidine) and *B. ravenspurgense* 5401308[T] (alanine). Phenylalanine, tyrosine, methionine and arginine degradation pathways were identified only in the strains of the phylogenetic groups *aurantiacum/sandarakinum/antiquum*, *linens/siliguriense/iodinum* and *casei*. Gamma-aminobutyrate is a four-carbon non-protein amino acid produced from glutamate by lactic acid bacteria during the ripening of cheeses [54]. All the genomes encode a 4-aminobutyrate transaminase [EC 2.6.1.19] and a succinate semialdehyde dehydrogenase [EC 1.2.1.16]. These enzymes convert Gamma-aminobutyrate into succinate, an intermediate of the TCA cycle.

Iron acquisition

Genes encoding putative Mn^{2+} or Fe^{2+} transporters are present in all of the 23 *Brevibacterium* genomes (Additional file 8). The EfeUOB transporter, which is a high-affinity uptake system for both Fe^{2+} and Fe^{3+} [55], is present in seven strains. Five genomes encode a putative ABC-type iron transport system whose closest homolog in *Haemophilus influenza* (FbpABC) is required for acquiring iron from transferrin [56]. Fe^{3+}/siderophore transport components are present in all of the 23 genomes, varying from nine components for *B. senegalense* JC43[T], up to 31 components for *B. sp.* Mu109. Comparison of the abundance of these components based on the analysis of the COG1120, COG4604, COG0609, COG4605, COG4606, COG4779, COG0614, COG4592 and COG4607, showed that the average number was higher for *Brevibacterium* (17.6 genes per genome) than for the other *Actinobacteria* (8.4 genes per genome, calculated from 5726 *Actinobacteria* genomes). The *Brevibacterium* genomes also encode siderophore interacting proteins (mean number of 3.6 genes per genome), which are required for iron release from Fe^{3+}/siderophore complexes. A putative hydroxamate-type siderophore biosynthesis cluster, which encodes a lysine N6-hydroxylase [EC 1.14.13.59], a siderophore synthetase component and, occasionally, a L-2,4-diaminobutyrate decarboxylase [EC 4.1.1.86], is present in 15 strains. All these strains belong to the phylogenetic groups *aurantiacum/sandarakinum/antiquum*, *linens/siliguriense/iodinum* and *casei* (Additional file 8). The genome of strain Mu109 contains a cluster (locus tag Ga0063700_02161 to Ga0063700_02193) encoding eight Fe^{3+}/siderophore transport components, one siderophore interacting protein and four proteins that are probably involved in the biosynthesis of a catecholate or a mixed catecholate/hydroxamate siderophore: 4'-phosphopantetheinyl transferase EntD, MbtH protein, non-ribosomal siderophore peptide synthetase component and L-ornithine N5-oxygenase [EC 1.14.13.195]. The closest homologs of the 4'-phosphopantetheinyl transferase EntD and the non-ribosomal siderophore peptide synthetase component are found in *Streptomyces* species. The genome of strain ATCC 9174 contains a cluster (locus tag BlinB01002486 to BlinB01002496) encoding three Fe^{3+}/siderophore transport components, one siderophore interacting protein, one siderophore exporter and seven genes that are probably involved in the biosynthesis of a catecholate siderophore: non-ribosomal siderophore peptide synthetase component, glycosyltransferase IroB, 2,3-dihydro-2,3-dihy-droxybenzoate dehydrogenase [EC 1.3.1.28], isochorismate synthase [EC 5.4.4.2], 2,3-dihydroxybenzoate-AMP lig-ase [EC 2.7.7.58], isochorismatase [EC 3.3.2.1], and a putative transferase component of siderophore synthe-tase. These siderophore biosynthesis genes probably re-sult from HGTs since many of their closest homologs are present, either in Gram-negative species or in *Streptomyces* or *Paenibacillus* species. In addition, comparison of the flanking regions of the siderophore biosynthesis cluster of strain ATCC 9174 to the CNRZ 920 genome (in which the cluster is absent) revealed that the cluster corresponded to an insertion that oc-curred in an ancestor of strain ATCC 9174 at the end of a tRNA-gly gene (locus tag BlinB_R0152 in ATCC 9174 and Ga0063691_00673 in CNRZ 920), which is followed in CNRZ 920 by a protein of unknown func-tion (locus tag Ga0063691_00672, which is an ortholog of BlinB01002497 in ATCC 9174). For strains ATCC 9172^T and Mu101, there is little evidence for sidero-phore biosynthesis, even if it cannot be excluded, since their genome encodes a lysine decarboxylase that is clustered with a siderophore interacting protein and a Fe^{3+}/siderophore binding component. With the excep-tion of isochorismate synthase, which is an enzyme that is also involved in menaquinone biosynthesis, no genes involved in siderophore biosynthesis were identified in the genomes of strains DSM 18261^T, SJ5-8^T, 5401308^T, ATCC 49030^T and JC43^T. In summary, the genomic analyses indicate that 16 of the 23 *Brevibacterium* strains are probably able to produce siderophores. Hydroxamate-type siderophore genes are found in 14 strains, catecholate-type siderophore genes in strain Mu109, and both types in strain ATCC 9174. Sidero-phore biosynthesis is predicted to occur in all of the nine strains of the phylogenetic group *aurantiacum/ sandarakinum/antiquum*, in four of the six strains of the group *linens/siliguriense/iodinum*, in the two strains of the group *casei*, but only in one of the six other *Brevibac-terium* strains, which corresponded to the cheese isolate Mu109. Interestingly, analysis of the genomes present in the IMG database also revealed that four gene clusters in-volved in iron acquisition were shared between *Brevibacter-ium* strains isolated from cheeses and cheese isolates belonging to other genera (*Glutamicibacter, Microbacter-ium* and *Corynebacterium*). This corresponded to recent HGT events since the percentages of identity at the amino acid level between the genes in *Brevibacterium* and the genes in the other genera were typically ~95–100%. In most cases, transposase genes were located close to the clusters. These clusters were denoted Iron-Brev1, Iron-Brev2, Iron-Brev3 and Iron-Brev4 (Additional file 9). The clus-ter Iron-Brev1 corresponds to the ActinoRUSTI region that was recently described [17].

Osmotic stress tolerance

Cheeses are salted by applying salt to their surface or by submerging them in saturated brine. One mechanism to overcome osmotic stress that results from high salt concentration is the accumulation of osmoprotectants. *Brevibacterium* strains are known to produce the osmo-protectant ectoine [57]. It is synthesized from L-aspartate-semialdehyde by the action of three enzymes: diaminobutyrate aminotransferase (EctB) [EC 2.6.1.76], diaminobutyrate acetyltransferase (EctA) [EC 2.3.1.178] and ectoine synthase (EctC) [EC 4.2.1.108]. Ectoine can be further converted to hydroxyectoine by the action of ectoine dioxygenase (EctD). Except for strains ATCC 49030^T and 5401308^T, all the investigated *Brevibacter-ium* genomes contained the *ectABC* cluster (Additional file 10). The *ectD* gene was present in 13 of the 15 strains from the phylogenetic groups *aurantiacum/san-darakinum/antiquum* and *linens/siliguriense/iodinum*, but it was absent in the two strains of the group *casei*, and only present in one of the six other *Brevibacterium* strains. Glycine-betaine is another osmoprotectant that can be synthesized by *Brevibacterium*. The genomic ana-lysis revealed the presence of two possible pathways. In the first one, choline is converted to glycine betaine by the combined action of a choline dehydrogenase [EC 1.1.99.1] and a betaine aldehyde dehydrogenase [EC 1.2.1.8] whereas in the second one, this conversion is catalyzed by a single enzyme, choline oxidase [EC 1.1.3.17]. Twelve of the strains have both pathways, nine have only the choline oxidase pathway, one has only the choline dehydrogenase pathway, and one has none of them. When present, the choline oxidase gene is located in a cluster containing a betaine/carnitine/choline trans-porter (BCCT) and a betaine-aldehyde dehydrogenase. Choline is present in milk and cheese [58], making it available for glycine-betaine biosynthesis. Interestingly the *Brevibacterium* genomes generally exhibit a greater number of BCCT than the other *Actinobacteria* ge-nomes (6.3 vs. 1.1 genes per genome, based on assigna-tions to COG1292) (Additional file 10). However, they do not have more ABC transport components involved in the transport of glycine betaine or related osmolytes (4.7 vs. 6.1). Trehalose is a non-reducing sugar that plays a physiological role in energy storage and also as a compatible solute [59, 60]. The genome of the 23 strains encodes a trehalose synthase [EC 5.4.99.16], which catalyzes the synthesis of trehalose from maltose (Additional file 10). Except for strain ATCC 49030^T, tre-halose can also be produced from UDP-glucose and glucose-6-phosphate via the trehalose-6-phosphate synthase [EC 2.4.1.15]/trehalose-6-phosphate phosphat-ase [EC 3.1.3.12] pathway. Two operons encoding multi-subunit (Na^+)(K^+)/proton antiporters (Mrp systems) were identified in all the *Brevibacterium* genomes. These

Mrp systems are composed of six subunits and the genes are organized as "group 2" *mrp* operons [61]. In other *Actinobacteria* genomes, the mean number of Mrp systems, based on assignation to COG1006, is about four times lower (0.5 vs. 2.0). The abundance of the other $(Na^+)(K^+)$/proton antiporters was similar in *Brevibacterium* and in the other *Actinobacteria* genomes (mean values of 3.9 and 4.0 genes per genome, respectively, based on assignations to the COG0025, COG1055, COG1757, COG3004, COG3067 and COG3263). It is noteworthy that *Brevibacterium* strains have a greater number of transporters from the Sodium Solute Symporter (SSS) family than other *Actinobacteria* strains (7.3 vs. 1.7 genes per genome, based on the number of proteins matching the PF00474 Hidden Markov Model).

Bacteriocines and phenazines

In cheese, bacteriocin producers can inhibit other microbial groups that share the same ecological niche, which confers them a selective advantage. Genomic analysis predicted the production of several bacteriocins in *Brevibacterium* strains (Additional file 11). Linocin M18-related bacteriocins (PF04454) were identified in 15 genomes, including five of the nine strains of the phylogenetic group *aurantiacum/sandarakinum/antiquum*, all six strains of the group *linens/siliguriense/iodinum* and four of the eight other strains (*B. album* DSM 18261[T], *B. senegalense* JC43[T], *B. jeotgali* SJ5-8[T] and *B. sp.* Mu109). Five groups of ribosomally synthesized and posttranslationally modified peptides (RiPPs) [62] were predicted in the 23 *Brevibacterium* strains. The first group corresponds to lanthipeptides, which are characterized by the presence of lanthionine (Lan) and/or methyl-lanthionine (MeLan) and/or labionin (Lab) residues [62, 63]. Many lanthipeptides have an antimicrobial activity, mainly against Gram-positive bacteria [64], and are referred to as lantibiotics. Lanthipeptide gene clusters were predicted in seven strains, of which six are cheese-associated (Additional file 11). Even if they have different structures, these clusters contain one or two putative lanthipeptide synthetases, one or two putative precursor peptides, one to three putative ABC transport system components and, in some cases, a prolyl oligopeptidase (PF00326) (Fig. 3A). Protein sequence analysis of the predicted lanthipeptide synthetases suggests that all of them belong to class III, corresponding to a trifunctional enzyme LanKC [65]. The sequences of all putative precursor peptides (Fig. 3B) contain two characteristic and conserved Ser/Ser(Thr)/Cys motifs necessary for the formation of (Me)Lan and/or (Me)Lab. One of them also contains a B-A-C-Leu-Gln motif in its N-terminal part (where B is Ile, Leu or Val; A is Phe or Leu; and C is Glu or Asp), which is highly conserved in class III precursor peptides and essential for the enzymatic processing of

the labyrinthopeptin A2 [66]. To our knowledge, as of this time, the *Kribbella flavida* DSM 17836 protein FlaP, which is a proline-specific oligopeptidase, is the only characterized protease involved in the removal of the leader peptide of a class III lanthipeptide [67]. Genes with a weak homology with *flaP* were identified in six of the seven predicted lanthipeptide gene clusters, but only one of the six predicted precursor peptides contains a Pro residue in the leader peptide, which could serve as the primary cleavage site for the predicted peptidase (Fig. 3B). Interestingly, the putative lanthipeptide gene cluster from the four cheese-associated strains *B. antiquum* CNRZ 918, *B. antiquum* P10, *B. aurantiacum* ATCC 9174 and *B. linens* ATCC 9172[T] is located in a ~96 kb genomic island, which is absent in the other *Brevibacterium* genomes investigated, but which is present in the genome of *Corynebacterium casei* LMG S-19264, a strain isolated from a smear-ripened cheese [28]. This genomic island, which we denoted as BreLI (**Bre**vibacterium **L**anthipeptide **I**sland), probably corresponds to an Integrative and Conjugative Element (ICE, or conjugative transposon). ICEs are typically composed of three core genetic modules involved in: (i) integration and excision; (ii) conjugation; and (iii) regulation [68]. Functional prediction of genes in BreLI revealed assignations to these three functions (Fig. 3C and Additional file 11). Integration of this ICE occurred at the 3′ end of a gene encoding a class Ib ribonucleotide reductase beta subunit, resulting in a 12-bp perfect repeat sequence, present at the two borders of BreLI. It is noteworthy that, in comparison to the four other strains, a segment containing three genes of the lanthipeptide gene cluster region is lacking in the ATCC 9174 BreLI Island (compare clusters (a) and (b) in Fig. 3A). The second group of predicted RiPPs is related to lactococcin 972 (PF09683), which has been characterized in *Lactococcus lactis* IPLA 972 [69, 70]. Genes with a weak homology with the lactococcin 972 precursor peptide were identified in six cheese-associated strains from the phylogenetic group *aurantiacum/sandarakinum/antiquum*, corresponding to two different putative prepeptides, one of them being present only in strain 239c, and the other in five *B. aurantiacum* strains (Additional file 11). In all these six strains, the bacteriocin structural gene forms an operon with three genes encoding two transmembrane proteins and a putative ABC transport systeme ATP-binding protein. Comparison of the flanking regions of the lactococcin 972-related bacteriocin biosynthesis gene cluster in strains 6(3) and 8(6) with the CNRZ 918 genome (in which the cluster is absent) revealed that the gene cluster is located in a genomic island that is inserted at the end of a tRNA-val gene (locus tag Ga0063697_02943 in strain 6(3), Ga0063698_00389 in strain 8(6) and Ga0063689_00412 in strain CNRZ 918), which is followed in CNRZ 918 by an

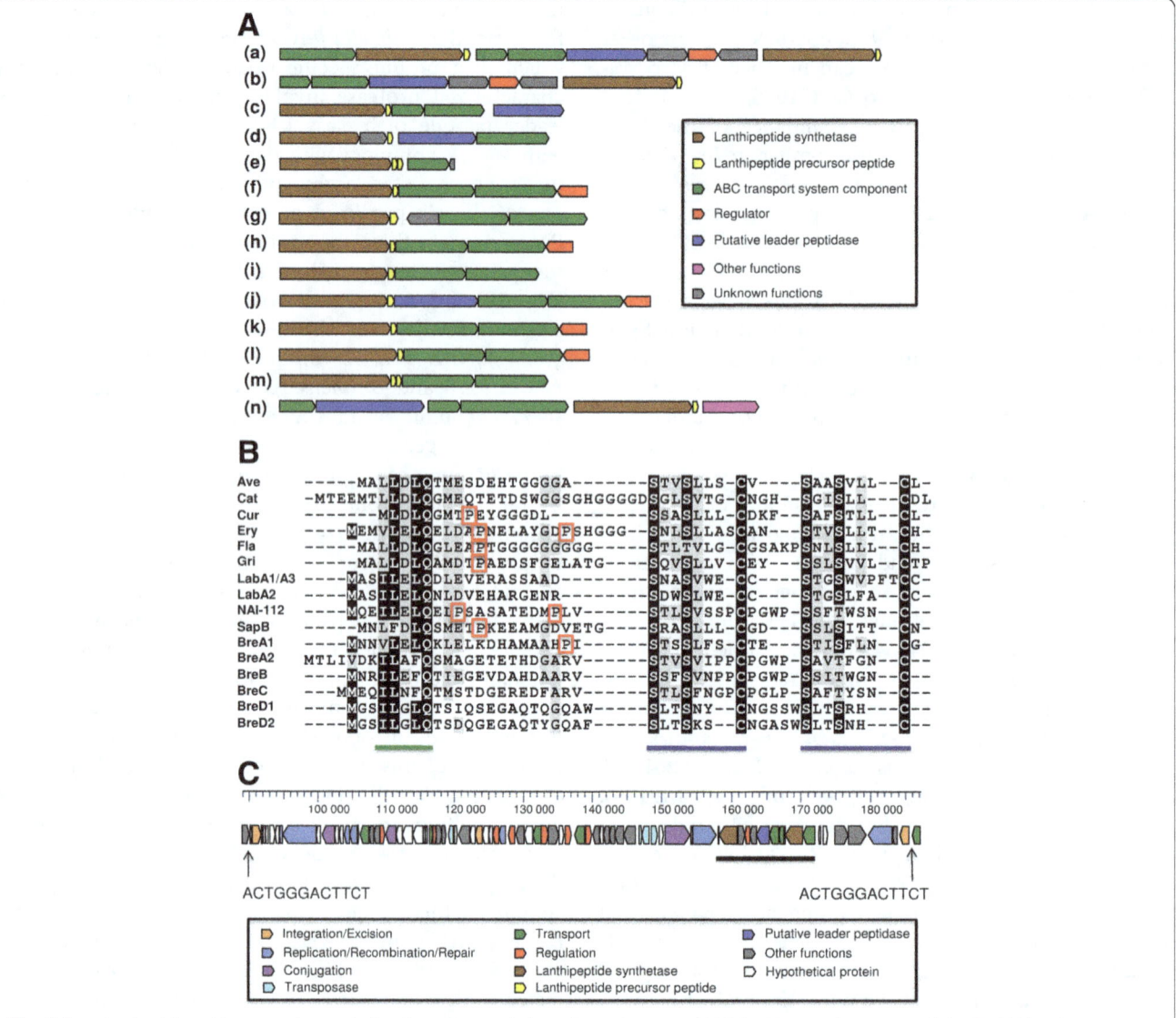

Fig. 3 Putative lanthipeptide gene clusters in *Brevibacterium* and *Corynebacterium casei* LMG S-19264, and structure of the BreLI Island. **a** Organization of class III lanthipeptide gene clusters. (a) *B. antiquum* P10, *B. antiquum* CNRZ 918, *B. linens* ATCC 9172^T and *Corynebacterium casei* LMG S-19264; (b) *B. aurantiacum* ATCC 9174; (c) *B. linens* Mu101; (d) *B. mcbrellneri* ATCC 49030^T; (e) *B. aurantiacum* CNRZ 920; (f) *Streptomyces avermitilis* MA-4680 (Avermipeptin, Ave); (g) *Catenulispora acidiphila* DSM 44928 (Catenulipeptin, Cat); (h) *Thermomonospora curvata* DSM 43183 (Curvopeptin, Cur); (i) *Saccharopolyspora erythraea* NRRL 2338 (Erythreapeptin, Ery); (j) *Kribbella flavida* DSM 17836 (Flavipeptin, Fla); (k) *Streptomyces griseus* NBRC 13350 (Griseopeptin, Gri); (l) *Streptomyces coelicolor* A3(2) (SapB); (m) *Actinomadura namibiensis* DSM 6313 (Labyrinthopeptin, Lab); (n) *Actinoplanes sp.* NAI112 (NAI-112). **b** Sequences of class III lanthipeptide precursors. (Me)Lan/(Me)Lab rings are indicated by a blue line, recognition motifs in the leader peptides are indicated by a green line, Pro residues in the leader are indicated by a red box. BreA1 corresponds to the protein product of Ga00637 01_00554 (strain P10), Ga0063689_03239 (CNRZ 918), Ga0063694_02885 (ATCC 9172^T), ORF1 (ATCC 9174, see Additional file 11 for more details), ORF25 (*Corynebacterium casei* LMG S-19264); BreA2 corresponds to the protein product of Ga0063701_00562 (P10), Ga0063689_03231 (CNRZ 918), ORF2 (ATCC 9172^T) and ORF26 (*Corynebacterium casei* LMG S-19264); BreB, BreC, BreD1 and BreD2 correspond to the protein products of Ga0063699_00711 (Mu101), HMPREF0183_0898 (ATCC 49030^T), Ga0063691_02687 and Ga0063691_02686 (CNRZ 920), respectively. **c** Schematic map of the BreLI Island in *B. antiquum* P10. Nucleotide position values refer to contig Ga0063701_102. The lanthipeptide gene cluster region (corresponding to BreA1 and BreA2 for strains P10, CNRZ 918, ATCC 9172^T and *Corynebacterium casei* LMG S-19264, and BreA1 for ATCC 9174) is underlined. The 12-bp sequences are the perfect repeat delimiting the BreLI Island

uracil-xanthine permease (locus tag Ga0063689_00413, which is an ortholog of Ga0063697_02919 in 6(3) and Ga0063698_00365 in 8(6)). The lactococcin 972-related bacteriocin biosynthesis gene cluster in strains ATCC 9174 and CNRZ 920 is also flanked by a tRNA-val gene.

The third group of predicted RiPPs corresponds to linear azol(in)e-containing peptides (LAPs), which are characterized by the presence of multiple thiazole and (methyl)oxazole heterocycles, and sometimes by their corresponding 2-electron reduced azoline state. LAP gene clusters were predicted in four *B. aurantiacum* cheese-associated strains (Additional file 11). One cluster is present in strains 6(3), ATCC 9175^T and ATCC 9174, and another is present only in strain 8(6). All these clusters contain putative precursor peptides, a SagB-type dehydrogenase domain-containing protein (TIGR03605), one or two YcaO cyclodehydratases (PF02624) and a cyclodehydratase-docking fusion protein (TIGR03882), representing the critical components in LAP biosynthesis [62, 71]. In addition, the gene cluster in strains ATCC 9175^T, ATCC 9174 and 6(3) also contains a methyltransferase, which may be involved in the methylation of the peptide [72]. Comparison of the flanking regions of the predicted LAP gene cluster in strain 6(3) with the CNRZ 920 genome (in which the cluster is absent) revealed that the gene cluster is located in a genomic island whose insertion is between an aspartate racemase (orthologs with locus tag Ga0063697_02599 in strain 6(3) and Ga0063691_01440 in strain CNRZ 920) and a short-chain dehydrogenase (orthologs with locus tag Ga0063697_02614 in strain 6(3) and Ga0063691_01438 in strain CNRZ 920). The LAP gene cluster in strain 8(6) contains two putative ABC transport system components. This type of component is frequently associated with LAP biosynthesis gene clusters [62, 72]. The fourth group of predicted RiPPs corresponds to the recently discovered linaridin family, which is characterized by the presence of thioether crosslinks, like for the lanthipeptide family, but whose post-translational modifications are unrelated to the lanthipeptide biosynthesis pathway [73]. Linaridin precursor peptides were predicted in the two cheese-associated strains 239c and 8(6). These two prepeptides showed significant similarity to the characterized cypemycin (51 to 53% identity) and grisemycin (58% identity) prepeptides [73, 74]. The putative linaridin gene clusters of strains 239c and 8(6) are very similar (Additional file 11). They contain homologs of the cypemycin biosynthetic gene cluster, i.e. *cypH, cypL, cypD* and *cypT* [73]. The fifth group of predicted RiPPs is related to the sporulation delaying protein (SDP), which has been characterized in *Bacillus subtilis* [75]. SDP is produced by the *sdpABC* operon in which *sdpC* encodes the SDP precursor peptide (pro-SdpC) and *sdpA* and *sdpB* are essential for the production of the active SDP toxin [76]. Genes with a weak homology with *sdpABC* were identified in three *B. aurantiacum* cheese-associated strains (Additional file 11). Comparison of the corresponding flanking regions in strains ATCC 9175^T and 6(3) with the 8(6) genome (in which the cluster is absent) revealed that this gene cluster is located in a genomic island that is inserted at the end of a NAD(P)H-dependent flavin oxidoreductase (locus tag Ga0063690_01635 in strain ATCC 9175^T, Ga0063697_00339 in strain 6(3) and Ga0063698_01092 in strain 8(6)), which is followed in 8(6) by a LuxR family transcriptional regulator (locus tag Ga0063698_01093, which is an ortholog of

Ga0063690_01646 in ATCC 9175^T and Ga0063697_00350 in strain 6(3)).

Phenazines are heterocyclic compounds that are substituted at different points around their rings [77]. They have various colors and biological activities, such as antibiotic and intercellular signaling activities. *Brevibacterium iodinum* ATCC 49514^T, isolated from milk, is known to produce purple extracellular crystals of the phenazine iodinin [78]. Its genome contains the *phzGFEDCB* gene cluster, which encodes one enzyme involved in chorismate synthesis (PhzG) and the five enzymes required for the generation of the "core" phenazines, phenazine-1,6- dicarboxylic acid or phenazine-1-carboxylic acid, which are precursors for strain-specific phenazine derivates (Additional file 12). This cluster is also present in four other strains, all isolated from cheeses (ATCC 9175^T, ATCC 9172^T, P10, and ATCC 9174), which are thus probably able to produce phenazine derivatives.

Discussion

Microbial cultures are widely used in the cheesemaking industry. There is a high degree of expertise today in terms of the design of lactic starter cultures. This expertise includes several important properties such as the ability to grow in milk, the resistance to bacteriophages, and the generation of adequate acidifying and texturizing activities. The situation is more problematic for ripening cultures for surface-ripened cheese, which contain strains that sometimes do not grow well in cheese. This may be explained by an insufficient fitness in the cheese surface habitat in comparison to the resident "house flora". Indeed, contrary to lactic starter cultures where growth is very fast (several hours) and facilitated by a massive inoculation in milk that contains only a low concentration of microbial cells, growth at the surface of cheeses takes a longer time (several weeks) and therefore offers more opportunities for the development of well-adapted adventitious strains.

Cheese-associated strains of *Brevibacterium* belong to different species, showing that adaptation to the cheese habitat occurred independently in different lineages. The strains sequenced in the present study belong to the species *B. aurantiacum, B. antiquum, B. linens, B. casei*, as well as to two not-yet described species (*B. sp.* Mu109 and 239c). In some cases, the genomes of cheese-associated strains are closely related to other strains, suggesting a recent adaptation. For example, the ANI value between strains S18 (human skin) and CIP 102111^T (cheese) is higher than the values between the different cheese isolates belonging to the species *B. aurantiacum* (98.22% vs. 97.48–98.04%). Interestingly, a complete prophage was found in *B. sp.* Mu109, which indicates that cheese-associated *Brevibacterium* strains may undergo phage attacks, even if, to our knowledge, there is no study in the scientific literature concerning the sensitivity to phages of ripening cultures containing *Brevibacterium* strains. It is possible that the impact of phages on these cultures has been overlooked.

In this study, genomic analyses revealed that *Brevibacterium* strains show differences in the ability to use energy compounds present in cheeses. Indeed, the galactose catabolism pathway is predicted to occur only in the phylogenetic groups *aurantiacum/sandarakinum/antiquum* and *linens/siliguriense/iodinum*, and the D-galactonate catabolism pathway only in the phylogenetic groups *aurantiacum/sandarakinum/antiquum*, *linens/siliguriense/iodinum* and *casei*. The strains from the *aurantiacum/sandarakinum/antiquum* and *linens/siliguriense/iodinum* groups have more lactate dehydrogenases than the strains from the other groups, suggesting that they have a better ability to consume lactate, which may be useful during growth in cheese. A predicted extracellular triacylglycerol lipase was found in 14 of the 15 genomes from the *aurantiacum/sandarakinum/antiquum* and *linens/siliguriense/iodinum* groups and in the cheese-associated strain Mu109, suggesting that many *Brevibacterium* strains are likely to contribute to cheese lipolysis. The resulting fatty acids constitute a potential energy source for the lipolitic *Brevibacterium* strains as well as for other microbial populations living on the cheese surface. A cluster of two genes encoding two secreted proteases with LPXTG motif is present in 16 of the 23 investigated *Brevibacterium* genomes. In *B. aurantiacum* ATCC 9174, one of these two proteases has been purified and characterized [53]. This protease is active on casein and secreted during the growth of the strain, which indicates that it probably contributes to cheese proteolysis [79, 80]. The amino acids produced from caseins and the non-protein amino acid 4-aminobutyrate are major energy substrates for the bacteria growing at the cheese surface. The 17 strains belonging to the groups *aurantiacum/sandarakinum/antiquum*, *linens/siliguriense/iodinum* and *casei* are well equipped in amino acid catabolism pathways and have a similar gene content. The six other *Brevibacterium* strains have a lower catabolic potential since they do not encode the enzymes involved in arginine, phenylalanine, methionine and tyrosine degradation.

Microorganisms growing at the cheese surface have to be able to withstand the osmotic stress due to the presence of salt. Most *Brevibacterium* genomes encode the pathways for the production of the osmoprotectants ectoine or hydroxyectoine from aspartate, glycine-betaine from choline, and trehalose. In addition, in comparison to the other *Actinobacteria* genomes, they also encode a large number of Betaine/Carnithine/Choline family

Transporters (BCCT), of multisubunit $(Na^+)(K^+)$/proton antiporters, and of Sodium Solute Symporters (SSS). The two last systems take advantage of Na^+ gradients to regulate the intracellular pH and to import nutrients. These properties probably contribute to the good resistance of *Brevibacterium* strains to the cheese salt and to their stimulation in the presence of 4% NaCl [81]. No major differences in the gene content concerning osmotic stress resistance was observed between the cheese-associated or the other strains, and between the investigated phylogenetic groups, except for *B. ravenspurgense* 5401308[T] and *B. mcbrellneri* ATCC 49030[T], which have a smaller number of BCCT and SSS systems. These two strains have the smallest genome sizes of the 23 investigated strains and have probably undergone a reductive genome evolution.

The present study confirms the central role of iron metabolism in bacteria from cheese surface microbial communities. *Brevibacterium* strains produce and utilize siderophores [82], and the growth of typical aerobic bacteria at the cheese surface is limited by the availability of iron [83]. The low iron content of milk, the presence of sequestering compounds such as lactoferrin, the presence of oxygen and the high pH of the cheese matrix during the growth of the acido-sensitive bacteria contribute to restricting the availability of iron. One strategy for improving iron acquisition in the cheese habitat is to produce siderophores, and this capacity can be acquired or improved by HGT. Hydroxamate-type siderophore biosynthesis seems to be widespread in *Brevibacterium* strains since gene clusters involved in their production were found in most genomes of the *aurantiacum/sandarakinum/antiquum*, *linens/siliguriense/iodinum* and *casei* groups. Clusters encoding catecholate-type siderophores were present only in two strains and resulted from horizontal transfers from other genera. Interestingly, both strains were isolated from cheese, and one of them (ATCC 9174) also possesses a gene cluster involved in the production of an acetohydroxamate-type siderophore. At least one other cheese-associated strain, *Glutamicibacter arilaitensis* Re117, has two siderophore biosynthesis clusters, including a catecholate siderophore resulting from HGT [16]. It is likely that siderophore production by cheese microorganisms also results in biotic interactions. In a recent study, it was observed that a *Staphylococcus equorum* strain, which was a weak competitor against other closely related *Staphylococcus* species in model cheese experiments, became dominant in the presence of the fungus *Scopulariopsis* [84]. This effect was attributed to fungal siderophore production, which may relieve *S. equorum* of the costly production of the siderophore staphyloferrin B and potentially provide an iron source through cross-feeding. This type of cross-feeding may also occur in microbial communities containing *Brevibacterium* strains that do not have the potential to produce siderophores, such as *B. linens* ATCC 9172[T] and *B. linens* Mu101. The Fe^{3+}/siderophore complexes are imported into the cells by ABC transport systems, whether or not these siderophores are produced by the same strain. Numerous ABC-type Fe^{3+}/siderophore components are present in *Brevibacterium* strains. These systems allow the cells to take advantage of different types of siderophores available in the medium. Horizontal gene transfers concerning ABC-type Fe^{3+}/siderophore components were observed in the cheese-associated strains *Glutamicibacter arilaitensis* Re117 [16], in *Corynebacterium variabile* DSM 44702 [15], and a recent study provided evidence of extensive HGTs concerning Fe^{3+}/siderophore acquisition in very diverse cheese-associated bacteria [17]. One of these regions, known as ActinoRUSTI, was present in two of the 23 *Brevibacterium* strains investigated in the present study. Three other islands involved in iron acquisition that are present both in cheese-associated *Brevibacterium* strains and in cheese-associated strains belonging to other genera (*Glutamicibacter arilaitensis* Re117, *Corynebacterium casei* UCMA 3821, *Corynebacterium variabile* DSM 44702 and *Microbacterium gubbeenense* DSM 15944) were detected in the investigated genomes. Interestingly, three of these four islands were present in the cheese-associated strain *B. sp.* Mu109, which is also the strain with the highest number of ABC-type Fe^{3+}/siderophore components (31 components).

Bacteriocin production is thought to play a critical role in mediating the microbial population or community interactions [85]. It may thus be assumed that it could have a significant impact on the cheese surface where microbial density may exceed 10^{10} cells per cm^2. In the present study, we detected several putative bacteriocin gene clusters in *Brevibacterium* genomes. They corresponded to six groups of bacteriocins, i.e., linocin M18-related bacteriocins and five groups of RiPPs, and were highly variable among strains. Linocin M18, which has been characterized in the red smear cheese bacterium *B. linens* M18, is an antilisterial and wide-spectrum bacteriocin [86]. Its activity against *Listeria spp.* has also been demonstrated in a model cheese [87]. In our study, linocin M18-related bacteriocins were detected in 15 of the 23 investigated genomes, belonging to different phylogenetic groups and isolated from different habitats. This result is consistent with the fact that the structural gene *lin* encoding linocin M18 is widely distributed in coryneform bacteria [88]. To our knowledge, up until now, there has been no experimental evidence about the production of RiPPs in *Brevibacterium*. However, it cannot be excluded that some of the antibacterial substances characterized from *Brevibacterium* strains, such as the Linecin A from *B. aurantiacum* ATCC 9175[T] [89] and/or the Linenscin OC2 from *B. linens* OC2 [90], are in fact RiPPs. Almost all the RiPP gene clusters detected in

our study were found in cheese-associated strains, except for a lanthipeptide gene cluster in *B. mcbrellneri* ATCC 49030[T], which is a human-associated strain. Interestingly, many of them seem to result from HGTs. In this study, we identified the BreLI island, which is probably a ~96-kb integrative and conjugative element (ICE) encoding for class III lanthipeptides. This island is present in four cheese-associated *Brevibacterium* strains as well as in a cheese-associated strain from another genus (*Corynebacterium casei* LMG S-19264). Taken together, our results consolidate the hypothesis that bacteriocin production may provide an ecological advantage to cheese-associated bacteria. It would be interesting in further studies to examine whether these bacteriocin gene clusters are functional, to investigate the activity spectra of these bacteriocins and to determine the influence of environmental conditions on their biosynthesis. Such information would be useful for the design of surface-ripened cheese cultures in order to improve their competitiveness against adventitious strains or to prevent growth of pathogens and spoilage microorganisms. It would also be interesting to investigate the role of phenazines, which have a broad-spectrum antibiotic activity and whose biosynthesis is predicted to occur in four cheese-associated strains. These strains belong to three different species, but the presence of the corresponding gene clusters cannot be explained by recent HGT events in view of the divergence in gene sequences.

The present study confirms that there are HGT events between microorganisms growing at the surface of cheeses. Acquisition of genes involved in siderophore biosynthesis, iron import and bacteriocin production can probably improve the fitness of the strains in the cheese habitat. It can be hypothesized that these gene transfers exert an influence on the balance between the resident "house flora" and the strains from the ripening cultures. For the latter type of strains, it is, in most cases, the same strains that are massively inoculated in all the manufacturing runs. For the adventitious "house flora", it would be beneficial to acquire genes that improve their competitiveness to the detriment of the inoculated strains, such as genes governing bacteriocin production or iron acquisition. It would thus be interesting to examine the extent to which gene acquisition by the adventitious strains impacts the growth and stability of the components from ripening cultures used for the production of surface-ripened cheeses.

Conclusion

Some properties deduced from genome analyses are similar in all the investigated strains, such as the ability to catabolize ethanol or 4-aminobutyrate. This is also observed for the ability to catabolize glycerol and for osmotolerant biosynthesis, except for strains 5401308^T and ATCC 49030^T, which are two strains that have a smaller genome size. Other properties are mainly correlated to the phylogenetic position of the strains, whether they were isolated or not from cheese, such as the ability to catabolize galactose, lactate, amino acids, or to secrete triacylglycerol lipases. The ability to catabolize D-galactonate is present in part of the strains, and this property does not seem to be correlated to the phylogenetic position or to the habitat of the strains. There are great differences in the number of Fe^{3+}/siderophore ABC transport components. Some of these genes are in clusters that are also present in cheese-associated bacteria belonging to other genera, indicating that these genes are disseminated by HGTs among strains living on the cheese surface. Two *Brevibacterium* strains isolated from cheeses also acquired a catecholate-type siderophore biosynthesis gene cluster by HGT. Bacteriocin biosynthesis genes are present in most of the strains, and one of the corresponding gene clusters is located in a probable conjugative transposon of ~96 kb (BreLI), which is present in four cheese-associated *Brevibacterium* strains as well as in *Corynebacterium casei* LMG S-19264, a strain isolated from a smear-ripened cheese. *Brevibacterium* strains thus show differences in genetic determinants involved in the growth on the cheese surface. Some of them are correlated to the phylogenetic position and others are the result of gene transfers. Part of these properties contributes to biotic interactions between strains. In the future, it would be interesting to take this information into account in order to improve the screening and selection of *Brevibacterium* strains and their association with other ripening culture components.

Additional files

Additional file 1: Genome statistics. General features of the Brevibacterium genomes. (XLSX 15 kb)

Additional file 2: 16S–ANI. (XLSX 15 kb)

Additional file 3: Orthology. (XLSX 6663 kb)

Additional file 4: Orthology (Fig). Number of orthologous gene clusters in the 23 Brevibacterium genomes. (PDF 17 kb)

Additional file 5: Galactose-galactonate-lactate-acetate-ethanol-citrate. (XLSX 36 kb)

Additional file 6: Lipid-glycerol. (XLSX 27 kb)

Additional file 7: Excreted proteases-aminoacids. (XLSX 43 kb)

Additional file 8: Iron. (XLSX 53 kb)

Additional file 9: Iron (Fig). (PDF 98 kb)

Additional file 10: Osmotolerance. (XLSX 24 kb)

Additional file 11: Bacteriocines. (XLSX 140 kb)

Additional file 12: Phenazines. (XLSX 17 kb)

Abbreviations

ANI: Average Nucleotide Identity; BCCT: Betaine/Carnithine/Choline family Transporter; BreLI: *Brevibacterium* Lanthipeptide Island; CDS: Coding DNA sequence; CRISPR: Clustered Regularly Interspaced Short Palindromic Repeats; DDH: DNA-DNA Hybridization; HGT: Horizontal Gene transfer; HMM: Hidden Markov Model; ICE: Integrative and Conjugative Element; iLDH: NAD-independent lactate dehydrogenase; Lab: Labionin; Lan: Lanthionine; LAP: Linear Azol(in)e-containing Peptides; MeLan: MethylLanthionine; nLDH: NAD-dependant lactate dehydrogenase; RiPP: Ribosomally synthesized and Post-translationally modified Peptide; SDP: Sporulation Delaying Protein; SSS: Sodium Solute Symporter

Acknowledgments

We are grateful to the INRA MIGALE bioinformatics platform (http://migale.jouy.inra.fr) for providing computational resources, and the IMG-ER system (https://img.jgi.doe.gov/cgi-bin/mer/main.cgi) for generating annotations.

Funding

N-P-P. is the recipient of a doctoral fellowship from the AgroParisTech Foundation and the INRA MEM metaprogram. This work was supported by France Génomique National infrastructure, funded as part of "Investissement d'avenir" program managed by Agence Nationale pour la Recherche (contrat ANR-10-INBS-09).

Authors' contributions

SL and FI initiated the project. SL and MV supervised sequencing of the 13 *Brevibacterium* genomes and SL performed their assembly. CM supervised the genomic analyses. All the authors contributed to the analyses regarding the general genomic properties or the metabolism of energy compounds. Analyses dealing with iron metabolism and osmoresistance were performed by CM, those regarding proteases and lipases by NPP, and those regarding bacteriocins by NPP and CM. NPP, EDB and CM wrote the paper. All authors read and approved the final manuscript.

Competing interests

The authors declare that they have no competing interests

Author details

[1]UMR GMPA, AgroParisTech, INRA, Université Paris-Saclay, 78850 Thiverval-Grignon, France. [2]US 1426, GeT-PlaGe, Genotoul, INRA, 31326 Castanet-Tolosan, France.

References

1. Bockelmann W. Secondary cheese starter cultures. In: law BA, Tamime AY, editors. Technol. Cheesemaking second Ed. Oxford: Wiley-Blackwell; 2010. p. 193–230.
2. Irlinger F, Layec S, Hélinck S, Dugat-Bony E. Cheese rind microbial communities: diversity, composition and origin. FEMS Microbiol Lett. 2015;362:1–11.
3. Monnet C, Landaud S, Bonnarme P, Swennen D. Growth and adaptation of microorganisms on the cheese surface. FEMS Microbiol Lett. 2015;362:1–9.
4. Brennan NM, Ward AC, Beresford TP, Fox PF, Goodfellow M, Cogan TM. Biodiversity of the bacterial flora on the surface of a smear cheese. Appl Environ Microbiol. 2002;68:820–30.
5. Feurer C, Vallaeys T, Corrieu G, Irlinger F. Does smearing inoculum reflect the bacterial composition of the smear at the end of the ripening of a French soft, red-smear cheese? J Dairy Sci. 2004;87:3189–97.

6. Goerges S, Mounier J, Rea MC, Gelsomino R, Heise V, Beduhn R, et al. Commercial ripening starter microorganisms inoculated into cheese milk do not successfully establish themselves in the resident microbial ripening consortia of a south german red smear cheese. Appl Environ Microbiol. 2008;74:2210–7.

7. Gori K, Ryssel M, Arneborg N, Jespersen L. Isolation and identification of the microbiota of Danish farmhouse and industrially produced surface-ripened cheeses. Microb Ecol. 2013;65:602–15.

8. Mounier J, Gelsomino R, Goerges S, Vancanneyt M, Vandemeulebroecke K, Hoste B, et al. Surface microflora of four smear-ripened cheeses. Appl Environ Microbiol. 2005;71:6489–500.

9. Rea MC, Görges S, Gelsomino R, Brennan NM, Mounier J, Vancanneyt M, et al. Stability of the biodiversity of the surface consortia of Gubbeen, a red-smear cheese. J Dairy Sci. 2007;90:2200–10.

10. Rattray FP, Fox PF. Aspects of enzymology and biochemical properties of Brevibacterium linens relevant to cheese ripening: a review. J Dairy Sci. 1999;82:891–909.

11. Forquin-Gomez M-P, Weimer BC, Sorieul L, Kalinowski J, Vallaeys T. The family Brevibacteriaceae. In: Rosenberg E, DeLong EF, Lory S, Stackebrandt E, Thompson F, editors. Prokaryotes Actinobacteria. 4th ed. Berlin, Heidelberg: Springer; 2014. p. 141–53.

12. Onraedt A, Soetaert W, Vandamme E. Industrial importance of the genus Brevibacterium. Biotechnol Lett. 2005;27:527–33.

13. Gavrish EI, Krauzova VI, Potekhina NV, Karasev SG, Plotnikova EG, Altyntseva OV, et al. Three new species of brevibacteria, Brevibacterium antiquum sp. nov., Brevibacterium aurantiacum sp. nov. and Brevibacterium permense sp. nov. Mikrobiologiia. 2004;73:218–25.

14. Montel M-C, Buchin S, Mallet A, Delbes-Paus C, Vuitton DA, Desmasures N, et al. Traditional cheeses: rich and diverse microbiota with associated benefits. Int J Food Microbiol. 2014;177:136–54.

15. Schröder J, Maus I, Trost E, Tauch A. Complete genome sequence of Corynebacterium Variabile DSM 44702 isolated from the surface of smear-ripened cheeses and insights into cheese ripening and flavor generation. BMC Genomics. 2011;12:545.

16. Monnet C, Loux V, Gibrat J-F, Spinnler E, Barbe V, Vacherie B, et al. The arthrobacter arilaitensis Re117 genome sequence reveals its genetic adaptation to the surface of cheese. PLoS One. 2010;5:e15489.

17. Bonham KS, Wolfe BE, Dutton RJ. Extensive horizontal gene transfer in cheese-associated bacteria. elife. 2017;6

18. Markowitz VM, I-MA C, Chu K, Pati A, Ivanova NN, Kyrpides NC. Ten years of maintaining and expanding a microbial genome and metagenome analysis system. Trends Microbiol. 2015;23:730–41.

19. Magoč T, Salzberg SL. FLASH: fast length adjustment of short reads to improve genome assemblies. Bioinforma Oxf Engl. 2011;27:2957–63.

20. Bankevich A, Nurk S, Antipov D, Gurevich AA, Dvorkin M, Kulikov AS, et al. SPAdes: a new genome assembly algorithm and its applications to single-cell sequencing. J. Comput. Biol. J Comput Mol Cell Biol. 2012;19:455–77.

21. Markowitz VM, I-MA C, Palaniappan K, Chu K, Szeto E, Grechkin Y, et al. IMG: the integrated microbial genomes database and comparative analysis system. Nucleic Acids Res. 2012;40:D115–22.

22. Huntemann M, Ivanova NN, Mavromatis K, Tripp HJ, Paez-Espino D, Palaniappan K, et al. The standard operating procedure of the DOE-JGI microbial genome annotation pipeline (MGAP v.4). Stand Genomic Sci. 2015;10:86.

23. Arndt D, Grant JR, Marcu A, Sajed T, Pon A, Liang Y, et al. PHASTER: a better, faster version of the PHAST phage search tool. Nucleic Acids Res. 2016;44:W16–21.

24. Weber T, Blin K, Duddela S, Krug D, Kim HU, Bruccoleri R, et al. antiSMASH 3. 0-a comprehensive resource for the genome mining of biosynthetic gene clusters. Nucleic Acids Res. 2015;43:W237–43.

25. van Heel AJ, de Jong A, Montalbán-López M, Kok J, Kuipers OP. BAGEL3: Automated identification of genes encoding bacteriocins and (non-)bactericidal posttranslationally modified peptides. Nucleic Acids Res. 2013; 41:W448–53.

26. Hammami R, Zouhir A, Le Lay C, Ben Hamida J, Fliss I. BACTIBASE second release: a database and tool platform for bacteriocin characterization. BMC Microbiol. 2010;10:22.

27. Mende DR, Sunagawa S, Zeller G, Bork P. Accurate and universal delineation of prokaryotic species. Nat Methods. 2013;10:881–4.

28. Walter F, Albersmeier A, Kalinowski J, Rückert C. Complete genome sequence of Corynebacterium Casei LMG S-19264T (=DSM 44701T), isolated from a smear-ripened cheese. J Biotechnol. 2014;189:76–7.

29. Almeida M, Hébert A, Abraham A-L, Rasmussen S, Monnet C, Pons N, et al. Construction of a dairy microbial genome catalog opens new perspectives for the metagenomic analysis of dairy fermented products. BMC Genomics. 2014;15:1101.

30. Altschul SF, Gish W, Miller W, Myers EW, Lipman DJ. Basic local alignment search tool. J Mol Biol. 1990;215:403–10.

31. Notredame C, Higgins DG, Heringa J. T-coffee: a novel method for fast and accurate multiple sequence alignment. J Mol Biol. 2000;302:205–17.

32. Edgar RC. MUSCLE: Multiple sequence alignment with high accuracy and high throughput. Nucleic Acids Res. 2004;32:1792–7.

33. Price MN, Dehal PS, Arkin AP. FastTree 2–approximately maximum-likelihood trees for large alignments. PLoS One. 2010;5:e9490.

34. Kumar S, Stecher G, Tamura K. MEGA7: Molecular Evolutionary Genetics Analysis Version 7.0 for Bigger Datasets. Mol Biol Evol. 2016;33:1870–4.

35. Contreras-Moreira B, Vinuesa P. GET_HOMOLOGUES, a versatile software package for scalable and robust microbial pangenome analysis. Appl Environ Microbiol. 2013;79:7696–701.

36. Li L, Stoeckert CJ, Roos DS. OrthoMCL: identification of ortholog groups for eukaryotic genomes. Genome Res. 2003;13:2178–89.

37. Tatusov RL, Galperin MY, Natale DA, Koonin EV. The COG database: a tool for genome-scale analysis of protein functions and evolution. Nucleic Acids Res. 2000;28:33–6.

38. Stackebrandt E, Frederiksen W, Garrity GM, Grimont PAD, Kämpfer P, Maiden MCJ, et al. Report of the ad hoc committee for the re-evaluation of the species definition in bacteriology. Int J Syst Evol Microbiol. 2002;52:1043–7.

39. Konstantinidis KT, Tiedje JM. Genomic insights that advance the species definition for prokaryotes. Proc Natl Acad Sci U S A. 2005;102:2567–72.

40. Goris J, Konstantinidis KT, Klappenbach JA, Coenye T, Vandamme P, Tiedje JM. DNA-DNA hybridization values and their relationship to whole-genome sequence similarities. Int J Syst Evol Microbiol. 2007;57:81–91.

41. Richter M, Rosselló-Móra R. Shifting the genomic gold standard for the prokaryotic species definition. Proc Natl Acad Sci U S A. 2009;106:19126–31.

42. Stackebrandt E, Goebel BM. Taxonomic note: a place for DNA-DNA Reassociation and 16S rRNA sequence analysis in the present species definition in bacteriology. Int J Syst Evol Microbiol. 1994;44:846–9.

43. Michel V, Martley FG. Streptococcus Thermophilus in cheddar cheese–production and fate of galactose. J Dairy Res. 2001;68:317–25.

44. Mounier J, Rea MC, O'Connor PM, Fitzgerald GF, Cogan TM. Growth characteristics of Brevibacterium, Corynebacterium, microbacterium, and staphylococcus spp. isolated from surface-ripened cheese. Appl Environ Microbiol. 2007;73:7732–9.

45. Jiang T, Gao C, Ma C, Microbial XP. Lactate utilization: enzymes, pathogenesis, and regulation. Trends Microbiol. 2014;22:589–99.

46. Pinchuk GE, Rodionov DA, Yang C, Li X, Osterman AL, Dervyn E, et al. Genomic reconstruction of Shewanella oneidensis MR-1 metabolism reveals a previously uncharacterized machinery for lactate utilization. Proc Natl Acad Sci U S A. 2009;106:2874–9.

47. Jolkver E, Emer D, Ballan S, Krämer R, Eikmanns BJ, Marin K. Identification and characterization of a bacterial transport system for the uptake of pyruvate, propionate, and acetate in Corynebacterium glutamicum. J Bacteriol. 2009;191:940–8.

48. Auchter M, Arndt A, Eikmanns BJ. Dual transcriptional control of the acetaldehyde dehydrogenase gene ald of Corynebacterium glutamicum by RamA and RamB. J Biotechnol. 2009;140:84–91.

49. Hajri T, Abumrad NA. Fatty acid transport across membranes: relevance to nutrition and metabolic pathology. Annu Rev Nutr. 2002;22:383–415.

50. Muñoz-Elías EJ, Upton AM, Cherian J, McKinney JD. Role of the methylcitrate cycle in mycobacterium tuberculosis metabolism, intracellular growth, and virulence. Mol Microbiol. 2006;60:1109–22.

51. Bott M, Niebisch A. The respiratory chain of Corynebacterium glutamicum. J Biotechnol. 2003;104:129–53.

52. Rawlings ND, Barrett AJ, Finn R. Twenty years of the MEROPS database of proteolytic enzymes, their substrates and inhibitors. Nucleic Acids Res. 2016; 44:D343–50.

53. Rattray FP, Bockelmann W, Fox PF. Purification and characterization of an extracellular proteinase from Brevibacterium linens ATCC 9174. Appl Environ Microbiol. 1995;61:3454–6.

54. Nomura M, Kimoto H, Someya Y, Furukawa S, Suzuki I. Production of gamma-aminobutyric acid by cheese starters during cheese ripening. J Dairy Sci. 1998;81:1486–91.

55. Miethke M, Monteferrante CG, Marahiel MA, van Dijl JM. The Bacillus Subtilis EfeUOB transporter is essential for high-affinity acquisition of ferrous and ferric iron. Biochim Biophys Acta. 2013;1833:2267–78.

56. Khan AG, Shouldice SR, Kirby SD, Yu R, Tari LW, Schryvers AB. High-affinity binding by the periplasmic iron-binding protein from Haemophilus influenzae is required for acquiring iron from transferrin. Biochem J. 2007;404:217–25.

57. Bernard T, Jebbar M, Rassouli Y, Himdi-Kabbab S, Hamelin J, Blanco C. Ectoine accumulation and osmotic regulation in Brevibacterium linens. Microbiology. 1993;139:129–36.

58. Zeisel SH, Mar M-H, Howe JC, Holden JM. Concentrations of choline-containing compounds and betaine in common foods. J Nutr. 2003;133:1302–7.

59. Frings E, Kunte HJ, Galinski EA. Compatible solutes in representatives of the genera Brevibacterium and Corynebacterium: occurrence of tetrahydropyrimidines and glutamine. FEMS Microbiol Lett. 1993;109:25–32.

60. Kempf B, Bremer E. Uptake and synthesis of compatible solutes as microbial stress responses to high-osmolality environments. Arch Microbiol. 1998;170:319–30.

61. Swartz TH, Ikewada S, Ishikawa O, Ito M, Krulwich TA. The Mrp system: a giant among monovalent cation/proton antiporters. Extremophiles. 2005;9:345–54.

62. Arnison PG, Bibb MJ, Bierbaum G, Bowers AA, Bugni TS, Bulaj G, et al. Ribosomally synthesized and post-translationally modified peptide natural products: overview and recommendations for a universal nomenclature. Nat Prod Rep. 2013;30:108–60.

63. Meindl K, Schmiederer T, Schneider K, Reicke A, Butz D, Keller S, et al. Labyrinthopeptins: a new class of carbacyclic lantibiotics. Angew Chem Int Ed Engl. 2010;49:1151–4.

64. Bierbaum G, Sahl H-G. Lantibiotics: mode of action, biosynthesis and bioengineering. Curr Pharm Biotechnol. 2009;10:2–18.

65. Zhang Q, Yu Y, Vélasquez JE, van der Donk WA. Evolution of lanthipeptide synthetases. Proc Natl Acad Sci U S A. 2012;109:18361–6.

66. Müller WM, Ensle P, Krawczyk B, Süssmuth RD. Leader peptide-directed processing of labyrinthopeptin A2 precursor peptide by the modifying enzyme LabKC. Biochemistry (Mosc). 2011;50:8362–73.

67. Völler GH, Krawczyk B, Ensle P, Süssmuth RD. Involvement and unusual substrate specificity of a prolyl oligopeptidase in class III lanthipeptide maturation. J Am Chem Soc. 2013;135:7426–9.

68. Bi D, Xu Z, Harrison EM, Tai C, Wei Y, He X, et al. ICEberg: a web-based resource for integrative and conjugative elements found in bacteria. Nucleic Acids Res. 2012;40:D621–6.

69. Martínez B, Fernández M, Suárez JE, Rodríguez A. Synthesis of lactococcin 972, a bacteriocin produced by Lactococcus lactis IPLA 972, depends on the expression of a plasmid-encoded bicistronic operon. Microbiol Read Engl. 1999;145(Pt 11):3155–61.

70. Sánchez C, Hernández de Rojas A, Martínez B, Argüelles ME, Suárez JE, Rodríguez A, et al. Nucleotide sequence and analysis of pBL1, a bacteriocin-producing plasmid from Lactococcus lactis IPLA 972. Plasmid. 2000;44:239–49.

71. Haft DH, Basu MK, Mitchell DA. Expansion of ribosomally produced natural products: a nitrile hydratase- and Nif11-related precursor family. BMC Biol. 2010;8:70.

72. Lee SW, Mitchell DA, Markley AL, Hensler ME, Gonzalez D, Wohlrab A, et al. Discovery of a widely distributed toxin biosynthetic gene cluster. Proc Natl Acad Sci U S A. 2008;105:5879–84.

73. Claesen J, Bibb M. Genome mining and genetic analysis of cypemycin biosynthesis reveal an unusual class of posttranslationally modified peptides. Proc Natl Acad Sci U S A. 2010;107:16297–302.

74. Claesen J, Bibb MJ. Biosynthesis and regulation of grisemycin, a new member of the linaridin family of ribosomally synthesized peptides produced by Streptomyces Griseus IFO 13350. J Bacteriol. 2011;193:2510–6.

75. Liu W-T, Yang Y-L, Xu Y, Lamsa A, Haste NM, Yang JY, et al. Imaging mass spectrometry of intraspecies metabolic exchange revealed the cannibalistic factors of Bacillus Subtilis. Proc Natl Acad Sci U S A. 2010;107:16286–90.

76. Pérez Morales TG, Ho TD, Liu W-T, Dorrestein PC, Ellermeier CD. Production of the cannibalism toxin SDP is a multistep process that requires SdpA and SdpB. J Bacteriol. 2013;195:3244–51.

77. Price-Whelan A, Dietrich LEP, Newman DK. Rethinking "secondary" metabolism: physiological roles for phenazine antibiotics. Nat Chem Biol. 2006;2:71–8.

78. Whitman W, Goodfellow M, Kämpfer P, Busse H-J, Trujillo M, Ludwig W, et al. Bergey's manual of systematic bacteriology (second edition) volume 5: the Actinobacteria. New York: Springer Science & Business Media; 2012.

79. Rattray FP, Fox PF, Healy A. Specificity of an extracellular proteinase from Brevibacterium linens ATCC 9174 on bovine alpha s1-casein. Appl Environ Microbiol. 1996;62:501–6.

80. Rattray FP, Fox PF, Healy A. Specificity of an extracellular proteinase from Brevibacterium linens ATCC 9174 on bovine beta-casein. Appl Environ Microbiol. 1997;63:2468–71.

81. Masoud W, Jakobsen M. The combined effects of pH, NaCl and temperature on growth of cheese ripening cultures of Debaryomyces Hansenii and coryneform bacteria. Int Dairy J. 2005;15:69–77.

82. Noordman WH, Reissbrodt R, Bongers RS, Rademaker JLW, Bockelmann W, Smit G. Growth stimulation of Brevibacterium sp. by siderophores. J Appl Microbiol. 2006;101:637–46.

83. Monnet C, Back A, Irlinger F. Growth of aerobic ripening bacteria at the cheese surface is limited by the availability of iron. Appl Environ Microbiol. 2012;78:3185–92.

84. Kastman EK, Kamelamela N, Norville JW, Cosetta CM, Dutton RJ, Wolfe BE. Biotic interactions shape the ecological distributions of staphylococcus species. MBio. 2016;7:e01157–16.

85. Riley MA, Wertz JE. Bacteriocins: evolution, ecology, and application. Annu Rev Microbiol. 2002;56:117–37.

86. Valdés-Stauber N, Scherer S. Isolation and characterization of Linocin M18, a bacteriocin produced by Brevibacterium linens. Appl Environ Microbiol. 1994;60:3809–14.

87. Eppert I, Valdés-Stauber N, Götz H, Busse M, Scherer S. Growth reduction of listeria spp. caused by undefined industrial red smear cheese cultures and bacteriocin-producing Brevibacterium linens as evaluated in situ on soft cheese. Appl Environ Microbiol. 1997;63:4812–7.

88. Valdes-Stauber N, Scherer S. Nucleotide sequence and taxonomical distribution of the bacteriocin gene lin cloned from Brevibacterium linens M18. Appl Environ Microbiol. 1996;62:1283–6.

89. Kato F, Eguchi Y, Nakano M, Oshima T, Murata A. Purification and characterization of Linecin-a, a Bacteriocin of Brevibacterium linens. Agric Biol Chem. 1991;55:161–6.

90. Maisnier-Patin S, Richard J. Activity and purification of linenscin OC2, an antibacterial substance produced by Brevibacterium linens OC2, an orange cheese coryneform bacterium. Appl Environ Microbiol. 1995;61:1847–52.

91. Kämpfer P, Schäfer J, Lodders N, Busse H-J. Brevibacterium sandarakinum sp. nov., isolated from a wall of an indoor environment. Int J Syst Evol Microbiol. 2010;60:909–13.

92. Maizel D, Utturkar SM, Brown SD, Ferrero MA, Rosen BP. Draft genome sequence of Brevibacterium linens AE038-8, an extremely arsenic-resistant bacterium. Genome Announc. 2015;3:e00316–5.

93. Roux V, Robert C, Gimenez G, Raoult D. Draft genome sequence of Brevibacterium massiliense strain 541308T. J Bacteriol. 2012;194:5151–2.

94. Kokcha S, Ramasamy D, Lagier J-C, Robert C, Raoult D, Fournier P-E. Non-contiguous finished genome sequence and description of Brevibacterium senegalense sp. nov. Stand Genomic Sci. 2012;7:233–45.

95. Ganesan B, Seefeldt K, Weimer BC. Fatty acid production from amino acids and alpha-keto acids by Brevibacterium linens BL2. Appl Environ Microbiol. 2004;70:6385–93.

8

RNA-seq-based genome annotation and identification of long-noncoding RNAs in the grapevine cultivar 'Riesling'

Zachary N. Harris[1,3]*, Laszlo G. Kovacs[1] and Jason P. Londo[2]*

Abstract

Background: The technological advances of RNA-seq and de novo transcriptome assembly have enabled genome annotation and transcriptome profiling in highly heterozygous species such as grapevine (*Vitis vinifera* L.). This work is an attempt to utilize a de novo-assembled transcriptome of the *V. vinifera* cultivar 'Riesling' to improve annotation of the grapevine reference genome sequence.

Results: Here we show that the transcriptome assembly of a single *V. vinifera* cultivar is insufficient for a complete genome annotation of the grapevine reference genome constructed from *V. vinifera* PN40024. Further, we provide evidence that the gene models we identified cannot be completely anchored to the previously published *V. vinifera* PN40024 gene models. In addition to these findings, we present a computational pipeline for the de novo identification of lncRNAs. Our results demonstrate that, in grapevine, lncRNAs are significantly different from protein coding transcripts in such metrics as length, GC-content, minimum free energy, and length-corrected minimum free energy.

Conclusions: In grapevine, high-level heterozygosity necessitates that transcriptome characterization be based on cultivar-specific reference genome sequences. Our results strengthen the hypothesis that lncRNAs have thermodynamically different properties than protein-coding RNAs. The analyses of both coding and non-coding RNAs will be instrumental in uncovering inter-cultivar variation in wild and cultivated grapevine species.

Keywords: Transcriptome, Genome re-annotation, RNA-seq, lncRNA, Minimum free energy, Riesling, *Vitis vinifera*

Background

RNA sequencing (RNA-seq) has emerged as a powerful technology for an in-depth, genome-wide view of the transcriptome. In grapevine, *Vitis vinifera* L., RNA-seq-based transcriptome profiling has been used to interrogate the molecular underpinning of such diverse biological phenomena as photosynthesis [1], berry development [2–4], tissue maturation [5], plant-pathogen interactions [6–8] and environmental effects [9]. A thorough understanding of grape biology is important for the development of new cultivars and the fine-tuning of cultural practices to meet the challenges of the changing climate. Furthermore, grapevine is becoming the model species for woody perennial plants, therefore grape genomic information can be leveraged well beyond its application in viticulture.

To transform RNA-seq data into biologically meaningful information, raw sequence reads are assembled into transcripts. These transcripts are, in turn, anchored to transcripts of the same or a different organism. This process is facilitated by a well-annotated reference genome sequence that can be used in varieties and accessions across a species. In grapevine, the comprehensive annotation of the reference genome sequence has proven difficult. Recent reports have demonstrated that a large portion of gene-models in the reference genome sequence cannot be anchored to newly assembled genomic or transcriptomic data. For example, the recently sequenced 'Cabernet Sauvignon' genome identified only 57% of the gene models

* Correspondence: zachary.n.harris@slu.edu; Jason.Londo@ARS.USDA.GOV
[1]Missouri State University, Biology Department, 901 S. National Ave, Springfield, MO, USA
[2]United States Department of Agriculture, Agricultural Research Service, Grape Genetics Research Unit, 630 W. North Street, Geneva, NY, USA
Full list of author information is available at the end of the article

in the 12Xv1 'Pinot Noir'-derived PN40024 reference genome annotation [10], and transcriptome analyses in 'Corvina' identified only 51% of the 12Xv1 gene models [2]. A likely reason for this is the high-level heterozygosity of grapevine which encompasses a broad genetic diversity even among its most commonly grown varieties [11]. Thus, most RNA-seq reads from a given grape variety are allelic variants of the reference genome sequence, which represents a single haplotype of a single 'Pinot Noir'-derived genotype. Similarly, most RNA-seq reads are allelic variants of annotated transcripts when the annotation is based on the transcriptome of a different variety. In instances where the extent of allelic divergence is substantial, there is an increased probability that RNA-seq reads are assembled into false chimeric transcripts or an assembly can be erroneously identified as a novel paralog in the genome. Alternatively, allelic variations may be interpreted as sequencing errors, particularly in low-level expressed transcripts. The increased probability of such artifacts can lead to the construction of an incorrect transcriptome.

RNA-seq is not limited by previously identified genetic information, but has the capacity to detect all transcribed sequences, including non-protein-coding transcripts. This lead to the discovery, in both plants and animals, of an entire new class of long non-coding RNA species (lncRNAs) in addition to the known ribosomal, transfer, short nuclear and short cytoplasmic RNA species. The new class of lncRNAs are operationally defined as 200-nt or longer transcripts that can be capped, spliced, and poly-adenylated, but that do not typically contain an open reading frame. Plants express thousands of lncRNAs, but only a handful of them have been experimentally validated [12–16]. The few validated lncRNAs were found to play a role in the regulation of such processes as vernalization, photomorphogenesis, phosphate homeostasis, and auxin-mediated gene expression regulation in *Arabidopsis thaliana* (reviewed by [17, 18]). Even fewer lncRNAs have been associated with a regulatory role in other plant species [18], and nothing is known about the function of lncRNAs in grapevine.

Here, we present a reannotation of the PN40024 reference genome sequence based on the transcriptome of a single *Vitis vinifera* cultivar. In an attempt to mitigate the problems associated with high-level heterozygosity, and to increase the probability of identifying novel transcripts, we constructed this annotation by taking a de novo transcriptome assembly approach. To catalog as many of the grape transcripts as possible, we used 'Riesling' RNA-seq libraries that collectively represented a broad range of grapevine tissues. Moreover, we present a pipeline for the de novo identification of long non-coding RNA entirely independent of a reference genome. This pipeline is then applied to *V. vinifera* cv. 'Riesling' for the first characterization of lncRNAs in the cultivar.

Results

Genome annotation

The *Vitis vinifera* cv. 'Riesling' transcriptome was assembled from RNA-seq reads derived from two accessions of this cultivar, 588,673 and Ventosa. Quality filtering and trimming resulted in 14,190,809 paired end reads for 588,673. Following quality control, reads were re-paired using the program pairfq_lite [19]. Pairfq_lite returned 6,679,255 reads with a paired read on both strands, 514,591 reads unpaired on the forward strand, and 317,708 reads unpaired on the reverse strand. Of the paired reads, 91.45% aligned to the *Vitis vinifera* PN40024 12Xv2 reference genome sequence hosted at Unité de Recherche Génomique Info (URGI) [20]. Of the unpaired reads, 79.45% aligned to the reference genome, leading to a total alignment rate of 79.74%. Both paired and unpaired reads were assembled using the de novo transcript assembler Trinity (v2.0.6) [21, 22]. Trinity assembled 62,745 contigs with an average contig length of 859 nt and a median contig length of 551 nt. The contig N50 for the assembly was 1325 nt. The 62,745 contigs assembled were represented by 49,330 clusters (putatively labeled as "genes" by Trinity). Quality control and trimming resulted in 103,677,027 reads for the Ventosa accession. Pairfq_lite [19] returned 48,639,916 reads paired on both strands, 4,393,048 reads unpaired on the forward strand, and 2,004,147 reads unpaired on the reverse strand. Of the paired and unpaired reads, 91.14% and 77.06% aligned, respectively, to the reference genome, resulting in a total alignment rate of 80.64%. Trinity identified 157,779 contigs with an average length of 840 nt and median length of 373 nt. These 157,779 contigs were clustered into 109,215 Trinity-identified clusters. The N50 for the assembly was 1434 nt. Additional statistics of the Trinity assemblies are presented in Table 1.

All transcripts from both accessions were used for a complete genome annotation of the *V. vinifera* PN40024 12Xv2 reference genome sequence, using the program Maker [23, 24]. Using these transcripts and the entire Uniprot-Swiss-Prot reference protein database, 65,342 putative gene models were identified. Gene models first identified by Maker were then assigned to proteins in the Uniprot-Swiss-Prot reference database [25] using the blastp algorithm of the BLAST v.2.2.29 software suite [26]. This operation linked 1680 gene models to proteins, 1004 of which were carried forward from the Uniprot database itself in the annotation.

Using the combined accession output from Maker, gene models were used to train the SNAP gene prediction algorithm [27]. Raw output was used to train the first pass of SNAP, and this output was used to train the second pass. This resulted in a statistical model, Riesling.hmm [see Additional file 1]. This model was then coupled to the statistical model derived from the

Table 1 Additional metrics describing the Trinity assemblies of the accessions Ventosa and 588,673

Metric	'Riesling' Accession	
	Ventosa	588,673
N10	3337	2953
N20	2599	2301
N30	2130	1912
N40	1762	1603
N50	1434	1325
E10N50	1413	1008
E20N50	1349	1119
E30N50	1290	1149
E40N50	1413	1259
E50N50	1520	1345
E60N50	1671	1480
E70N50	1815	1612
E80N50	1947	1620
E90N50	1922	1499
E100N50	1437	1328

gene predictor algorithm Augustus [28] trained with *Arabidopsis thaliana* data. These models were then reintroduced to the genome sequence, this time in the context of the raw Trinity transcripts. The combination of Maker, SNAP, and Augustus predicted the presence of 19,446 gene models that were supported by RNA-seq evidence [see Availability of Data and Materials]. The sequence of steps for the annotation of the genome is depicted in Fig. 1. These annotated gene-models had a Benchmarking Universal Single-Copy Ortholog (BUSCO) score of 59.4%, indicating a ~60% recovery of the predicted *V. vinifera* transcriptome.

Functional annotation of the gene models

Predicted protein domains were searched against the reference protein domain database Pfam31.0 [29] using the program hmmscan in the HMMER v3.1b2 software suite [30]. Hits were considered significant if they matched with an expected value (E-value) of less than or equal to 1×10^{-05}. Under this threshold, 26,287 protein domains were identified. This cohort was composed of 3721 unique domains identified across 13,942 unique gene models. In many cases, Pfam domains could be tied to Gene Ontology (GO) [31] terms, classes, and functions using a custom boilerplate SQLite database. In total, 1742 Pfam domains could be tied to 3713 unique GO terms for a total of 27,823 tied instances. In total, 16,598 (59.7%), 8645 (31.1%), and 2582 (9.3%) instances of the molecular function, the biological process, and the cellular component classes were tied, respectively [see Additional file 2]. Proteins were further functionally annotated using blastp

against the Uniprot-Swiss-Prot and Uniprot-Uniref90 [32] databases. Annotation against Uniprot-Swiss-Prot anchored 14,866 and 12,572 proteins with E-values of less than or equal to 1×10^{-05} and 1×10^{-20}, respectively. Annotation against Uniprot-Uniref90 anchored 18,535 and 17,970 proteins with an E-value of 1×10^{-05} and $1 \times$

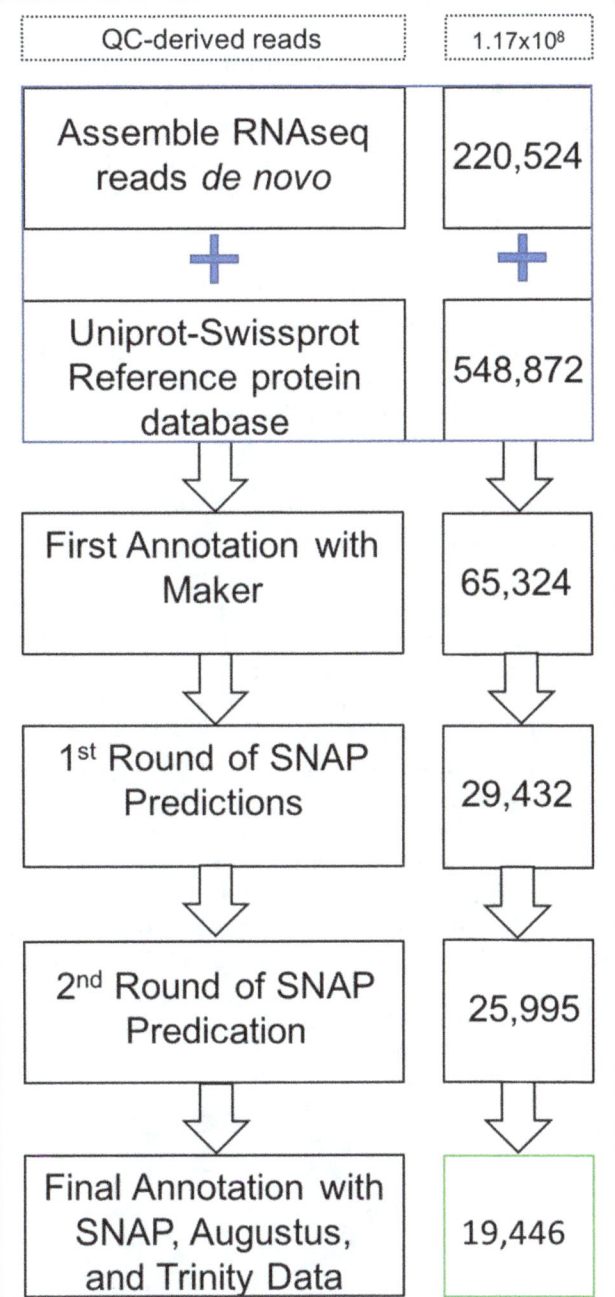

Fig. 1 Diagrammatic representation of the steps taken to annotate the 12Xv2 PN40024 reference genome. The numbers on the right indicate the number of transcripts (or proteins in the case of Uniprot-Swiss-Prot) that were fed into or derived from the step to the left. The final number, framed in green, shows the number of gene models in the final annotation

10^{-20}, respectively [see Additional file 3]. The most frequent species from which the BLAST homologies were identified are listed in Fig. 2.

Anchoring gene models to the legacy transcriptomes

In order to anchor newly annotated transcripts to the legacy *V. vinifera* PN40024 v2.1 transcriptome [33] (filtered to only use the top isoform for each transcript), we devised an iterative approach for reciprocal best hit (iRBH) analysis based on the assumption that each gene is present in the same number of copies in both 'Riesling' and PN40024. Though this working assumption may not be correct for all gene families, it is likely to be correct for the majority of genes in light of the pronounced karyotypic conservation and interfertility across *V. vinifera* accessions [34]. This analysis sorted the forward and backward blastp results in such a way that each transcript only matched each unique target one time. The following processes then occurred iteratively: (1) BLAST results were sorted such that only the highest scoring hit for each query was kept, (2) RBHs were identified, and 3) a gene model labeled as a RBH (either target or query) was removed from further analysis at every incidence of the BLAST output.

This process was executed iteratively 25 times, whereby 16,600 gene models were putatively anchored to the legacy transcriptome with a threshold E-value of 1×10^{-05}. Of the models anchored to the v2.1 transcriptome, 15,364 were identified on the same chromosome in both annotations. Over half of the models that differed in chromosomal location (677) were assigned to chrUkn (a compilation of scaffolds that cannot be assigned to any of the 19 grape chromosomes) in either the legacy or the present annotation. Those models that differed in chromosomal location, but were not assigned to chrUkn in either annotation, had overall a lower bit-score/length (v2.1 model) ratios (0.82 vs. 1.64) and marginally higher E-values (7.67×10^{-23} vs. 6.75×10^{-24}) than those that found anchors on the same chromosome [see Additional file 3].

Gene-models from the v2.1 annotation that returned no "reciprocal good hits" were labeled as unsupported models. Gene models that were annotated by either Maker, SNAP, or Augustus, but that lacked RNA-seq support, were scanned for these models using a single RBH and chromosomal location. In total, 15,245 v2.1 gene models lacked Maker anchors. Of the 9558 models not supported by RNA-seq, 8886 models found RBHs at an E-value threshold of 1×10^{-05}, and 8047 RBHs occurred on the same chromosome. Of those 1505 models that differed in chromosomal location, 559 models were assigned to chrUkn in either the v2.1 or Maker annotation.

Gene duplication

Using MCScanX [35], duplicated genes were identified using a self-BLAST-based collinear approach at various E-values. Gene duplication was initially classified using the MCScanX tool duplicate_gene_classifier, whereby 17,115 genes were considered the results of whole genome duplication. Furthermore, 1360, 226, and 715 genes were considered dispersed, proximal, and tandem duplicates, respectively, and 30 genes were considered singletons [see Additional file 3]. Because genes can represent more than one of these gene types, all evidence of duplication was explicitly searched in the gene models at various E-value thresholds. At a threshold E-value of 1×10^{-20}, evidence was found for 20,563 gene duplication pairs across 11,925 genes. MCScanX was also used to detect all tandemly arrayed genes (TAGs). Regardless of threshold E-value, 2281 tandem duplications were identified across 3480 (17.9%) unique genes. Of these, 1928 (55%) appeared in arrays of at least 3 genes, and 1552 in 2-gene arrays.

Design of a de novo pipeline for lncRNA identification

A computational pipeline was constructed to glean lncRNAs from assembled transcriptomes. To make the pipeline broadly applicable, it was designed to identify

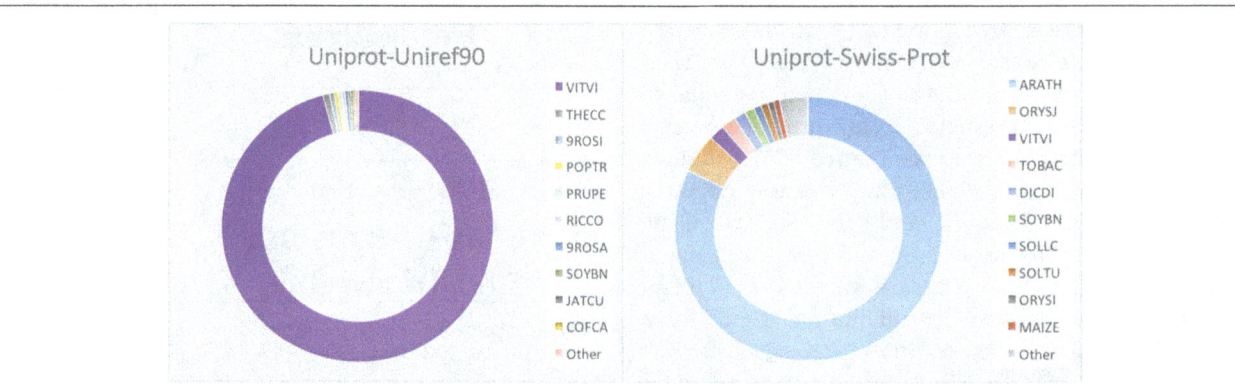

Fig. 2 Most common BLAST targets in protein comparisons of the gene-model annotation to two reference protein databases (Uniprot-Swiss-Prot and Uniprot-Uniref90) at E-value threshold of 1×10^{-20}. The species codes are Uniprot species IDs (http://www.uniprot.org/docs/speclist). Color selection is random with the excpetion of VITVI and SOYBN which are repesented by purple and green, respectively, in both figures

lncRNAs from raw RNA-seq reads in a reference genome-independent manner. The essential function of the pipeline was to remove protein-coding transcripts and short non-coding RNA sequences. First, Trinity [21, 22] was used to assemble raw RNA-seq reads into a set of transcripts, which was then purged of redundantly identified transcripts using CD-HIT [36]. Clustered transcripts were then filtered by expression level via RSEM [37], and the remaining set was further filtered to remove known protein coding genes identified by BLAST using Trinotate v2.0.2 [38]. Finally, various sets of non-protein-coding transcripts extracted from independent RNA-seq data of the same grape genotype were juxtaposed retaining only transcripts present in multiple sets to ensure a low false positive identification rate [39]. These final transcripts were then searched against the reference RNA database Rfam [40] using Infernal [41] to remove known ncRNAs and validated using the Coding Potential Calculator (CPC) [42]. Fig. 3 is a diagrammatic representation of this pipeline.

lncRNA identification pipeline

Due to the highly redundant nature of de novo assembled transcriptome builds, the Trinity output was clustered for both 'Riesling' accessions 588,673 and Ventosa using the cd-hit-est. algorithm implemented by the CD-HIT software suite. Clustering in accessions 588,673 and Ventosa resulted in 48,769 and 110,900 contigs, respectively. To further reduce the complexity of the data sets, clustered transcript sets were filtered by an expression level threshold of FPKM $> = 1.50$. Accessions 588,673 and Ventosa resulted in 46,699 and 31,103 contigs, respectively.

To remove all transcripts putatively annotated as protein-coding from the clustered and expression-level filtered transcript sets, we employed the Trinotate pipeline. In preparation, all transcripts in both data sets were translated to proteins using the tool TransDecoder [43]. RNA sequences were then searched using the blastx algorithm and translated proteins were searched using the blastp algorithm against the UniProt-Swiss-Prot and Uniprot-Uniref90 reference protein databases using an E-value threshold of 1×10^{-20}. Only the top BLAST hit for each sequence in the accessions was accepted based on bit score, E-value, and percent identity. Contigs were binned into categories of Viridiplantae proteins, non-plant proteins, and proteins for which no homologous hit was found. For 588,673, the number of transcripts identified to encode Viridiplantae proteins, non-plant proteins, and proteins with no homology were 41,686, 3340, and 13,755, respectively. For Ventosa, 34,321 transcripts were identified to encode Viridiplantae proteins, 682 to encode non-plant proteins, and 10,292 to encode no homologous protein in the database.

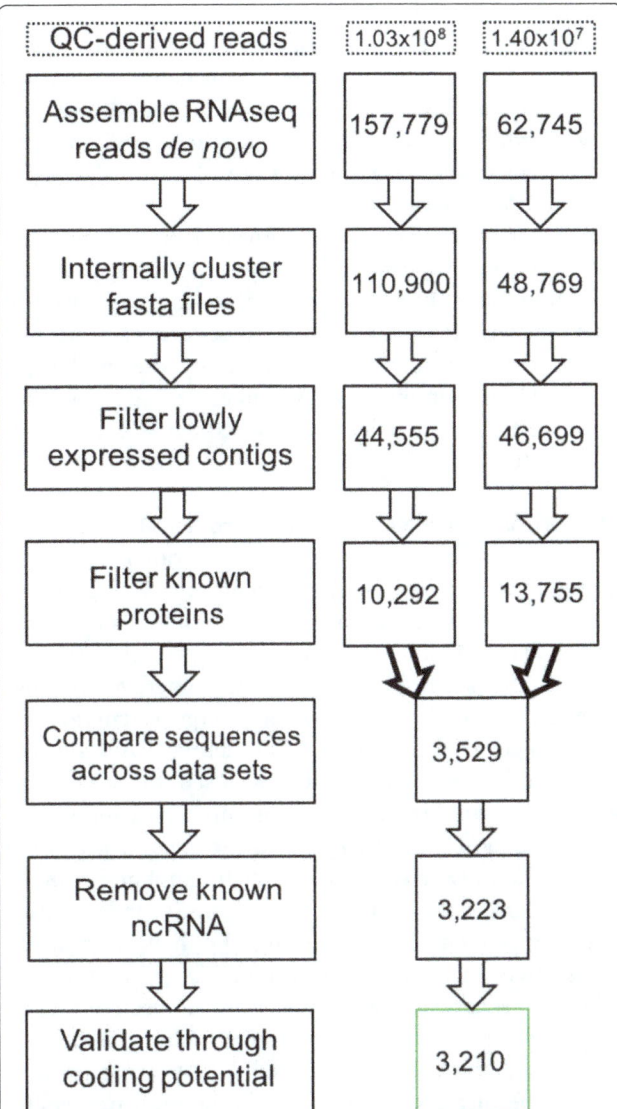

Fig. 3 Diagram of the long non-coding RNA identification pipeline. The numbers on the left of the figure indicate the number of transcripts remaining after the filtering step shown on the right. Through the 'Compare sequences across data sets' step, numbers are shown for accessions Ventosa and 588,673 respecitvely. The final number, framed in green, shows the number of psuedo-validated lncRNAs through the Coding Potential Calculator

To identify RNA sequences that occurred in both accessions, blastn was used with default parameters. Only the top BLAST hit for each contig in the Ventosa accession was accepted based on bit score, E-value, and percent identity. Only matches that had an alignment length of at least 200 nt were carried forward in the analysis. In matches of at least 200 nt, the longest transcript from either 588,673 or Ventosa was taken. This resulted in 3529 sequences.

These 3529 putatively identified non-coding RNAs then were filtered for the presence of known non-coding

RNAs housed in the RFAM [40] v12.0 database via the cmscan tool in the Infernal suite v1.1.1 [41]. Using cmscan, 196 sequences were considered significant based on the E-value threshold of 0.01 and were removed from the data set. This resulted in 3223 putatively labeled long non-coding RNAs [see Availability of Data and Materials].

In order to validate the putatively labeled long non-coding RNAs, we used the program Coding Potential Calculator (CPC) [42]. Using this tool against the UniProt-Swiss-Prot database, 3210 sequences were predicted to be non-coding, substantiating the predictions generated by our pipeline. Alignment of the predicted lncRNAs to the reference genome sequence lead to 3049 mapped transcripts.

Comparison of lncRNAs and protein coding RNAs

It has been previously observed by others that the secondary structure of lncRNAs tend to have higher free energy (less stable conformation) than protein-coding mRNAs [12, 44, 45]. To examine if grape lncRNAs identified in this study have a higher free-energy level than mRNAs, we used the RNA free energy calculator and folding algorithm RNAfold of the ViennaRNA-2.2.5 software package [46]. RNAfold was used to predict the secondary structure and the minimum free energy of all putative lncRNAs that aligned to the reference genome and a randomly selected set of 3049 annotated protein-coding transcripts identified by Maker. Secondary structures of sequences representing the highest and the lowest free energies are shown in Fig. 4. Free energy values of secondary structures were corrected for the length of the sequence. The corrected minimum free energy distribution of all analyzed RNAs are shown in Fig. 4. The mean length-corrected minimum free energy for annotated protein coding genes was –0.276 kcal/mol/nt with a standard deviation of 0.026 kcal/mol/nt.

The mean length-corrected minimum free energy content of the putatively annotated lncRNAs was –0.210 kcal/mol/nt with a standard deviation of 0.041 kcal/mol/nt. The means of these two groups were found to be significantly different using a two-tailed Welch's t-test whereby $p < 2.2 \times 10^{-16}$. To assess whether or not this difference was substantive, the effect size (Cohen's d) was derived from the t-statistic. The resulting d value of 2.08 indicated a 'large' effect size of the difference [47]. The result of a Kolmogorov-Smirnov test demonstrated that the two groups were sampled from different distributions at $p < 2.2 \times 10^{-16}$. Additionally, both length of the transcripts and the GC-content were found to be significantly different (t-test, $p < 2.2 \times 10^{-16}$) and to be sampled from different populations (KS-test, $p < 2.2 \times 10^{-16}$). Thus, our data confirm previous reports that lncRNAs fold into secondary structures of lower free energy than protein-coding mRNAs do. Genomic distribution of lncRNAs and raw free energy computations are shown in Fig. 5.

Discussion

The purpose of this work was to create a novel annotation of the 12Xv2 *V. vinifera* PN40024 reference genome sequence using expression support exclusively from *V. vinifera* cv. 'Riesling'. To mitigate the problems associated with the high-level heterozygosity of grapevine, we have taken a de novo approach to assembling the 'Riesling' transcriptome, and coupled it with sequence information from the 12Xv2 reference genome sequence to generate a novel annotation for *V. vinifera*. Furthermore, to detect as many transcripts as possible, we used 'Riesling' RNA-seq libraries that collectively represented a broad range of grapevine tissues, including root, dormant bud, leaf, tendril, rachis, flower, and unripe and post-veraison berry. We identified 19,446 gene models that had various levels of RNA-seq support. In attempted functional annotation, we found that 13,942 (71.7%) of these gene models

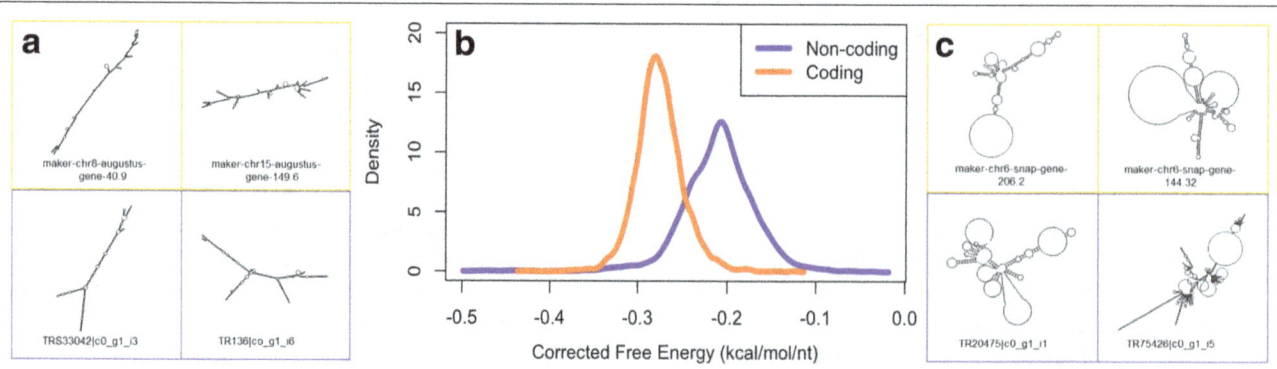

Fig. 4 Minimum free energy structures and free energy distributions for non-coding and coding transcripts. **a** Characteristically stable (low free energy) minimum free energy structures for predicted coding (orange) and non-coding (purple) transcripts. **b** Free energy distributions for long non-coding and coding transcripts ($p < 2.2 \times 10^{-16}$; d = 2.05). **c** Characteristically unstable (high free energy) structures for predicted coding and non-coding transcripts

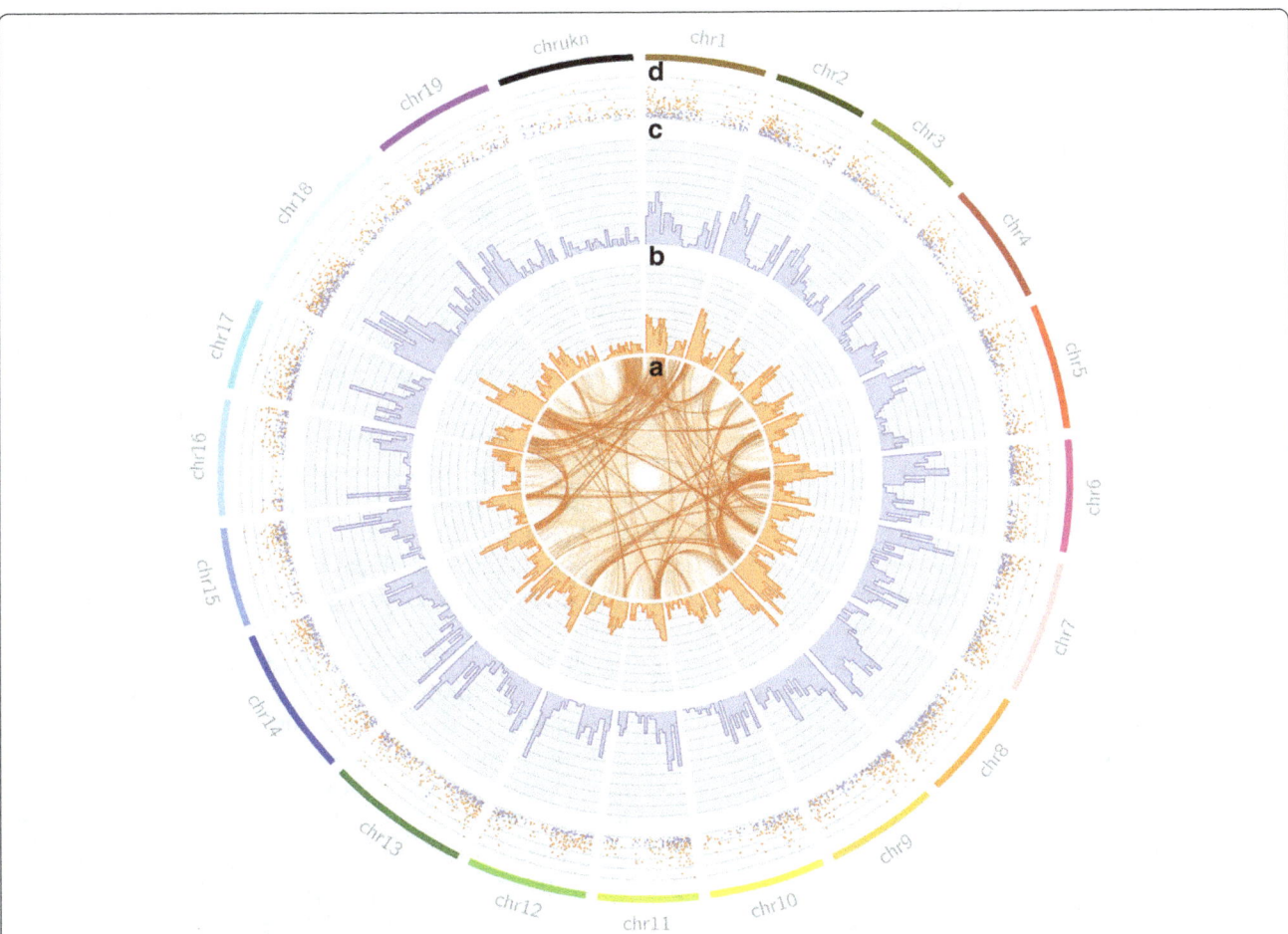

Fig. 5 Duplications, gene density, and free energy for genes anchored to genomic locations. **a** Evidence of gene duplications in the gene annotation at two E-value thresholds: 1×10^{-20} and 1×10^{-50} (light and dark orange, respectively). **b** Frequency distributions of 3049 randomly selected protein coding genes using a weighted sampling schema binned into 1 Mbp bins. **c** Long non-coding RNA frequency of all lncRNAs that aligned to the reference genome binned into 1Mbp bins. **d** Raw (uncorrected) free energy values for 3049 lncRNAs that aligned to the reference genome (purple) and 3049 randomly selected protein coding genes (orange) ($p < 2.2 \times 10^{-16}$)

contain at least one Pfam domain and that 14,886 (76.4%) models have significant homology to a protein in the Uniprot-Swiss-prot database. These proportions are similar to those reported for the v2.1 reference transcriptome indicating that the function of a large segment of grapevine transcriptome cannot be predicted based on currently available data. This is not a grapevine-specific problem: results of recent plant transcriptome annotation efforts indicated that our knowledge of gene function in higher plants is still limited. For example, a recent transcriptome analysis in rose-scented geranium (*Pelargonium graveolens*) and giant cane (*Arundo donax*) found protein homology to only 66% and 55% of the transcripts [48, 49], respectively. Collectively, plant de novo transcriptome analysis studies confirmed that, despite the wealth of genomics and bioinformatics resources accumulated, our understanding of plant biology is still hindered by our inability to assign even tentative function to a large number of plant genes.

Validation of the transcriptome, using BUSCO, identified 59.4% of the expected 1440 single-copy embryophyta orthologs. This result suggested that we have recovered about 60% of the 'Riesling' transcriptome. This is in agreement with previous estimates that the grapevine genome contained about 30,000 genes. Even though only a subset of the v2.1 gene models were identified, several quantitative features of the 'Riesling' transcriptome were similar to those of other plants. For example, of the genes predicted, 17.9% (3480 genes) occurred in tandemly duplicated gene arrays (TAGs), a percentage consistent with *Arabidopsis thaliana* (16.6%) [50] and even such distantly related organisms as human, mouse, and rat (14–17%) [51].

The fraction of the transcriptome we captured is similar in range to that reported by Venturini et al. [2] who characterized a transcriptome for *V. vinifera* cv. 'Corvina', and was able to recover only 51% of the v1 reference annotation's 29,971 gene-models. These results raised the question of why such a low percentage of RNA-seq-supported *V. vinifera* transcripts could be anchored to the gene models of the reference transcriptome of the same species. Venturini et al. [2] speculated that this might be due to different assortment of genes expressed in the reference and the de novo transcriptomes. While differential expression certainly plays a part, we contend that the high-level genetic diversity within *V. vinifera* is an even greater source of the difficulty in anchoring transcripts among varieties. Strong support for the this contention was provided by a recent genome assembly by Chin and co-workers [10] who assembled the genome of *Vitis vinifera* cv. 'Cabernet Sauvignon'. Chin et al. were able to align only 16,981 (57%) of the v1 reference transcriptome's 29,971 gene-models to the 'Cabernet Sauvignon' genome. As they have worked with genomic sequence alignments only, their anchoring could not have been influenced by differential gene expression. Nonetheless, their anchoring success rate of 57% was similar to that (51%) obtained by Ventirini et al. [2], who worked with de novo assembled transcripts. Importantly, both Venturini et al. [2] and Chin et al. [10] achieved these results by attempting to identify corresponding genes between varieties of the same species, *V. vinifera*.

The high level of heterozygosity within cultivated *V. vinifera* varieties is well documented and is attributed primarily to the large effective population size of the ancestral *Vitis vinifera* ssp. *sylvestris*, and the resultant extensive chromosomal recombination Using a grapevine genotyping array, Myles et al. [52] showed that the *V. vinifera* genome was composed of short haplotype blocks which reflect chromosomal recombination over a long evolutionary time [52]. Myles et al. also demonstrated that nearly a quarter of the polymorphisms for which *V. vinifera* varieties segregate are shared with the North American *Vitis* species. This suggested vast effective population sizes which lead to the maintenance of genetic diversity from deep ancestry. We propose that this genomic diversity within *V. vinifera* manifests itself in such high-level allelic diversity that the accurate anchoring of many gene models is not possible across varieties.

This problem could have been exacerbated by the inclusion into the v2.1 annotation of transcript information from two rootstock cultivars, the genomes of which had been derived to a great extent from non-vinifera grape genotypes. These rootstocks were collectively bred from three North American *Vitis* species, namely *Vitis berlandieri*, *Vitis rupestris*, and *Vitis riparia*, in addition to *V. vinifera*. The divergence time between the three North American and the Eurasian vinifera clades was estimated to be 11.12 (16.58–6.59) million years, whereas the divergence within the ancestral *V. vinifera* ssp. *sylverstris*, the wild progenitor of *V. vinifera*, occurred

during a considerably shorter evolutionary time [53]. Thus, the inclusion of *V. berlandieri*, *V. rupestris*, and *V. riparia* transcripts in the v2.1 annotation likely introduced genetic divergence well beyond the already great divergence in *V. vinifera* itself.

Results of the transcriptome analysis presented here, supported by previous transcriptome and genotyping work by others, suggest that the level of genetic diversity in grapevine prevents the creation of a well-annotated transcriptome based on a single *V. vinifera* reference genome sequence [2, 10, 54]. We propose that a much more complete and accurate annotation can be constructed only on the basis of a cultivar-specific transcriptome assembly and a cultivar-specific genome sequence. Such genotype-specific annotations will be essential for comparative genomics of grape varieties. Only then will we be able to truly define varietal differences in grapevine on the transcriptional level.

Anchoring gene models to reference genomes and transcriptomes may be fraught with difficulties in other perennial crops as well. Although artificial selection during domestication resulted in self-compatibility in most fruit species, many of them have been derived from obligate outcrossing progenitors and, consequently, represent broad genetic diversity. Thus, variety-specific genome annotation may also become necessary to elucidate genotypic differences in such economically important fruit species as apple [55], plum [56] and sweet cherry [57]. Other highly heterozygous woody perennials for which genomics tools have been developed are cocoa [58] and poplar [59].

As our insight into the regulation of protein coding genes improves, there is mounting evidence that long non-coding RNAs (lncRNAs) play a part in regulatory processes (reviewed in [17, 60, 61]). Identification of these transcripts is paramount for understanding the role of these RNA species. We present a standardized computational pipeline for the identification of lncRNAs, which is particularly useful in non-model species. This pipeline represents a logical sequence of processes for removing known protein-coding genes and other non-coding RNAs using the most effective computational methods available to date. The pipeline predicts lncRNAs, then attempts to validate them using a pseudo-independent software, the Coding Potential Calculator. We consider this validation pseudo-independent, because both the pipeline and the Calculator incorporate BLAST results, albeit to varying levels of confidence. These transcripts are, at best, predictions, and only experimental evidence will validate their true function. This, in fact, is the major limitation of this pipeline. For example, we cannot assess the sensitivity or the specificity of the pipeline due to a lack of validated non-coding RNA sequences. This problem is inherent not only to this study, but in all current studies seeking to classify long RNA transcripts. We can, nonetheless, attempt to demonstrate that the putative lncRNAs look fundamentally different from protein coding RNA molecules. For example, our lncRNAs were overall enriched for shorter transcripts ($p < 2.2 \times 10^{-16}$) and lower GC content ($p < 2.2 \times 10^{-16}$). These two factors are both consistent with the previously proposed idea that lncRNAs have differential stability as compared to protein-coding transcripts [12, 45]. This result was validated by tests to determine if there was a difference in means (Welch's t-test) and if the two metric populations were sampled from the same distributions (KS-test). At this large sample size, it was not unexpected that the results of these tests produced statistically significant differences. Nonetheless, the effect size indicated that we may be looking at true functional differences.

This pipeline is optimal for species that lack well-established reference genomes. Moreover, the pipeline works well for species that suffer from the problems presented previously, namely high heterozygosity that interferes with genome-guided methodologies. While this pipeline is well suited for such species, there is no inherent limitation of its use in other species. It is constructed in a way that it maximizes the retention of transcripts that apparently do not code for proteins regardless of the initial data. As our understanding of lncRNA biology deepens, we will return to this and other pipelines to test their efficacy in calling functionally validated lncRNAs.

Conclusions

These and previous results of grapevine transcriptome assembly projects suggest that RNA-seq and predictive method-based genome annotation will be greatly improved by the availability of cultivar-specific genome sequences and corresponding cultivar-specific transcriptomes. This is especially necessary for the development of gene models and inter-cultivar analyses of variations. The data presented here strengthen the hypothesis that lncRNAs have thermodynamic properties that differ from those of protein-coding RNAs. The analysis of both coding and non-coding RNAs will be instrumental in uncovering inter-cultivar variation in wild and cultivated grapevine species.

Methods
Plant material and tissue extraction
The two accessions of *Vitis vinifera* cv. 'Riesling' used in this work were 588,673, a clone maintained at the USDA-ARS cold hardy grape germplasm repository in Geneva, NY and Ventosa, a commercially grown Johannisburg clone collected from Ventosa Vineyards, also in Geneva, NY. In total, tissues of seven different organs were collected from these vines representing young leaf, tendril, rachis, flower, berry (unripe and post-veraison), dormant bud, and root tissue. All tissues were collected

between 10 am and 12 pm during sunny, dry conditions. In addition to field collected leaf tissue, young leaf tissue from an ongoing temperature stress experiment was collected from cuttings grown in a growth chamber (25 °C) as well as cuttings exposed to chilling temperatures (4 °C, 48 h) and freezing temperatures (–3 °C, 30 min). mRNA was isolated from each tissue-type separately using a commercially available extraction kit (Sigma Spectrum RNA kit). Following mRNA isolation and quantification, RNA pools were constructed for each genotype, equilibrated and barcoded prior to sequencing. Tissue was not available from all organs for both accessions. Thus, RNA pools for 588,673 included dormant bud, leaf, tendril, flower, rachis, unripe berry, and post-veriason berry. The Ventosa Vineyard sample included tendril, flower, dormant bud and root tissues as well as field-collected leaf tissue and chilling/freezing exposed leaf tissue. Barcoded RNA libraries were then sequenced as 150 bp, paired-end reads on a HiSeq2000 at Cornell University.

Genome annotation
Raw RNA-seq reads from each sequenced library were assorted into bins corresponding to a directional, trial specific barcode using the FASTX tool fastx_barcode_splitter. Quality filtering was accomplished with fastq_quality_filter with the following parameters: -Q33 −q 25 −p 25. Barcodes were removed using fastx_trimmer with the following parameters: -Q33 -f 7. Further trimming of the adapter sequences.

(rcprAC = "AGATCGGAAGAGCGTCGTGTAGGGA AAGAGTGTAGATCTCGGTGGTCGCCGTATCATT", rcprBC = "AGATCGGAAGAGCGGTTCAGCAGGA ATGCCGAGACCGATCTCGTATGCCGTCTTCTGC TTG") was performed with cutadapt [62] with the following parameters: −minimum-len 25 −O 3. The script for this processing is available in the listed GitHub repository. Reads from different sequencing lanes were concatenated into one FASTQ file representing all libraries sequenced from the left terminus and all libraries sequenced from the right terminus. Corresponding reads were paired with the tool pairfq_lite [19] using default parameters. Transcripts were de novo assembled with the program Trinity using the following parameters: −seqType fq −max_memory 22G −SS_lib_type FR −CPU 1. Flags for −left and −right were given both paired and unpaired reads from pairfq_lite delimited by a comma.

The unmasked *V. vinifera* PN40024 12Xv2 reference genome, all Trinity-assembled transcripts, and the Uniprot-Swiss-Prot database were passed as FASTA files to Maker for the first round of annotation. All settings were kept at default, with the exception of the "hidden setting" est_forward = 1. FASTA and gff3 files were merged across the entire genome using the *fasta_merge* and *gff3_merge* tools in the Maker suite to generate a preliminary transcriptome.

Using the preliminary Maker annotation, gene prediction models were trained for the final annotation of 'Riesling' data. Using all Maker-generated gene-models, we trained SNAP [27] to generate a gene prediction model following the methods described in the Maker Wiki [63]. The SNAP model was trained twice, iteratively. The final annotation was performed by passing the SNAP-trained Riesling.hmm (see Additional file 1), the Arabidopsis-derived Augustus model [28], and all Trinity-derived RNA-seq evidence. Options for SNAP and Augustus were defined in the Maker control file as instructed in the Wiki.

Functional annotation of the gene models

Gene-model protein domains were identified against the Pfam31.0 [29] reference domain database using the program hmmscan (output declared as tab-delimited, –E 1e-05) of the HMMER v3.1b software suite [30]. Pfam to GO anchors were downloaded from the Gene Ontology project's website (http://geneontology.org/external2go/pfam2go). the. Gene model searches against the Uniprot-Uniref90 and the Uniprot-Swiss-Prot protein databases were done using the blastp algorithm implemented by the BLAST suite v2.29. Results were filtered for the top hit of each query based on bit score, E-value, and percent alignment. BUSCO analyses were executed against the empryophyta_odb9 reference single-copy ortholog data set.

Gene duplication

Duplicate genes were identified using an all-by-all self-blastp using the program MCScanX [35] with the following parameters: $path/to/MCScanX. /self_blast -e $i, where *i* represented the threshold E-value. Gene were classified into categories using the duplicate_gene_classifier algorithm implemented by the MCScanX tool. Tandem duplications were further identified from the .tandem output file. From these, TAGs were identified using the second gene in a tandem pair as an anchor for the identification of 3-gene arrays.

lncRNA identification pipeline

The output from Trinity was filtered for redundant transcripts using the cd-hit-est. algorithm of CD-HIT [36] with the following parameters: -i Trinity.fasta -n 5 -o clust_Trinity.fasta -c 0.90 -m 8000 -T 6. Filtering by expression was executed with RSEM [37] implemented by the Trinity-provided script align_and_estimate_abundance.pl with the following flags: −seqType fq −transcripts clust_Trinity.fasta −SS_lib_type FR −est_method RSEM −ali_method bowtie −trinity_mode −prep_reference.

Comparison of lncRNAs and protein coding RNAs

The minimum free energy of each transcript was calculated using the rnafold algorithm implemented by the ViennaRNA-2.2.5 software package [46] using the following options: -p –d2 −noLP. Results are output to a tab delimited file in which the name of the sequence, the minimum free energy (MFE), Centroid free energy, and ensemble diversity are reported. The minimum free energies of the transcripts were then compared to the minimum free energy of a randomly selected set of putative protein coding genes as annotated by Maker. Random genes were selected to be reproducible in R using set.seed (1992), and the genes were selected to coincide with occurrence patterns of lncRNAs (146 genes from chr1, 139 genes from chr2, etc.). All statistical analyses were completed in R v3.2.3 using the standard t.test () and ks-test () functions. Effect size was calculated using the formula $d = 2t/\sqrt{df}$.

Additional files

Additional file 1: Riesling-trained Hidden Markov Model for Gene Prediction. This file is a statistical model that can be used with Maker. (HMM 45 kb)

Additional file 2: Pfam to GO Anchors in the Maker Annotation. This table details the genes identified by Maker that contained Pfam domains that could be anchored to Gene Ontology terms. (CSV 4992 kb)

Additional file 3: Predicted Functional Annotation of the Maker Annotation. This table details the best hits of the Maker-derived genes to the Uniprot-swissprot and the Uniprot-Uniref90 reference protein databases. Additional data is shown for the best match gene in the reference PN40024 annotation and the type of predicted duplication classification. (CSV 1661 kb)

Abbreviations
BUSCO: Benchmarking universal single copy orthologs; GO: Gene ontology; lncRNA: Long non-coding RNA; nt: nucleotides; RBH: Reciprocal best hit; SOYBN: Soy bean; VITVI: *Vitis vinifera*

Acknowledgments
We would like to thank Matthew R. Siebert and Ridwan Sakidja, for providing advice and resources for computational methods and Sean Maher for his guidance in statistical analyses. We would like to thank Jaquelyn Lillis and Mikhail Osipovitch for providing bioinformatic support.
This work utilized the Extreme Science and Engineering Discovery Environment (XSEDE), which is supported by National Science Foundation grant number OCI-1053575. Specifically, we acknowledge the Texas Advanced Computing Center (TACC) at The University of Texas at Austin for providing HPC and grid resources that have contributed to the research results reported within this paper. Further, we acknowledge the Bridges system, which is supported by NSF award number ACI-1445606, at the Pittsburgh Supercomputing Center (PSC).

Funding
This work was conducted and supported as part of the USDA-ARS-GGRU CRIS:8060–21,220-006-00D. Funding was also provided by the Graduate College at Missouri State University. Funding sources had no role in study design, data collection, computation, or data interpretation.

Authors contributions

All authors contributed to the conception and design of this study. JL collected tissue and performed RNA extraction/ sequencing. ZH performed all computational work with guidance form JL. ZH and LK were major contributors to the writing of the manuscript. All authors read and approved the final manuscript.

Competing interests

The authors declare that they have competing interests.

Author details

[1]Missouri State University, Biology Department, 901 S. National Ave, Springfield, MO, USA. [2]United States Department of Agriculture, Agricultural Research Service, Grape Genetics Research Unit, 630 W. North Street, Geneva, NY, USA. [3]Present address: Saint Louis University, Department of Biology, 1 N. Grand Blvd, Saint Louis, MO, USA.

References

1. Pervaiz T, Haifeng J, Haider MS, Cheng Z, Cui M, Wang M, et al. Transcriptomic analysis of grapevine (cv. Summer black) leaf, using the illumina platform. PLoS One. 2016;11:1–20.
2. Venturini L, Ferrarini A, Zenoni S, Tornielli GB, Fasoli M, Dal Santo S, et al. De novo transcriptome characterization of Vitis Vinifera cv. Corvina unveils varietal diversity. BMC Genomics. 2013;14:41.
3. Shangguan L, Mu Q, Fang X, Zhang K, Jia H, Li X, et al. RNA-sequencing reveals biological networks during table grapevine ('Fujiminori') fruit development. PLoS One. 2017;12:1–23.
4. Zenoni S, Ferrarini A, Giacomelli E, Xumerle L, Fasoli M, Malerba G, et al. Characterization of transcriptional complexity during berry development in Vitis Vinifera using RNA-Seq. Plant Physiol. 2010;152:1787–95.
5. Fasoli M, Dal Santo S, Zenoni S, Tornielli GB, Farina L, Zamboni A, et al. The grapevine expression atlas reveals a deep Transcriptome shift driving the entire plant into a maturation program. Plant Cell. 2012;24:3489–505.
6. Blanco-Ulate B, Amrine KC, Collins TS, Rivero RM, Vicente AR, Morales-Cruz A, et al. Developmental and metabolic plasticity of white-skinned grape berries in response to Botrytis Cinerea during noble rot. Plant Physiol. 2015; 169:2422–43.
7. Blanco-Ulate B, Hopfer H, Figueroa-Balderas R, Ye Z, Rivero RM, Albacete A, et al. Red blotch disease alters grape berry development and metabolism by interfering with the transcriptional and hormonal regulation of ripening. J Exp Bot. 2017;68:1225–38.
8. Massonnet M, Morales-Cruz A, Figueroa-Balderas R, Lawrence DP, Baumgartner K, Cantu D. Condition-dependent co-regulation of genomic clusters of virulence factors in the grapevine trunk pathogen Neofusicoccum parvum. Mol Plant Pathol. 2016. doi:10.1111/mpp.12491.
9. Pastore C, Dal Santo S, Zenoni S, Movahed N, Allegro G, Valentini G, et al. Whole plant temperature manipulation affects Flavonoid metabolism and the Transcriptome of grapevine berries. Front Plant Sci. 2017;8:929.
10. Chin C-S, Peluso P, Sedlazeck FJ, Nattestad M, Concepcion GT, Clum A, et al. Phased diploid genome assembly with single molecule real-time sequencing. Nat Methods. 2016;13:1050-54.
11. This P, Martínez Zapater J, Péros J-P, Lacombe T. Natural variation in Vitis. In: Adam-Blondon A-F, Martínez-Zapater JM, Kole C, editors. Genetics, genomics, and breeding of grapes. 1st ed. Enfield, New Hampshire: Science Publishers; 2010. p. 30–67.
12. Mohammadin S, Edger PP, Pires JC, Schranz ME. Positionally-conserved but sequence-diverged: identification of long non-coding RNAs in the Brassicaceae and Cleomaceae. BMC Plant Biol. 2015;15:217.
13. Li L, Eichten SR, Shimizu R, Petsch K, Yeh C-T, Wu W, et al. Genome-wide discovery and characterization of maize long non-coding RNAs. Genome Biol. 2014;15:R40.
14. Wang H, Niu QW, HW W, Liu J, Ye J, Yu N, et al. Analysis of non-coding transcriptome in rice and maize uncovers roles of conserved lncRNAs associated with agriculture traits. Plant J. 2015;84:404–16.
15. Lu X, Chen X, Mu M, Wang J, Wang X, Wang D, et al. Genome-wide analysis of long noncoding RNAs and their responses to drought stress in cotton (Gossypium Hirsutum L). PLoS One. 2016;11:e0156723.
16. Kwenda S, Birch PRJ, Moleleki LN. Genome-wide identification of potato long intergenic noncoding RNAs responsive to Pectobacterium carotovorum subspecies brasiliense infection. BMC Genomics. 2016;17:614.
17. Heo JB, Lee Y-S. Molecular functions of long noncoding transcripts in plants. J Plant Biol. 2015;58:361–5.
18. Shafik S, Li J, Sun Q. Functions of plants long non-coding RNAs. Biochim biophys Acta. 2015.
19. Pairfq. https://github.com/sestaton/Pairfq. Accessed 8 Feb 2017.
20. University of Padua. Unive. http://www.cribi.unipd.it. Accessed 8 Feb 2017.
21. Grabherr MG, Haas BJ, Yassour M, Levin JZ, Thompson DA, Amit I, et al. Full-length transcriptome assembly from RNA-Seq data without a reference genome. Nat Biotechnol. 2011;29:644–52.
22. Haas BJ, Papanicolaou A, Yassour M, Grabherr M, Blood PD, Bowden J, et al. De novo transcript sequence reconstruction from RNA-seq using the trinity platform for reference generation and analysis. Nat Protoc. 2013;8:1494–512.
23. Cantarel BL, Korf I, Robb SMC, Parra G, Ross E, Moore B, et al. MAKER: an easy-to-use annotation pipeline designed for emerging model organism genomes. Genome Res. 2008;18:188–96.
24. Holt C, Yandell M. MAKER2: an annotation pipeline and genome-database management tool for second-generation genome projects. BMC Bioinformatics. 2011;12:491.
25. Uniprot-Swiss-Prot. http://www.uniprot.org/uniprot/. Accessed 8 Feb 2017.
26. Camacho C, Coulouris G, Avagyan V, Ma N, Papadopoulos J, Bealer K, et al. BLAST plus : architecture and applications. BMC Bioinformatics. 2009;10
27. Korf I. Gene finding in novel genomes. BMC Bioinformatics. 2004;5:59.
28. Stanke M, Steinkamp R, Waack S, Morgenstern B. AUGUSTUS: a web server for gene finding in eukaryotes. Nucleic Acids Res. 2004;32 WEB SERVER ISS: 309–12.
29. Bateman A. The Pfam protein families database. Nucleic Acids Res. 2004;32:138D–141.
30. Eddy SR. A new generation of homology search tools based on probabilistic inference. Genome Inform. 2009;23:205–11.
31. Gene T, Consortium O. The gene ontology project in 2008. Nucleic Acids Res. 2008;36(Database issue):D440–4.
32. Uniprot-Uniref90. http://www.uniprot.org/uniref/. Accessed 8 Feb 2017.
33. Vitulo N, Forcato C, Carpinelli EC, Telatin A, Campagna D, D'Angelo M, et al. A deep survey of alternative splicing in grape reveals changes in the splicing machinery related to tissue, stress condition and genotype. BMC Plant Biol. 2014;14:99.
34. Terral JF, Tabard E, Bouby L, Ivorra S, Pastor T, Figueiral I, et al. Evolution and history of grapevine (Vitis Vinifera) under domestication: new morphometric perspectives to understand seed domestication syndrome and reveal origins of ancient European cultivars. Ann Bot. 2010;105:443–55.
35. Wang Y, Tang H, Debarry JD, Tan X, Li J, Wang X, et al. MCScanX: a toolkit for detection and evolutionary analysis of gene synteny and collinearity. Nucleic Acids Res. 2012;40:1–14.
36. Li W, Godzik A. Cd-hit: a fast program for clustering and comparing large sets of protein or nucleotide sequences. Bioinformatics. 2006;22:1658–9.
37. Li B, Dewey CN. RSEM accurate transcript quantification from RNA-Seq data with or without a reference genome. BMC Bioinformatics. 2011;12:323.
38. Trinotate: Transcriptome Functional Annotation and Analysis. http://trinotate.github.io. Accessed 8 Feb 2017.
39. Al-Tobasei R, Paneru B, Salem M. Genome-wide discovery of long non-coding RNAs in rainbow trout. PLoS One. 2016;11:e0148940.
40. Nawrocki EP, Burge SW, Bateman A, Daub J, Eberhardt RY, Eddy SR, et al. Rfam 12.0: updates to the RNA families database. Nucleic Acids Res. 2015;43: D130–7.
41. Nawrocki EP, Eddy SR. Infernal 1.1: 100-fold faster RNA homology searches. Bioinformatics. 2013;29:2933–5.
42. Kong L, Zhang Y, Ye ZQ, Liu XQ, Zhao SQ, Wei L, et al. CPC: assess the protein-coding potential of transcripts using sequence features and support vector machine. Nucleic Acids Res. 2007;35(SUPPL.2):345–9.

43. Haas B. TransDecoder (Find Coding Regions Within Transcripts). trinotate. github.io. Accessed 3 Apr 2017.

44. Mu C, Wang R, Li T, Li Y, Tian M, Jiao W, et al. Long Non-Coding RNAs (lncRNAs) of Sea Cucumber: Large-Scale Prediction, Expression Profiling, Non-Coding Network Construction, and lncRNA-microRNA-Gene Interaction Analysis of lncRNAs in *Apostichopus japonicus* and *Holothuria glaberrima* During LPS Challenge and Radial Organ Complex Regeneration. Mar Biotechnol. 2016;18:485-99.

45. Yang JR, Zhang J. Human long noncoding RNAs are substantially less folded than messenger RNAs. Mol Biol Evol. 2015;32:970-7.

46. Lorenz R, Bernhart SH, zu Siederdissen C, Tafer H, Flamm C, Stadler PF, et al. {ViennaRNA} Package 2.0. Algorithms Mol Biol. 2011;6:26.

47. Cohen J. Statistical power analysis for the behavioural sciences New York. NY Acad. 1988.

48. Narnoliya LK, Kaushal G, Singh SP, Sangwan RS. De novo transcriptome analysis of rose-scented geranium provides insights into the metabolic specificity of terpene and tartaric acid biosynthesis. BMC Genomics. 2017;18:74.

49. Evangelistella C, Valentini A, Ludovisi R, Firrincieli A, Fabbrini F, Scalabrin S, et al. De novo assembly, functional annotation, and analysis of the giant reed (Arundo Donax L.) leaf transcriptome provide tools for the development of a biofuel feedstock. Biotechnol Biofuels. 2017;10:138.

50. Rizzon C, Ponger L, Gaut BS. Striking similarities in the genomic distribution of tandemly arrayed genes in Arabidopsis and rice. PLoS Comput Biol. 2006; 2:0989-1000.

51. Shoja V, Zhang LA. Roadmap of tandemly arrayed genes in the genomes of human, mouse, and rat. Mol Biol Evol. 2006;23:2134-41.

52. Myles S, Chia JM, Hurwitz B, Simon C, Zhong GY, Buckler E, et al. Rapid genomic characterization of the genus Vitis. PLoS One. 2010;5

53. Wan Y, Schwaninger HR, Baldo AM, Labate J a, Zhong G-Y, Simon CJ. A phylogenetic analysis of the grape genus (Vitis L.) reveals broad reticulation and concurrent diversification during neogene and quaternary climate change. BMC Evol Biol. 2013;13:141.

54. Da Silva C, Zamperin G, Ferrarini a, Minio a, Dal Molin A, Venturini L, et al. The high Polyphenol content of grapevine cultivar Tannat berries is conferred primarily by genes that are not shared with the reference genome. Plant Cell. 2013;25:4777-88.

55. Velasco R, Zharkikh A, Affourtit J, Dhingra A, Cestaro A, Kalyanaraman A, et al. The genome of the domesticated apple (Malus × domestica Borkh.). Nat Genet. 2010;42:833-9.

56. Topp BL, Russell DM, Neumuller M, Dalbo M, Plum LW. In: Badenes ML, Byrne DH, editors. Fruit Breeding. New York: SPringer; 2012. p. 571-623.

57. Shirasawa K, Isuzugawa K, Ikenaga M, Saito Y, Yamamoto T, Hirakawa H, et al. The genome sequence of sweet cherry (Prunus Avium) for use in genomics-assisted breeding. DNA Res. 2017;24:499-508.

58. Trognitz B, Scheldeman X, Hansel-Hohl K, Kuant A, Grebe H, Hermann M. Genetic population structure of cacao plantings within a young production area in Nicaragua. PLoS One. 2011;6

59. Tuskan G a, Difazio S, Jansson S, Bohlmann J, Grigoriev I, Hellsten U, et al. The genome of black cottonwood, Populus Trichocarpa (Torr. \& gray). Science (80-). 2006;313:1596.

60. Bhat SA, Ahmad SM, Mumtaz PT, Malik AA, Dar MA, Urwat U, et al. Long non-coding RNAs: mechanism of action and functional utility. Non-coding RNA Res. 2016;

61. Bai Y, Dai X, Harrison AP, Chen M. RNA regulatory networks in animals and plants: a long noncoding RNA perspective. Brief Funct Genomics. 2015;14: 91-101.

62. Martin M. Cutadapt removes adapter sequences from high-throughput sequencing reads. EMBnet. Journal. 2011;17

63. MAKER Tutorial for GMOD Online Training 2014. http://weatherby.genetics. utah.edu/MAKER/wiki/index.php/MAKER_Tutorial_for_GMOD_Online_ Training_2014. Accessed 28 Nov 2017.

9

Alternative patterns of sex chromosome differentiation in *Aedes aegypti* (L)

Corey L. Campbell[1]* , Laura B. Dickson[1], Saul Lozano-Fuentes[1], Punita Juneja[2], Francis M. Jiggins[2] and William C. Black

Abstract

Background: Some populations of West African *Aedes aegypti*, the dengue and zika vector, are reproductively incompatible; our earlier study showed that divergence and rearrangements of genes on chromosome 1, which bears the sex locus (*M*), may be involved. We also previously described a proposed cryptic subspecies SenAae (PK10, Senegal) that had many more high inter-sex F_{ST} genes on chromosome 1 than did *Ae.aegypti aegypti* (Aaa, Pai Lom, Thailand). The current work more thoroughly explores the significance of those findings.

Results: Intersex standardized variance (F_{ST}) of single nucleotide polymorphisms (SNPs) was characterized from genomic exome capture libraries of both sexes in representative natural populations of Aaa and SenAae. Our goal was to identify SNPs that varied in frequency between males and females, and most were expected to occur on chromosome 1. Use of the assembled AaegL4 reference alleviated the previous problem of unmapped genes. Because the *M* locus gene *nix* was not captured and not present in AaegL4, the male-determining locus, per se, was not explored. Sex-associated genes were those with F_{ST} values ≥ 0.100 and/or with increased expected heterozygosity (H_{exp}, one-sided T-test, $p < 0.05$) in males. There were 85 genes common to both collections with high inter-sex F_{ST} values; all genes but one were located on chromosome 1. Aaa showed the expected cluster of high inter-sex F_{ST} genes proximal to the *M* locus, whereas SenAae had inter-sex F_{ST} genes along the length of chromosome 1. In addition, the Aaa *M*-locus proximal region showed increased H_{exp} levels in males, whereas SenAae did not. In SenAae, chromosomal rearrangements and subsequent suppressed recombination may have accelerated X-Y differentiation.

Conclusions: The evidence presented here is consistent with differential evolution of proto-Y chromosomes in Aaa and SenAae.

Keywords: Population genetics, Arbovirus vector, Dimorphic traits, Genomics, Evolution of reproductive proteins, Sex determination

Background

The dengue, yellow fever, chikungunya and zika vector, *Aedes aegypti*, has at least two major subspecies in tropical and subtropical regions; these consist principally of forest and peridomestic types [1–3]. Although morphological features such as abdominal scale patterns have been used to differentiate these groups, definitive molecular markers for subspecies identification are not yet available [1, 2, 4, 5]. Population-specific differences in west African population vector competence for flaviviruses have been described [6, 7]; and a trend toward reproductive isolation [8] may contribute toward these differences, as well as other traits [6, 7, 9]. *Ae. aegypti* has a dominant male-determining sex locus (*M*) on chromosome 1, for which males are heterozygous (*Mm*). This locus is primarily responsible for sex determination [10], however male and female chromosomes are also cytologically distinct [11]. The male-determining factor (M factor) *nix*, an M-linked myosin heavy chain gene, *myo-sex*, and two sex determination transcription factors have been characterized [10, 12–15], but little else is known about the specific genes contributing dimorphic phenotypes in aedine mosquitoes.

* Correspondence: corey.campbell@colostate.edu
[1]Department of Microbiology, Immunology and Pathology, Colorado State University, Campus Delivery 1692, Fort Collins, CO 80523, USA
Full list of author information is available at the end of the article

Metazoan proteins involved in mating and reproduction evolve more rapidly than genes in other functional groups, and this phenomenon may contribute to reproductive isolation and subsequent speciation (reviewed in [16–18]). The opposing evolutionary forces of male sexual selection and female conflict may be involved in this process [19, 20]. Rapid sex-associated gene evolution has been described in *Anopheles* mosquitoes [21] and drosophilids [22]. Haerty et al. showed rapid divergence of sex-associated genes in drosophilid males [22]. Such rapid evolution is also supported in taxa without a hemizygous X, as is the case in *Ae. aegypti* [8, 23], and has been attributed to sexual selection acting mostly on males [24]. It is expected that alleles with sexually antagonistic effects on fitness would accumulate on sex chromosomes, where they would be expressed predominantly or exclusively in the sex where they are advantageous (reviewed in [25]). In a species, such as *Ae. aegypti*, with recombining homomorphic sex chromosomes, these genes are expected to be enriched in regions tightly linked to the *M* locus. Because recombination should be suppressed in the *M* locus proximal region, differentiation of males and females likely occurs by genetic drift or possibly by selection of specific genes. For these reasons, analysis of sex-specific genetic variation in reproductively isolated mosquito populations could reveal gene diversity contributing to reproductive isolation and speciation [26].

A Senegalese sylvatic population (PK10, SenAae) has increased genetic and structural diversity at chromosome 1 compared to the type form *Ae. aegypti aegypti* (Aaa), possibly due to chromosomal rearrangements [26, 27]. In addition, PK10 showed reproductive incompatibility when mated to PK10 males with different abdominal banding patterns [26]. Interestingly, this strain also lacked the expected genetic linkage of the *white-eye* and the *M* locus in 26% of genetic families [27], which was consistent with the observations of sex chromosome structural diversity. Further, high throughput sequencing (HTS) showed that overall standardized variance (F_{ST}) was greater in SenAae than the representative type form, Aaa [27]. These unusual attributes in SenAae sex chromosomal structure and reproductive isolation led us to further explore sex-specific genomic polymorphisms in order to increase understanding of sex-specific differences in *Aedes* subspecies.

Therefore the over-arching goal of this study was to extend our earlier study [27] and use population genomics analyses of SenAae and Aaa to characterize sex-specific allele frequency differences. Our hypothesis was that genes with high sex-specific or inter-sex F_{ST} values would be located proximal to the *M* locus on chromosome 1 [28]. We used orthology information to predict whether these genes would be involved in in sex determination, reproduction and/or sexual dimorphic traits. Exome capture [29] genomic DNA (gDNA) HTS data from independent replicate pools ($n = 12$) of adult *Ae.aegypti* males and females were compared for two geographically and genetically distinct populations, with subsequent analysis of sex-specific single nucleotide polymorphisms (SNPs). The collections, SenAae and the type form Aaa from Thailand, have been highlighted in previous studies [26, 27, 30]. Standardized variance in SNP frequencies (F_{ST}) was used to compare sex-specific differences [31]. Thus, in the context of this work, high inter-sex F_{ST} values revealed SNPs that differed in frequency between males and females. We also expected that genes linked to the *M* locus would be more heterozygous in males [8, 32]. The Hardy-Weinberg expected heterozygosity (H_{exp}) score indicated the predicted level of sex-specific genetic diversity. Genes with high H_{exp} and/or F_{ST} levels may play roles in mating, sex determination, dimorphic development or trends in reproductive isolation.

Results

Exome-wide analysis of sex-specific polymorphisms

Exome-captured HTS libraries were sequenced from pools of Aaa and SenAae males and females; Table 1 shows library-specific information and overall polymorphism statistics. Two biological replicates for each pool (12 mosquitoes per pool) of males and females from each location produced a total of eight libraries (SenAae: 2 male PK10, 2 female PK10; Aaa: 2 male Thai, 2 female Thai). Roughly 34-38 million trimmed reads were produced from SenAae, and 18-25 million reads were obtained from the Aaa collection (Additional file 1). The chromosome-length assembled Aaa genome was used as a reference for all alignments (AaegL4) [33–35]; and 90-92% of trimmed reads aligned in each population (Additional file 1). Sex-specific polymorphisms were identified at each nucleotide site (at least 15 read counts per site) using the F_{ST} calculation (see Methods); sex-specific H_{exp} scores were also calculated [31]. SNPs that were completely fixed for both sexes but different from the reference, also known as monomorphic SNPs, were removed. The Aaa collection had about 1.9 million sex-specific polymorphisms and SenAae had about 3.0 million (Table 1). To rule out the possibility that population-specific differences arose from dissimilarities in sequencing coverage, the ratio of variant sites (* 1000) per aligned nucleotide were calculated on each chromosome (Additional file 1). In SenAae, the variant/aligned ratio ranged from 2.0-3.6 per chromosome, while in Aaa, they ranged from 2.6-4.8. Therefore, the overall relative number of variants per aligned nucleotide in Aaa was higher than that of SenAae, indicating that the features described below were not due to library size differences.

Table 1 Polymorphisms and Coverage

	Aaa		SenAae	
Monomorphic SNPs -Excluded	21,849,618		23,861,997	
Number of variant sites	1,901,845		3,044,292	
Coverage per nucleotide[a]-	-Min	60		60
	-Max	3564		3745
	-Mean	180		261.4
	-Median	152		214
Allele frequency Statistics	Aaa		SenAae	
	Female	Male	Female	Male
H_{exp} across all genes				
Mean (95% cl)	0.097 +/− 0.084	0.098 +/− 0.084	0.113 +/− 0.090	0.112 +/− 0.089
Median	0.024	0.024	0.029	0.026
Mean sample variance	0.002	0.002	0.002	0.002
Standard deviation	0.043	0.043	0.046	0.046
Chr 1 H_{exp}				
Mean (95% cl)	0.099 +/− 0.084	0.102 +/− 0.086	0.112 +/− 0.089	0.111 +/− 0.089
Median	0.025	0.027	0.031	0.027
Mean sample variance	0.002	0.002	0.002	0.002
Standard deviation	0.043	0.044	0.046	0.045
Increased male H_{exp}, T-test p value		2.20E-16		ns
M locus proximal region				
Mean (95% cl)	0.074 +/− 0.074	0.082 +/- 0.078	0.082 +/− 0.078	0.081 +/− 0.034
Median	0.018	0.02	0.016	0.015
Mean sample variance	0.001	0.002	0.002	0.0003
Standard deviation	0.038	0.04	0.04	0.0176
Increased male H_{exp}, T-test p value		2.20E-16		ns

ns not significant (one-sided T test, $p < 0.05$)
[a]Collection-wide total coverage

Polymorphisms were examined to identify gene-wise inter-sex F_{ST} values (Fig. 1, Materials and Methods), which were expected to follow a *beta* distribution (Additional file 2). We chose a cut-off of $F_{ST} \geq 0.100$ to identify genes of interest for this study. This cut-off was chosen rather than a percent cut-off, such as the upper 5%, because Aaa showed many fewer genes with F_{ST} greater than 0.100 in the upper 5% than did SenAae. For example, in the upper 5% subset, Aaa had 441 genes with $F_{ST} < 0.100$, while SenAae had none. The chromosome-length reference allowed us to examine inter-sex F_{ST} averages per gene relative to each physical location. The number of high inter-sex genes was significantly higher on chromosome 1 than either chromosome 2 or 3 (Fisher's Exact test, SenAae, $p < 0.0001$, Aaa, $p < 0.0001$). Interestingly, Aaa had a distinct cluster of high inter-sex F_{ST} genes on chromosome 1, with a few in distal locations ($n = 171$). This region overlaps a similar region of high inter-sex F_{ST} reported for the Liverpool Aaa strain [28]. This is consistent with retention of the

sex locus in Aaa, with a cluster of inter-sex F_{ST} genes proximal to *nix* (Fig. 1). Importantly, the male-specific M locus *nix* was not included in the AaegL4 reference or in our capture probes [12, 13, 36], however, predictions from AaegL4 indicate that *nix* is located between AAEL015064-RA and AAEL014760-RA at the location of the Aaa high F_{ST} cluster [35].

SenAae had high inter-sex F_{ST} genes across most of the chromosome ($n = 1233$). This pattern is consistent with extensive chromosome length X-Y differentiation, which is different from findings of other aedine populations [28]. The high level of reported SenAae genetic diversity may have contributed to the chromosome-wide pattern [26, 27], as mosquito pools rather than individuals were evaluated in this study.

Organisms with a single sex-determining locus, such as *Aedes spp.*, would be expected to bear sexually dimorphic heterozygosity proximal to the sex locus, and males should have greater heterozygosity at these sites. We assessed H_{exp} values along the length of

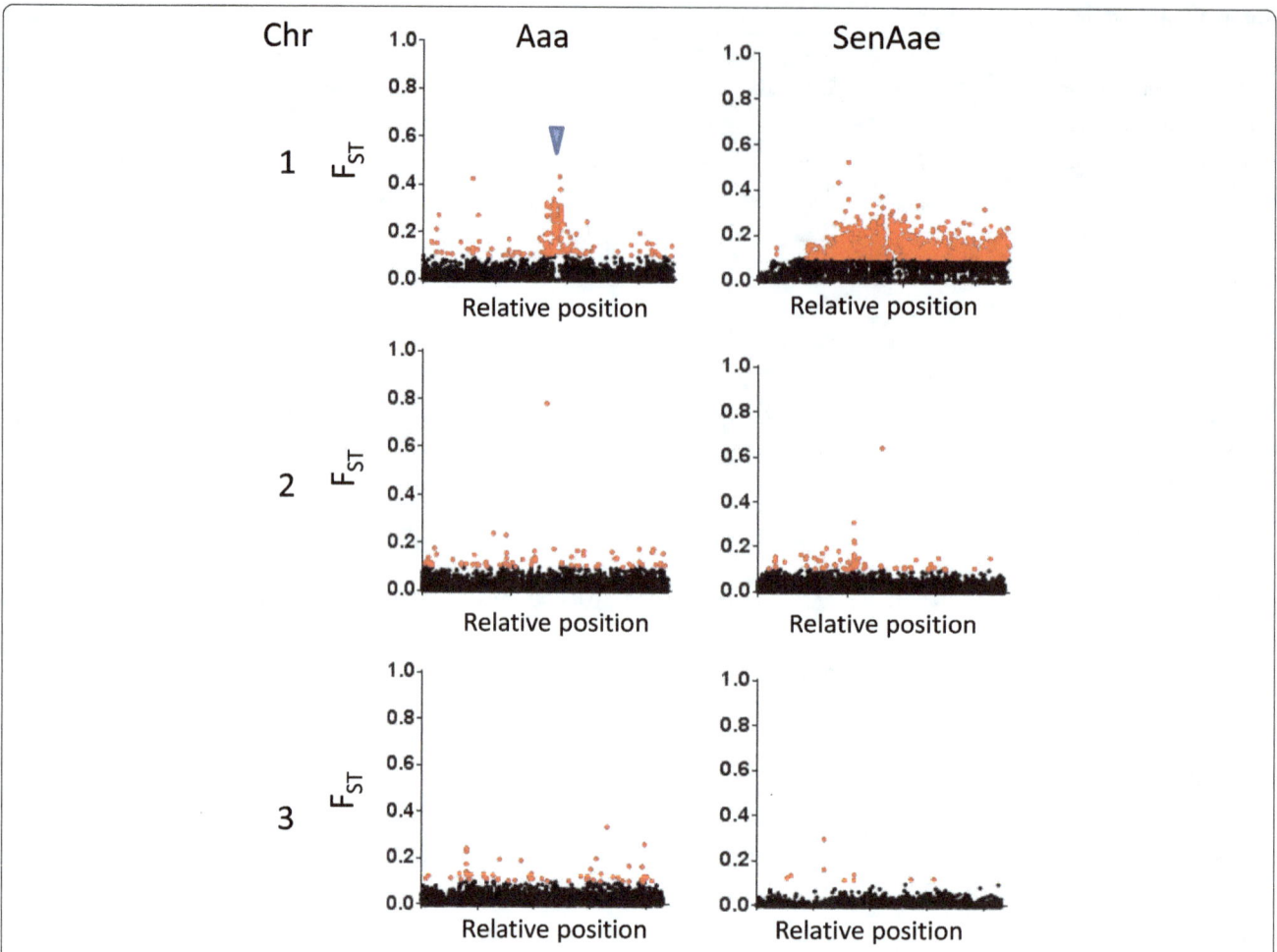

Fig. 1 Inter-sex F$_{ST}$ values vary among *A.aegypti* populations. Relative position of gene-wise F$_{ST}$ values per chromosome. Red dots indicate genes with F$_{ST}$ values ≥ 0.100 (Aaa, (Thai) collection, $n = 304$; SenAae (PK10) $n = 1310$); black dots indicate F$_{ST}$ values below the threshold. Blue carat indicates predicted location of *nix* at the *M* locus

chromosome 1, testing specifically for higher average H$_{exp}$ values in males over females (one-sided T-test). Along the entire length of chromosome 1, Aaa males had increased heterozygosity levels; this was especially marked in the central third of chromosome 1 (Fig. 2 and Table 1, one-sided T-test, $p < 2.2E\text{-}16$), which also corresponds to an area of reduced recombination reported by Fontaine et al. [28]. Curiously, increased male heterozygosity was not observed in any portion of Chromosome 1 in SenAae.

Though high inter-sex F$_{ST}$ genes were expected only on chromosome 1, genes with association to sex were present on all three chromosomes in both populations. High inter-sex F$_{ST}$ genes on chromosomes 2 and 3 could be involved in processes other than sex determination, such as reproduction, sexual dimorphic development or behavior. Alternatively, this category could also include genes that contribute to sex distortion phenotypes [37–39].

We expected that F$_{ST}$ calculations of female-vs-female comparisons from each population should be reduced

proximal to the *M* locus. Indeed, graphs of female-vs-female and male-vs-male comparisons indicate a marked reduction in F$_{ST}$ values proximal to the M locus in females but not in males (Additional file 3). The high number of F$_{ST}$ values > 0.100 shows the high level of differentiation between Aaa and SenAae.

Common features of X-Y differentiation

Our gene-by-gene F$_{ST}$ calculations provided a unique opportunity to explore specific high inter-sex F$_{ST}$ genes that were shared among the two populations. The premise of this line of inquiry was to identify specific genes that may contribute to male-female differentiation. Indeed, a study of humans showed that high F$_{ST}$ genes were enriched on X chromosomes relative to autosomes [40]. The intersection of high inter-sex F$_{ST}$ genes (≥ 0.100, $n = 85$) among Aaa and SenAae was assessed (Fig. 3a, Additional file 4). As expected, chromosome 1 was most represented in this subset (Fisher's Exact test, $p < 0.001$); just a single gene

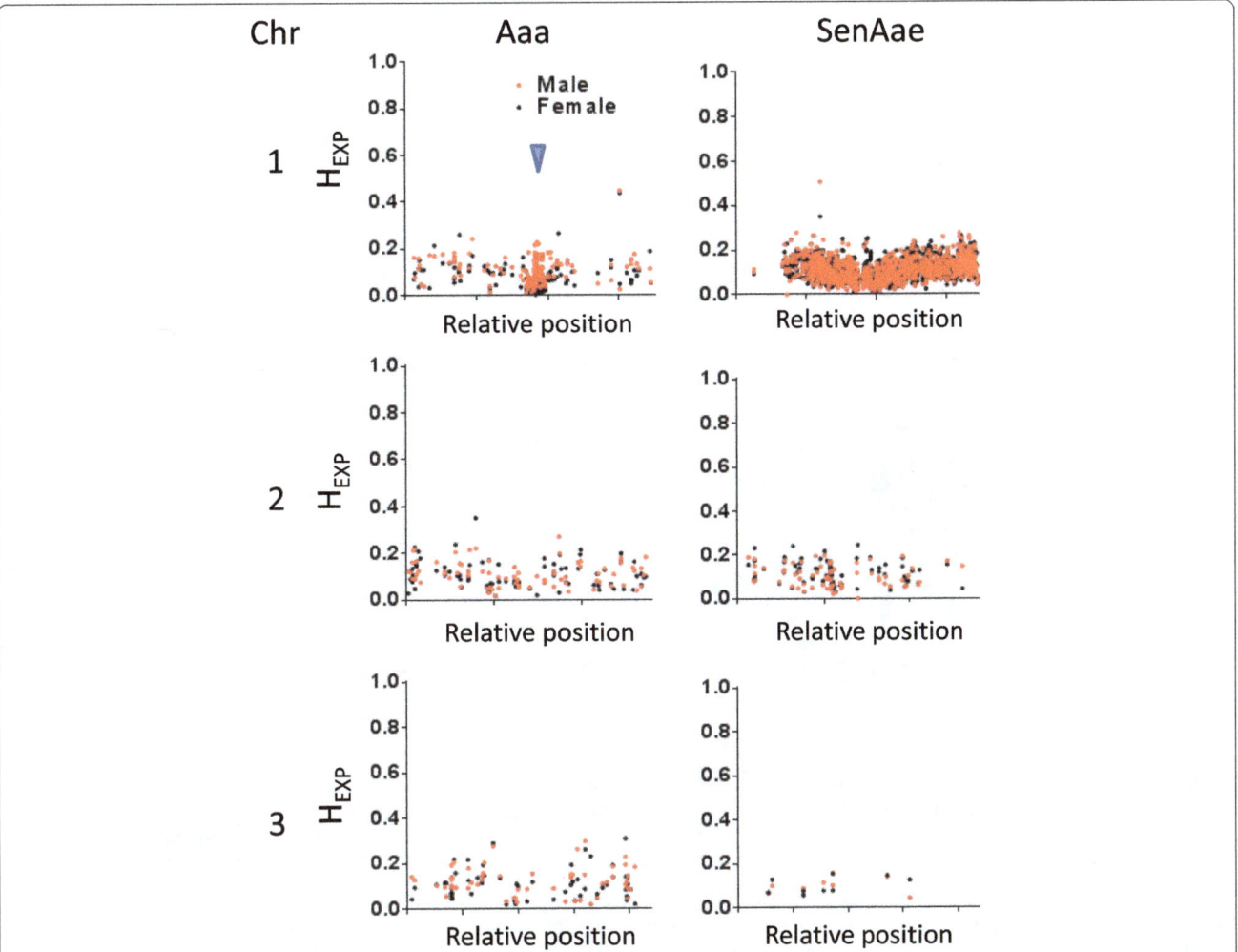

Fig. 2 Inter-sex H_{exp} values vary among *A.aegypti* populations. Relative position of gene-wise H_{exp} values for those genes in the high F_{ST} group ($F_{ST} \geq 0.100$). Red dots indicate male gene-wise H_{exp} values; black dots indicate female gene-wise H_{exp} values. Blue carat indicates predicted location of *nix* at the *M* locus

(AAEL001298) on chromosome 2 was present. The data are consistent with an overall lack of common autosomal high inter-sex F_{ST} genes, indicating that most autosomal high F_{ST} genes are due to population-specific trends. Importantly, high F_{ST} values on X-Y chromosomes could be due to genetic drift or sex-specific selection, therefore both are possible explanations for these high F_{ST} values [41]. Nevertheless, coordinated cis-regulation of gene expression on sex chromosomes has also been described [42] and provides support for the hypothesis that sex differentiation genes, other than the *M* locus, are present within the high F_{ST} clusters.

To further explore the common gene set, the genes were assigned to functional categories by orthology (BLAST, E^{-20} cut-off) to other dipterans [43] (Fig. 3b, Additional file 4). Excluding the diverse and uncharacterized subsets, the largest subset contained genes involved in DNA repair/replication/transcription/translation, which accounted for 18% of the total and was over-represented in this subset (hypergeometric analysis, $p < 0.003$). Nine genes had domains consistent with transcriptional activation or suppression activities. Sex-linked genes could also be those that contribute to sexually dimorphic phenotypes. For example, genes predicted to be involved in the chemosensory response (3.8%) were also represented, though not significantly over-represented. A possible sexual dimorphic bias in chemosensory function was expected, given that males and females have distinct food sources and mating behaviors [44, 45].

In insects, sex determination mechanisms are highly variable across genera and species, and sometimes vary within a single species, as occurs in *Musca domestica* [46–48].

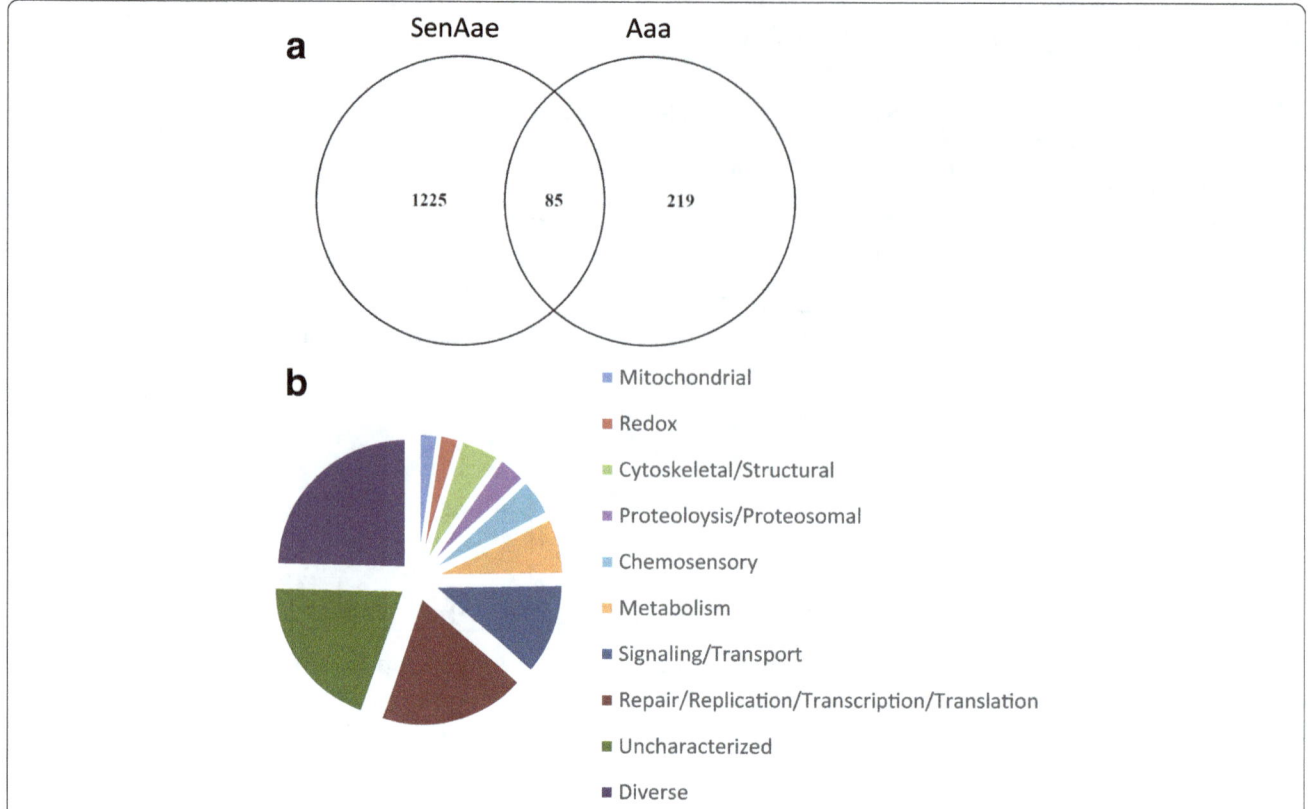

Fig. 3 Functional categories of genes showing sex-specific polymorphisms. **a** The intersection of genes among both populations with high inter-sex F_{ST} values (≥ 0.100). **b** The resulting 85 common genes were classified by functional category and are shown as a portion of the pie chart. Legend: the list of functional groups, arranged from top to bottom, is represented in the pie chart clock-wise, starting at the top-most slice

One common mechanism of dipteran sex determination, which also holds true for *Ae.aegypti* [14, 15], is dimorphic regulation of alternative RNA splicing mechanisms (reviewed in [49]). Just two RNA processing/splicing or sex determination genes were in the high F_{ST} group common to both SenAae and Aaa populations, AAEL017421 and AAEL006713. Although a specific function for nucleolar protein 56 (AAEL017421) has not been identified, other nucleolar proteins are important for tissue-specific development and maintenance of heterochromatin and ribosomal RNA [50]. In addition, a U2 snRNP auxiliary factor subunit (AAEL006713) was also in this group. U-type snRNPs make up the canonical RNA-splicing machinery (reviewed in [49]).

Each collection was further interrogated to identify male seminal fluid genes [51]. 2 genes were common to both populations; they code for seminal fluid proteins AAEL010935, a gamma glutamyl transpeptidase, and AAEL014053, a vacuolar ATPase. In SenAae alone, AAEL003746, a gene with predicted 4-hydroxybutyrate CoA-transferase activity, and AAEL005790, a predicted malate dehydrogenase were identified. Also, in Aaa alone, AAEL008489, a predicted calcium ion binding protein, was identified.

Discussion

Proto-Y chromosomes evolve from autosomes upon the acquisition of a male-determining factor (reviewed in [52, 53]). The evidence presented here is consistent with differential evolution of proto-Y chromosomes in Aaa and SenAae. We showed that Aaa had a cluster of high inter-sex F_{ST} genes ($F_{ST} > 0.100$) proximal to the *M* locus (Fig. 1). In contrast, SenAae showed high inter-sex F_{ST} along the majority of chromosome 1. In Aaa but not SenAae, the *M* locus proximal region had significantly higher male H_{exp} levels (Fig. 2), which is consistent with reduced recombination. Suppressed recombination is a necessary prelude to the development of heteromorphic sex chromosomes. The reason for the absence of these features in SenAae is unknown. It could be due to a high level of genetic diversity in this population but is also consistent with accelerated X-Y differentiation. For example, chromosomal rearrangements in SenAae [26] could have contributed to accelerated X-Y differentiation of chromosome 1. Indeed, chromosomal inversions can also reduce recombination rates in proto-sex chromosomes [54].

Both groups showed population-specific trends for high inter-sex F_{ST} genes on chromosomes 2 and 3. The identification of high intersex F_{ST} genes on autosomal chromosomes is consistent with previous studies that identified multiple independent loci contributing to the sex phenotype in other culicine species [55]. In addition, it is also consistent with the presence of high F_{ST} autosomal genes in flies that arose from sex-specific selection due to sexual-antagonistic mechanisms [41]. Alternatively, these could also be due to sex distortion trends, though we were unable to test this hypothesis in this study. In an organism heterozygous for a given pair of alleles, we expect equal recovery of each allele in the gametes. Loci in which this fails to occur constitute "meiotic drive" (MD) or "segregation distortion" systems. Because of the ease of detection, sex ratio distortion has been the best-studied system. In Aedes aegypti, [10] the male parent determines the sex ratio in progeny and, given normal segregation, equal numbers of males and females should occur. However, departures from a 1:1 sex ratio are often observed in culicine mosquitoes and have been best studied in Aedes aegypti, wherein 35 to 45% females are found in field collected populations [56]. A study of sex ratio in 19 laboratory strains revealed that some strains had ~50% females, others had a slight excess of males (~40% females) and a few showed distinct deviations in sex ratio (< 30% female) [57]. In 1976, a meiotic drive (MD) gene product that is tightly linked to and acts in trans with the M allele was observed to cause breakage of the m allele (female)-carrying chromosome [58]. It was proposed that the m allele carrying chromosome is sensitive (ms) or insensitive (mi) to MD. Additionally, some m alleles vary in their sensitivity to distortion over a range of haplotypes [37, 59–62].

Most recently, investigators selected for a strain in which only 14.7% of progeny are females [63]. This distortion is due to an inherited factor that causes a predominance of males in certain strains and for the progeny of single pair matings. The factor is transmitted only by males [64]. Several modifiers of MD have been identified. The tolerance of Distorter locus is near the re locus at 47 cM on chromosome 1 and results in a reduction in sex ratio distortion [39]. Another suppressor of MD is linked with the spot abdomen (s) locus on 29 cM on chromosome 2, and an enhancer of MD was linked with the black tarsus (blt) locus at 28 cM on chromosome 3 [65]. The actual genes associated with these genetic loci have not been identified.

Conclusion

In Aaa, increased male heterozygosity levels and high intersex F_{ST} genes are consistent with the presence of a proto-Y chromosome (reviewed in [25]). In contrast, chromosomal rearrangements and subsequent suppressed recombination in SenAae may have accelerated X-Y differentiation, as the features observed in Aaa were absent. Our approach also allowed us to identify additional genes associated with sex, which may include candidates for M locus modifiers. However, further characterization will be required to confirm possible mechanisms. Taken together, these data could inform transgenic strategies for vector control and the overall understanding of evolution of sex-associated genes in aedine mosquitoes.

Methods
Samples and sequencing
SenAae (PK10) and Aaa (Thai) collections were processed as follows. Deep sequencing libraries were made from pools of F1 individuals collected from the PK10 forest, Senegal in 2011 and Ae. aegypti aegypti from a collection in Pai Lom, Thailand in 2002 [7, 66, 67]. For both comparisons, mosquitoes were collected as larvae, reared to adulthood and frozen until DNA extractions. Two biological replicates for each pool (12 mosquitoes per pool) of males and females from each location produced a total of eight libraries (2 male PK10, 2 female PK10, 2 male Thai, 2 female Thai). Prior to pooling, DNA in individual mosquitoes was quantified using Pico Green (Life Technologies, Thermo Fisher Scientific Inc.) and equal amounts of DNA per mosquito were pooled. A Covaris S2 sonicator (Covaris Ltd., Brighton UK) sheared pooled DNA to an average size of 500 bp. Sonication conditions were: duty cycle 10%, Intensity 5.0, Cycles per burst 200, Duration 40 s, Mode Frequency sweeping, Displayed Power 23 W, Temperature 5.5° to 6 °C. Each TruSeq DNA LT (v.2) library was prepared using 1 μg of sheared genomic DNA following manufacturer's recommendations. Equimolar quantities of prepared libraries were pooled and enriched for coding sequences by exome capture using custom SeqCap EZ Developer probes (Nimblegen) [29]. In total, 26.7 Mb of the genome (2%) was targeted for enrichment, as described elsewhere [29]. Overlapping probes covering the protein coding sequence (not including UTRs) in the AaegL1.3 gene annotations (https://www.vectorbase.org/organisms/aedes-aegypti/liverpool-lvp/AaegL1.3) were produced by Nimblegen. Enrichment followed the Nimblegen SeqCap EZ protocol. Briefly, pooled TruSeq libraries were hybridized to the probes for 64 h, unbound DNA was washed away, and the targeted DNA was eluted and amplified. These were then sequenced on 2 lanes of a HiSeq2000 (Illumina) for paired-end 2×100 nt sequencing. TruSeq library preparation, exome capture and sequencing were performed by the High-Throughput Genomics Group at the Wellcome Trust Centre for Human Genetics (Oxford, UK) and produced reads with quality scores > 30.

Bioinformatics

Alignments and population genetics pipeline

All raw reads were trimmed of adapters and filtered using cutadapt (v. 1.14) [68]. The AaegL4 genome build [35] of 18,769 transcripts was used, including all 5'UTRs, exons, introns, 3'UTRs. The 5′ and 3′ non-transcribed regions in previously reported alignments were excluded [27]. Individual replicate fastq files were aligned to the AaegL4 genome using GSNAP (version 2017-02-25), allowing 10% divergence [69]. Using SAMtools "mpileup" command [70], GSNAP outputs were converted to *.mpileup files. The "readcounts" command in Varscan2 (v2.3.5) [71] was used to convert *.mpileup files to readcounts output, using the following options: −min-coverage 15 −min-base-qual 30. The readcounts output listed each SNP as a single row and A, C, G, T, in/del in columns.

To address possible sequencing errors, the following steps were taken: 1) a minimum of 15 variants per SNP site were required for a site to be considered; 2) only reads with Q30 passed trimming (cutadapt); this quality score was also required at each base upon alignment to the reference; 3) only those SNP sites that were present in both replicate libraries were included in F_{ST} calculations. PCR duplicates were not removed, because of the evidence that removal does not significantly alter variant calls [72].

For each SNP, in-house FORTRAN (F77) scripts (available on request) used the variant coverage per SNP site to calculate the Fumagalli F_{ST}. Between-group component (a_s), a within-group component (b_s) and F_{ST} calculated from a_s and b_s following Fumagalli [31] where:

$$a_s = \frac{4n_i\left(\widehat{p}_{(i,s)}-\widehat{p}_s\right)^2 + 4n_j\left(\widehat{p}_{(j,s)}-\widehat{p}_s\right)^2 - b_s}{2\left(2n_in_j/\left(n_i+n_j\right)\right)}$$

and $\quad b_s = \frac{n_i\alpha_{(i,s)}+n_j\alpha_{(j,s)}}{n_i+n_j-1}$

where $\alpha_{(i,s)} = 2\widehat{p}_{(i,s)}\left(1-\widehat{p}_{(i,s)}\right)$ and $\alpha_{(j,s)} = 2\widehat{p}_{(j,s)}\left(1-\widehat{p}_{(j,s)}\right)$.

$\widehat{p}_{(i,s)}$ is the coverage of a nucleotide at SNP site (s) divided by the total coverage of s in collection (i). n_i and n_j are the number of mosquitoes sampled in collections i and j, and $\widehat{p}_s$ is the coverage of a nucleotide at s in both i and j collections divided by the total coverage of s in both i and j collections. The estimate of F_{ST} for s is:

$$F_{ST}(s) = \frac{a_s}{a_s+b_s}$$

and for an entire gene (g) with m SNPs is:

$$F_{ST}(g) = \frac{\sum_{s=1}^{m} a_s}{\sum_{s=1}^{m}(a_s+b_s)}$$

Genes were annotated using Gene Ontology terms and SwissProt functional annotation data listed in AegyXcel (http://exon.niaid.nih.gov/transcriptome.html#aegyxcel), using a cut-off e value of E^{-20}.

Hardy-Weinberg expected heterozygosity (H_{exp}) values were calculated for SNP sites that were present in both males and females using the following formula, $\alpha_{(i,s)} = 2\widehat{p}_{(i,s)}\left(1-\widehat{p}_{(i,s)}\right)$; $\alpha_{(i,s)}$ is expected heterozygosity (H_{exp}). $\widehat{p}_{(i,s)}$ is the coverage of a variant at a SNP(s) site divided by the total coverage of s in the collection (i).

Female-vs-female and male-vs-male comparisons

Similar to the inter-sex comparisons, F_{ST} was also calculated for replicate female SenAae and Aaa libraries (Pk10 female vs Thai female). F_{ST} was also calculated for male-vs-male libraries to obtain the plots shown in Additional file 3.

Statistics

Descriptive statistics were calculated in R (version 3.0.2). We evaluated H_{exp} values along the length of chromosome 1 using a one-sided T-test ($p\ value < 0.05$) that tested specifically for higher average H_{exp} values in males over females. The ratio of variant sites per nucleotide of aligned reads was calculated as follows. Using flagstat (SAMtools), the number of aligned reads was determined in the reads aligned to reference *.bam files, and multiplied by the read-length (100nts) to achieve the total nucleotides aligned. The number of variants per chromosome was multiplied by 1000 and divided by the estimated total nucleotides aligned (Additional file 1).

Studies of *Ae. aegypti* RAD-tag and SNP-CHIP analyses allowed just 2 alternate alleles per locus [73, 74], whereas, here, loci with 3 or more alternate alleles were included in the final analysis. Moreover, our approach is not subject to ascertainment bias, as occurs when a small number of SNPs from the entire genome are analyzed [75, 76]. This systematic bias occurs when limited loci are analyzed rather than complete genotypic profiles.

Additional files

Additional file 1: Sequencing Statistics. SenAae (PK10) and Aaa (Thai) HTS details. Total trimmed reads aligned to the AaegL4 reference; percent reads mapped; percent properly paired; number of variant sites; ratio of variant sites per aligned nucleotide*1000. (XLSX 16 kb)

Additional file 2: F_{ST} frequency distributions. SenAae (PK10) and Aaa (Thai). (PDF 19 kb)

Additional file 3: F_{ST} frequency distributions for female-vs-female and male-vs-male comparisons. SenAae (PK10) and Aaa (Thai). Black dots indicate average F_{ST} values < 0.100; red dots indicate FST values ≥ 0.100. (PDF 699 kb)

Additional file 4: Genes with significant sex-association values common to both populations. Vectorbase number (VBN), transcript, Chr, function, Function_description_Vectorbase. (XLSX 166 kb)

Abbreviations

Aaa: *Aedes aegypti aegypti*; AaegL4: Chromosome-length *Ae. aegypti aegypti* reference sequence; *blt*: *black tarsus* locus; DNA: Deoxyribonucleic acid; F_{ST}: Standardized variance, a measure of genetic association; H_{exp}: Expected heterozygosity; *m* allele: Female recessive locus; *M* locus: Male determining locus; MD: Meiotic drive; PK10: *Aedes aegypti* collection from southeast Senegal; *s*: *Spot* locus; SenAae: Senegal *Aedes aegypti* collection; SNP: Single-nucleotide polymorphisms

Acknowledgments

We thank Karen Fleming for mosquito work and collections, as well as lab support. We thank Tyler Eike and Thomas Harrison for help with bioinformatics details. We also thank Mariangela Bonizzoni for helpful advice on the manuscript. The authors wish to thank the Franklin Graybill Statistical Laboratory at Colorado State University for statistical consulting.

Funding

This work was supported by National Institutes of Health (NIH) R01AI0833680 to WCB4. The funding body had no role in the study, collection, analysis, interpretation of data or in writing the manuscript.

Authors' contributions

CLC, LBD, and WCB4 designed the study. PJ and FJ provided the capture probe and gave technical advice on preparation details, probe design and capture probe details. CLC and LBD prepared the samples. SLF wrote scripts and provided guidance on data analysis. CLC and WCB4 analyzed the data. CLC and WCB4 wrote the manuscript. All authors read and approved the final manuscript.

Competing interests

The authors declare that they have no competing interests.

Author details

[1]Department of Microbiology, Immunology and Pathology, Colorado State University, Campus Delivery 1692, Fort Collins, CO 80523, USA. [2]Department of Genetics, University of Cambridge, Downing Street, Cambridge CB2 3EH, UK.

References

1. Brown JE, McBride CS, Johnson P, Ritchie S, Paupy C, Bossin H, Lutomiah J, Fernandez-Salas I, Ponlawat A, Cornel AJ, Black WC IV, Gorrochotegui-Escalante N, Urdaneta-Marquez L, Sylla M, Slotman M, Murray KO, Walker C, Powell JR. Worldwide patterns of genetic differentiation imply multiple 'domestications' of Aedes aegypti, a major vector of human diseases. Proc Biol Sci. 2011;278(1717):2446–54.
2. McClelland GAH. A worldwide survey of variation in scale pattern of the abdominal tergum of Aedea aegypti (L.) (Diptera:Culicidae). Trans Royal Entomol Soc Lond. 1974;126:239–59.
3. Tabachnick WJ, Powell JR. A world-wide survey of genetic variation in the yellow fever mosquito, Aedes aegypti. Genet Res. 1979;34(3):215–29.
4. Mattingly PF. Taxonomy of Aedes aegypti and related species. Bull World Health Organ. 1967;36(4):552–4.
5. Rasic G, Filipovic I, Weeks AR, Hoffmann AA. Genome-wide SNPs lead to strong signals of geographic structure and relatedness patterns in the major arbovirus vector, Aedes aegypti. BMC Genomics. 2014;15:275.
6. Dickson LB, Sanchez-Vargas I, Sylla M, Fleming K, Black WC IV. Vector competence in West African Aedes aegypti Is Flavivirus species and genotype dependent. PLoS Negl Trop Dis. 2014;8(10):e3153.
7. Sylla M, Bosio C, Urdaneta-Marquez L, Ndiaye M, Black WC IV. Gene flow, subspecies composition, and dengue virus-2 susceptibility among Aedes aegypti collections in Senegal. PLoS Negl Trop Dis. 2009;3(4):e408.
8. Presgraves DC, Orr HA. Haldane's rule in taxa lacking a hemizygous X. Science. 1998;282(5390):952–4.
9. Mattingly PF. Genetical aspects of the Aedes aegypti problem. I. Taxonom: and bionomics. Ann Trop Med Parasitol. 1957;51(4):392–408.
10. McClelland GA. Sex-linkage at two loci affecting eye pigment in the mosquito Aedes aegypti (diptera: culicidae). Can J Genet Cytol. 1966;8(2):192–8.
11. Newton ME, Southern DI, Wood RJ. X and Y chromosomes of Aedes aegypti (L.) distinguished by Giemsa C-banding. Chromosoma 1974, 49(1):41-49.
12. Hall AB, Basu S, Jiang X, Qi Y, Timoshevskiy VA, Biedler JK, Sharakhova MV, Elahi R, Anderson MA, Chen XG, Sharakhov IV, Adelman ZN, Tu Z. A male-determining factor in the mosquito Aedes aegypti. Science. 2015;348:1268.
13. Hall AB, Timoshevskiy VA, Sharakhova MV, Jiang X, Basu S, Anderson MA, Hu W, Sharakhov IV, Adelman ZN, Tu Z. Insights into the preservation of the homomorphic sex-determining chromosome of Aedes aegypti from the discovery of a male-biased gene tightly linked to the M-locus. Genome Biol Evol. 2014;6(1):179–91.
14. Salvemini M, D'Amato R, Petrella V, Aceto S, Nimmo D, Neira M, Alphey L, Polito LC, Saccone G. The orthologue of the fruitfly sex behaviour gene fruitless in the mosquito Aedes aegypti: evolution of genomic organisation and alternative splicing. PLoS One. 2013;8(2):e48554.
15. Salvemini M, Mauro U, Lombardo F, Milano A, Zazzaro V, Arca B, Polito LC, Saccone G. Genomic organization and splicing evolution of the doublesex gene, a drosophila regulator of sexual differentiation, in the dengue and yellow fever mosquito Aedes aegypti. BMC Evol Biol. 2011;11:41.
16. Clark NL, Aagaard JE, Swanson WJ. Evolution of reproductive proteins from animals and plants. Reproduction. 2006;131(1):11–22.
17. Coyne J, Orr HA. Speciation. Sunderland: Sinauer Associates; 2004.
18. Swanson WJ, Vacquier VD. The rapid evolution of reproductive proteins. Nat Rev Genet. 2002;3(2):137–44.
19. Hosken DJ, Stockley P. Sexual selection and genital evolution. Trends Ecol Evol. 2004;19(2):87–93.
20. Krzywinska E, Kokoza V, Morris M, de la Casa-Esperon E, Raikhel AS, Krzywinski J. The sex locus is tightly linked to factors conferring sex-specific lethal effects in the mosquito Aedes aegypti. Heredity. 2016;117(6):408–16.
21. Krzywinska E, Krzywinski J. Analysis of expression in the Anopheles gambiae developing testes reveals rapidly evolving lineage-specific genes in mosquitoes. BMC Genomics. 2009;10:300.
22. Haerty W, Jagadeeshan S, Kulathinal RJ, Wong A, Ravi Ram K, Sirot LK, Levesque L, Artieri CG, Wolfner MF, Civetta A, Singh RS. Evolution in the fast lane: rapidly evolving sex-related genes in Drosophila. Genetics. 2007;177(3):1321–35.
23. CI W, Davis AW. Evolution of postmating reproductive isolation: the composite nature of Haldane's rule and its genetic bases. Am Nat. 1993;142(2):187–212.
24. Ellegren H, Parsch J. The evolution of sex-biased genes and sex-biased gene expression. Nat Rev Genet. 2007;8(9):689–98.
25. Bachtrog D. Y-chromosome evolution: emerging insights into processes of Y-chromosome degeneration. Nat Rev Genet. 2013;14(2):113–24.
26. Dickson LB, Sharakhova MV, Timoshevskiy VA, Fleming KL, Caspary A, Sylla M, Black WC IV. Reproductive incompatibility involving Senegalese Aedes aegypti (L) is associated with chromosome rearrangements. PLoS Negl Trop Dis. 2016;10(4):e0004626.
27. Dickson LB, Campbell CL, Juneja P, Jiggins FM, Sylla M, Black WC IV. Exon-enriched libraries reveal large genic differences between Aedes aegypti from Senegal, West Africa, and populations outside Africa. G3. 2017;7(2):571–82.
28. Fontaine A, Filipovic I, Fansiri T, Hoffmann AA, Cheng C, Kirkpatrick M, Rasic G, Lambrechts L. Extensive genetic differentiation between Homomorphic sex chromosomes in the mosquito vector, Aedes aegypti. Genome Biol Evol. 2017;9(9):2322–35.
29. Juneja P, Ariani CV, Ho YS, Akorli J, Palmer WJ, Pain A, Jiggins FM. Exome and Transcriptome sequencing of Aedes aegypti identifies a locus that confers resistance to Brugia malayi and alters the immune response. PLoS Pathog. 2015;11(3):e1004765.

30. Crawford JE, Alves JM, Palmer WJ, Day JP, Sylla M, Ramasamy R, Surendran SN, Black WC IV, Pain A, Jiggins FM. Population genomics reveals that an anthropophilic population of Aedes aegypti mosquitoes in West Africa recently gave rise to American and Asian populations of this major disease vector. BMC Biol. 2017;15(1):16.

31. Fumagalli M, Vieira FG, Korneliussen TS, Linderoth T, Huerta-Sanchez E, Albrechtsen A, Nielsen R. Quantifying population genetic differentiation from next-generation sequencing data. Genetics. 2013;195(3):979–92.

32. Haldane JS. Sex ratio and unisexual sterility in hybrid animals. J Genetics. 1922;12:101–9.

33. Barreiro LB, Laval G, Quach H, Patin E, Quintana-Murci L. Natural selection has driven population differentiation in modern humans. Nat Genet. 2008;40(3):340–5.

34. VectorBase: a home for invertebrate vectors of human pathogens (http://www.vectorbase.org/).

35. Dudchenko O, Batra SS, Omer AD, Nyquist SK, Hoeger M, Durand NC, Shamim MS, Machol I, Lander ES, Aiden AP, Aiden EL. De novo assembly of the Aedes aegypti genome using Hi-C yields chromosome-length scaffolds. Science. 2017;356(6333):92–5.

36. Nene V, Wortman JR, Lawson D, Haas B, Kodira C, ZJ T, Loftus B, Xi Z, Megy K, Grabherr M, Ren Q, Zdobnov EM, Lobo NF, Campbell KS, Brown SE, Bonaldo MF, Zhu J, Sinkins SP, Hogenkamp DG, Amedo P, Arsenburger P, Atkinson PW, Bidwell S, Biedler J, Birney E, Bruggner RV, Costas J, Coy MR, Crabtree J, Crawford M, Debruyn B, Decaprio D, Eiglmeier K, Eisenstadt E, El-Dorry H, Gelbart WM, Gomes SL, Hammond M, Hannick LI, Hogan JR, Holmes MH, Jaffe D, Johnston SJ, Kennedy RC, Koo H, Kravitz S, Kriventseva EV, Kulp D, Labutti K, Lee E, Li S, Lovin DD, Mao C, Mauceli E, Menck CF, Miller JR, Montgomery P, Mori A, Nascimento AL, Naveira HF, Nusbaum C, O'Leary SB, Orvis J, Pertea M, Quesneville H, Reidenbach KR, Rogers YH, Roth CW, Schneider JR, Schatz M, Shumway M, Stanke M, Stinson EO, Tubio JM, Vanzee JP, Verjovski-Almeida S, Werner D, White O, Wyder S, Zeng Q, Zhao Q, Zhao Y, Hill CA, Raikhel AS, Soares MB, Knudson DL, Lee NH, Galagan J, Salzberg SL, Paulsen IT, Dimopoulos G, Collins FH, Bruce B, Fraser-Liggett CM, Severson DW. Genome sequence of Aedes aegypti, a major Arbovirus vector. Science. 2007;316:1718.

37. Hickey WA, Craig GB Jr. Distortion of sex ratio in populations of Aedes aegypti. Can J Genet Cytol. 1966;8(2):260–78.

38. Hoang KP, Teo TM, Ho TX, Le VS. Mechanisms of sex determination and transmission ratio distortion in Aedes aegypti. Parasit Vectors. 2016;9:49.

39. Wood RJ, Newton ME. Sex-ratio distortion caused by meiotic drive in mosquitoes. Am Nat. 1991;137(3):379–91.

40. Lucotte EA, Laurent R, Heyer E, Segurel L, Toupance B. Detection of allelic frequency differences between the sexes in humans: a signature of sexually antagonistic selection. Genome Biol Evol. 2016;8(5):1489–500.

41. Cheng C, Kirkpatrick M. Sex-specific selection and sex-biased gene expression in humans and flies. PLoS Genet. 2016;12(9):e1006170.

42. Coolon JD, Stevenson KR, McManus CJ, Yang B, Graveley BR, Wittkopp PJ. Molecular mechanisms and evolutionary processes contributing to accelerated divergence of gene expression on the drosophila X chromosome. Mol Biol Evol. 2015;32(10):2605–15.

43. AegyXcel; http://exon.niaid.nih.gov/transcriptome.html#aegyxcel.

44. McBride CS, Baier F, Omondi AB, Spitzer SA, Lutomiah J, Sang R, Ignell R, Vosshall LB. Evolution of mosquito preference for humans linked to an odorant receptor. Nature. 2014;515(7526):222–7.

45. Sparks JT, Bohbot JD, Dickens JC. The genetics of chemoreception in the labella and tarsi of Aedes aegypti. Insect Biochem Mol Biol. 2014;48:8–16.

46. Dubendorfer A, Hediger M, Burghardt G, Bopp D. Musca domestica, a window on the evolution of sex-determining mechanisms in insects. Int J Dev Biol. 2002;46(1):75–9.

47. Franco MG, Rubini PG, Vecchi M. Sex-determinants and their distribution in various populations of Musca domestica L. of Western Europe. Genet Res. 1982;40(3):279–93.

48. Thompson PE, Bowen JS. Interactions of differentiated primary sex factors in Chironomus Tentans. Genetics. 1972;70(3):491–3.

49. Black DL. Mechanisms of alternative pre-messenger RNA splicing. Annu Rev Biochem. 2003;72:291–336.

50. Marinho J, Martins T, Neto M, Casares F, Pereira PS. The nucleolar protein Viriato/Nol12 is required for the growth and differentiation progression activities of the Dpp pathway during Drosophila eye development. Dev Biol. 2013;377(1):154–65.

51. Sirot LK, Hardstone MC, Helinski ME, Ribeiro JM, Kimura M, Deewatthanawong P, Wolfner MF, Harrington LC. Towards a semen proteome of the dengue vector mosquito: protein identification and potential functions. PLoS Negl Trop Dis. 2011;5(3):e989.

52. Charlesworth B. The evolution of chromosomal sex determination and dosage compensation. Curr Biol. 1996;6(2):149–62.

53. Lahn BT, Pearson NM, Jegalian K. The human Y chromosome, in the light of evolution. Nat Rev Genet. 2001;2(3):207–16.

54. Lemaitre C, Braga MD, Gautier C, Sagot MF, Tannier E, Marais GA. Footprints of inversions at present and past pseudoautosomal boundaries in human sex chromosomes. Genome Biol Evol. 2009;1:56–66.

55. Graham DH, Holmes JL, Black WC IV. Identification of quantitative trait loci affecting sex determination in the eastern treehole mosquito (Ochlerotatus triseriatus). J Hered. 2004;95(1):35–45.

56. Christophers S. Aedes aegypti (L.), the yellow-fever mosquito; its life history, bionomics and structure. London: Cambridge University Press; 1960.

57. Craig GB Jr, Vandehey RC, Hickey WA. Genetic variability in populations of Aedes aegypti. Bull World Health Organ. 1961;24:527–39.

58. Newton ME, Wood RJ, Southern DI. A cytogenetic analysis of meiotic drive in the mosquito, Aedes aegypti (L.). Genetica. 1976;46(3):297–318.

59. Hickey WA, Craig GB Jr. Genetic distortion of sex ratio in a mosquito, Aedes aegypti. Genetics. 1966;53(6):1177–96.

60. Owusu-Daaku KO, Wood RJ, Butler RD. Selected lines of Aedes aegypti with persistently distorted sex ratios. Heredity. 1997;79(Pt 4):388–93.

61. Suguna SG, Wood RJ, Curtis CF, Whitelaw A, Kazmi SJ. Resistance to meiotic drive at the MD locus in an Indian wild population of Aedes aegypti. Genet Res. 1977;29(2):123–32.

62. Wood RJ. Between-family variation in sex ratio in the Trinidad (T-30) strain of Aedes aegypti (L.) indicating differences in sensitivity to the meiotic drive gene MD. Genetica. 1976;46(3):345–61.

63. Mori A, Chadee DD, Graham DH, Severson DW. Reinvestigation of an endogenous meiotic drive system in the mosquito, Aedes aegypti (Diptera: Culicidae). J Med Entomol. 2004;41(6):1027–33.

64. Craig GB Jr, Hickey WA, Vandehey RC. An inherited male-producing factor in Aedes aegypti. Science. 1960;132(3443):1887–9.

65. Wood RJ, Ouda NA. The genetic basis of resistance and sensitivity to the meiotic drive gene D in the mosquito Aedes aegypti L. Genetica. 1987;72(1):69–79.

66. Bosio CF, Harrington LC, Jones JW, Sithiprasasna R, Norris DE, Scott TW. Genetic structure of Aedes aegypti populations in Thailand using mitochondrial DNA. Am J Trop Med Hyg. 2005;72(4):434–42.

67. Huber K, Ba Y, Dia I, Mathiot C, Sall AA, Diallo M. Aedes aegypti in Senegal: genetic diversity and genetic structure of domestic and sylvatic populations. Am J Trop Med Hyg. 2008;79(2):218–29.

68. Martin M. Cutadapt removes adapter sequences from high-throughput sequencing reads. EMBnet J. 2011;17(1):10–2.

69. TD W, Nacu S. Fast and SNP-tolerant detection of complex variants and splicing in short reads. Bioinformatics. 2010;26(7):873–81.

70. Li H, Handsaker B, Wysoker A, Fennell T, Ruan J, Homer N, Marth G, Abecasis G, Durbin R, Genome Project Data Processing S. The sequence alignment/map format and SAMtools. Bioinformatics. 2009;25(16):2078–9.

71. Koboldt DC, Zhang Q, Larson DE, Shen D, McLellan MD, Lin L, Miller CA, Mardis ER, Ding L, Wilson RK. VarScan 2: somatic mutation and copy number alteration discovery in cancer by exome sequencing. Genome Res. 2012;22(3):568–76.

72. Ebbert MT, Wadsworth ME, Staley LA, Hoyt KL, Pickett B, Miller J, Duce J, Alzheimer's Disease Neuroimaging I, Kauwe JS, Ridge PG. Evaluating the necessity of PCR duplicate removal from next-generation sequencing data and a comparison of approaches. BMC Bioinf. 2016;17(Suppl 7):239.

73. Brown JE, Evans BR, Zheng W, Obas V, Barrera-Martinez L, Egizi A, Zhao H, Caccone A, Powell JR. Human impacts have shaped historical and recent evolution in Aedes aegypti, the dengue and yellow fever mosquito. Evolution. 2014;68(2):514–25.

74. Evans BR, Gloria-Soria A, Hou L, McBride C, Bonizzoni M, Zhao H, Powell JR. A multipurpose, high-throughput single-nucleotide polymorphism chip for the dengue and yellow fever mosquito, Aedes aegypti. G3. 2015;5(5):711–8.

75. Frascaroli E, Schrag TA, Melchinger AE. Genetic diversity analysis of elite European maize (Zea mays L.) inbred lines using AFLP, SSR, and SNP markers reveals ascertainment bias for a subset of SNPs. Theor Appl Genet. 2013;126(1):133–41.

76. Seeb JE, Carvalho G, Hauser L, Naish K, Roberts S, Seeb LW. Single-nucleotide polymorphism (SNP) discovery and applications of SNP genotyping in nonmodel organisms. Mol Ecol Resour. 2011;11(Suppl 1):1–8.

10

Isogenic mice exhibit sexually-dimorphic DNA methylation patterns across multiple tissues

Helen McCormick[1,2], Paul E. Young[1], Suzy S. J. Hur[1], Keith Booher[3], Hunter Chung[3], Jennifer E. Cropley[1,2], Eleni Giannoulatou[1,2*] and Catherine M. Suter[1,2*]

Abstract

Background: Cytosine methylation is a stable epigenetic modification of DNA that plays an important role in both normal physiology and disease. Most diseases exhibit some degree of sexual dimorphism, but the extent to which epigenetic states are influenced by sex is understudied and poorly understood. To address this deficit we studied DNA methylation patterns across multiple reduced representation bisulphite sequencing datasets (from liver, heart, brain, muscle and spleen) derived from isogenic male and female mice.

Results: DNA methylation patterns varied significantly from tissue to tissue, as expected, but they also varied between the sexes, with thousands of sexually dimorphic loci identified. The loci affected were largely autonomous to each tissue, even within tissues derived from the same germ layer. At most loci, differences between genders were driven by females exhibiting hypermethylation relative to males; a proportion of these differences were independent of the presence of testosterone in males. Loci harbouring gender differences were clustered in ontologies related to tissue function.

Conclusions: Our findings suggest that gender is underwritten in the epigenome in a tissue-specific and potentially sex hormone-independent manner. Gender-specific epigenetic states are likely to have important implications for understanding sexually dimorphic phenotypes in health and disease.

Keywords: Epigenetics, DNA methylation, Sexual dimorphism, Gender, Tissue-specific methylation, RRBS

Background

Sexual dimorphism, in which the two sexes exhibit different characteristics, affects a range of traits in animals. Almost all human diseases exhibit some component of sexual dimorphism, which can manifest as differences in disease incidence, age of onset, or severity of symptoms [1]. The discordance cannot be completely explained by genetic (i.e. sex chromosome) differences or the actions of sex hormones [2]. Despite increasing efforts in understanding the epigenetic contribution to complex disease [3], the contribution of epigenetic factors (such as DNA methylation) to sexual dimorphism is relatively underexplored.

There is evidence for gender differences in DNA methylation in various tissues in eutherian mammals. Unsurprisingly, the majority of differentially methylated loci occur on the X-chromosome, where they are hypermethylated in females relative to males [4]. However, sex-specific methylation also occurs on autosomes. A recent meta-analysis of Infinium 450 K array data from 76 individual studies identified gender-specific methylation at about 200 autosomal CpG sites in peripheral blood [4]. Gender-specific methylation differences have also been reported using 450 K array in human prefrontal cortex [5], saliva [6], and pancreatic islets [7], and by reduced representation bisulphite sequencing (RRBS) in mouse liver [8]. A notable recent study used RRBS to find gender-specific methylation at 160 autosomal loci in mouse liver [9]. In this study, a bias towards hypomethylation in males was dependent on testosterone exposure during puberty, which was coincident with loss of

* Correspondence: e.giannoulatou@victorchang.edu.au;
c.suter@victorchang.edu.au
[1]Victor Chang Cardiac Research Institute, 405 Liverpool Street, Darlinghurst, NSW 2010, Australia
Full list of author information is available at the end of the article

methylation. This study suggests that there is an inter-action between sex hormones and epigenetic marks, but whether gender-specific marks might exist independent of hormonal or other postnatal factors is unknown.

Given that the bulk of de novo DNA methylation in mammals takes place during the very early stages of embryogenesis, it is possible that at least some gender methylation differences are specified early in development, or are even innate. Such gender differences would be consistent across different tissues, possibly even tissues derived from different germ layers. Here we have explored this possibility by performing RRBS on male and female mouse tissues derived from each of the three germ layers: liver (endoderm), heart (mesoderm), and brain (ectoderm). We find sexually dimorphic, differentially methylated loci in all tissues. By combining our analyses with publically available data, we find that in the liver, a proportion of loci exhibit gender-specific methylation in the absence of testosterone. Very few gender-specific differences are shared among tissues, even those from the same germ layer.

Results
Gender is predicted by DNA methylation patterns in the mouse liver
We generated snapshots of genome-wide cytosine methylation patterns with reduced representation bisul-phite sequencing (RRBS). RRBS captures only ~1% of the mammalian genome, but the captured fraction is highly enriched for functional regions such as CpG islands, shores, and other gene regulatory elements. Here we used an enhanced RRBS protocol to produce methylomes from the livers of six female and six male isogenic C57BL/6 J mice. Overall, data quality was very high, as were inter-sample correlations (Additional file 1: Figure S1). For all analyses we considered only those CpGs that were present at >10× coverage in all 12 samples. To combine statistical evidence of neighbour-ing CpGs, we calculated DNA methylation levels in tiles of 100 bp across the genome (resulting in 820,388 tiles covered across all 12 samples).

Unsupervised hierarchical clustering separated samples by gender, even when data from sex chromosomes was removed (Fig. 1a). Principal components analysis (PCA) also showed a distinct spatial clustering of samples by gender irrespective of whether sex chromosomes were included (Fig. 1b) or not (Fig. 1c). We identified regions of difference between genders using methylKit [10]. We found 1093 tiles that were differentially methylated (q-value < 0.01 and ≥ 25% methylation difference) be-tween males and females (Fig. 1d; Additional file 2: Table S1). Despite the huge enrichment for CpG islands in our RRBS libraries, the majority of differentially methylated tiles (DMTs) were found outside of CpG islands, in both

genic and intergenic regions (Fig. 1e). Around a third of the DMTs overlapped with ENCODE liver H3K4me1 and H3K27Ac peaks, a highly significant enrichment (both $p < 1 \times 10^{-4}$). This is consistent with many of the intergenic DMTs residing in active enhancers.

We chose a subset of ten DMTs to validate by COBRA [11]. All but one of the ten loci showed a difference in methylation levels between males and females as pre-dicted by the RRBS signal (Additional file 3: Figure S2). This experimental validation indicates that our RRBS and informatics strategy detects gender-specific differ-ences with high confidence.

Gender DMTs in the liver can be testosterone independent
We next reanalysed data from Reizel et al.. [9] (GEO Accession GSE60012), who previously identified 160 gender DMTs in the adult mouse liver. Using our bio-informatics pipeline with their liver dataset, we identified 83 autosomal gender DMTs, of which 48 (~58%) over-lapped with the gender DMTs from our data (Fig. 2a, Additional file 2: Table S2). The difference in the num-ber of DMTs identified can be at least partially attributed to the difference in dataset size (our liver dataset con-tained 820,388 tiles, and Reizel et al's contained 167,462 tiles). Like Reizel and colleagues, we found that the ma-jority of gender DMTs in the liver could be attributed to hypermethylation in females relative to males; this was true across all autosomes (Fig. 2b). Interestingly, the gender DMTs identified in the liver by Reizel et al. were absent from the liver of males who were castrated when young, and could be reconstituted in castrated males with exogenous testosterone administration [9]. This implies that gender DMTs in the liver are testosterone-dependent. We extended these observations by perform-ing an unbiased comparison of the Reizel et al. females with the castrated males, using our own informatics pipeline. In doing so we were able to identify 228 gender DMTs, despite the absence of testosterone in the males (Additional file 2: Table S3). These testosterone-independent gender DMTs were also heavily skewed towards hypermethylation in females (Fig. 2c). Only four of these DMTs are also found in our set of 1093 DMTs (Fig. 2d), although almost all the tiles in the castrated male dataset were also represented in our dataset. This suggests that there are additional factors beyond testosterone that are able to specify gender-specific methylation patterns.

Heart and brain also harbour gender DMTs
The robust gender differences observed in the mouse liver in this study and others, along with reports of gen-der bias in human non-liver tissues such as peripheral blood leukocytes, prompted us to ask whether the

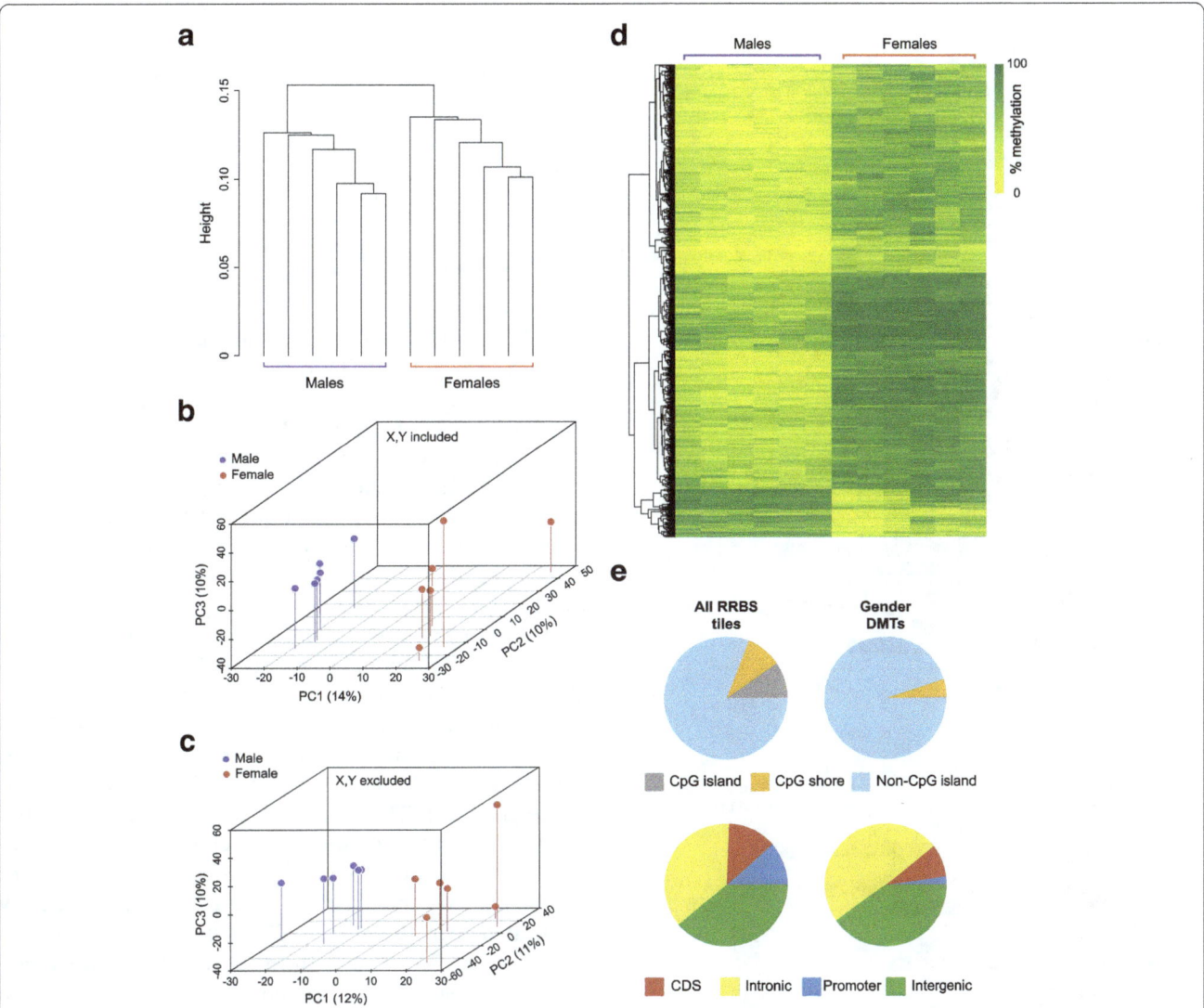

Fig. 1 Autosomal DNA methylation patterns in the liver distinguish gender. **a** Dendrogram showing results of unsupervised hierarchical clustering of liver RRBS data from six males and six females with sex chromosomes excluded. **b, c** Pseudo-3D principal component analysis (PCA) plots of the first three principal components of liver RRBS data as in (a) with sex chromosome data included (b), or removed (c); males are shown in blue, females in red. **d** Clustered heat map of differentially methylated tiles (DMTs) identified in the liver of males versus females. **e** CpG island (top) and genomic annotation (bottom) of CpG tiles present in all liver RRBS data (left) and gender DMTs (right)

gender DMTs might be conserved across tissues. Given that some DMTs in liver appear to be sex hormone-independent, it is possible that these differences in methylation arise in the germline, or very early in development. To address this possibility we interrogated the methylomes of brain and heart from three of the males and three of the females used for liver analysis. We chose these tissues as they derive from different germ layer origins (heart, mesoderm; brain, ectoderm; liver, endoderm). Unlike the liver methylome, analysis of the brain methylome by hierarchical clustering or PCA did not separate the genders, even when sex chromosomes were included (Fig. 3a); likewise in the heart we observed no distinct clustering, although the genders

separate very slightly on the first principal component (PC1; Fig. 3b). Despite the lack of unsupervised clustering by gender, both heart and brain harboured gender DMTs. Applying the same stringent parameters as used for the liver analysis, we identified 957 autosomal DMTs in the brain (Fig. 3c; Additional file 2: Table S4), and 145 in the heart (Fig. 3d; Additional file 2: Table S5). While the brain gender DMTs exhibited, like the liver, a clear bias towards hypermethylation in females (Fig. 3f), this bias was absent from the gender DMTs in the heart (Fig. 3g). The gender DMTs in both tissues were again mostly outside CpG islands, in intronic and intergenic regions (Fig. 3e). Like the liver DMTs, the brain DMTs were significantly associated with enhancer regions ($p <$

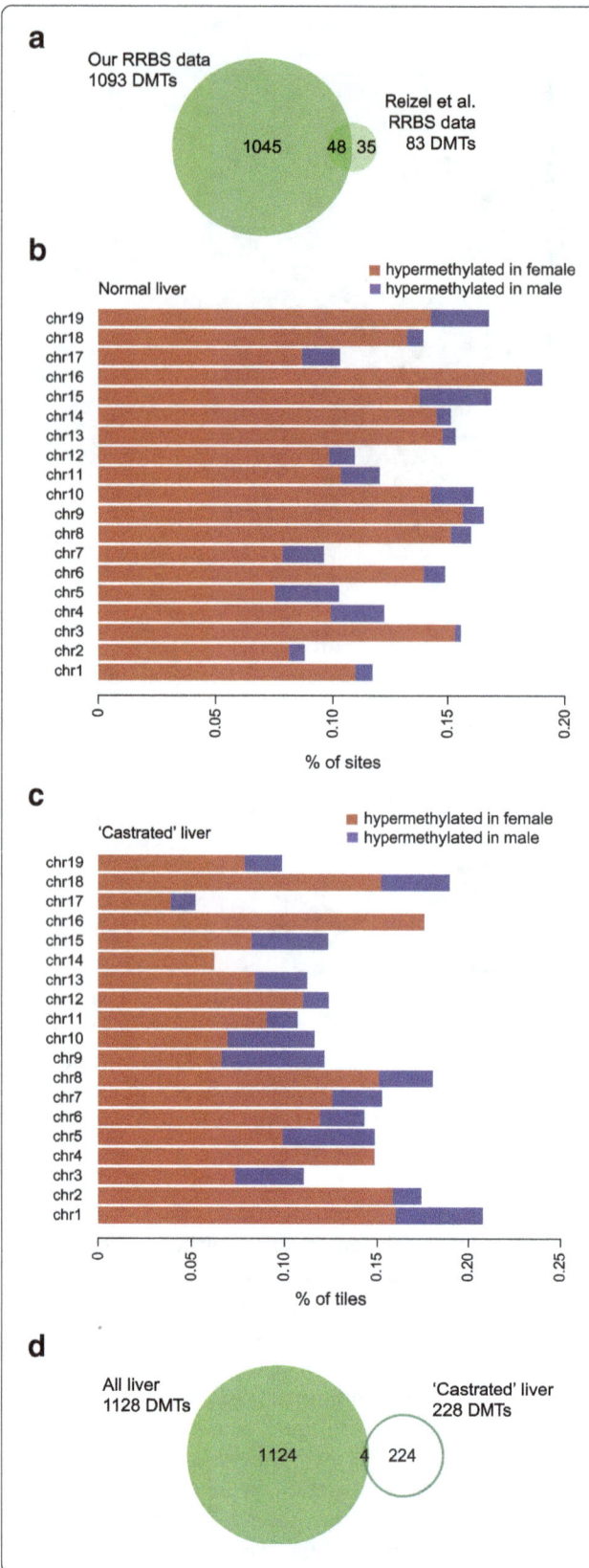

Fig. 2 Gender-specific methylation in the liver does not require testosterone. **a** Venn diagram showing overlap of gender DMTs in the liver identified by this study and our reanalysis of Reizel et al. [9]. **b** Bar plot of gender DMTs from our liver RRBS data showing % of tiles hypermethylated in females (red) and males (blue) across all autosomes. **c** Bar plot of gender DMTs between normal females and castrated males showing % of tiles hypermethylated in females (red) and males (blue) across all autosomes. **d** Venn diagram showing overlap of all liver gender DMTs (both those identified here and those from Reizel et al. [9]) and testosterone independent DMTs (i.e. differentially methylated between normal females and castrated males)

1×10^{-4}), albeit with less fraction of overlap (~5% of DMTs); but the heart DMTs were not ($p = 0.74$).

Gender DMTs are tissue specific

We then asked whether there was any overlap of the gender DMTs we identified among tissues of distinct germ layer origin that might suggest that they were inborn. Despite the large number of DMTs identified across the three tissues in total we found almost no overlap (Fig. 4a). Only two liver gender DMTs were common to the brain, and no liver DMT was common to the heart (although 11 of the 145 heart DMTs were also differentially methylated in the brain). We determined functional pathways associated with the DMTs in each tissue; in most cases, significantly enriched pathways were functionally related to the relevant tissue (Fig. 4b). This may not be surprising given that the DMTs tend to be over-represented in enhancer regions. PCA of all of our RRBS data across all tissues confirmed that methylation patterns overall are tightly associated with tissue type (Fig. 4c); it is interesting to note that methylation patterns in the brain are much more variable among isogenic individuals than the other tissues examined.

While this analysis suggests that gender DMTs do not arise in the germline, we considered whether gender DMTs might be common to tissues of the same germ layer origin (i.e. set early in development). We identified DMTs from additional mesodermal tissues studied by Reizel et al. (skeletal muscle and spleen from adult mice) to compare with our mesodermal (heart) tissue. We identified 354 and 48 gender DMTs in muscle and spleen respectively (Additional file 2: Tables S6 and S7), but these DMTs were exclusive to both each other and to the heart (Fig. 4d); this is despite almost all tiles from the muscle and spleen datasets being represented in our heart dataset. Taken together, these comparisons indicate that differential cytosine methylation between genders is almost entirely tissue-specific.

Discussion

Here we confirm the existence of sex-specific cytosine methylation in the mouse liver and find that tissues

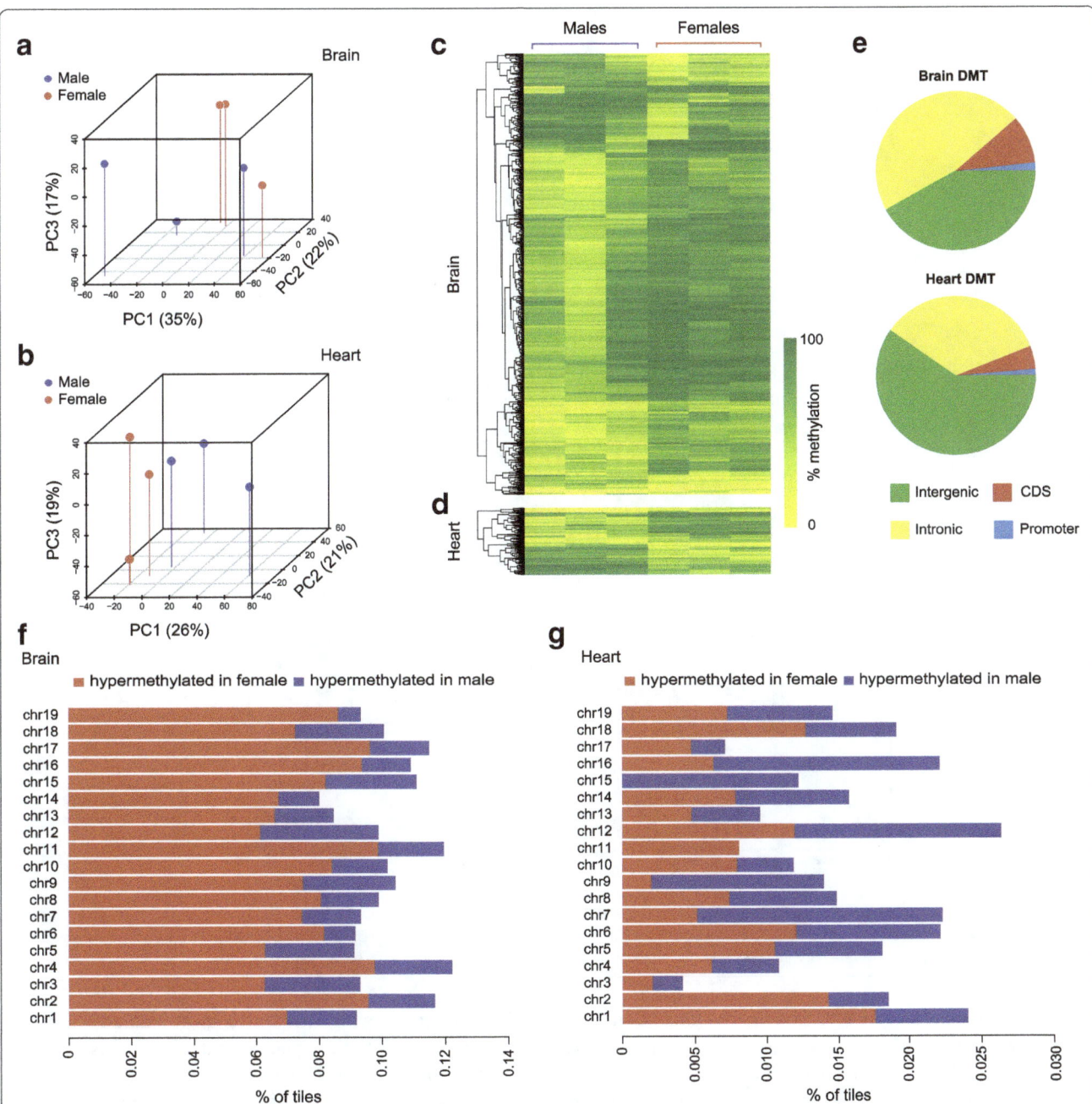

Fig. 3 Mouse heart and brain also harbour gender-specific methylation. **a, b** Pseudo-3D PCA plots of the first three principal components of RRBS data from brain (a) and heart (b); males are shown in blue, females in red. **c, d** Heat maps of differentially methylated tiles (DMTs) identified in the brain (**c**) and heart (**d**) of males versus females. **e** Genomic annotations of brain and heart gender DMTs. **f, g** Bar plot of gender DMTs from brain (**f**) and heart (**g**) showing % of tiles hypermethylated in females (in red) and males (blue) across all autosomes

derived from all three embryonic germ layers also exhibit sexually dimorphic patterns of methylation that are essentially idiosyncratic to a tissue. In the majority of instances, sex differences were manifest as a strong female bias towards hypermethylation, and this was the case in the liver even when the demethylating action of testosterone in males was removed. Our findings suggest that the DNA methylome undergoes gender differentiation in multiple tissues at some point after lineage specification, in response to tissue-dependent mechanisms.

Tissues of different origins displayed varying extents of gender-specific methylation: the brain harboured more than a thousand differentially methylated regions whereas the heart had only a few hundred. Sex-specific methylation in the brain was remarkable also because it occurred on a very high background of inter-individual variation.

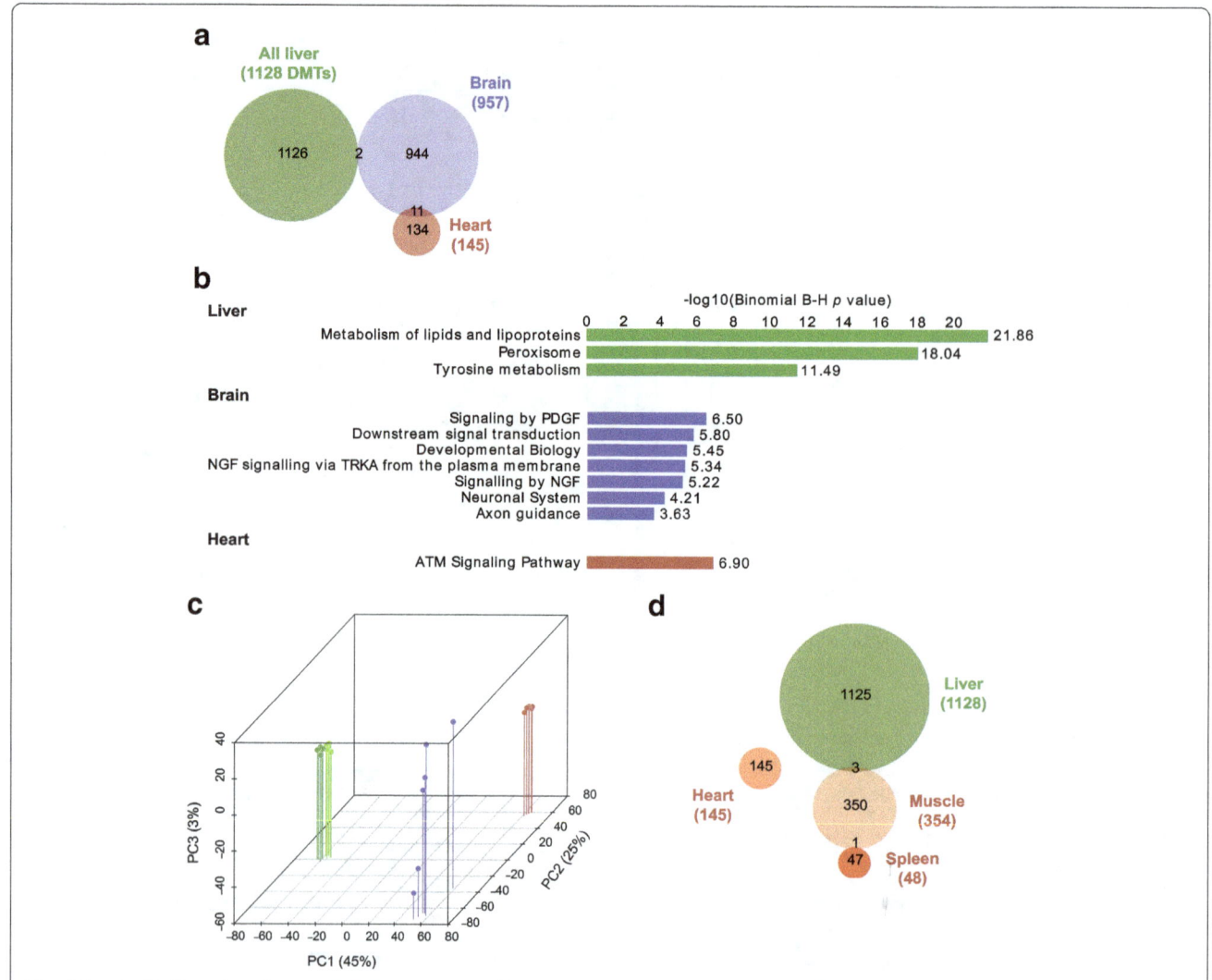

Fig. 4 Gender DMTs are largely tissue autonomous. **a** Venn diagram showing overlap of gender DMTs from liver, brain and heart. **b** Molecular functions overrepresented by regions harbouring gender DMTs in liver, brain and heart. **c** Pseudo-3D PCA plot of RRBS data from liver (green), brain (blue) and heart (red). **d** Venn diagram showing overlap of gender DMTs from spleen, heart, skeletal muscle and liver

Extensive inter-individual variation in the methylome of the human brain has been recognised only recently, and attributed largely to a secondary influence of genetic variation [12]. However in our study all samples were derived from isogenic mice, indicating that at least some proportion of inter-individual variation in methylation patterns in the brain is purely epigenetic in nature, and sex-specific epigenetic signatures are overlaid on this. As our criteria for calling sex-specific methylation was strict ($q < 0.01$, methylation difference $\geq 25\%$), it is probable that thousands more sex-specific differences exist in the brain, albeit smaller in magnitude or restricted to specific cell subtypes.

Despite identifying many thousands of gender differences in cytosine methylation overall, we found very few that were in common among tissues. Even tissues originating from the same germ layer displayed a completely autonomous set of gender methylation differences.

Taken together with the ontological analysis showing enrichment for tissue-specific pathways, this strongly suggests that gender-specific epigenetic differences are likely to have a tissue-specific function. This idea is supported by our finding that gender-specific methylation differences overlap significantly with histone markers of enhancer regions. It is also consistent with the recent finding that gender-specific methylation in the liver correlates with hepatic gene expression, particularly when the methylation differences occur in regions corresponding to tissue-specific enhancers [9]. The expression of more than a thousand genes in the mammalian liver is sex-dependent, which is not surprising given that sexual dimorphism in both steroid and drug metabolism underpin normal liver physiology [13]. Whether gender biases in methylation in heart and brain also influence their respective transcriptomes is not known, however the tissue-specific functional pathways we find suggest that this may well be the case.

The mechanisms underpinning gender-specific methylation are generally unexplored. However two recent studies have implicated testosterone in the process. In the mouse forebrain, male-specific methylation patterns can be induced in females by neonatal administration of exogenous testosterone [14]. In the liver, testosterone can trigger male-specific demethylation of certain regions; like females, castrated males maintain these regions as methylated [9]. However, not all gender-specific methylation can be attributed to the actions of testosterone: in an unbiased comparison we found more than 300 gender differences between the livers of females and castrated males. While these testosterone-independent gender differences account for a minority of all liver differences, they show the same strong tendency for hypermethylation in females, suggesting that other sex-specific *trans*-acting factors are also capable of effecting methylation dimorphism. Such factors may involve early and indirect actions of the sex chromosomes, as has been shown for early embryonic gene expression [15]. Female hormones such as estradiol are also capable of skewing gene expression patterns [16], and so may also be a contributor.

While the weight of our evidence supports widespread tissue-specific methylation differences between genders, our study is not without limitations and there are many unknowns that warrant further investigation. Comparative analysis of many tissues across multiple developmental stages (including embryonic, fetal, neonatal, and adult) would be required to establish the precise timing of establishment of gender differences, and the relationship to the changing hormonal milieu. Furthermore, RRBS allows the interrogation of methylation levels at only a representative fraction of the genome, and while this proportion is enriched for functionally relevant regions, whole genome bisulphite sequencing would be required to capture the full extent of gender-specific DNA methylation. Such studies will only become more feasible with the increasing affordability of whole genome sequencing.

Tissue-specific gender-bias in DNA methylation represents yet another aspect of sexual dimorphism but its function is currently unclear. While likely to reflect normal tissue physiology, such gender bias in epigenetic states may also have broader implications. Intrinsic gender bias in epigenetic state holds the potential to influence disease susceptibility and disease course, and modulate the response to environmental stressors. Gender differences in environmental epigenetic programming are pervasive [17] and our findings here suggest that at least some of the gender bias in induced phenotypes derives from baseline gender differences in epigenetic state.

Conclusion

Our study of the DNA methylomes of multiple tissues of male and female mice indicates that sex significantly influences DNA methylation patterns in a tissue-specific manner. These findings provide a platform to better understand the role of DNA methylation in health and disease and have important implications for the study of complex and programmed phenotypes. Our findings underscore the need to consider both genders in epigenome-wide association studies, and reinforce the requirements related to the choice of tissue to study.

Methods

Mice and tissues

All animals in this study were generated contemporaneously. Mice were handled in accordance with good practice as defined by the National Health and Medical Research Council (Australia) Statement on Animal Experimentation, and requirements of state government legislation. The study was approved by the Garvan/St Vincent's Animal Ethics Committee (#13/35). Six male and six female C57BL/6 J littermate pairs were selected at weaning and maintained on NIH-31 chow until 24 weeks of age. Tissues (brain, heart, liver) were collected and snap frozen prior to storage at −80°C until DNA extraction. DNA was extracted from a 5 mm coronal section of the rostral end of the brain, a 5 mm apical section of the heart, and the extreme caudal section of the left lobe of the liver.

DNA extraction and RRBS

Frozen tissue sections were homogenised in lysis buffer (50 mM Tris-HCl pH 8.0, 100 mM EDTA, 1% SDS, 100 mM NaCl) and incubated with 400 µg/ml Proteinase K (Roche) overnight at 55°C, followed by phenol:chloroform extraction and ethanol precipitation. Genomic DNA was used for RRBS library preparation and sequencing through the Methyl-MiniSeq service of Zymo Research (Irvine, CA, USA).

Differential methylation analysis of RRBS data

RRBS data from liver heart and brain generated for this study have been deposited in the Gene Expression Omnibus (GEO) under accession number GSE84573. eRRBS and mapping was performed by Zymo Research using proprietary methods. Bed data supplied was reformatted, and identification of differentially methylated cytosines and annotation was carried out in R with the methylKit package, v0.99.2 [10]. To combine statistical evidence of neighbouring CpGs, we calculated DNA methylation levels in tiles of 100 bp across the genome. For all analyses we considered only tiles that were present at > 10× coverage in all replicate samples and < 99.8th percentile of coverage values across all samples in the comparison.

Hierarchical clustering was performed using correlation and the Ward's minimum variance method. Identification of differentially methylated tiles was carried out with methylKit; thresholds for calling a difference were set at a methylation difference of $\geq 25\%$, and $q \leq 0.01$. Genomic annotations of differentially methylated tiles were performed using the mm10 refgene table from UCSC Table Browser. Gene ontology analysis was performed with GREAT [18] using all default parameters with the exception that distal gene regulatory domains were set at a maximum of 100 kb. Histone modification enrichment analysis was performed by permutation of overlaps with replicated H3K4me1 and H3K27Ac peak files from ENCODE (liver, ENCSR000CDH and ENCSR000CAO; brain, ENCSR000CDF and ENCSR000CAE; heart, ENCSR000CDD and ENCSR000CAI); the overlap of DMTs was compared with the overlap observed using the same number of tiles randomly chosen from the dataset, over 10,000 iterations.

Differential methylation analysis of RRBS data obtained from Reizel et al.

Datasets from liver, castrated liver, muscle and spleen from adult mice (20 weeks) were obtained from GEO Accession GSE60012. Sequencing reads were aligned to the mm10 genome with Bismark [19] and methylation percentage calls for each CpG site calculated using MethylKit. The sex chromosomes were removed. Our quality control based on pairwise correlations of all samples identified five liver outliers (three female and two male samples) that were excluded from further analysis. Similarly to the analysis of our RBBS data, to combine statistical evidence of neighbouring CpGs, we calculated DNA methylation levels in tiles of 100 bp across the genome, and considered only tiles that were present at $> 10\times$ coverage in replicate samples and < 99.8th percentile of coverage values. Given the different number of replicate samples used by Reizel et al. compared to our experimental design, we set this coverage restriction in at least across six samples in each group for liver female vs. male samples comparison, five samples in each group for females vs. castrated liver samples, six samples in each group for muscle female vs. male samples and three samples in each group for the spleen female vs. male samples comparison. Identification of differentially methylated tiles and annotations was carried out with methylKit; thresholds were set at a methylation difference of $\geq 25\%$, and $q \leq 0.01$.

Combined bisulphite restriction analysis (COBRA) validations

Ten loci identified as differentially methylated by RRBS were randomly chosen for independent verification by COBRA, with the additional criterion that they were in the vicinity of coding genes, long non-coding RNA or retrotransposon repeats. Primers were designed using MethPrimer [20] and sequences are available on request. Following bisulphite PCR and digestion with the relevant restriction enzyme, amplicons representing the fraction of methylated and unmethylated template were quantified using densitometry (Multi Gauge V2.3).

Additional files

> **Additional file 1: Figure S1.** Correlation matrices and corresponding correlation coefficients for each liver RRBS dataset. (PDF 251 kb)
>
> **Additional file 2: Table S1.** DMTs between females and males in mouse liver. P-values and q-values were calculated using MethylKit. (XLSX 217 kb)
>
> **Additional file 3: Figure S2.** COBRA vadlidation of candidate liver gender-specific DMRs. Bs-seq plots (top panels) show average percentage methylation values across each differentially methylated locus as assessed by RRBS for females (red) and males (blue). The CpG/s interrogated by COBRA are indicated by arrows. Graphs (bottom panels) show average percentage methylation in females (red) and males (blue) as assessed by densitometry of COBRA bands. Error bars represent SEM; *** $p < 0.0001$; NS, non significant. (PDF 634 kb)

Acknowledgements
Not applicable

Funding
This work was supported by an Australian Research Council Discovery Project grant (DP120100825). HM was supported by an Australian Postgraduate Award. JEC is supported by an Australian Research Council DECRA (DE120100723). CMS was supported by an Australian Research Council Future Fellowship (FT120100097).

Authors' contributions
JEC and CMS conceived and oversaw the study; JEC, EG and CMS designed the analytical strategy; KB, HC, JEC and HM performed experiments; EG, HM, PEY analysed data; JEC, EG, HM, SH and CMS interpreted data; JEC, EG, SH, HM and CS drafted the manuscript. All authors read and approved the final manuscript.

Authors' information
Not applicable

Ethics approval
All animals in this study were handled in accordance with good practice as defined by the National Health and Medical Research Council (Australia) Statement on Animal Experimentation, and requirements of state government legislation. The study was approved by the Garvan/St Vincent's Animal Ethics Committee (#13/35).

Competing interests
Keith Booher and Hunter Chung are employees of Zymo Research; The authors declare that they have no competing interests.

Author details
[1]Victor Chang Cardiac Research Institute, 405 Liverpool Street, Darlinghurst, NSW 2010, Australia. [2]St Vincents Clinical School, Faculty of Medicine, University of New South Wales, Kensington, NSW 2052, Australia. [3]Zymo Research, Murphy Ave, Irvine, CA 92614, USA.

References
1. Ober C, Loisel DA, Gilad Y. Sex-specific genetic architecture of human disease. Nat Rev Genet. 2008;9(12):911–22.
2. Morrow EH. The evolution of sex differences in disease. Biol Sex Differ. 2015;6:5.
3. Rakyan VK, Down TA, Balding DJ, Beck S. Epigenome-wide association studies for common human diseases. Nat Rev Genet. 2011;12(8):529–41.
4. McCarthy NS, Melton PE, Cadby G, Yazar S, Franchina M, Moses EK, Mackey DA, Hewitt AW. Meta-analysis of human methylation data for evidence of sex-specific autosomal patterns. BMC Genomics. 2014;15:981.
5. Xu H, Wang F, Liu Y, Yu Y, Gelernter J, Zhang H. Sex-biased methylome and transcriptome in human prefrontal cortex. Hum Mol Genet. 2014; 23(5):1260–70.
6. Liu J, Morgan M, Hutchison K, Calhoun VD. A study of the influence of sex on genome wide methylation. PLoS One. 2010;5(4):e10028.
7. Hall E, Volkov P, Dayeh T, Esguerra JL, Salo S, Eliasson L, Ronn T, Bacos K, Ling C. Sex differences in the genome-wide DNA methylation pattern and impact on gene expression, microRNA levels and insulin secretion in human pancreatic islets. Genome Biol. 2014;15(12):522.
8. Orozco LD, Rubbi L, Martin LJ, Fang F, Hormozdiari F, Che N, Smith AD, Lusis AJ, Pellegrini M. Intergenerational genomic DNA methylation patterns in mouse hybrid strains. Genome Biol. 2014;15(5):R68.
9. Reizel Y, Spiro A, Sabag O, Skversky Y, Hecht M, Keshet I, Berman BP, Cedar H. Gender-specific postnatal demethylation and establishment of epigenetic memory. Genes Dev. 2015;29(9):923–33.
10. Akalin A, Kormaksson M, Li S, Garrett-Bakelman FE, Figueroa ME, Melnick A, Mason CE. methylKit: a comprehensive R package for the analysis of genome-wide DNA methylation profiles. Genome Biol. 2012;13(10):R87.
11. Xiong Z, Laird PW. COBRA: a sensitive and quantitative DNA methylation assay. Nucleic Acids Res. 1997;25(12):2532–4.
12. Illingworth RS, Gruenewald-Schneider U, De Sousa D, Webb S, Merusi C, Kerr AR, James KD, Smith C, Walker R, Andrews R, et al. Inter-individual variability contrasts with regional homogeneity in the human brain DNA methylome. Nucleic Acids Res. 2015;43(2):732–44.
13. Waxman DJ, Holloway MG. Sex differences in the expression of hepatic drug metabolizing enzymes. Mol Pharmacol. 2009;76(2):215–28.
14. Ghahramani NM, Ngun TC, Chen PY, Tian Y, Krishnan S, Muir S, Rubbi L, Arnold AP, de Vries GJ, Forger NG, et al. The effects of perinatal testosterone exposure on the DNA methylome of the mouse brain are late-emerging. Biol Sex Differ. 2014;5:8.
15. Lowe R, Gemma C, Rakyan VK, Holland ML. Sexually dimorphic gene expression emerges with embryonic genome activation and is dynamic throughout development. BMC Genomics. 2015;16:295.
16. Sakakibara M, Uenoyama Y, Minabe S, Watanabe Y, Deura C, Nakamura S, Suzuki G, Maeda K, Tsukamura H. Microarray analysis of perinatal-estrogen-induced changes in gene expression related to brain sexual differentiation in mice. PLoS One. 2013;8(11):e79437.
17. Gabory A, Attig L, Junien C. Sexual dimorphism in environmental epigenetic programming. Mol Cell Endocrinol. 2009;304(1-2):8–18.
18. McLean CY, Bristor D, Hiller M, Clarke SL, Schaar BT, Lowe CB, Wenger AM, Bejerano G. GREAT improves functional interpretation of cis-regulatory regions. Nat Biotechnol. 2010;28(5):495–501.
19. Krueger F, Andrews SR. Bismark: a flexible aligner and methylation caller for Bisulfite-Seq applications. Bioinformatics. 2011;27(11):1571–2.
20. Li LC, Dahiya R. MethPrimer: designing primers for methylation PCRs. Bioinformatics. 2002;18(11):1427–31.

11

Deciphering the genetic regulation of peripheral blood transcriptome in pigs through expression genome-wide association study and allele-specific expression analysis

T. Maroilley[1]* (iD), G. Lemonnier[1], J. Lecardonnel[1], D. Esquerré[2], Y. Ramayo-Caldas[1], M. J. Mercat[3], C. Rogel-Gaillard[1]* and J. Estellé[1]*

Abstract

Background: Efforts to improve sustainability in livestock production systems have focused on two objectives: investigating the genetic control of immune function as it pertains to robustness and disease resistance, and finding predictive markers for use in breeding programs. In this context, the peripheral blood transcriptome represents an important source of biological information about an individual's health and immunological status, and has been proposed for use as an intermediate phenotype to measure immune capacity. The objective of this work was to study the genetic architecture of variation in gene expression in the blood of healthy young pigs using two approaches: an expression genome-wide association study (eGWAS) and allele-specific expression (ASE) analysis.

Results: The blood transcriptomes of 60-day-old Large White pigs were analyzed by expression microarrays for eGWAS (242 animals) and by RNA-Seq for ASE analysis (38 animals). Using eGWAS, the expression levels of 1901 genes were found to be associated with expression quantitative trait loci (eQTLs). We recovered 2839 local and 1752 distant associations (Single Nucleotide Polymorphism or SNP located less or more than 1 Mb from expression probe, respectively). ASE analyses confirmed the extensive *cis*-regulation of gene transcription in blood, and revealed allelic imbalance in 2286 SNPs, which affected 763 genes. eQTLs and ASE-genes were widely distributed on all chromosomes. By analyzing mutually overlapping eGWAS results, we were able to describe putative regulatory networks, which were further refined using ASE data. At the functional level, genes with genetically controlled expression that were detected by eGWAS and/or ASE analyses were significantly enriched in biological processes related to RNA processing and immune function. Indeed, numerous distant and local regulatory relationships were detected within the major histocompatibility complex region on chromosome 7, revealing ASE for most class I and II genes.

Conclusions: This study represents, to the best of our knowledge, the first genome-wide map of the genetic control of gene expression in porcine peripheral blood. These results represent an interesting resource for the identification of genetic markers and blood biomarkers associated with variations in immunity traits in pigs, as well as any other complex traits for which blood is an appropriate surrogate tissue.

Keywords: Blood, Pig, Transcriptome, eQTL, eGWAS, RNA-Seq, Allele-Specific Expression

* Correspondence: tatiana.maroilley@inra.fr; claire.rogel-gaillard@inra.fr; jordi.estelle@inra.fr
[1]GABI, INRA, AgroParisTech, Université Paris-Saclay, 78350 Jouy-en-Josas, France
Full list of author information is available at the end of the article

Background

Modern livestock breeding programs aim to foster sustainability and improve production by selecting animals for health- and environment-related phenotypes. However, complex health traits such as immune capacity, disease resistance, and robustness are difficult to directly measure in living animals. There is a strong need for the identification of intermediate phenotypes that can be used as proxies for these traits. In pigs, data on immune function are regularly obtained by scoring innate and adaptive immunity traits measured in blood and serum (e.g., blood cell counts, specific and non-specific antibodies, and acute phase proteins) or by studying the effect of in vitro stimulation of total blood or leucocytes (e.g., production of chemokines and cytokines, phagocytosis) [1, 2]. We and others have reported that many immunity traits are heritable in swine [1–3]. We have also shown that changes in the peripheral blood transcriptome are correlated to variation in immune traits [4]. In this context, then, peripheral blood appears to be a relevant tissue with which to phenotype immunity traits, as well as a relevant surrogate tissue for the quantification of other physiological traits.

Variation in transcript abundance has been studied as an inherited quantitative trait [5] and association mapping of loci for expression traits has been performed in several species for a wide range of tissues [6]. Indeed, as demonstrated by Schadt et al. [7], characterizing the relationships between DNA sequence and RNA expression is an appropriate step in understanding the links between genotype and phenotype. In the first global analysis of expression quantitative trait loci (eQTLs) conducted in mammals, these authors showed that gene expression traits in mice were connected with complex phenotypes like obesity. eQTLs can control the expression levels of local and/or distant transcripts [8] and are nowadays identified by expression genome-wide association studies (eGWASs). Local and distant associations are often referred to as *cis-* and *trans-*associations and correspond to putative *cis-* and *trans-*regulatory relationships, which act, respectively, in an allele-specific and non-allele-specific manner. Complementarily to gene expression microarrays, an RNA-Seq-based approach provides sequencing data for the detection of both transcript expression levels and single-nucleotide polymorphisms (SNPs). By integrating these two types of data, allele-specific expression (ASE) analyses provide information on the relative expression levels of two alleles of a gene from the same individual, thus targeting *cis-*acting regulation [9]. An approach that combines eGWAS and ASE techniques is highly effective in characterizing candidate genetic variants acting as *cis-* and *trans-*regulators, as has been shown in the mouse for liver [10] and adipose tissue [11].

In humans, results on the genetic control of the blood transcriptome were reported to be robust and reproducible despite possible confounding effects due to variations in white and red cell counts [33]. These results highlighted the utility of blood eQTL mapping for improving the interpretability of GWAS results for complex phenotypes, especially by helping with the prioritization of candidate genes. Similarly, Schramm et al. [31] showed that eQTL analysis of whole blood is reliable and may be used to identify biomarkers and to enhance understanding of the molecular mechanisms underlying associations between genes and disease. Together, all these eQTL data have provided new types of resources and associated databases for the study of relationships between gene expression and phenotype of interest [12].

Few eGWASs have been reported in pigs. Liaubet et al. [13] analyzed gene expression in pig skeletal muscle sampled shortly after slaughtering, and showed an overrepresentation of genes that encoded proteins involved in processes induced during muscle development and metabolism, cell morphology, stress response, and apoptosis. More recently, eGWASs have been integrated with phenotyping studies in order to identify candidate genes and causative mutations associated with phenotypic variations in several tissues: for example, liver gene expression linked to blood and lipid traits [41], gene expression in *longissimus dorsi* or *gluteus medius* muscles associated with growth, fatness, yield, and meat quality [14–19], and hypothalamus gene expression connected to coping behavior [20].

To our knowledge, no eGWAS of total blood has yet been performed in pigs. In this study, our aim was to build a genome-wide map of the genetic control of gene expression in the blood of 60-day-old Large White pigs by combining eGWAS and ASE results. Such a resource will be valuable for guiding integrated studies of complex phenotypes, as has been demonstrated in studies of humans [12], especially those linked to health and immune response in young pigs, which are prone to be subjected to infection challenges. After detecting the eQTL and ASE effects, we performed *in silico* functional analyses in order to illustrate the usefulness of our data in not only prioritizing candidate blood biomarkers and SNPs associated with immunity and physiological trait variations but also in deciphering the complexity of blood gene transcriptional regulation in pigs.

Results

eGWAS revealed numerous cases of local and distant regulation of gene transcription in blood

We performed an eGWAS by combining expression data from customized single-channel 8X60K Agilent arrays and SNP genotypes from Illumina PorcineSNP60K genotyping chips taken from 242 French Large White pigs. We searched for significant associations between all expressed probes that mapped to a unique position on the Sscrofa10.2 reference genome (39,649 probes from a total of 59,774 expressed probes) and all SNPs on the reference genome that met the quality control threshold (42,234 SNPs from a total of 61,557). For these association analyses, all probes, including those annotated to the same gene, were considered independent.

Overall, we identified 4591 associations, which involved 3195 eQTL-SNPs (see below) and 3419 probes that were retained for further analysis (Fig. 1 and Additional file 1: Table S1). Initially, 48,536 significant associations (FDR < 0.05) were detected between 18,541 SNPs and 3419 expressed probes mapping to an unique genomic position. In order to reduce redundancy in SNPs that was due to linkage disequilibrium, we represented the signals of all associated SNPs that harbored the same effect on probe expression by the most significant SNP in a window of 5 Mb. We used this SNP to represent the eQTL in further analyses and designated it as an eQTL-SNP. From this subset of eQTL-SNPs, 77% of the 3419 associated probes could be assigned to 1901 pig genes. Probe annotation revealed that gene assignment included alignments of sense (1828 genes), antisense (73 genes), or both sense and antisense (244 genes)

probe sequences (Additional file 2: Table S2). In addition, 421 probes could not be assigned to any known gene but were found associated with at least one eQTL-SNP.

A large majority of eQTL-SNPs ($N = 2454$) was associated with variation in the expression of a single probe. The other 965 eQTL-SNPs were each associated with between two (483 eQTL-SNPs) and 51 probes (one eQTL-SNP) (Table 1). Reciprocally, probes were associated with up to six eQTL-SNPs, although most were linked to only one or two (Table 1).

Among all associations, we characterized 2839 as local and 1752 as distant. In local associations, candidate eQTL-SNPs were located within 1 Mb of the start position of the associated probe and were expected to act on that probe's expression in an allele-specific manner, meaning a *cis*-regulatory action. All other associations were classified as distant associations and the corresponding eQTL-SNPs are referred to as distant eQTL-SNPs, expected to act in *trans* by regulating the associated gene in a non-allele-specific manner. Among the 3195 eQTL-SNPs, 2124 were local and 1187 were distant, while 116 were involved in both local and distant associations. Furthermore, 2407 probes were affected by at least one local eQTL-SNP and 1427 probes by at least one distant eQTL-SNP, with 415 probes associated with at least one local and one distant

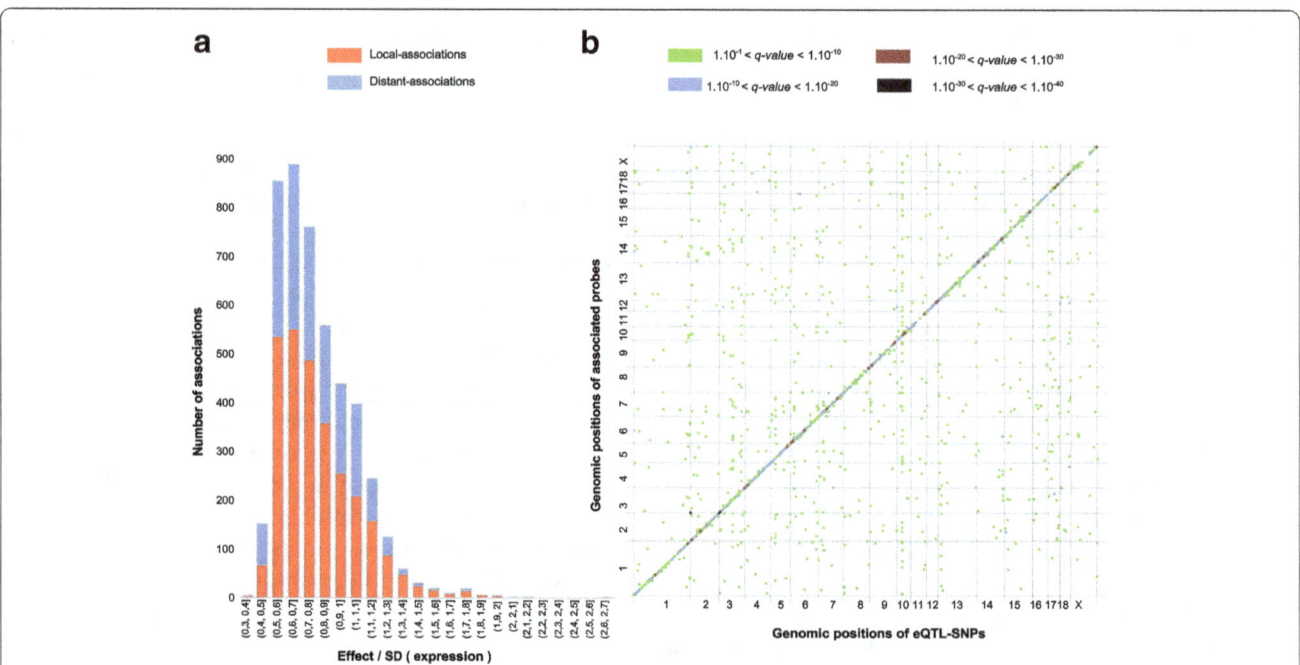

Fig. 1 Overview of associations identified by eGWAS. **a**: The distribution of additive effects of local-acting (red bars) or distant-acting (blue bars) eQTL-SNPs. eQTL-SNP effects are expressed as the ratio between the effect of the eQTL-SNP on probe expression and the standard deviation of variation in probe expression. **b**: Genomic locations of expressed probes versus genomic positions of associated eQTL-SNPs. Each significant association is represented by a dot; the color of dots indicates the significance (q-value) of the association (green: $1 \times 10^{-1} < q < 1 \times 10^{-10}$, dark red: $1 \times 10^{-10} < q < 1 \times 10^{-20}$, blue: $1 \times 10^{-20} < q < 1 \times 10^{-30}$, black: $1 \times 10^{-30} < q < 1 \times 10^{-40}$). Dots on the diagonal represent local associations; off-diagonal dots correspond to distant associations

Table 1 Numbers of significant associations between probes and eQTL-SNPs, and vice versa

	Number of associated eQTL-SNPs or probes											
	1	2	3	4	5	6	7	8	9	10	11–51	Total
Probes	2423	839	141	14	1	1						3419
eQTL-SNPs	2454	483	146	44	23	16	7	7	2	4	9	3195

eQTL-SNP. For 2382 local associations (> 83%), the eQTL-SNP was located less than 500 kb from the start position of the probe, and for a subset of 1334 associations, the eQTL-SNP was within 100 kb of the associated probe.

In order to evaluate the degree of the eQTL-SNP effect with respect to global variation in probe expression, we calculated the ratio of the eQTL-SNP effect to the standard deviation of the expression of its associated probe. The ratio distribution is represented in Fig. 1a and showed that eQTL-SNPs had a wide range of additive effect sizes on the expression variation of associated probes. The means of local and distant eQTL-SNP effects were significantly different (unilateral Student test <0.05), with local associations showing larger effects. Globally, the median ratio was equal to 0.8, meaning

that for a majority of associations, the effect of the eQTL-SNP was nearly equal to the standard deviation of the gene probe expression. For 930 associations, the effect of the eQTL-SNP was larger than the standard deviation (ratio higher than 1; Fig. 1a). In eight extreme cases, the effect was at least two times higher than the standard deviation of gene expression. The most extreme association had a ratio of 2.68; this involved the *IFITM2* gene on SSC2 and was also the most significant association found by eGWAS (q-value = 1.25×10^{-39}).

The analysis of eQTL genomic locations revealed that both local and distant associations were widely distributed on all chromosomes, although there were significant inter-chromosomal differences in the density of eGWAS signals (Fig. 2a to d). Chromosomes SSC1, SSC8, SSC11, SSC13, SSC15, and SSC16 had significantly lower proportions of eQTL-SNPs than the genome as a whole, while SSC4, SSC7, SSC12, and SSC14 were relatively enriched in eQTL-SNPs (one-sided Fisher <0.05). In addition, the relative proportions of local, distant, and local / distant eQTL-SNPs differed among chromosomes, especially on SSC1, SSC7, SSC12, and SSC15 (Chi² < 0.05). We also observed that the relative percentage of associated genes varied among

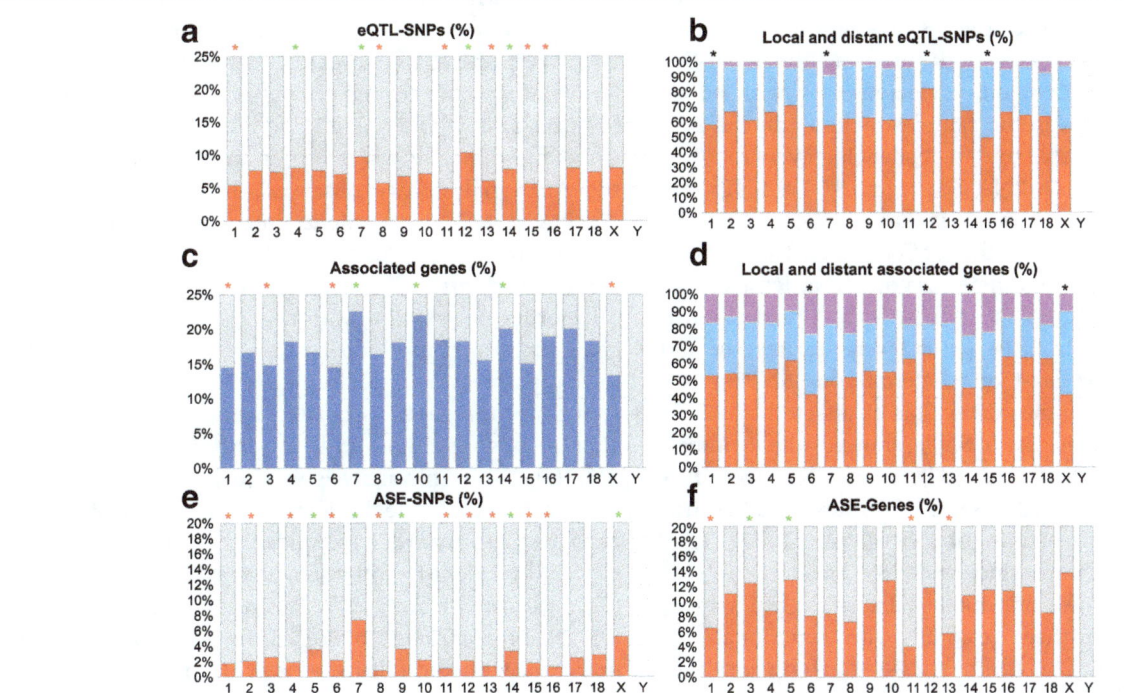

Fig. 2 Genome-wide distribution of associations identified by eGWAS (A-D), and ASE-SNPs or -genes (E-F). **a**: Per-chromosome percentage of eQTL-SNPs detected by eGWAS (red) in the larger SNP set. **b**: Relative proportions of SNPs that were detected as local-acting (red), distant-acting (blue) or both local- and distant-acting (purple) eQTL-SNPs by eGWAS. **c**: Per-chromosome percentage of genes whose transcription levels were linked to eQTL-SNPs. This percentage was calculated from the whole set of genes detected as expressed in blood. **d**: Percentages of genes associated with local-acting (red), distant-acting (blue) or both local- and distant-acting (purple) eQTL-SNPs. **e**: Per-chromosome percentage of ASE-SNPs of the whole set of 80,939 heterozygous SNPs in the group of 38 pigs included in the ASE analysis. **f**: Per-chromosome percentage of genes harboring at least one ASE-SNP in the set of genes with heterozygous SNPs in the group of 38 pigs

chromosomes (Fig. 2c). In particular, SSC1, SSC3, SSC6, and SSCX had lower percentages of associated genes than the genome as a whole (Fisher <0.05), while SSC7, SSC10, and SSC14 were found to be enriched in associated genes (Fisher <0.05). We also observed inter-chromosomal variation in the percentage of genes associated with local, distant, or local + distant eQTL-SNPs, as represented in Fig. 2d. In particular, SSC6, SSC12, SSC14, and SSCX presented a significantly different ratio between local- and distant-associated genes (Chi^2 < 0.05) in comparison to the proportion in the whole genome (Additional file 3: Figure S1). From these results, we targeted a subset of the most promising regions and networks, according to their level of significance and connectivity or their relevance to immune response, for further analyses.

ASE analysis confirmed extensive *cis*-regulation of gene expression in blood and revealed additional *cis*-regulatory events

ASE analysis was performed on whole-blood RNA-Seq data from a group of 38 Large White pigs. We found 12,811 Ensembl genes (release 82) expressed in blood; of these, 9501 were assigned a HUGO gene symbol. To be considered expressed, a gene had to be represented by a minimum of 10 reads in at least one sample. Since genome re-sequencing of the 38 pigs was not available, SNP identification was carried out using the RNA-Seq data. Thus, this analysis only detected SNPs with different expressed alleles in the samples included in the ASE study, and was not able to detect silencing events.

After SNP calling on RNA-Seq reads, 82,419 SNPs met the quality control criteria, including 71,365 (86.2%) SNPs already known in the dbSNP database. Among these SNPs, 80,939 were heterozygous in at least one of the 38 samples, and 86.3% of these were already known in dbSNP. Overall, 6824 genes were expressed and overlapped with at least one heterozygous SNP in one individual, and were thus detectable as affected by ASE.

From our analysis, 2286 SNPs showed allelic imbalance (Additional file 4: Table S3). These SNPs were further referred to as ASE-SNPs. Of these, 63.5% had already been described in dbSNP (1452 SNPs). They overlapped 1312 transcripts and 989 Ensembl (release 82) annotated genes, of which 763 had a gene symbol and could be further explored with functional enrichment analysis. Overall, these *cis*-regulated genes represented roughly 11% of all expressed genes that overlapped at least one heterozygous SNP. The genome localization of ASE-SNPs revealed that the candidate *cis*-regulated genes were widely distributed along chromosomes, with a great deal of inter-chromosomal variability. Specifically, enrichment in ASE-SNPs was detected on SSC5, SSC7, SSC9, SSC14, and SSCX, with the highest

levels found on SSC7 (Fig. 2e). In addition, SSC1, SSC11, and SSC13 were significantly depleted in genes affected by ASE-SNPs, while SSC3 and SSC5 were significantly enriched (Fig. 2f, one-sided Fisher <0.05). The distribution of ASE-SNPs along each chromosome was not homogeneous (Additional file 5: Figure S2).

In total, 149 genes were detected by both the eGWAS and ASE analysis (Additional file 4: Table S3); this corresponded to 7% of the associated genes found by eGWAS, 20% of genes affected by ASE, and 2.7% of all genes detected by both approaches. Among these 149 genes, 109 were represented by a sense-probe and associated with a local eQTL-SNP by eGWAS. This result provided validation of the genetic *cis*-regulation of these genes. Fifty-two genes that were affected by ASE were also found by eGWAS to be associated with a distant eQTL-SNP, and of these, 16 had a distant association with an eQTL-SNP on another chromosome. These 16 genes were thus candidates for being both *cis*- and *trans*-regulators.

Functional annotation of eQTL-SNPs, ASE-SNPs, and candidate regulated genes

The effects of eQTL-SNPs and ASE-SNPs were predicted using the Variant Effect Predictor (VEP) [21] and are summarized in Table 2. Of the eQTL-SNPs, 50.5% and 34% corresponded to intergenic and intronic variants, respectively. However, there were significantly fewer intergenic variants among eQTL-SNPs than in the SNP set as a whole (*i.e.* all SNPs that met QC criteria), while the opposite was true for intronic variants. With the ASE analysis, we observed that a majority of the detected SNPs with allelic imbalances corresponded to 3'UTR (24.5%), missense (23.6%), or synonymous (18.1%) variants.

In order to explore whether the genetic control of gene expression in blood has an impact on specific biological pathways, we performed a functional enrichment analysis of the genes associated with local and/or distant eQTL-SNPs (Fig. 3a) and genes that contained an ASE-SNP (Fig. 3b). This analysis was carried out by first converting the pig gene lists into human gene lists using Biomart [22], in order to upload them into the GOrilla tool [23] to identify and visualize enriched gene ontology (GO) terms. For both eGWAS and ASE approaches, the top 20 most-enriched biological functions are presented in Fig. 3, with the full lists for eQTL in Additional files 6, 7 and 8: Tables S4, S5 and S6 and for ASE in Additional file 9: Table S7.

Among the top 20 most-significant biological processes that were enriched among genes associated by eGWAS with local and/or distant eQTL-SNPs (Fig. 3a), 12 were related to metabolic and catabolic processes and three to RNA processing (RNA and non-coding RNA).

Table 2 Summary of the effects of SNPs identified by eGWAS and ASE, and their relative proportions in the larger population of a given SNP category (%)

SNP effects	eGWAS			ASE
	QC-SNPs[a]	eQTL-SNPs	P-values[b]	ASE-SNPs
3'UTR variant	274 (0.65%)	47 (1.47%)	1.71×10^{-6}*	560 (24.5%)
5'UTR variant	51 (0.12%)	5 (0.16%)	5.95×10^{-1}	51 (2.2%)
Downstream gene variant	1278 (3%)	150 (4.7%)	9.03×10^{-7}*	125 (5.5%)
Intergenic variant	27,211 (64.7%)	1606 (50.5%)	4×10^{-56}*	150 (6.6%)
Intronic variant	10,893 (25.9%)	1083 (34%)	1.4×10^{-22}*	209 (9.1%)
Missense variant	187 (0.44%)	28 (0.87%)	1.79×10^{-3}*	539 (23.6%)
Non-coding exon variant	18 (0.04%)	3 (0.09%)	1.80×10^{-1}	127 (5.6%)
Splice region variant	68 (0.16%)	7 (0.21%)	3.70×10^{-1}	14 (0.6%)
Stop lost	7 (0.016%)	1 (0.03%)	4.42×10^{-1}	5 (0.2%)
Stop retained variant	16 (0.038%)	4 (0.13%)	4.78×10^{-2}*	1 (0.04%)
Synonymous variant	553 (1.31%)	68 (1.19%)	3.45×10^{-4}*	414 (18.1%)
Upstream gene variant	1526 (3.63%)	182 (5.7%)	1.92×10^{-8}*	51 (2.2%)
Mature miRNA variant	1 (0.002%)	0		1 (0.04%)
Splice acceptor variant	1 (0.002%)	0		10 (0.4%)
Stop gained	5 (0.012%)	0		17 (0.7%)
Splice donor variant				12 (0.5%)

[a]SNPs that met QC criteria

[b]P-values for a Fisher test. For $P < 0.05$, the proportion of the SNP effect is significantly different between that of SNPs that passed the QC and that of eQTL-SNPs. Significant differences are labeled by a *

Other significantly enriched processes included the tricarboxylic acid cycle, the cellular catabolic process, mitochondrial transport, and the organophosphate biosynthetic process. The two immunity-related functions on the list involved the "interferon-gamma-mediated signaling pathway" and the "presentation of exogenous peptide antigen via MHC class I, TAP-independent", which was also among the top 20 enriched processes in the ASE analysis (Additional file 9: Table S7). In order to prevent bias in the enrichment analysis due to the inclusion of both local and distant associations, we also performed a separate enrichment study for each association type (Additional files 7 and 8: Tables S5 and S6, Fig. 3a). There were 18 GO terms enriched among genes that were associated with distant eQTL-SNPs, while 7 terms appeared on the list generated by an analysis of genes linked to local eQTL-SNPs. The only function present in both lists was the ncRNA metabolic process. No immunity-related functions were shared by both local and distant associations; instead, the "interferon-gamma-mediated signaling pathway" was enriched in local-associated genes and the "antigen processing and presentation of exogenous peptide antigen via MHC class I" function was identified only by combining all data. Three biological processes related to RNA processing were enriched among distant-associated genes. Overall, although a few processes were shared between genes associated with local or distant eQTL-SNPs, for the most

part, the two groups of genes seemed to be involved in different biological pathways (Additional files 7 and 8: Tables S5 and S6 and Additional file 10: Figure S3). The high degree of overlap in enriched GO terms between the global analysis and the local-associated genes could be mostly due to the over-representation in the former analysis of genes associated with local eQTL-SNPs.

The genes that overlapped ASE-SNPs were significantly linked to a much higher number of biological functions than that identified from eGWAS data (Additional file 9: Table S7). In total, the ASE-genes corresponded to 105 enriched functions, with many of these related to immunity. Indeed, the top function listed among the 20 most-significant biological processes (Fig. 3b) was the immune system process (GO:0002376), followed by 12 other processes associated with immunity and defense, response to stimulus, or viral transcription, and 5 processes associated with adhesion and signaling pathways, which included both type I interferon and interferon-gamma-mediated signaling pathways. In addition, *cis*-regulated genes revealed by the ASE analysis were also found to be enriched for processes related to translational initiation (GO:0006413), nuclear-transcribed mRNA catabolic process nonsense-mediated decay (GO:0000184), and regulation of translation (GO:0006417), which could all be linked to the genetic control of gene expression, especially by master regulators. Two functions linked to heme metabolism and

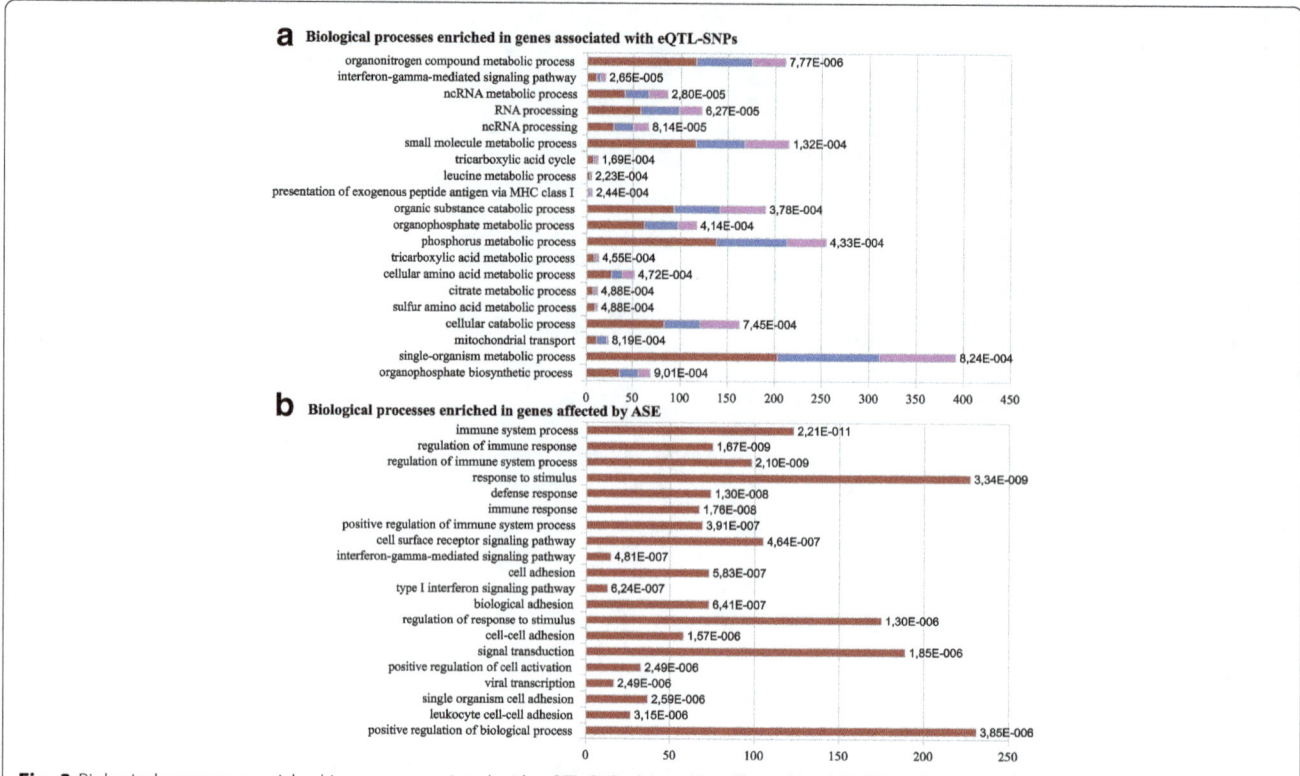

Fig. 3 Biological processes enriched in genes associated with eQTL-SNPs (**a**) and/or affected by ASE (**b**). **a:** The top 20 biological processes enriched in genes associated with eQTL-SNPs by eGWAS. Horizontal bars represent the number of associated genes per biological process, with the corresponding q-values. Bars were split in order to represent the respective numbers of local-associated genes (dark red), distant-associated genes (blue), and local + distant-associated genes (purple). **b:** Biological processes enriched in genes affected by ASE. The horizontal bars represent the number of ASE-genes involved in each biological process, with the corresponding q-values. The two enrichment analyses were performed with the GOrilla tool

biosynthesis were also significantly enriched among the *cis*-regulated ASE genes (GO:0006783: "Heme biosynthetic process", GO:0006783: "Heme metabolic process"), which is unsurprising for genes expressed in blood. Only two GO functions were shared between the list based on ASE genes and that produced using the eGWAS approach (Additional file 10: Figure S3): the "interferon-gamma-mediated signaling pathway" (GO:0060333) and "antigen processing and presentation of exogenous peptide antigen via MHC class I, TAP-independent" (GO:0002480).

Construction of association networks revealed that genes are functionally linked and under shared genetic control
Since several eQTL-SNPs were associated with variation in the expression of more than one gene, and many genes were also linked to more than one eQTL-SNP (Table 1), we were able to construct potential regulatory networks by overlapping the eQTL signals. The whole set of SNP-probe associations was visualized in Cytoscape v3.2.1 [24], which identified 831 independent networks involving more than two nodes (Table 3). A network of three nodes represented either two eQTL-SNPs associated with the same gene or two genes

associated with the same eQTL-SNP. Most networks were simple (655 consisted of one eQTL-SNP affecting several genes). When all networks were taken together, nodes had on average 1.5 neighbors, and we found 329 multi-edge node pairs, which represented associations between a single eQTL-SNP and multiple probes assigned to the same gene.

The largest network (Fig. 4a) was centered on eQTL-SNP rs81422644 (H3GA0029721), which mapped onto SSC10 at 30,039,892 bp (Additional file 1: Table S1). This eQTL-SNP was associated with the distant regulation of transcription in 51 probes, which were annotated to 47 distinct genes (three probes could not be annotated to any known porcine genes and for the *LNPEP* gene there were two probes). These 47 genes were widely distributed on all chromosomes, with the exception of SSC3 and SSCY (Additional file 11: Figure S4A).

Table 3 Number and size of putative regulatory networks constructed from eGWAS-based associations

Number of nodes per network	3	4	5	6	7	8–154	Total
Number of networks	488	166	73	42	13	41	831

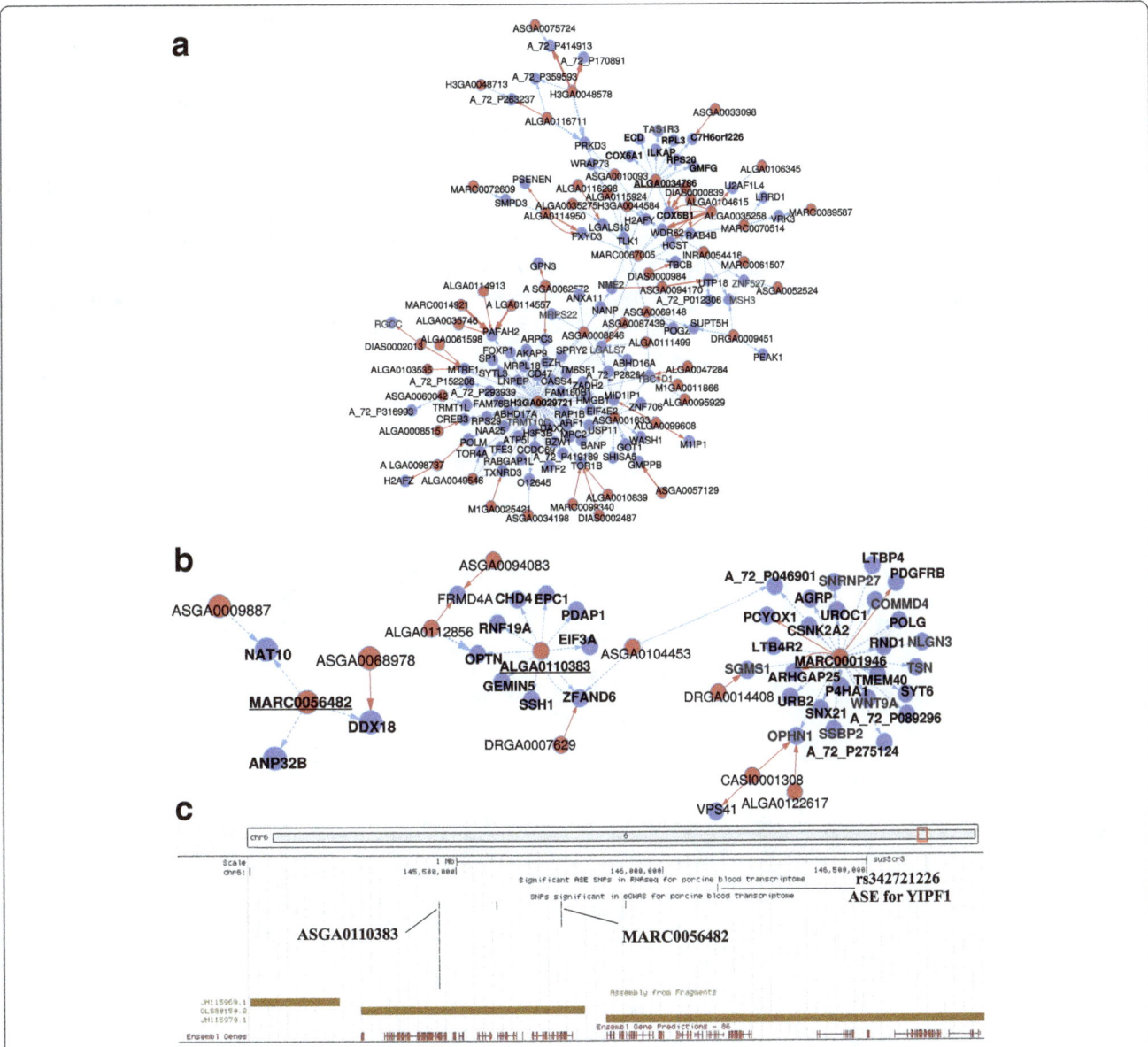

Fig. 4 Visualization of association networks revealed by eGWAS (**a**, **b**) and links with ASE (**c**). Association networks were constructed based on multiple associations between eQTL-SNPs or genes, and were visualized by Cytoscapev3.2.1. eQTL-SNPs and genes are represented by red and blue dots, respectively. Associations are depicted with directed edges (oriented arrows) to show regulation by the eQTL-SNP genotype on gene transcription and not the reverse. Local associations are drawn with red arrows and distant associations with blue dotted arrows. All nodes correspond either to a gene or an eQTL-SNP, with the names indicated. Names in bold are further described in the results. **a**: The biggest association network drawn in this study comprises 154 nodes (eQTL-SNPs or genes). It is centered on two eQTL-SNPs: H3GA0029721 (rs81422644), which was distant-associated with 51 probes representing 47 genes, and ALGA0034786 (rs81393122). **b**: Visualization of the association networks centered on eQTL-SNPs MARC0001946, ALGA0110383, and MARC0056482. **c**: Map of a segment of SSC6 that comprises the ASE-SNP rs242721226 and the eQTL-SNPs ALGA0110383 and MARC0056482. The distances between the position of ASE-SNP rs342721226 and the eQTL-SNPs ALGA0110383 and MARC0056482 are 676 and 381 Kb, respectively

Using the R package PCIT [25], we applied the Partial Correlation and Information Theory algorithm [26] on expression correlations; this showed that variations in transcription levels of these genes were positively correlated regardless of SNP genotype (average correlation coefficient = 0.75; Additional file 11: Figure S4B). These analyses suggested that the 47 genes are genetically controlled by a common regulator in close vicinity to eQTL-SNP rs81422644. However, only two individuals in our analyses were homozygous for the alternative allele of this eQTL-SNP, which means that this potentially interesting master regulatory SNP must be confirmed using another pig population in which this allele is present at a higher minimum allele frequency (MAF).

This large network was connected to another network that was centered around eQTL-SNP rs81393122 (ALGA0034786), which was located on SSC6 at 19,761,160 bp (Additional file 1: Table S1). This eQTL-SNP was linked to the distant regulation of 11 probes, annotated for ten genes on six distinct chromosomes (Fig. 5a, Additional file 1: Table S1): ribosomal protein L3 (*RPL3*), ribosomal protein S20 (*RPS20*), cytochrome c oxidase subunit VIb polypeptide 1 (*COX6B1*), cytochrome c oxidase subunit VIa polypeptide 1 (*COX6A1*), non-metastatic cells 2 protein (*NME2*), taste receptor, type 1, member 3 (*TAS1R3*), integrin-linked kinase-

associated serine/threonine phosphatase (*ILKAP*), glia maturation factor gamma (*GMFG*), uncharacterized protein C7H6orf226, and ecdysoneless cell cycle regulator (*ECD*). PCIT analysis revealed positive correlations among transcription variations in the 11 probes, with an average correlation coefficient of 0.61 (Fig. 5b). We also observed that this eQTL-SNP genotype affected the transcription of the ten genes in the same direction (Fig. 5c). Strikingly, nine of the ten associated genes were included in a unique IPA-predicted network related to the functions "cellular assembly and organization", "cell death and survival", "endocrine

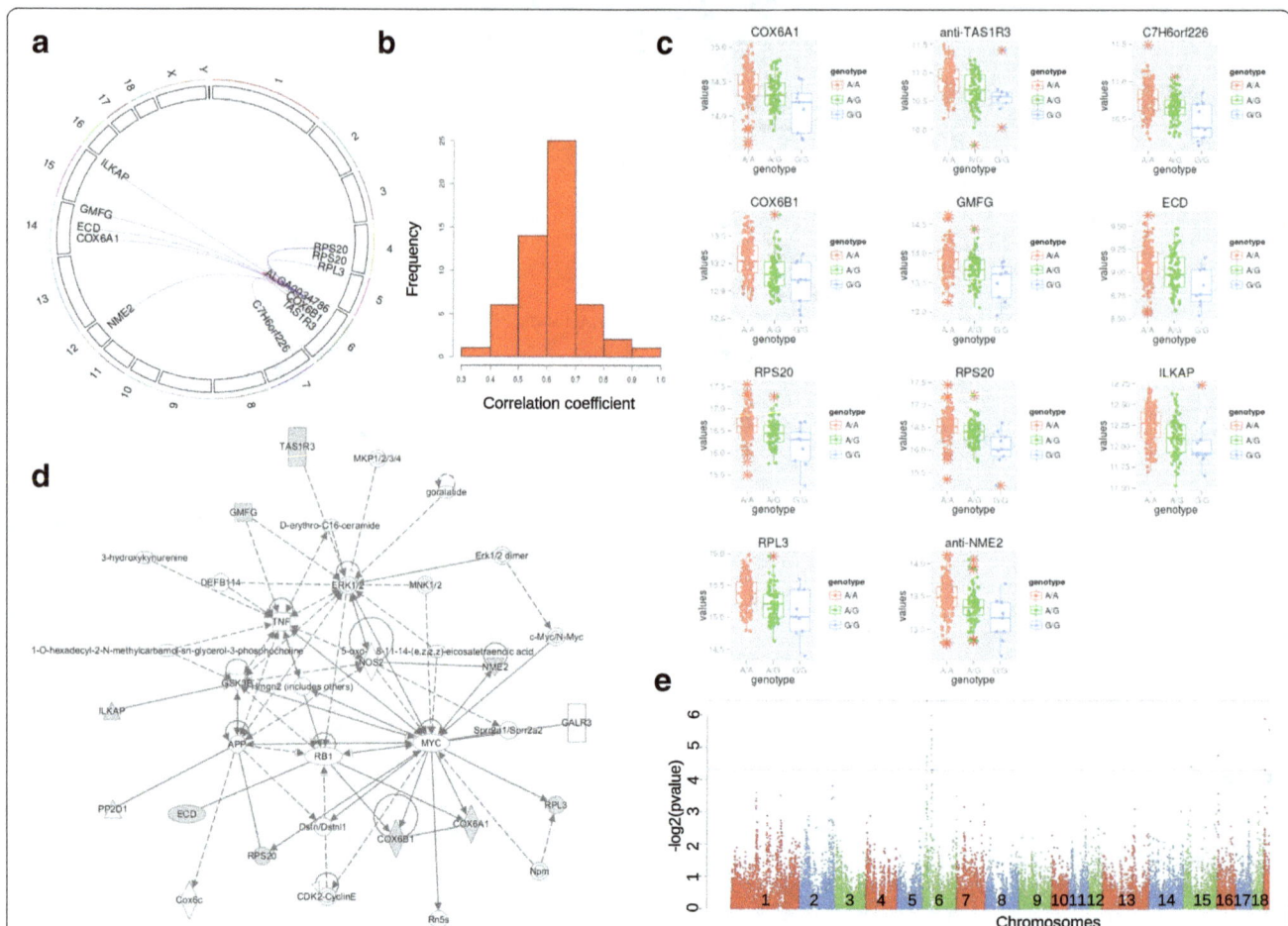

Fig. 5 Detailed analysis of the association network centered on eQTL-SNP ALGA0034786 on SSC6. **a**: Circo plot produced by R package RCirco which maps eQTL-SNPs and associated genes according to their genomic positions. Links in red correspond to associations between the eQTL-SNP ALGA0034786 and genes mapping to SSC6, and links in blue represent associations with genes on other chromosomes. **b**: The distribution of all (grey) and significant (red) correlation coefficients calculated from pairwise comparisons of expression variation in probes associated with ALGA0034786, performed by PCIT. Red areas of the histogram correspond to coefficient correlations involving significant edges, as determined by PCIT, and gray areas represent non-significant edges. **c**: Boxplots of expression variations in probes annotated for porcine genes, depending on the eQTL-SNP genotype. Each dot represents one animal. Pink dots correspond to pigs homozygous for the most-frequent allele (genotype A/A), green dots to heterozygous pigs (genotype A/G), and blue dots to animals homozygous for the less-frequent allele (genotype G/G). **d**: Ingenuity Pathway Analysis of the genes included in the association network. The functional network comprised nine of the ten genes that were distant-associated to eQTL-SNP ALGA0034786. The symbols of these nine genes are colored in gray. The IPA-drawn network revealed functions linked to "cellular assembly and organization", "cell death and survival", and "endocrine system disorder". **e**: eGWAS results for probe A_72_P211467, which was annotated for the gene *COX6B1*, visualized with a Manhattan plot. The association peak on SSC6 corresponds to the eQTL represented by eQTL-SNP ALGA0034786. The red horizontal line corresponds to an FDR threshold <0.01 and the horizontal blue line to an FDR threshold <0.05

system disorders" (Fig. 5d). One of the most significant associations in the network was with the gene *COX6B1* (Fig. 5e).

As represented in Fig. 4b and c, a candidate master regulatory SNP (rs81288717, or MARC0001946) was detected in an intergenic hotspot on SSC3, at 76,408,626 bp (Additional file 1: Table S1). This eQTL-SNP was associated with the distant-regulation of transcription variation in 26 genes, distributed on 14 different chromosomes (Additional file 12: Figure S5A). The eQTL-SNP genotype affected the transcription of the 26 genes in different directions, but with additive effects. As an example, allele G was associated with increased transcription of the gene *UROC1*, but with decreased transcription of the gene *ARHGAP2* (Additional file 12: Figure S5B). Complementarily, PCIT analysis revealed that transcription variations in these 26 genes were strongly correlated, either positively (average correlation coefficient = 0.74) or negatively (average correlation coefficient = 0.47) (Additional file 12: Figure S5C). IPA highlighted that the 26 genes could be included in three different networks associated with immunity-related processes: ten genes were linked to "cellular development, hematological system development and function, hematopoiesis", nine genes to "inflammatory response, cell signaling, molecular transport", and five genes to "developmental disorder, hereditary disorder, organismal injury and abnormalities" (Additional file 12: Figure S5D). Thus, eQTL-SNP rs81288717 could be an interesting candidate genetic marker for the regulation of immunity-related genes in blood, but further work is needed to identify the causal genetic variant.

For a few networks, the central eQTL-SNP was located within 1 Mb of an ASE-SNP. As an example, we present the case of two eGWAS signals that were 295 Kb apart, both within 500 Kb of the same ASE-SNP (Fig. 4c). The eQTL-SNP rs81338631 (ALGA0110383) maps onto SSC6 at 145,457,494 bp, inside an intron of *PCSK9*, while eQTL-SNP rs81244817 (MARC0056482) is located on the same chromosome at 145,752,528 bp (Additional file 1: Table S1). The nearby ASE-SNP, rs342721226, is contained within the 3'UTR region of the *YIPF1* gene, at 146,133,708 bp (Fig. 4c). Association networks revealed that eQTL-SNP rs81338631 was linked to the distant-regulation of nine different genes located on seven distinct chromosomes, while eQTL-SNP rs81244817 was associated with three genes located on three different chromosomes (Additional file 13: Figure S6A). In a PCIT analysis, the transcription levels of genes associated with these two eQTL-SNPs were strongly positively correlated (0.63 on average; Additional file 13: Figure S6B). Moreover, IPA revealed that the whole set of genes associated with the two eQTL-SNPs, as well as the ASE-gene *YIPF1* and *PCSK9*, the gene that contained the eQTL-SNP

rs81338631, could be included in a unique functional network related to "cell death and survival", "organismal injury and abnormalities" and "tissue morphology" (Additional file: Figure S6C). Both eQTL-SNPs affected the transcription of their associated genes in the same direction (Additional file 13: Figure S6D). Overall, these two regulatory networks seem to be related and the ASE affecting *YIPF1* designed this gene as an interesting candidate for understanding the molecular mechanism of transcriptome regulation of associated genes (*YIPF1* is associated to GO annotations that include RNA binding and ribonuclease activity).

A high density of local- and distant-acting eQTLs in the swine Leucocyte antigen (SLA) complex on SSC7

Chromosome SSC7 was enriched in both eQTL- and ASE-SNPs (Figure 2, Additional files 3 and 5: Figures S1 and S2). In particular, we observed numerous local and distant associations involving genes and SNPs within the MHC region, as has already been reported for human MHC [27–29].

The MHC in pigs is referred to as the SLA complex and maps to SSC7 on both sides of the centromere; the class I and III subregions cover 1.8 Mb of 7p1.1 band while the class II subregion spans 0.6 Mb of 7q1.1 band, with the size of the in-between centromere as-yet unknown [30]. Within this relatively small region, the SLA complex contains 151 annotated gene loci, and due to this high gene density, our previous definition of local/distant associations (inside/outside a 2-Mb window centered on the expressed probe) was not appropriate. Indeed, for the majority of eQTLs that mapped to the SLA complex, we could not distinguish between local and distant associations. Figure 6 illustrates the associations identified between gene expression levels and eQTL-SNPs, as well as genes affected by ASE-SNPs that have already been described in dbSNP. According to our eGWAS results, the expression levels of 51 genes within the SLA complex were found to be genetically controlled by 78 eQTL-SNPs, which included 11 SNPs located on chromosomes others than SSC7 and 31 SNPs on SSC7 but outside the SLA region. We identified 36 non-redundant eQTL-SNPs within the SLA complex and their VEP annotation is summarized in Fig. 6. Of these 36 eQTL-SNPs, 29 mapped to annotated genes, which included four SLA genes (*SLA-7*, *SLA-11*, *SLA-DRB3*, and *SLA-DOA*) and 23 non-SLA genes. The remaining two were considered to be intergenic when only the most severe consequence of each SNP was considered in VEP analysis.

Twelve SLA genes and 39 non-SLA genes were found to be regulated by between one and five distinct eQTL-SNPs. Among those affected were four functional SLA class I genes, the classical class Ia genes *SLA-1* and *SLA-*

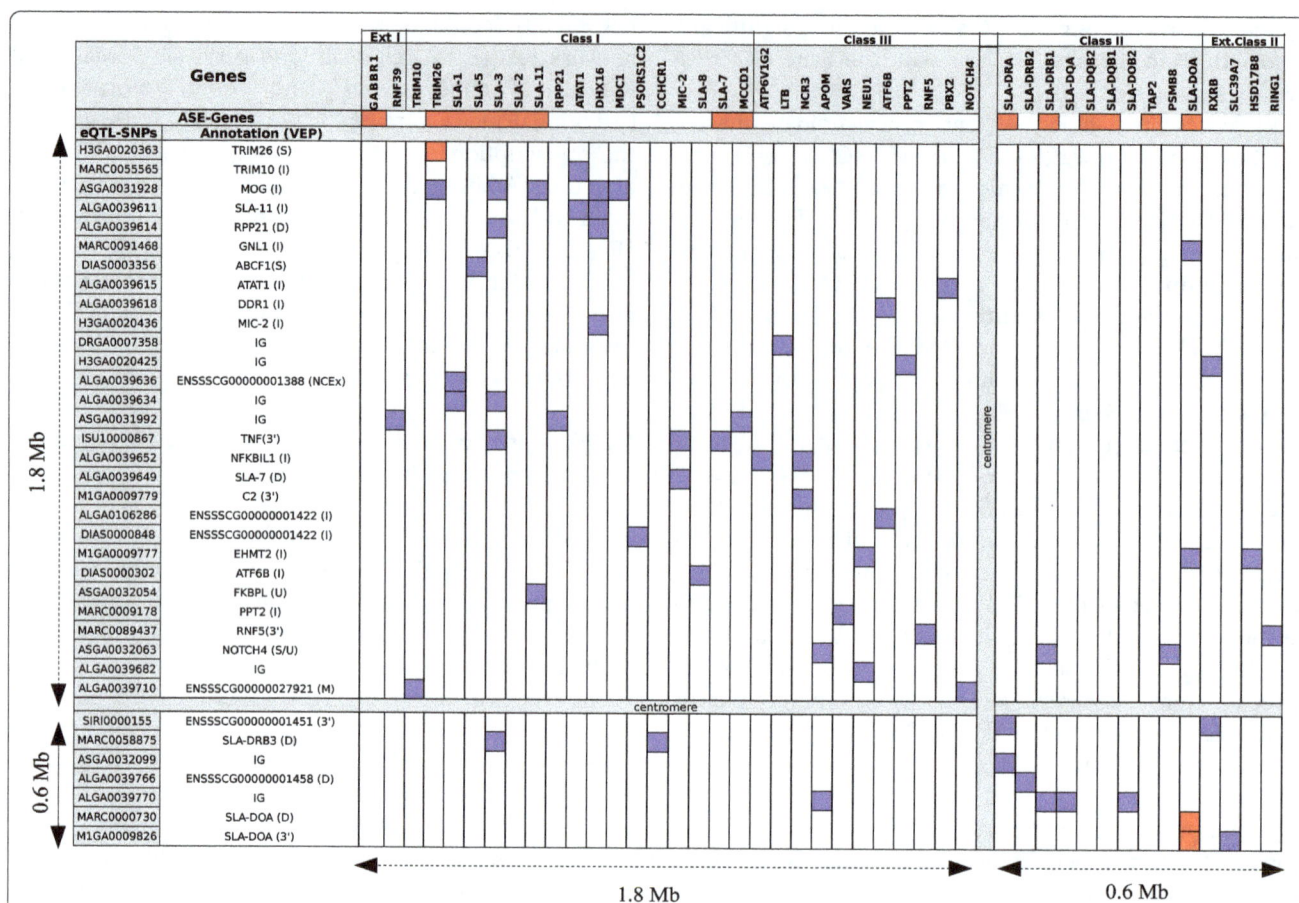

Fig. 6 Representation of associations and ASE-SNPs mapping to genes within the SLA region. SNPs detected as eQTL-SNPs by eGWAS are listed on the y-axis and their VEP annotation is indicated. These SNPs are ordered according to their physical position on SSC7, using the SusScrofa.10.2 genome assembly as a reference sequence. The genes listed along the x-axis are those that were found to harbor ASE-SNPs already referenced in dbSNP or that were associated with eQTL-SNPs by eGWAS. These genes are ordered using the map of the Hp1a.1 haplotype as a reference sequence [30]. All genes affected by ASE are indicated with red boxes and all distant associations are indicated with blue boxes. The centromere known to map between the class III and II sub-regions is shown. I = Intronic variant; IG = Intergenic variant; D = Downstream variant; U = Upstream variant; 3' = 3'UTR variant; S = Synonymous variant; NCEx = Non-coding exonic variant; M = Missense variant

3, as well as two non-classical class Ib genes, *SLA-7* and *SLA-8*. In the class II subregion, the expression levels of most genes were found to be genetically regulated, including *SLA-DRA*, *SLA-DRB1*, *SLA-DRB2*, *SLA-DQA*, *SLA-DOB2*, and *SLA-DOA*, as well as the non-SLA genes *TAP2* and *PSMB8*. All these data revealed that transcriptional activity is both dense and locally regulated. eQTL-SNPs were found to act either within a sub-region or more distantly, on genes mapping to both sides of the centromere. Only *SLA-DOA* and *TRIM26* were associated with an eQTL-SNP that had been annotated as a variant for these genes: a 3'UTR variant for *SLA-DOA* and a synonymous variant for *TRIM26*. These two genes were thus identified as good candidates for *cis*-regulation. We did not find eQTL-SNPs outside of SSC7 that were associated with changes in the expression of SLA genes; the only associated eQTL-SNPs were located either within the SLA complex or in close

vicinity. We detected 32 eQTL-SNPs specifically associated with a single gene (*SLA-1*, *SLA-3*, *SLA-5*, *SLA-8*, *SLA-11*, *SLA-DRA*, *SLA-DRB2*, or *SLA-DOA*). These SNPs could thus be considered genetic markers that specifically target the expression of one of these genes.

Through ASE analysis, we identified nine genes within the class I subregion and six genes within the class II subregion that harbored putative ASE-SNPs (Fig. 6). Interestingly, each SLA gene was found to be affected by specific ASE-SNPs. *SLA-DQB1* and *SLA-DQB2* shared the same three ASE-SNPs because of their close vinicity. The number of ASE-SNPs per SLA gene varied. *SLA-3* harbored 84 SNPs with allelic imbalance, while *SLA-2* and *SLA-DRB1* showed 27 and 28, respectively. *SLA-DOA* and *SLA-DRA* harbored only one ASE-SNP. This observed variation in the number of SNP variants per SLA gene is consistent with the expected allelic variability according to each gene (www.ebi.ac.uk/ipd/mhc/group/SLA).

The functional classification of ASE-SNPs for SLA genes is presented in Table 4.

Overlap of eQTL-SNPs and ASE-SNPs with blood biomarkers previously identified in pigs

We compared the lists of genes found using the ASE or eGWAS datasets to lists of genes that we previously reported to be differentially expressed (DE) in the blood of pigs; this included results from in vitro production of IL-2 and IL-10 after stimulation of total blood, in vitro phagocytosis tests, and counts of T lymphocytes CD4-CD8+ [4]. For each immunity-related trait, numerous DE genes were associated with eQTLs or ASE (Table 5), indicating that expression levels of blood biomarkers for these traits are genetically controlled. For example, local eQTL-SNPs were associated with expression changes in the *ALOX12* gene, which has been identified as a blood biomarker for phagocytosis, as well as the genes *GZMB* or *KLRG1*, which are correlated CD4-CD8+ cell count. Instead, two distant eQTL-SNPs were implicated in the transcription variation of *GATM*, which has been identified as a candidate biomarker for phagocytosis level; these eQTL-SNPs mapped less than 3 Mb from *GATM*. Finally, for CD4-CD8+ cell count, the candidate blood biomarkers *TNL1*, *GNLY*, *GZMB*, and *KLRG1* were found to be affected by a SNP with allelic imbalance.

Discussion
eGWAS and ASE analysis provide complementary results for mapping genetic regulation of gene transcription in blood

In the current study we investigated the genetic control of the blood transcriptome in 60-day-old pigs by combining two high-throughput methods: i) an eGWAS which analyzed transcription in 242 pigs (130 uncastrated males and 112 females) using expression

Table 5 Correspondence between sets of eQTL- and ASE-SNPs detected in this study and gene sets that were differentially expressed (DE) in the blood of pigs studied for levels of immunity traits

	IL-2[a]	IL-10[b]	Phagocytosis[c]	CD4-CD8 +[d]
DE genes [3]	850	733	1195	52
DE local-associated genes	69	64	107	5
DE distant-associated genes	56	47	74	0
DE genes with ASE signal	48	29	61	7

[a]Production of IL-2 after in vitro stimulation of total blood with phorbolmyristate acetate (PMA) and ionomycin for 48 h
[b]Production of IL-10 after in vitro stimulation of total blood with PMA and ionomycin for 48 h
[c]Quantification of phagocytosis capacity via in vitro tests on total blood
[d]CD4-CD8+ lymphocyte count measured by flow cytometry

microarrays and genome-wide SNP genotyping, and ii) an evaluation of ASE in 38 castrated males based on RNA-Seq of total blood RNA. The groups of animals used for the eGWAS and ASE analysis were independent from each other but selected from the same larger experimental population of Large White pigs, and partly generated from common boars. In doing this, we aimed to prevent discrepancies in the results due to heterogeneity in genetic backgrounds. Each approach had a different focus: eGWAS maps eQTLs as genomic loci that influence levels of gene transcripts through local and/or distant actions [8], while ASE analysis identifies allelic imbalance among transcripts. In our study, eQTL-SNPs were defined as acting locally or distantly based on whether the SNP was less or more than 1 Mb from either side of the start position of the associated probe, respectively. Globally, we found 2124 local eQTL-SNPs which affected the expression of 1363 genes, and 1187 distant eQTL-SNPs affecting 1010 genes. In parallel, ASE analysis enabled us to detect 2286 candidate SNPs that showed allelic imbalance, and which affected 763 genes.

Table 4 Number and annotation of ASE-SNPs affecting MHC genes

	SLA-1	SLA-2	SLA-3	SLA-5	SLA-7	SLA-11	SLA-DRA	SLA-DRB1	SLA-DOA	SLA-DQB1/2
3'UTR variant	13	15				6	1			3
5'UTR variant						1				
Intron variant	17	5	13		2			2		
Misssense variant	28	2			1	7		14		
Non-coding transcript exon variant	2		61	12						
Splice region variant	12	1	1							
Synonymous variant						3		6	1	
Downstream gene variant		4	12		3	3		6		
Intergenic variant										
Total[a]	72	27	84	12	3	20	1	28	1	3

[a] Total of ASE-SNPs per gene identified in the 38 pigs included in the analysis. The total numbers do not correspond to the sum of annotated SNPs because a few SNPs were annotated differently depending on the transcript they overlapped

Both eGWAS and ASE analyses were performed using stringent parameters to ensure robust mapping. For eGWAS, we chose to simultaneously analyze local and distant eQTLs [31–33], while previous studies had performed two independent association analyses which separated local- and distant-acting SNPs [27, 29, 34]. Due to the large amount of transcript information and the number of SNPs involved, eGWASs require extremely extensive calculations, which can raise issues related to statistical power [32]. To address this, we used stringent quality control thresholds, and corrected local and distant associations for multiple testing with an FDR procedure at two levels: i) for all association tests for each probe, and ii) for all association tests for all probes. For ASE analyses we filtered out SNPs with fewer than 10 reads and retained only ASE-SNPs with significant allelic imbalance in their read abundances in at least one-third of heterozygous animals.

For both eGWAS and ASE approaches, we observed that the genetic control of gene transcription was heterogeneously distributed along the genome, with important variations among chromosomes in both numbers of eQTL- or ASE-SNPs as well as in the relative proportions of genes that demonstrated ASE or that were associated with eQTLs. However, we found only limited overlap between the local associations that were identified by eGWAS and the *cis*-associations that were identified by the ASE analyses, which is consistent with results from other studies [10, 11]. Indeed, this was expected due to inherent differences between the approaches and their respective limits. For eGWAS, only eQTLs that corresponded to SNPs represented on the 60K SNP chip could be detected and in this design, intergenic SNPs are overrepresented [35]. Consequently, the functional annotation of eQTL-SNPs remains limited, and they mainly reflect cases of linkage disequilibrium with nearby causal variants and not direct causal links. The ASE analysis was limited by the lack of whole-genome sequence data for the animals included in this analysis, which meant that SNPs were called using RNA-Seq data. Therefore, in order to limit errors in SNP discovery, only ASE-SNP positions for which both alleles were expressed (Minor Allele Count >3) were considered in the final analysis, and transcripts with low read counts were discarded. In addition, due to the lack of parental information, we could not differentiate between allelic imbalance and parental imprinting. However, among the 763 ASE genes found in our study, only three are described as imprinted genes in the Geneimprint database (http://www.geneimprint.com/): *SNRPN*, which is imprinted by the paternal allele in *Homo sapiens*, *Macaca mulatta*, and *Bos taurus*; and *SNORD64* and *DMTN*, which are reported as imprinted only in *Homo sapiens*.

Overall, all these limitations suggest that these eQTL-SNPs and ASE-SNPs are not causal variants but SNPs in linkage disequilibrium with the causal variants that could be used as genetic markers for gene expression in the pig population included in our study. We did not address the question of sex effect on blood transcriptome variation but we included the sex as a cofactor in the eGWAS models and used only castrated males for the ASE analyses.

In a comparison of ASE and eGWAS results, 149 associated genes in eGWAS overlapped with ASE-SNPs. For genes found *cis*-associated in eGWAS, ASE analysis validated the putative *cis*-regulation of their expression. By combining eGWAS and ASE analyses, we were able to more extensively map the genetic control of the blood transcriptome and this approach will likely contribute to a more fine-scale mapping of the candidate causal polymorphisms. The heterogeneity of associated SNPs along the genome can be observed with greater precision using ASE approaches than with eGWAS, and this precision enabled us to pinpoint the most relevant genomic regions, as illustrated by the example of the SLA region. At the functional level, we found that ASE-genes were more enriched in biological functions specifically related to immunity than the genes locally or distantly associated with eQTL-SNPs were. However, the few enriched GO terms that were shared between the two analyses were immune-related. The GO results were complemented by the fact that the ASE analysis pinpointed immune functions as one of particular phenotypes affected by gene expression found under genetic control. These findings suggest that ASE results are more precise and accurate than eQTL results. In our study, this difference could be related to the relatively low density of SNPs in the chip used for eQTL mapping. Finally, ASE can provide information relevant in refining the mechanism of genetic regulation of gene expression by completing for instance putative regulatory networks.

Results are consistent with previous eQTL studies of human whole blood

Globally, our results provided general patterns of eQTL mapping in blood that were consistent with previous results on whole blood in humans. First, we detected 1.6 times more local associations (2839) than distant associations (1751), an imbalance that has already been reported in previous studies [12, 31, 33, 36]. This could be because distantly acting regulators have weaker effects than locally acting ones, resulting in associations that require greater statistical power for detection. Moreover, *trans*-eQTLs are more cell specific than *cis*-eQTLs [29] and because the heterogeneous nature of blood, *trans*-signals will be weaker. Although we cannot preclude the explanation that the larger number of local associations recovered here is due to a lack of statistical power to detect distant/*trans*-regulation, our results suggest that the number of local/*cis*-regulators truly does exceed the number of distant/*trans*-regulatory SNPs.

We observed that local eQTLs, distant eQTLs, and *cis*-regulators were widely distributed across all chromosomes, as has been previously described [31, 33]. Interestingly, nine genes that were associated with local eQTL-SNPs, two of which belonged to the *TMEM* family, had also been listed among the top 25 *cis*-associated genes reported by Joehanes et al. [12]. Likewise, a set of 324 *cis*-acting genes reported by Mehta et al. [33] in a study on whole blood also contained 32 of the *cis*-associated genes found here. Moreover, as has also been reported in humans, we were able to identify putative candidate master regulatory variants, despite the heterogeneous nature of peripheral blood [28, 29, 31]. We detected at least 13 master regulatory loci, as represented by eQTL-SNPs that were associated with between 10 and 51 probes. These eQTL-SNPs influenced the expression of various genes, mostly in the same direction, as illustrated here by the examples of two SNPs, ASGA0110383 and MARC0056482. However, most master regulators were not associated with local genes, which was consistent with the results of a previous study [31].

We paid special attention to the SLA region on chromosome SSC7 because it harbored the greatest concentration of genetic regulatory elements in both eGWAS and ASE analyses. Previous studies in humans have reported similar findings. For example, Fairfax et al. [29] found that several loci in the MHC showed significant *trans*-associations. Examination of the MHC also revealed numerous eQTL-SNPs linked to complex diseases [27]. In addition, an eGWAS performed on blood from cohorts of healthy women and breast cancer survivors showed an association between ten HLA genes (*HLA-C, HLA-E, HLA-F, HLA-G, HLA-H, HLA-DPB1, HLA-DQA1, HLA-DQB1, HLA-DRB3, HLA-DRB4*) and SNPs in 100 genes located on human chromosome 6 [37]. In particular, *CIITA* has been identified as a master *trans*-activator, with an essential role in initiating transcription of MHC class II genes in human stimulated B cells and monocytes [38]. However, our eGWAS did not generate any evidence of *CIITA* acting as a master regulator, despite the presence of one *CIITA*-related SNP on the array. In the ASE analysis, *CIITA* was expressed in the RNA-Seq data, but had no significant allelic imbalance. A possible explanation for this discrepancy is that such associations could be specific to conditions of immune stimulation and are not significant in a basal health state. Strikingly, we observed that almost all functional SLA class I and II genes harbored allelic imbalance, suggesting that fine-scale modulations of the expression levels of histocompatibility class I and II molecules may depend on the relative abundance of alleles. More in-depth analyses are required to determine whether this allelic imbalance affects peptide presentation efficiency and is associated with different immunity phenotypes.

Our results seem to be globally consistent with studies on the whole blood transcriptome in humans [28, 29, 31]. However, additional studies are needed on whole blood in pigs to assess the robustness of the list of local eQTL-SNPs reported here and to start to establish comparative eQTL maps where relevant. Thus, our work is a first step in generating useful resources for the study of eQTLs in pigs, as has been established in humans [12].

A step forward in the characterization of blood biomarkers, particularly for immune capacity

Peripheral blood contains various types of important immune cells. Schramm et al. [31] demonstrated reasonable concordance between the *cis*-eQTLs found in their own study of whole blood and those reported in prior studies of primary monocytes or blood-derived lymphoblastoid cell lines. They thus confirmed that whole-blood eQTL studies are a good resource for the discovery of biomarkers, especially in the context of disease in humans. With the goal of improving selection for health traits in pigs, blood is also a relevant surrogate tissue for phenotyping immune capacity [39]. As shown in human studies, whole blood can be used for robust eQTL analysis despite its heterogeneous cell composition. Indeed, Mehta et al. [33] demonstrated that variations in cell count would exert only a minor effect on expression levels, as eQTLs are consistently replicated between whole blood and independent human cohorts.

We showed a notable degree of overlap between the genes found in this study to be genetically controlled and the QTLs available in the AnimalQTL database [40] for pigs. More precisely, we found overlap between the lists of eQTL- and ASE-SNPs generated here and QTLs of the AnimalQTL database that had been associated with immune-related phenotypes, including interferon-gamma level, IgG level, lymphocyte count and other traits related to blood function (Additional files 14 and 15: Table S8 and Table S9). In addition, we identified the genetic regulation of several previously published blood biomarkers for immunity traits (e.g., *ALOX12* for phagocytosis, *GZMB* and *KLRG1* for CD4-CD8+ cell count) [4], which suggests that these biomarkers could be used as molecular expression phenotypes in genetic selection efforts.

The integration of eGWAS- and ASE-based approaches, together with GWAS data when relevant, facilitates the identification of candidate genes and polymorphisms that are related to complex traits. Similar integrated approaches in pigs have been recently published for complex traits related to meat properties [14, 16], as well as for blood lipid traits linked with cardiovascular diseases [41] and traits associated with coping behavior [20]. In addition, this approach could help to pinpoint the biological and molecular bases of phenotype-genotype links that are highlighted by a GWAS-based approach. Integrated methods are also expected to be more accurate in the detection of loci under genetic control as the link between genotype and transcriptome is more direct than the link between genotype and end phenotype. These techniques can thus help to prioritize the study of candidate variants linked to changes in the transcriptome and phenotype.

Conclusions

This study is, to the best of our knowledge, the first genome-wide analysis of the genetic control of gene expression in blood in the pig, based on a cohort of 60-day-old French Large White pigs. The combination of eGWAS and ASE results showed extensive genetic regulation of gene transcription, and provides a relevant resource for the study of genomic regions and variants linked to gene expression variation in pig blood. All results are available in a public Track Hub entry that can be explored using the genome browsers of Ensembl and UCSC. Additional studies are needed to cover more ages, breeds and environments in order to build an even more extensive map of genetic regulations of gene expression in blood. Overall, these data and approaches should contribute to future efforts to decipher the ultimate molecular mechanisms behind the genetic determination of pig phenotypes, especially for immunity- and health-related traits.

Methods
Animals and blood sampling

Two groups of pigs from the same French Large White selected line were bred either on a test farm at Le Rheu (38 pigs, IMMOPIG project) or on an INRA experimental farm at le Magneraud (243 pigs, SUS_FLORA project). The animals were partly generated from common boars and were weaned at 28 days of age. The group of 38 pigs belongs to a population of castrated males that was described in Flori et al. (2011) [2]. The group of 243 pigs comprised 131 uncastrated males and 112 females from a larger cohort that is described in Ramayo-Caldas et al. (2016) [42]. Details on the pig feed diet are available in Mach et al. (2015) [43] and correspond to standard nutritional practices in pig production. Peripheral blood (jugular vein) was sampled from 60-day-old pigs using EDTA-coated tubes and PAXgene Blood RNA tubes (PreAnalystiX, Qiagen, Germany). Blood samples

were stored at −20 °C (EDTA tubes) or −80 °C (PAX-Gene tubes) prior to DNA or RNA extraction procedures, respectively. Complete blood cell counts were recorded for a subset of 195 pigs from the group of 243 pigs (Additional file 16: Table S10), in agreement with our observations that the animals showed no clinical signs of infection. Information on production traits are available in Sanchez et al., (2014) [44] for the ASE pig group and in Ramayo-Caldas et al. (2016) [42] for the eQTL pig group.

Total RNA extraction from blood

Total RNA from blood was isolated using the PAXgene Blood RNA Kit (Qiagen, Germany) following the manufacturer's instructions. RNA purity and concentration were determined using a NanoDrop 1000 spectrophotometer (Thermo Scientific, USA), and RNA integrity was assessed using the Bioanalyzer 2100 (Agilent Technologies, USA). RNA samples used for RNA-Seq had RINs ranging from 6.7 to 8.6 with 28S:18S ratios ranging from 0.8 to 1.6. RNA samples used for microarrays had RINs between 6.1 and 9.7.

SNP genotyping

DNA was extracted from blood samples and genotyped using the Illumina PorcineSNP60 DNA chip which contains over 60,000 SNPs [35]. Genomic DNA extraction and SNP genotyping were carried out with the Labogena-DNA platform (Jouy-en-Josas, France). The initial quality control (QC) step of genotyping was performed using the R package GenABEL [45] and included heterozygosity testing and multidimensional scaling to identify population outliers. For SNP filtering, we used sample and SNP call rates >95%, a MAF > 5%, a Hardy-Weinberg equilibrium threshold corrected by FDR of 1%, and a XXY call of 0.8. We removed one individual that did not meet these QC criteria, resulting in 242 individuals (130 males and 112 females) with data available for 44,281 SNPs.

Analysis of blood transcriptome using expression microarrays

Total blood RNA (200 ng) was retro-transcribed, Cy3-labeled, and hybridized onto customized single-channel 8X60K Agilent Technologies arrays (platform Agilent-037880/INRA Susscrofa60K v1), which were enriched in genes linked to immunity and muscle physiology, following the manufacturer's instructions and a protocol described in Jacquier et al. (2015) [46]. Because genomic sequencing data were not available, it was not possible to take a possible effect of within cDNA target polymorphisms on hybridization levels into account. Preprocessing and quality control were performed by the Limma R package [47] on 60,306 probes. In brief, we first normalized the background by the normexp method. Then, intensities quantified by median signals were normalized by the quantile method. Probes were filtered according to their intensities in order to eliminate unexpressed genes; to be retained, probes had to be expressed at levels at least 10% higher than those of the negative controls on at least four arrays. In total, 59,774 probes that passed the quality control and the filtering step were considered to be expressed in the blood samples included in this study.

Refinement of probe sequence annotation of the Agilent-037880/INRA *Susscrofa*60K v1 platform

Probe mapping and annotation were updated to refine all genomic locations. The 60-mer sequence probes included in the Agilent-037880/INRA *Susscrofa*60K v1 platform were mapped onto the pig reference genome (SusScrofa v10.2) with the TopHat aligner (v2.0.14) [48]. Allowing two mismatches and two gaps per probe sequence, we found 44,326 probes mapping to unique genomic positions. Among these, 41,839 probes were expressed in our samples. The annotation of these probes was performed by combining annotation information provided by the Ensembl (release 79) and NCBI (GCF_000003025.5) databases.

Expression genome-wide association study (eGWAS)

The eGWAS was carried out on data from 242 pigs and linked SNP genotypes, based on animal genotypes obtained with PorcineSNP60 DNA chips, to blood gene-transcript levels, as detected by transcriptome analysis using Agilent-037880/INRA *Susscrofa*60K v1 microarrays. The association studies were performed with the GenABEL package [45] using the family-based score test for association [49], a polygenic mixed model to estimate the SNP effect, and a *p*-value in two steps. First, the polygenic additive model and likelihood were estimated using available data, and then the FASTA test statistic was computed using the maximum-likelihood estimates of the intercept, the proportion of variance explained by the polygenic component, and the residual variance. The model was corrected for different co-factors, with batch and sex as fixed effects, and with the familial structure of the sample as a random effect, represented by the genomic kinship matrix estimated using autosomal data. The individual variability in blood cell counts was not included in the analysis model since the data were not available for all the animals. Nonetheless, we observed a limited variability of these blood parameters within the population (see Additional file 16: Table S10). All probes and SNPs were treated independently for the first step of the association study. Next, multiple testing corrections were conducted at two levels. First, for each probe, *p*-values of the associations between the expression

variation of that probe and all SNPs were corrected by an FDR procedure that controlled the overall type-I error rate at 5%. Secondly, to assess the significance of associations of all probes, a second significance threshold was calculated by FDR (5%) for the *p*-values of all associations. Only associations with a FASTA *p*-value lower than these two thresholds were considered significant.

Finally, in order to facilitate the global interpretation of results, we implemented additional association tests to limit redundancy. We reduced the definition of an eQTL to a representative SNP, i.e. for each eQTL region, we kept only the most significant associated SNP to represent the eQTL, referred to hereafter as an eQTL-SNP. In order to eliminate other significant associated SNPs in a 5 Mb window around a designated eQTL-SNP without missing an independent eQTL, we repeated the association analysis between the probe and these SNPs, correcting the polygenic mixed model by the effect of the most significant SNP, as has been reported in previous studies of human data [29, 50]. If the *p*-values associated with a given SNP did not reach the previously calculated significance threshold, the SNP was considered to have the same effect as the most significant SNP (the previously designated eQTL-SNP); it was thus considered redundant and removed from further analyses.

Allele-specific expression (ASE) analysis by RNA-Seq

Sequencing libraries from total blood RNA (1.5 µg) were prepared using the TruSeq RNA Sample Preparation Kit (Illumina, San Diego, USA) according to the manufacturer's instructions at INRA's GetPlage platform (Toulouse, France) and as previously reported in Mach et al. (2014) [51]. Tagged cDNA libraries were sequenced on 2.66 lanes of a HiSeq2000 (Illumina) in 100-bp single-end reads. Then, raw reads were aligned to the pig reference genome (v10.2) with the STAR 2.4.0 aligner using an approach that comprised two alignment steps, as recommended by the Broad Institute (https://software.broadinstitute.org/gatk/guide/article?id=3891). First, splice junctions were detected by a first pass of STAR mapping. Detected splice junctions were then used during a second round of mapping to take into account the discontinuous nature of the RNA sequencing data, allowing only one mapping position per read. STAR includes a soft-clipping step, which takes care of adapter contamination. After the two-pass procedure, we used a GATK (v3.4, [52]) tool called SplitNCigarReads, which split reads into exon segments and hard-clips any sequences overhanging into the intronic regions to reduce the number of called false variants. Then, reads were realigned around indels to limit artifacts due to the alignment algorithm. Finally, the standard GATK base recalibration procedure re-adjusted the quality scores assigned to the individual base calls in each sequence read, thus reducing the systematic technical error linked to the sequencing machines.

In order to prevent false positive ASE signals due to allelic mapping biases that could influence allele counts, variants were called in the dataset, filtered using vcftools (v.0.1.12a), and then nucleotides that were characterized as SNPs were masked in the reference genome with maskFastaFromBed (Bedtools v2–2.24.0). Reads were then realigned on the masked reference genome using the same procedure as described previously in order to have the same constraint of one mismatch for reads that had either of the two alleles at the SNP position.

Allele counts were performed using a GATK (v.3.4) tool called ASEReadCounter which calculates allele counts at SNP positions. A binomial test determined whether there was a significant difference in abundance of each allele of a SNP. We corrected p-values on a per-animal basis using the FDR procedure (5×10^{-2}). Only bi-allelic SNPs with a significant differential abundance in alleles in more than 1/3 of heterozygous animals for the considered SNP were analyzed further.

Analysis of function enrichment and SNP annotation

A functional enrichment analysis of genes associated with eQTL- or ASE-SNPs was performed with GOrilla (Gene Ontology enRIchment anaLysis and visuaLizAtion tool, Eden et al., 2009). The Gorilla database is periodically updated using the GO database (Gene Ontology Consortium) and other sources and we used the updated version of December 2016. The p-value threshold was equal to 1×10^{-3} and an FDR correction was performed on p-values. For gene networks, we also performed functional analyses that used QIAGEN's Ingenuity Pathway Analysis (IPA®, QIAGEN, Redwood City, CA, USA, http://www.qiagen.com/ingenuity, release date 2016–12–05) with default parameters. eQTL- and ASE-SNPs were annotated using the Variant Effect Predictor [21] to determine their known effects on gene, transcript, and protein sequences.

Visualization of results

To visualize eGWAS results, we used Cytoscape v3.2.1 [24], which enables the construction of association networks. We used genes and eQTL-SNPs as nodes and significant associations as edges. The R package RCirco [53] was used to illustrate master regulators, which we defined as SNPs that had associations with multiple genes. To generate Venn diagrams, we used jvenn, a plug-in for the jQuery Javascript library [54]. Boxplots were produced with the R package ggplot. The R package PCIT [25] was used to analyze the correlation between the expression variation of probes associated with the same eQTL-SNP.

Additional files

Additional file 1: Table S1. Associations detected by eGWAS. (XLSX 342 kb)

Additional file 2: Table S2. Genes annotated by sense and/or antisense probes found associated with local or distant eQTL-SNPs. (DOCX 11 kb)

Additional file 3: Figure S1. Distribution of eQTL-SNPs along pig autosomal chromosomes. The histograms represent the distribution of eQTL-SNPs along chromosomes, with each bar corresponding to a 5-Mb segment. The density of eQTL-SNPs is represented by a color gradient from green to red. The scale for the eQTL-SNP count is on the Y-axis and varies among chromosomes. On SSC7, the region between 20 Mb and 30 Mb that has a high density of eQTL-SNPs overlaps the MHC locus. (PDF 30 kb)

Additional file 4: Table S3. ASE analysis results. (XLSX 126 kb)

Additional file 5: Figure S2. Distribution of ASE-SNPs along pig autosomal chromosomes. The histograms represent the distribution of ASE-SNPs along chromosomes, with each bar corresponding to a 5-Mb segment. The density of ASE-SNPs is represented by a color gradient from green to red. The scale for the eQTL-SNP count is on the Y-axis and among chromosomes. The highest density of ASE-SNPs was found on SSC7 at the MHC locus. (PDF 29 kb)

Additional file 6: Table S4. Enrichment analysis of associated genes in eGWAS by GOrilla. (XLSX 39 kb)

Additional file 7: Table S5. Enrichment analysis of local-associated genes in eGWAS by GOrilla. (XLSX 28 kb)

Additional file 8: Table S6. Enrichment analysis of distant-associated genes in eGWAS by GOrilla. (XLSX 21 kb)

Additional file 9: Table S7. Enrichment analysis of cis-genes in ASE analysis by Gorilla. (XLSX 67 kb)

Additional file 10: Figure S3. Comparison of enriched functions for genes detected by eGWAS (total, local, distant), and ASE analysis. A: Venn diagram produced with jvenn [54] showing shared and specific gene ontology (GO) terms. The numbers of enriched GO terms are shown for all associated genes detected by eGWAS (green), distant- (blue) or local-associated genes (red) detected by eGWAS, and ASE-genes (yellow). B: Number of enriched GO terms detected for each analysis. (PNG 40 kb)

Additional file 11: Figure S4. Gene association network centered on eQTL-SNP H3GA0029721. A: Circo plot produced by R package RCirco which maps eQTL-SNP H3GA0029721 and its associated genes according to their genomic position. The blue links represent associations between this eQTL-SNP on SSC10 and 47 genes that map onto other chromosomes. B: The distribution of all (grey) and significant (red) correlation coefficients calculated from pairwise comparisons of expression variation in probes associated with H3GA0029721, performed by the PCIT algorithm. (PDF 131 kb)

Additional file 12: Figure S5. Detailed analysis of the association network centered on eQTL-SNP MARC0001946. A: Circo plot produced by R package RCirco which maps eQTL-SNP MARC0001946 and its associated genes according to their genomic positions. Red links represent associations between MARC0001946 and genes on SSC3, and blue links represent associations with genes on other chromosomes. B: The boxplots represent the expression variation of probes annotated for the porcine genes UROC1 and ARHGAP2, depending on eQTL-SNP genotype. Each dot represents one animal. Pink dots correspond to pigs homozygous for the more-frequent allele (genotype A/A), green dots to heterozygous pigs (genotype A/G), and blue dots to animals homozygous for the less-frequent allele (G/G). The boxplots show that the eQTL-SNP genotypes affected the transcription level of the two associated genes in opposite directions. C: The distribution of all (grey) and significant (red) correlation coefficients calculated from pairwise comparisons of expression variation in probes associated with MARC0001946, performed by the PCIT algorithm. D: Three functional networks produced by IPA, on which genes associated with eQTL-SNP MARC0001946 are represented by gray symbols. The functional network related to "cellular development", "hematological system development and function", and "hematopoiesis" includes ten associated genes; the network related to "inflammatory response", "cell signaling", and "molecular transport" includes nine associated genes; and the network related to "developmental disorder", "hereditary disorder", and "organismal injury and abnormalities" includes five associated genes. (PDF 608 kb)

Additional file 13: Figure S6. Detailed analysis of the association networks centered around eQTL-SNPs ALGA0110383 and MARC0056482. A: Circo plot produced by R package RCirco which maps the two eQTL-SNPs on SSC6 together with their associated genes, according to their genomic positions. Green links represent associations between genes and ALGA0110383. Purple links represent associations between genes and MARC0056482. B: The distribution of all (grey) and significant (red) correlation coefficients calculated from pairwise comparisons of expression variation in probes associated with ALGA0110383 and MARC0001946, performed by the PCIT algorithm. C: Functional network produced by IPA in which genes associated with eQTL-SNPs ALGA0110383 and MARC0001946, as well as the genes *YIPF1* and *PCSK9* (which harbor an ASE-SNP and an eQTL-SNP, respectively) are represented by gray-colored symbols. The functional network, which contains 14 associated genes, is related to "cell death and survival", "organismal injury and abnormalities", and "tissue morphology". D: The boxplots represent expression variations in probes annotated for porcine genes according to each eQTL-SNP genotype. Each dot represents an animal. Pink dots correspond to pigs homozygous for the most-frequent allele (A/A), green dots to heterozygous pigs (A/G), and blue dots to animals homozygous for the less-frequent allele (G/G). (PDF 594 kb)

Additional file 14: Table S8. QTLs in AnimalQTLdb (<10 Mb) for pigs harbouring at least one eQTL-SNP. (XLSX 90 kb)

Additional file 15: Table S9. QTLs in AnimalQTLdb for pigs (<10 Mb) harbouring at least one ASE-SNPs. (XLSX 44 kb)

Additional file 16: Table S10. Complete Blood Cell Counts for 195 animals among the 243 eGWAS pigs. (XLSX 14 kb)

Abbreviations
ASE: Allele-specific expression; eGWAS: expression genome-wide association study; eQTL: expression quantitative trait locus; FASTA: Family bAsed Score Test for Associations; GO: Gene ontology; GWAS: Genome-wide association study; IPA: Ingenuity pathway analysis; MHC: Major Histocompatibility Complex; PCIT: Partial correlation information theory; QC: Quality control; QTL: Quantitative trait locus; SLA: Swine Leucocyte Antigen; SNP: Single nucleotide polymorphism; VEP: Variant effect predictor

Acknowledgments
We are grateful to the Genotoul bioinformatics platform at INRA de Toulouse Midi-Pyrenees (BioinfoGenotoul) for providing computing and storage resources. We are grateful to the INRA MIGALE bioinformatics platform (http://migale.jouy.inra.fr) for help and support. We are grateful to Laurence Liaubet (INRA, France) for sharing information on the expression microarrays and to Maria Ballester (IRTA, Spain) for useful comments and suggestions during the data analysis. We are grateful to the INRA @bridge platform for the secure storage of samples and the personnel at GENESI's farm at Le Magneraud for their implication on the generation of animals and samples.

Funding
TM was financially supported by a PhD fellowship co-funded by the INRA Animal Genetics Division and BIOPORC (association composed of AXIOM, CHOICE GENETICS FRANCE, NUCLEUS, and IFIP). Experiments were funded by the Sus_Flora (ANR-10-GENM-016) and IMMOPIG (ANR-06-GANI-08) projects of the French National Agency and INRA's Animal Genetics Division.

Authors' contributions
CRG coordinated the Sus_Flora and IMMOPIG projects. JE, TM, and CRG designed the experiments and interpreted all results. TM performed all analyses and was supervised by JE for computational analyses. YRC contributed to network analyses. CRG, JE, and GL contributed to animal sampling; GL organized all procedures for animal sampling and safe sample storage. GL and JL performed RNA extraction and microarray hybridization on the @BRIDGe platform (http://abridge.inra.fr). DE performed RNA-Seq on the GetPlaGe platform. TM drafted the manuscript which was further completed by CRG and JE. All authors improved the manuscript. All authors read and approved the final manuscript.

Competing interests
The authors declare they have no competing interests.

Author details
[1]GABI, INRA, AgroParisTech, Université Paris-Saclay, 78350 Jouy-en-Josas, France. [2]GenPhySE, INRA, INPT, ENVT, Université de Toulouse, 31326 Castanet-Tolosan, France. [3]IFIP - Institut du porc/BIOPORC, La Motte au Vicomte, BP 35104, 35651 Le Rheu, France.

References
1. Clapperton M, Diack AB, Matika O, Glass EJ, Gladney CD, Mellencamp MA, et al. Traits associated with innate and adaptive immunity in pigs: heritability and associations with performance under different health status conditions. Genet Sel Evol. 2009;41:54.
2. Flori L, Gao Y, Laloë D, Lemonnier G, Leplat J-J, Teillaud A, et al. Immunity traits in pigs: substantial genetic variation and limited covariation. PLoS One. 2011;6:e22717.
3. Clapperton M, Glass EJ, Bishop SC. Pig peripheral blood mononuclear leucocyte subsets are heritable and genetically correlated with performance. Animal. 2008;2:1575–84.
4. Mach N, Gao Y, Lemonnier G, Lecardonnel J, Oswald IP, Estellé J, et al. The peripheral blood transcriptome reflects variations in immunity traits in swine: towards the identification of biomarkers. BMC Genomics. 2013;14:894.
5. Jin W, Riley RM, Wolfinger RD, White KP, Passador-Gurgel G, Gibson G. The contributions of sex, genotype and age to transcriptional variance in Drosophila Melanogaster. Nat Genet. 2001;29:389–95.
6. Rockman MV, Kruglyak L. Genetics of global gene expression. Nat Rev Genet. 2006;7:862–72.
7. Schadt EE, Monks SA, Drake TA, Lusis AJ, Che N, Colinayo V, et al. Genetics of gene expression surveyed in maize, mouse and man. Nature. 2003;422: 297–302.
8. Albert FW, Kruglyak L. The role of regulatory variation in complex traits and disease. Nat Rev Genet. 2015;16:197–212.
9. Pastinen T. Genome-wide allele-specific analysis: insights into regulatory variation. Nat Rev Genet. 2010;11:533–8.
10. Lagarrigue S, Martin L, Hormozdiari F, Roux P-F, Pan C, van Nas A, et al. Analysis of allele-specific expression in mouse liver by RNA-Seq: a comparison with Cis-eQTL identified using genetic linkage. Genetics. 2013; 195:1157–66.
11. Hasin-Brumshtein Y, Hormozdiari F, Martin L, van Nas A, Eskin E, Lusis AJ, et al. Allele-specific expression and eQTL analysis in mouse adipose tissue. BMC Genomics. 2014;15:471.
12. Joehanes R, Zhang X, Huan T, Yao C, Ying S-X, Nguyen QT, et al. Integrated genome-wide analysis of expression quantitative trait loci aids interpretation of genomic association studies. Genome Biol. 2017;18:16.

13. Liaubet L, Lobjois V, Faraut T, Tircazes A, Benne F, Iannuccelli N, et al. Genetic variability of transcript abundance in pig peri-mortem skeletal muscle: eQTL localized genes involved in stress response, cell death, muscle disorders and metabolism. BMC Genomics. 2011;12:548.

14. González-Prendes R, Quintanilla R, Cánovas A, Manunza A, Figueiredo Cardoso T, Jordana J, et al. Joint QTL mapping and gene expression analysis identify positional candidate genes influencing pork quality traits. Sci Rep. 2017;7:39830.

15. Martínez-Montes AM, Muiños-Bühl A, Fernández A, Folch JM, Ibáñez-Escriche N, Fernández AI. Deciphering the regulation of porcine genes influencing growth, fatness and yield-related traits through genetical genomics. Mamm Genome. 2016;28:130.

16. Ponsuksili S, Murani E, Trakooljul N, Schwerin M, Wimmers K. Discovery of candidate genes for muscle traits based on GWAS supported by eQTL-analysis. Int J Biol Sci. 2014;10:327–37.

17. Cinar MU, Kayan A, Uddin MJ, Jonas E, Tesfaye D, Phatsara C, et al. Association and expression quantitative trait loci (eQTL) analysis of porcine AMBP, GC and PPP1R3B genes with meat quality traits. Mol Biol Rep. 2012; 39:4809–21.

18. Ponsuksili S, Murani E, Schwerin M, Schellander K, Wimmers K. Identification of expression QTL (eQTL) of genes expressed in porcine M. Longissimus dorsi and associated with meat quality traits. BMC Genomics. 2010;11:572.

19. Puig-Oliveras A, Revilla M, Castelló A, Fernández AI, Folch JM, Ballester M. Expression-based GWAS identifies variants, gene interactions and key regulators affecting intramuscular fatty acid content and composition in porcine meat. Sci Rep. 2016;6:31803.

20. Ponsuksili S, Zebunke M, Murani E, Trakooljul N, Krieter J, Puppe B, et al. Integrated genome-wide association and hypothalamus eQTL studies indicate a link between the circadian rhythm-related gene PER1 and coping behavior. Sci Rep. 2015;5:16264.

21. McLaren W, Gil L, Hunt SE, Riat HS, Ritchie GRS, Thormann A, et al. The Ensembl variant effect predictor. Genome Biol. 2016;17:122.

22. Yates A, Akanni W, Amode MR, Barrell D, Billis K, Carvalho-Silva D, et al. Ensembl 2016. Nucleic Acids Res. 2016;44:D710–6.

23. Eden E, Navon R, Steinfeld I, Lipson D, Yakhini Z. GOrilla: a tool for discovery and visualization of enriched GO terms in ranked gene lists. BMC Bioinformatics. 2009;10:48.

24. Shannon P, Markiel A, Ozier O, Baliga NS, Wang JT, Ramage D, et al. Cytoscape: a software environment for integrated models of biomolecular interaction networks. Genome Res. 2003;13:2498–504.

25. Watson-Haigh NS, Kadarmideen HN, Reverter A. PCIT: an R package for weighted gene co-expression networks based on partial correlation and information theory approaches. Bioinformatics. 2010;26:411–3.

26. Reverter A, Chan EKF. Combining partial correlation and an information theory approach to the reversed engineering of gene co-expression networks. Bioinformatics. 2008;24:2491–7.

27. Fehrmann RSN, Jansen RC, Veldink JH, Westra H-J, Arends D, Bonder MJ, et al. Trans-eQTLs reveal that independent genetic variants associated with a complex phenotype converge on intermediate genes, with a major role for the HLA. PLoS Genet. 2011;7:e1002197.

28. Fairfax BP, Makino S, Radhakrishnan J, Plant K, Leslie S, Dilthey A, et al. Genetics of gene expression in primary immune cells identifies cell type-specific master regulators and roles of HLA alleles. Nat Genet. 2012;44:502–10.

29. Fairfax BP, Humburg P, Makino S, Naranbhai V, Wong D, Lau E, et al. Innate immune activity conditions the effect of regulatory variants upon Monocyte gene expression. Science. 2014;343:1246949.

30. Renard C, Hart E, Sehra H, Beasley H, Coggill P, Howe K, et al. The genomic sequence and analysis of the swine major histocompatibility complex. Genomics. 2006;88:96–110.

31. Schramm K, Marzi C, Schurmann C, Carstensen M, Reinmaa E, Biffar R, et al. Mapping the genetic architecture of gene regulation in whole blood. PLoS One. 2014;9:e93844.

32. Sasayama D, Hori H, Nakamura S, Miyata R, Teraishi T, Hattori K, et al. Identification of single nucleotide polymorphisms regulating peripheral blood mRNA expression with genome-wide significance: an eQTL study in the Japanese population. PLoS One. 2013;8:e54967.

33. Mehta D, Heim K, Herder C, Carstensen M, Eckstein G, Schurmann C, et al. Impact of common regulatory single-nucleotide variants on gene expression profiles in whole blood. Eur J Hum Genet. 2013;21:48–54.

34. Rotival M, Zeller T, Wild PS, Maouche S, Szymczak S, Schillert A, et al. Integrating genome-wide genetic variations and monocyte expression data reveals trans-regulated gene modules in humans. PLoS Genet. 2011;7: e1002367.

35. Ramos AM, Crooijmans RPMA, Affara NA, Amaral AJ, Archibald AL, Beever JE, et al. Design of a high density SNP genotyping assay in the pig using SNPs identified and characterized by next generation sequencing technology. PLoS One. 2009;4:e6524.

36. Petretto E, Mangion J, Dickens NJ, Cook SA, Kumaran MK, Lu H, et al. Heritability and tissue specificity of expression quantitative trait loci. PLoS Genet. 2006;2:e172.

37. Landmark-Høyvik H, Dumeaux V, Nebdal D, Lund E, Tost J, Kamatani Y, et al. Genome-wide association study in breast cancer survivors reveals SNPs associated with gene expression of genes belonging to MHC class I and II. Genomics. 2013;102:278–87.

38. Wong D, Lee W, Humburg P, Makino S, Lau E, Naranbhai V, et al. Genomic mapping of the MHC transactivator CIITA using an integrated ChIP-seq and genetical genomics approach. Genome Biol. 2014;15:494.

39. Schroyen M, Tuggle CK. Current transcriptomics in pig immunity research. Mamm Genome. 2015;26:1–20.

40. Hu Z-L, Fritz ER, Reecy JM. AnimalQTLdb: a livestock QTL database tool set for positional QTL information mining and beyond. Nucl Acids Res. 2007;35:D604–9.

41. Chen C, Yang B, Zeng Z, Yang H, Liu C, Ren J, et al. Genetic dissection of blood lipid traits by integrating genome-wide association study and gene expression profiling in a porcine model. BMC Genomics. 2013;14:848.

42. Ramayo-Caldas Y, Mach N, Lepage P, Levenez F, Denis C, Lemonnier G, et al. Phylogenetic network analysis applied to pig gut microbiota identifies an ecosystem structure linked with growth traits. ISME J. 2016;10:2973–7.

43. Mach N, Berri M, Estellé J, Levenez F, Lemonnier G, Denis C, et al. Early-life establishment of the swine gut microbiome and impact on host phenotypes. Environ Microbiol Rep. 2015;7:554–69.

44. Sanchez M-P, Tribout T, Iannuccelli N, Bouffaud M, Servin B, Tenghe A, et al. A genome-wide association study of production traits in a commercial population of large white pigs: evidence of haplotypes affecting meat quality. Genet Sel Evol. 2014;46:12.

45. Aulchenko YS, Ripke S, Isaacs A, van Duijn CM. GenABEL: an R library for genome-wide association analysis. Bioinformatics. 2007;23:1294–6.

46. Jacquier V, Estellé J, Schmaltz-Panneau B, Lecardonnel J, Moroldo M, Lemonnier G, et al. Genome-wide immunity studies in the rabbit: transcriptome variations in peripheral blood mononuclear cells after in vitro stimulation by LPS or PMA-Ionomycin. BMC Genomics. 2015;16:26.

47. Ritchie ME, Phipson B, Wu D, Hu Y, Law CW, Shi W, et al. Limma powers differential expression analyses for RNA-sequencing and microarray studies. Nucleic Acids Res. 2015;43:e47.

48. Trapnell C, Pachter L, Salzberg SL. TopHat: discovering splice junctions with RNA-Seq. Bioinformatics. 2009;25:1105–11.

49. Chen W-M, Abecasis GR. Family-based association tests for genomewide association scans. Am J Hum Genet. 2007;81:913–26.

50. Bryois J, Buil A, Evans DM, Kemp JP, Montgomery SB, Conrad DF, et al. Cis and trans effects of human genomic variants on gene expression. PLoS Genet. 2014;10:e1004461.

51. Mach N, Berri M, Esquerré D, Chevaleyre C, Lemonnier G, Billon Y, et al. Extensive expression differences along porcine small intestine evidenced by transcriptome sequencing. PLoS One. 2014;9:e88515.

52. DePristo MA, Banks E, Poplin RE, Garimella KV, Maguire JR, Hartl C, et al. A framework for variation discovery and genotyping using next-generation DNA sequencing data. Nat Genet. 2011;43:491–8.

53. Zhang H, Meltzer P, Davis S. RCircos: an R package for Circos 2D track plots. BMC Bioinformatics. 2013;14:244.

54. Bardou P, Mariette J, Escudié F, Djemiel C, Klopp C. jvenn: an interactive Venn diagram viewer. BMC Bioinformatics. 2014;15:293.

12

Validation of whole-blood transcriptome signature during microdose recombinant human erythropoietin (rHuEpo) administration

Guan Wang[1,10], Jérôme Durussel[2], Jonathan Shurlock[3], Martin Mooses[4], Noriyuki Fuku[5], Georgie Bruinvels[6], Charles Pedlar[6], Richard Burden[6], Andrew Murray[7], Brendan Yee[8], Anne Keenan[2], John D. McClure[2], Pierre-Edouard Sottas[9] and Yannis P. Pitsiladis[1,10]*

Abstract

Background: Recombinant human erythropoietin (rHuEpo) can improve human performance and is therefore frequently abused by athletes. As a result, the World Anti-Doping Agency (WADA) introduced the Athlete Biological Passport (ABP) as an indirect method to detect blood doping. Despite this progress, challenges remain to detect blood manipulations such as the use of microdoses of rHuEpo.

Methods: Forty-five whole-blood transcriptional markers of rHuEpo previously derived from a high-dose rHuEpo administration trial were used to assess whether microdoses of rHuEpo could be detected in 14 trained subjects and whether these markers may be confounded by exercise ($n = 14$ trained subjects) and altitude training ($n = 21$ elite runners and $n = 4$ elite rowers, respectively). Differential gene expression analysis was carried out following normalisation and significance declared following application of a 5% false discovery rate (FDR) and a 1.5 fold-change. Adaptive model analysis was also applied to incorporate these markers for the detection of rHuEpo.

Results: *ALAS2*, *BCL2L1*, *DCAF12*, *EPB42*, *GMPR*, *SELENBP1*, *SLC4A1*, *TMOD1* and *TRIM58* were differentially expressed during and throughout the post phase of microdose rHuEpo administration. The *CD247* and *TRIM58* genes were significantly up- and down-regulated, respectively, immediately following exercise when compared with the baseline both before and after rHuEpo/placebo. No significant gene expression changes were found 30 min after exercise in either rHuEpo or placebo groups. *ALAS2*, *BCL2L1*, *DCAF12*, *SLC4A1*, *TMOD1* and *TRIM58* tended to be significantly expressed in the elite runners ten days after arriving at altitude and one week after returning from altitude (FDR > 0.059, fold-change varying from 1.39 to 1.63). Following application of the adaptive model, 15 genes showed a high sensitivity (≥ 93%) and specificity (≥ 71%), with *BCL2L1* and *CSDA* having the highest sensitivity (93%) and specificity (93%).

(Continued on next page)

* Correspondence: y.pitsiladis@brighton.ac.uk
[1]Centre of Sports Medicine for Anti-Doping Research, University of Brighton, Eastbourne, UK
[10]Department of Movement, Human and Health Sciences, University of Rome "Foro Italico", Rome, Italy
Full list of author information is available at the end of the article

(Continued from previous page)

Conclusions: Current results provide further evidence that transcriptional biomarkers can strengthen the ABP approach by significantly prolonging the detection window and improving the sensitivity and specificity of blood doping detection. Further studies are required to confirm, and if necessary, integrate the confounding effects of altitude training on blood doping.

Keywords: Recombinant human erythropoietin, Whole blood, Transcriptome, Altitude, Exercise, Athlete biological passport

Background

The performance-enhancing drug recombinant human erythropoietin (rHuEpo) stimulates red blood cell production and although the World Anti-Doping Agency (WADA) prohibits its use, is frequently abused by athletes. The early anti-doping approach was to set an upper limit for haemoglobin and haematocrit levels in an attempt to discover rHuEpo abuse [1]. The first direct analytical procedure to detect rHuEpo was introduced in 2000 and exploited the differences in the charge profiling of endogenously and exogenously produced Epo in urine by isoeletric focusing (IEF) [2]. The main limitations of this direct approach are a variable and short detection window and low sensitivity [3]. There is now an improved direct analytical test to detect rHuEpo using sarcosyl polyacrylamide gel electrophoresis (SAR-PAGE, a modified sodium dodecyl sulfate polyacrylamide gel electrophoresis) with increased discriminatory capacity compared to IEF and detection window of 24 to 85 h using blood and urine samples [4]. According to the WADA technical document (i.e. TD2014EPO), SAR-PAGE is currently recommended for rHuEpo detection in both the initial and confirmation testing procedures [5]. Despite these advances, important limitations in detection of the direct approach have prompted a paradigm shift to the indirect identification of the effect of the prohibited method and/or substance.

One of the initial indirect methods involved the use of altered haematological markers such as reticulocyte haematocrit, haematocrit and percent macrocytes for the detection of current and recent rHuEpo intake using the ON- and OFF-statistical models, respectively [6]. This further advance culminated in the development of the Athlete Biological Passport (ABP) [1], introduced in 2009 by WADA. The ABP monitors changes in the blood matrix of an individual over time using Bayesian inference techniques to establish an individual's haematological profile that can reveal evidence of doping, not confined only to rHuEpo but also other forms of blood manipulation [7]. The stability of the ABP haematological parameters is limited to 48 h for reticulocytes

and 72 h for haemoglobin from blood collection to analysis when samples are handled at 4 °C and a limit of 36 h has been recommended by WADA for improved analytical quality [8]. A particular advantage of the ABP approach is the incorporation of other evidence of doping, such as longitudinal performance data, additional biomarkers yet to be discovered and validated and other nonanalytical evidence [1]. Despite this significant advance, the detection of rHuEpo and blood doping in general, using the current ABP approach, remains unsatisfactory [9, 10]. For example, the application of the ABP failed to reveal any evidence of rHuEpo microdosing (i.e. 20–30 $IU \cdot kg^{-1}$ body mass rHuEpo twice weekly for 8 weeks) in 10 healthy male subjects [11]. In addition, the haematological parameters of the ABP may be confounded with factors such as altitude exposure, since hypoxia may affect the blood variables and vascular volumes [12, 13]. Notwithstanding these limitations, the addition of other biomarkers to the ABP is envisaged to improve the sensitivity and specificity of the ABP model, and therefore substantially enhancing future doping detection strategies [1, 14, 15].

Current advances in omics technologies permit a global transcriptional, translational or epigenomic feature of a cell, tissue or organism under altered physiological or developmental conditions to be investigated. For example, a state-of-the-art omics approach has been successfully applied to epidemic diseases (e.g. [16, 17]) and cancer diagnosis [18]. Undoubtedly, the application of omics to the field of anti-doping will help reveal potential doping biomarkers. In a study by Varlet-Marie et al., the gene expression profile in response to darbepoetin alpha was determined using the serial analysis of gene expression (SAGE) by pooling whole-blood samples from 14 healthy, active subjects (50% male) into three SAGE libraries (before, during and after administration) [19]. The authors then confirmed the differential expression of 95 genes identified using SAGE in two well-trained male athletes by qPCR [19]. Five genes remained significantly expressed following a further high dose and microdose rHuEpo treatment in these athletes based on a fold change of 1.5 and a false discovery rate (FDR) of 0% [19]. This initial promise of improved discriminatory potential of the transcriptomic bio-markers to detecting doping, encouraged a number of other attempts to investigate the global gene expression patterns in whole blood or lymphocytes for the detection of testosterone, anabolic steroids, recombinant human growth hormone, gene doping, blood transfusions and rHuEpo (see review [14]). It should be noted that samples are often collected for anti-doping purposes at sporting or training venues after intense exercise [20] and prior intense exercise training has been reported to have significant impact on gene expression profiles of peripheral blood mononuclear cells [21] and white blood cells [22]. We previously carried out whole-blood gene expression profiling using a microarray-based approach to detect rHuEpo doping (i.e. 50 IU·kg^{-1} body mass every two days for 4 weeks). Briefly, 34 of 45 selected genes from two independent groups following the microarray analysis were validated using a different quantification platform [23]. The limitations of this study were the absence of a control/placebo group and the now out-dated rHuEpo dosing regimen involving near clinical doses of rHuEpo as opposed to the more commonly used strategy involving rHuEpo microdosing [23]. With these limitations in mind, the aims of the present study were to investigate 1) whether the previously identified 45 transcriptional markers could detect rHuEpo microdosing in a randomised, double-blind, placebo-controlled cross-over study; and 2) whether these gene responses differ among rHuEpo microdosing and potential confounders such as strenuous exercise training and moderate altitude exposure.

Methods

Forty-five candidate transcripts and five reference genes identified from the whole-blood transcriptome profiling in subjects administered with 50 IU·kg^{-1} body mass of rHuEpo for 4 weeks (i.e. the rHuEpo high-dose study, see Additional file 1) [23] were interrogated in subjects participating in the rHuEpo microdose study and the rHuEpo confounders studies involving exercise and altitude training (Fig.1).

Study design

Microdose study (MDS)

Fourteen endurance-trained healthy males (mean ± standard deviation (SD); age: 29.9 ± 4 yrs., height: 178.8 ± 4.5 cm, maximal aerobic capacity (V̇O$_2$max): 55.3 ± 4.7 ml·kg^{-1}·min^{-1}) at sea-level (Glasgow, Scotland) not involved in competition during the study period participated in a randomised, double-blind, placebo-controlled crossover microdose rHuEpo study. Written informed consents were obtained from all participants. The study was approved by the University of Glasgow Ethics Committee (Scotland, UK). The subjects received 20–40 IU·kg^{-1} body mass subcutaneous injections of rHuEpo (NeoRecormon, Roche, Welwyn Garden City, UK) or equivalent saline (NaCl 0.9%, Baymed Healthcare Limited, Glasgow, UK; placebo injection) twice a week for 7 weeks (Fig. 2). All subjects received daily iron tablets providing approximately 105 mg of elemental iron derived from 350 mg of dried ferrous sulphate (Almus, Barnstaple, UK), while lactose (Minerals-Water, Purfleet, UK) substituted daily iron during the placebo trial.

All subjects were also subjected to a modified Wingate test comprising of 10 sprints of 10 s at baseline and during the week after the last rHuEpo or placebo injections (Fig. 2; performance data not included here). Specifically, after a 5-min cycling warm-up at 100 W, with a flat-out sprint at 3 min for 5 s, followed by a 5-min rest [24], each subject performed a series of ten, maximal effort 10 s sprints, separated by a 50 s rest interval. The subject then underwent 10 min of active recovery at 100 W followed by 110 min of rest.

Altitude training study (ATS)

Twenty-one elite endurance runners (12 males and 9 females) were recruited (mean ± SD; age: 23.2 ± 2 yrs., height: 175.3 ± 8.5 cm, body mass index (BMI): 19.7 ± 1.2 kg·m^{-2}). Informed, written consents were obtained from all participants. This study was approved by the Ethics Committee of the University of Brighton (England, UK). Participants were randomly assigned into an altitude group (n = 12; trained at Sierra Nevada, Spain, 2320 m, for approximately 2–3 weeks, with one participant returning from altitude after 8 days and another after 28 days)

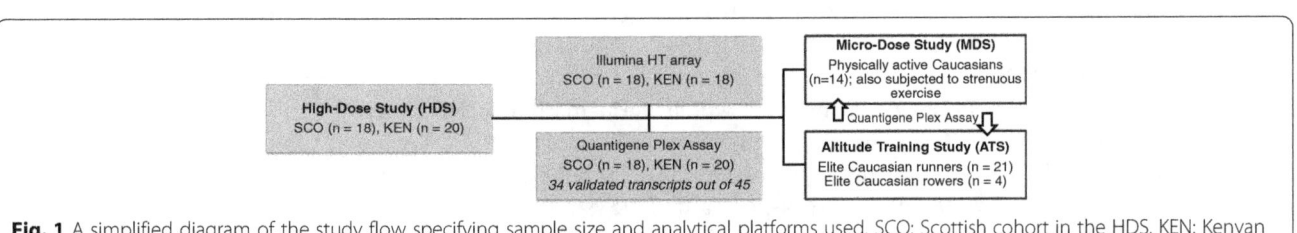

Fig. 1 A simplified diagram of the study flow specifying sample size and analytical platforms used. SCO: Scottish cohort in the HDS. KEN: Kenyan cohort in the HDS

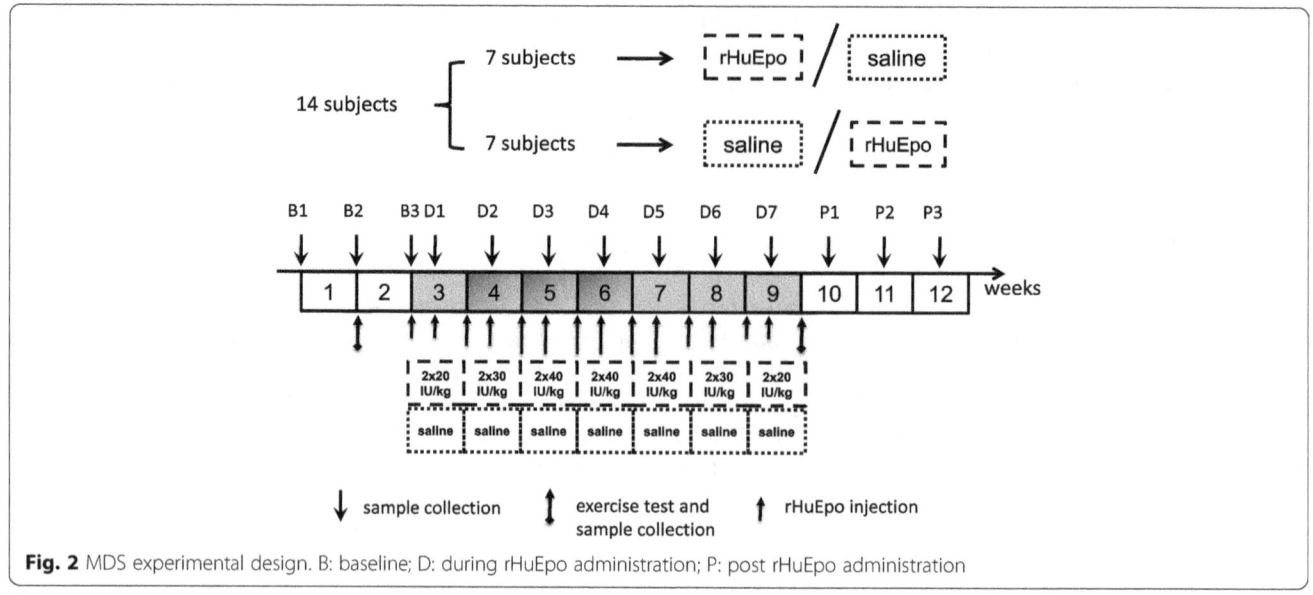

Fig. 2 MDS experimental design. B: baseline; D: during rHuEpo administration; P: post rHuEpo administration

and a control group ($n = 9$; trained at sea level for approximately 2 weeks). Both groups followed the same training programme in preparation for a major international athletics competition.

Four elite male rowers were also recruited (mean ± SD; age: 25.3 ± 3.6 yrs., height: 193.3 ± 3.6 cm, BMI: 26.3 ± 1.2 kg·m^{-2} and $\dot{V}O_2$max: 63.8 ± 4.3 ml·kg^{-1}·min^{-1}). These athletes underwent 2-week of altitude training (Santa Caterina di Valfurva, Italy, 1850 m) in preparation for an international competition. Prior to and after the altitude exposure, these athletes remained in a hypoxic chamber/ altitude room (16 m^2; excluding training and for meals) set to an oxygen concentration equivalent to 2500 m for 5 days. Informed and written consent was obtained from all participants and approved by the Ethics Committee of the University of Glasgow (Scotland, UK).

Blood sampling and RNA extraction
In the MDS, whole blood samples were collected in triplicate at baseline (approximately day −14, −7 and 0) and then once a week for 7 weeks during rHuEpo administration (approximately day 3, 10, 17, 24, 31, 38 and 45) and for 3 weeks post rHuEpo administration (approximately day 52, 59 and 66) (Fig. 2). Whole blood samples were also obtained before, immediately after the 10 sprints and 30 min after the last sprint following the modified Wingate test both at baseline and after the last rHuEpo or placebo injections (Fig. 2). In the ATS, whole blood samples were collected from the elite runners at baseline (approximately day −14), during (approximately day 10), 48-h-, 1-week-, 2-week- and 4-week-post altitude exposure, respectively. Samples were also obtained from the control group of elite runners at baseline

(approximately day −14), 2-week- and 4-week-post the sea-level training period. Whole blood samples were collected from the elite rowers at baseline (day −14 and −10), during (day 5 and 22) and post (day 29, 33 and 39) simulated and natural altitude exposure (days are relative to the first day of the simulated altitude). For all studies, 3 mL whole blood sample was collected using the Tempus™ Blood RNA tube (Life Technologies, Carlsbad, CA, USA) and mixed vigorously with 6 mL stabilising reagent immediately after collection. The whole sample was incubated at room temperature for approximately 3 h and then stored at −20 °C before subsequent analysis. Three-millilitre whole blood was collected in K$_3$EDTA tubes (Greiner Bio-One Ltd., Stonehouse, UK) for haematology analysis, and the whole blood was mixed thoroughly with the tube additive by gently inverting the tube 5–10 times as per the manufacturer's instructions.

Total RNA was isolated from the whole blood collected using Tempus tubes according to the manufacturer's instructions (Tempus™ Spin RNA Isolation Kit, Life Technologies, Carlsbad, CA, USA). The purified total RNA was eluted in 90 µL elution buffer and stored in three aliquots at −80 °C until further analysis. RNA quantity and purity was assessed by the Nanodrop™ ND-2000 Spectrophotometer (Thermo Fisher Scientific, Wilmington, DE, USA).

Haematological analysis
The mixed K$_3$EDTA blood tubes (Greiner Bio-One Ltd., Stonehouse, UK) was tested on the Sysmex XT-2000i (Sysmex, Norderstedt, Germany) for the MDS and ATS elite rowers samples, or the Advia 2120i system (Siemens, Worldwide) for the ATS elite runners samples.

Standard haematological parameters were measured (i.e. haemoglobin, HGB; haematocrit, HCT; and reticulocyte percent, RET%). These samples were measured in accordance with the WADA Athlete Biological Passport Operating Guidelines (version 4.0, 2013) [25]. The blood data was analysed using R lme4 (for applying the linear model) [26, 27] and phia (for post-hoc interaction analysis) [28] packages using a mixed design, with two within-subject variables (the rHuEpo/placebo trial and/or time covariates) as the fixed factors and subject as the random factor, in the MDS and ATS, respectively. Significance level was adjusted using the holm-bonferroni method.

QuantiGene Plex experiment and data analysis

Two hundred nanogram RNA was run in duplicate for quantification of the 45 selected RNA targets and 5 reference genes (ACTB, ACTR10, MRFAP1, PPIB and RAB11A) in subjects participating in the MDS and ATS, using the QuantiGene Plex Assay (Affymetrix, Santa Clara, CA, USA). The resulting fluorescence signal was measured on the MAGPIX (Luminex, Austin, TX, USA). The median fluorescence intensity (MFI) values were viewed and exported from the xPonent software (Luminex, Austin, TX, USA) for statistical analysis. The MFI data was background subtracted and normalised to the geometric mean of the five reference genes. The coefficient of variation (CV) for assay precision was calculated using the duplicated samples. The "limma" function implemented in the R limma package [29] was used to perform the differential expression analysis based on the adjusted and log2 transformed data. Gene expression values were compared to the averaged baseline or the baseline in the MDS (i.e. in the rHuEpo and placebo groups, respectively), MDS exercise samples (i.e. post vs. pre exercise as well as before-after treatment comparisons in the rHuEpo and placebo groups, respectively) and ATS (i.e. in the altitude and control groups, respectively), where appropriate. Gene expression changes over time were also examined in rHuEpo vs. placebo in the MDS and MDS exercise samples and in altitude vs. control in the elite runners. Transcripts expression exceeding a FDR [30] < 0.05 and a 1.5 fold-change were considered meaningful in the current context.

Adaptive model analysis of blood and molecular signatures

In the MDS, the adaptive Bayesian model was applied on the two primary markers (i.e. HGB and the stimulation index OFF-score) of the haematological module of the ABP. Parameters of the model, including within- and between-subject components of variance, were chosen to represent a modal population of Caucasian male athletes aged 20–40 yrs. [31]. The data were analysed by

an investigator without knowing whether the profile was obtained from a rHuEpo or placebo sample. Following the unblinding, sensitivity was calculated based on the portion of samples in the rHuEpo group that produced at least one atypical value out of individual limits (i.e. true positives/size of the rHuEpo group), and specificity was calculated given the portion of samples in the placebo group that presented no atypical value out of individual limits (i.e. true negatives/size of placebo group). For 41 of the 45 transcripts, within- and between-subject variances were estimated using the analysis of variance on the placebo data of log2 expression with subject as a random effect. A leave-one-out cross-validation procedure was used to minimise overfitting. The adaptive model was applied assuming universal components of variance and normality in the within-subject variations of the transcripts. A specificity level of 99% was chosen for all reference ranges returned by the adaptive model. Area under the Receiving Operating Characteristic (ROC) curve was computed on the percentiles at which HGB, OFF-score and all transcript sequences were falling in the distribution of sequences as returned by the adaptive model.

Results
Samples available for analysis

In the MDS, 343 out of 364 samples (i.e. 14 subjects × 2 trials × 13 time points), including 174 and 169 samples following rHuEpo and placebo injections, respectively, were collected and analysed using the QuantiGene Plex Assay (Affymetrix, Santa Clara, CA, USA). The average CV was 10.6% across all samples analysed (vs. 15% typical CV of the QuantiGene Plex Assay [32]). 171 (98.3% of 174) and 163 (96.4% of 169) samples under rHuEpo and placebo administration were available for the haematological analysis, respectively. 164 of 168 MDS exercise samples (i.e. 14 subjects × 2 trials × 6 exercise samples) were available for the QuantiGene Plex 2.0 Assay (Affymetrix, Santa Clara, CA, USA) analysis. The average CV was 16% across these samples. In the ATS elite runners, 66 of 72 samples (i.e. 12 subjects × 6 time points) obtained following altitude training and 22 of 27 samples (i.e. 9 subjects × 3 time points) obtained following sea-level training were processed using the QuantiGene Plex 2.0 Assay (Affymetrix, Santa Clara, CA, USA). 48 and 23 samples were available for HGB/HCT analysis in the elite runners and controls, and 37 and 21 samples for RET% calculation in the elite runners and controls, respectively. In the ATS elite rowers, 28 samples (i.e. 4 subjects × 7 time points) were available for the haematological and gene expression analyses. The average CVs were 8.4% and 9.8% across the samples from elite runners and rowers, respectively. Forty-one of the 45 transcripts exceeding the assay limit of detection were available for gene expression analysis in the MDS and ATS and 35 for MDS exercise-induced gene expression analysis, respectively.

Haematological analysis

In the MDS, both HGB concentration and HCT percentage were gradually increased throughout the rHuEpo administration relative to the baseline values and reached the maximum one week after the last injection (Holm-Bonferroni adjusted $p < 0.05$) (Fig. 3). RET% increased rapidly after the first injection and remained significantly elevated for 4 weeks (Holm-Bonferroni adjusted $p < 0.05$) (Fig. 3). The RET% was significantly lower compared to the baseline values (Holm-Bonferroni adjusted $p < 0.05$) throughout the post-rHuEpo phase (Fig. 3). The OFF-score showed an increasing trend during rHuEpo administration and significantly increased throughout the post phase (Holm-Bonferroni adjusted $p < 0.05$) (Fig. 3). No significant differences over time were found for the HGB and HCT parameters compared to baseline values in the placebo trial in the MDS, however, RET% were significantly increased at During 5 and Post 2 and while OFF-scores were significantly lower at During 4–7 and Post 1,2 in the placebo trial (Holm-Bonferroni adjusted $p < 0.05$). When comparing the blood data between the rHuEpo and placebo groups, similar trends and findings were obtained to those observed in the rHuEpo group. Haematological analysis in the ATS elite runners revealed a significant decrease in HGB during altitude (approximately 10 days at altitude), a significant increase in RET% and a significant decrease in OFF-scores post 2 weeks of sea-level training (Holm-Bonferroni adjusted $p = 0.036$, 0.02 and 0.002, respectively, see Additional files 2 and 3). No other significant differences were observed in these blood parameters over time at available time points (compared to the baseline), neither in the altitude group nor in the control group of the ATS elite runners (Holm-Bonferroni adjusted $p > 0.05$). Similarly, in the ATS elite rowers, no significant changes were observed over time in comparison to baseline values for the four haematological markers (see Additional file 4).

Gene expression analysis

In the group of subjects following the rHuEpo injection in the MDS, differentially expressed genes were firstly selected based on the moderated F-statistic computed by the "eBayes" function implemented in the R limma package. Thirty-six out of the 41 genes exceeded the overall test of significance at the FDR adjusted p of 0.05. Of

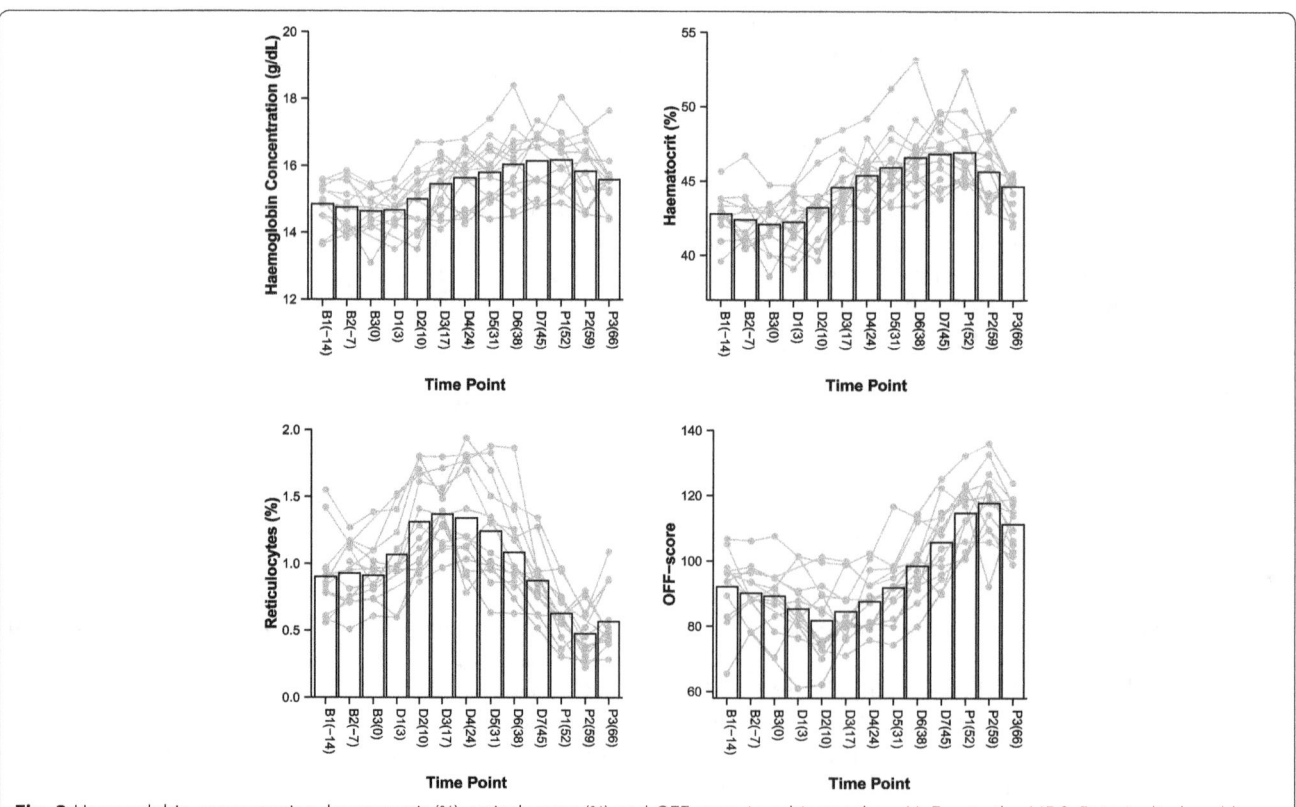

Fig. 3 Haemoglobin concentration, haematocrit (%), reticulocytes (%) and OFF-score in subjects taken rHuEpo in the MDS. Data is displayed by means with corresponding individual changes over time. B: baseline; D: during rHuEpo administration; P: post rHuEpo administration. The number in the parentheses indicates the blood sampling day relative to the day of the first rHuEpo injection (i.e. B3)

these 36 genes, 23 were selected when individual contrasts in gene expression revealed non-zero differences seven days after the last rHuEpo injection (i.e. Post 1) at FDR < 0.05. Subsequently, 17 of the 23 genes were found significantly altered in expression, exceeding a fold-change of 1.5 (FDR < 0.05), ten days (i.e. During 2) after the first injection. Among the 17 genes, 11 were consistently over-expressed from During 2 (Day 10) to During 5 (Day 31) and were then under-expressed throughout the post-administration stage, i.e. Post 1 (Day 52) to Post 3 (Day 66) (FDR < 0.05 with 1.5 fold-change threshold) (Table 1). Nine of the 11 genes, *ALAS2*, *BCL2L1*, *DCAF12*, *EPB42*, *GMPR*, *SELENBP1*, *SLC4A1*, *TMOD1* and *TRIM58* (Table 1 and Additional file 5), were common with the 34 transcripts previously identified and validated [23]. None of the 41 genes were differentially expressed in the placebo group over time (FDR > 0.15). When comparing the levels of gene expression in the rHuEpo group with that in the placebo group, 24 of the 41 genes were down-regulated at Post 1–3 (FDR < 0.05 with 1.5 fold-change threshold) (Table 2), overlapping

the 9 genes aforementioned and no genes were differentially expressed in the "During" stage between the rHuEpo group and the placebo group (FDR > 0.17).

Seven genes (1 up-, *CD247* and 6 down-regulated, *BPGM*, *FECH*, *SNCA*, *STRADB*, *TRIM58* and *YOD1*) of the 35 were significantly altered immediately following 10 sprints of 10 s compared to baseline (i.e. pre exercise) before the rHuEpo injection and 2 genes (1 up-, *CD247* and 1 down-regulated, *LOC100130562*) after the last rHuEpo injection (FDR < 0.05 with 1.5 fold-change threshold) (see Additional file 6). One (up-regulated, *CD247*) and 16 (1 up-, *CD247* and 15 down-regulated, *ADIPOR1*, *BCL2L1*, *BPGM*, *CA1*, *DCAF12*, *FAM46C*, *FBXO7*, *FECH*, *OSBP2*, *SNCA*, *STRADB*, *TRIM58*, *UBXN6*, *YBX3* and *YOD1*) genes of the 35 were found significantly altered immediately after the repeated sprint tests in comparison to pre-exercise gene expression before and after the placebo injection, respectively (FDR < 0.05 with 1.5 fold-change threshold) (see Additional file 6). No significant changes were found post 30 min of exercise vs. pre exercise in either rHuEpo or placebo trials. When

Table 1 11 genes differentially expressed at During 2 to 5 and throughout the post rHuEpo administration in the MDS

Gene	During 2	During 3	During 4	During 5	Post 1	Post 2	Post 3	During 2	During 3	During 4	During 5	Post 1	Post 2	Post 3
Log2 FC								FC						
ALAS2	0.86	0.95	0.80	0.70	−0.82	−1.32	−1.18	1.81	1.93	1.75	1.63	−1.77	−2.50	−2.26
BCL2L1	0.71	0.81	0.79	0.69	−0.67	−0.88	−0.89	1.63	1.75	1.73	1.61	−1.59	−1.84	−1.85
DCAF12	0.60	0.78	0.73	0.66	−0.59	−0.90	−0.89	1.52	1.72	1.66	1.58	−1.50	−1.87	−1.86
EPB42	0.86	0.98	0.88	0.76	−0.89	−1.25	−1.08	1.82	1.97	1.84	1.69	−1.85	−2.37	−2.11
GMPR	0.81	0.94	0.79	0.68	−0.72	−0.94	−0.84	1.75	1.92	1.73	1.60	−1.65	−1.92	−1.79
OSBP2	0.96	1.08	1.05	1.00	−0.89	−1.16	−1.03	1.95	2.12	2.07	2.00	−1.85	−2.23	−2.04
SELENBP1	0.98	1.09	0.97	0.83	−0.90	−1.19	−1.05	1.97	2.13	1.96	1.77	−1.87	−2.29	−2.07
SLC4A1	0.89	1.08	0.96	0.82	−0.87	−1.18	−1.10	1.86	2.12	1.94	1.77	−1.83	−2.26	−2.14
TMOD1	0.65	0.89	0.76	0.61	−0.71	−1.06	−1.00	1.57	1.85	1.70	1.53	−1.64	−2.08	−2.01
TNS1	0.84	1.10	0.98	0.87	−0.94	−1.22	−1.10	1.79	2.14	1.97	1.83	−1.91	−2.34	−2.14
TRIM58	0.73	0.82	0.72	0.67	−0.74	−0.94	−0.97	1.66	1.76	1.65	1.59	−1.67	−1.92	−1.96
Uncorrected *P* val.								FDR						
ALAS2	0.00181	0.00055	0.00337	0.01011	0.00274	0.000003	0.00002	0.00778	0.00173	0.00767	0.02030	0.01443	0.00006	0.00008
BCL2L1	0.00180	0.00038	0.00048	0.00230	0.00317	0.00013	0.00009	0.00778	0.00171	0.00367	0.01678	0.01443	0.00031	0.00025
DCAF12	0.00912	0.00072	0.00157	0.00434	0.01090	0.00014	0.00012	0.01558	0.00193	0.00430	0.01678	0.02482	0.00031	0.00031
EPB42	0.00488	0.00139	0.00416	0.01333	0.00378	0.00007	0.00045	0.01052	0.00300	0.00877	0.02277	0.01551	0.00022	0.00092
GMPR	0.00165	0.00026	0.00200	0.00793	0.00479	0.00034	0.00110	0.00778	0.00152	0.00492	0.01835	0.01679	0.00064	0.00196
OSBP2	0.00024	0.00004	0.00006	0.00014	0.00071	0.00002	0.00009	0.00778	0.00079	0.00252	0.00591	0.01443	0.00008	0.00025
SELENBP1	0.00118	0.00030	0.00129	0.00603	0.00283	0.00011	0.00053	0.00778	0.00152	0.00407	0.01835	0.01443	0.00030	0.00103
SLC4A1	0.00848	0.00148	0.00478	0.01499	0.01016	0.00069	0.00126	0.01511	0.00303	0.00933	0.02458	0.02451	0.00124	0.00215
TMOD1	0.00520	0.00015	0.00105	0.00849	0.00218	0.000010	0.00002	0.01052	0.00152	0.00388	0.01835	0.01443	0.00008	0.00008
TNS1	0.00531	0.00026	0.00114	0.00376	0.00181	0.00007	0.00027	0.01052	0.00152	0.00388	0.01678	0.01443	0.00022	0.00058
TRIM58	0.00190	0.00051	0.00204	0.00450	0.00172	0.00009	0.00004	0.00778	0.00173	0.00492	0.01678	0.01443	0.00026	0.00014

Log2 FC log2 transformed fold-change, *FC* fold-change, *FDR* false discovery rate adjusted significance

Table 2 24 genes responded differently (significantly down-regulated over Post 1, 2, 3) in the rHuEpo group relative to the placebo group in the MDS

Gene	Log2 FC			FC			Uncorrected P val.			FDR		
	Post 1	Post 2	Post 3	Post 1	Post 2	Post 3	Post 1	Post 2	Post 3	Post 1	Post 2	Post 3
ADIPOR1	−0.69	−1.17	−1.02	−1.61	−2.24	−2.03	0.00669	0.00001	0.00007	0.01958	0.00009	0.00048
ALAS2	−1.10	−1.78	−1.55	−2.14	−3.43	−2.93	0.00504	0.00001	0.00008	0.01720	0.00009	0.00048
BCL2L1	−0.97	−1.29	−1.12	−1.96	−2.44	−2.18	0.00269	0.00010	0.00050	0.01720	0.00026	0.00172
BPGM	−0.72	−1.11	−1.04	−1.64	−2.16	−2.05	0.00417	0.00002	0.00004	0.01720	0.00010	0.00048
CA1	−0.82	−1.28	−1.12	−1.76	−2.42	−2.18	0.01519	0.00022	0.00090	0.02967	0.00044	0.00223
CSDA	−0.80	−1.25	−1.12	−1.75	−2.39	−2.18	0.00791	0.00005	0.00022	0.01959	0.00020	0.00088
DCAF12	−0.77	−1.24	−1.05	−1.70	−2.36	−2.07	0.01941	0.00024	0.00144	0.03351	0.00045	0.00282
EPB42	−1.30	−1.76	−1.44	−2.47	−3.39	−2.72	0.00286	0.00009	0.00097	0.01720	0.00025	0.00223
FAM46C	−0.66	−1.12	−0.98	−1.58	−2.17	−1.98	0.00462	0.00000	0.00003	0.01720	0.00009	0.00048
FBXO7	−0.64	−1.02	−0.94	−1.56	−2.03	−1.91	0.00792	0.00003	0.00011	0.01959	0.00014	0.00049
FECH	−0.78	−1.20	−1.13	−1.72	−2.30	−2.18	0.00545	0.00003	0.00007	0.01720	0.00014	0.00048
GMPR	−1.06	−1.39	−1.11	−2.08	−2.62	−2.16	0.00391	0.00022	0.00249	0.01720	0.00044	0.00465
GUK1	−0.77	−1.26	−1.05	−1.71	−2.40	−2.07	0.01462	0.00010	0.00098	0.02967	0.00026	0.00223
KRT1	−1.16	−1.53	−1.22	−2.24	−2.89	−2.33	0.00181	0.00006	0.00110	0.01720	0.00021	0.00236
OSBP2	−0.78	−1.13	−0.99	−1.72	−2.19	−1.99	0.00820	0.00020	0.00086	0.01959	0.00043	0.00223
SELENBP1	−1.28	−1.68	−1.39	−2.43	−3.21	−2.62	0.00295	0.00014	0.00124	0.01720	0.00033	0.00254
SLC4A1	−1.27	−1.60	−1.37	−2.42	−3.03	−2.58	0.00860	0.00123	0.00476	0.01959	0.00209	0.00813
SNCA	−0.84	−1.47	−1.28	−1.79	−2.77	−2.42	0.00952	0.00001	0.00009	0.02054	0.00009	0.00048
STRADB	−0.85	−1.36	−1.20	−1.80	−2.56	−2.30	0.00438	0.00001	0.00006	0.01720	0.00009	0.00048
TMOD1	−0.99	−1.43	−1.22	−1.98	−2.69	−2.33	0.00303	0.00003	0.00025	0.01720	0.00014	0.00094
TNS1	−1.26	−1.63	−1.26	−2.40	−3.10	−2.39	0.00318	0.00020	0.00330	0.01720	0.00043	0.00589
TRIM58	−0.93	−1.25	−1.15	−1.91	−2.38	−2.22	0.00536	0.00027	0.00061	0.01720	0.00049	0.00181
UBXN6	−0.71	−1.23	−1.05	−1.64	−2.35	−2.07	0.01962	0.00009	0.00062	0.03351	0.00025	0.00181
YOD1	−0.59	−1.10	−1.04	−1.50	−2.14	−2.05	0.01831	0.00002	0.00004	0.03351	0.00011	0.00048

Log2 FC log2 transformed fold-change, *FC* fold-change, *FDR* false discovery rate adjusted significance

comparing exercise gene expression changes before-after rHuEpo or placebo injections, no differences were observed, neither immediately nor 30 min post exercise (FDR > 0.93). Furthermore, there were no significant changes in transcription following exercise when comparing the rHuEpo group vs. the placebo group (FDR > 0.70).

In the 12 elite runners, 28 out of the 41 genes exceeded an overall F-test significance at the FDR adjusted p of 0.05 following altitude training (see Additional file 7, green section). The remaining 13 genes demonstrated non-significant changes in gene expression (F-test FDR > 0.05; see Additional file 7). Following pairwise comparisons, trends towards 5% FDR were observed for 20 genes (out of the 28) down-regulated one week after returning from altitude and 13 genes (out of the 20) were up-regulated approximately 10 days after reaching altitude (FDR > 0.059 and 0.064, respectively; see Additional file 7). Of the 13 genes, *ALAS2, BCL2L1, DCAF12, SLC4A1 TMOD1* and *TRIM58* (FDR > 0.059 with the fold-change varied from 1.39 to 1.63, see

Additional files 7 and 8) were in common with the 9 MDS genes. In the elite runners, no genes were differentially expressed in the control group or responded differently over time in the altitude group relative to the control group post 2 and 4 weeks of altitude or sea-level training (FDR > 0.20 and 0.60, respectively). In the 4 elite rowers, no genes responded differently following 2-week altitude in conjunction with adaptation to simulated altitude when compared with the averaged baseline.

Adaptive model analysis

In the MDS, using 13 samples per subject collected at weekly intervals including 3 weekly samples post-supplementation, rHuEpo use was identified in 13 out of 14 subjects and without any false positives, for a specificity of 99% in the ABP using the haematological markers of HGB and OFF-score (Table 3); 41 out of the 45 were amenable to analysis using the adaptive model. A higher between- compared to within-subject variability was observed across the examined genes (averaged variance:

Table 3 The adaptive model analysis summarising the within- and between-subject variances, sensitivity, specificity and ROC area for the HGB concentration, OFF-score and 41 transcripts analysed in the MDS

	Mean	Within-subject variance	Between-subject variance	Sensitivity (%)	Specificity (%)	ROC area
HGB (g·L $^{-1}$)	150	29	36	79	93	0.87
OFF-score	92	51	62	93	93	0.93
ADIPOR1	3.33	0.098	0.085	86	79	0.97
ALAS2	3.09	0.204	0.282	100	79	0.97
BCL2L1	1.33	0.159	0.202	93	93	0.96
BPGM	2.74	0.107	0.087	93	71	0.98
CA1	1.28	0.159	0.190	93	79	0.97
CCR7	−0.15	0.092	0.155	29	93	0.78
CD247	−0.66	0.059	0.046	43	71	0.62
CD3D	0.30	0.061	0.048	43	64	0.63
CSDA	2.68	0.136	0.150	93	93	0.98
DCAF12	2.18	0.146	0.198	93	79	0.98
EEF1D	1.75	0.026	0.027	43	64	0.63
EPB42	−0.91	0.243	0.448	93	86	0.96
FAM46C	3.74	0.083	0.061	93	79	0.95
FBXO7	3.54	0.079	0.075	79	79	0.92
FECH	1.34	0.142	0.119	86	71	0.95
GMPR	0.23	0.181	0.281	86	86	0.93
GUK1	0.51	0.130	0.228	86	79	0.95
GYPE	−4.87	0.145	0.282	79	93	0.92
HBD	0.20	0.243	0.885	64	64	0.95
KRT1	−1.31	0.327	1.229	86	71	0.86
LEF1	0.37	0.083	0.129	36	93	0.68
LOC100130562	2.78	0.032	0.027	50	71	0.63
LOC286444	0.73	0.034	0.049	50	79	0.69
MIF	−0.15	0.043	0.021	36	71	0.62
OSBP2	0.61	0.205	0.206	93	79	0.97
PITHD1	0.20	0.125	0.183	86	86	0.92
RBM38	0.44	0.150	0.566	79	79	0.89
RNF213	0.76	0.102	0.062	36	79	0.42
SELENBP1	0.63	0.240	0.401	93	79	0.92
SGK223	−2.81	0.064	0.042	36	79	0.60
SKAP1	−1.52	0.055	0.052	43	79	0.61
SLC4A1	1.36	0.213	0.595	93	86	0.96
SNCA	0.87	0.172	0.154	86	71	0.96
STRADB	2.29	0.136	0.145	93	86	0.99
TMOD1	−0.57	0.188	0.163	93	71	0.95
TNS1	−2.05	0.234	0.306	93	86	0.96
TPRA1	−2.43	0.011	0.010	50	79	0.67
TRIM58	2.23	0.156	0.199	86	71	0.98
UBXN6	2.06	0.107	0.209	93	71	0.98
VEGFB	−1.62	0.045	0.039	57	64	0.67
YOD1	1.48	0.137	0.050	71	71	0.90

ROC receiving operating characteristic

0.21 vs. 0.13; Table 3). Fifteen transcripts showed a sensitivity of ≥93% and a specificity of ≥71%, with the ROC area ≥ 0.92 (Table 3). *BCL2L1* and *CSDA* were the genes with the highest sensitivity (93%) and specificity (93%), with the ROC area of 0.96 and 0.98, respectively (Table 3). In the MDS differential gene expression analysis, the *BCL2L1* gene was significantly expressed at During 2–5 and Post 1–3 in the rHuEpo group (Table 1, FDR < 0.05 and fold-change >1.5) and at Post 1–3 in the rHuEpo group vs. the placebo group (Table 2, FDR < 0.05 and fold-change >1.5); the *CSDA* gene was significantly down-regulated at Post 1–3 (FDR < 0.05 and fold-change >1.5) in the rHuEpo group vs. the placebo group (Table 2). HGB concentration (g·L^{-1}), OFF-score and gene expression changes (of the 41 transcripts) across 28 subjects (14 subjects × 2 trials) obtained from the adaptive model and ROC curves are provided in Additional files 9, 10 and 11.

Discussion

Twenty-four of the 41 genes showed a significant and long lasting down-regulation following the last rHuEpo injection in the MDS given a fold-change of 1.5 and 5% FDR spanning the post rHuEpo stage for 3 weeks (Table 2). This prolonged detection window in terms of the long lasting effect and the stability of the gene markers collected using the Tempus™ Blood RNA method is promising and will undoubtedly improve the efficiency of rHuEpo detection given the substantially shorter detection duration of 36 h for improved analytical quality when using the current haematological markers [8]. Fifteen of the 24 genes showed a high sensitivity (≥ 93%) with specificity equal to or above 71% (Table 3). Particularly, the *BCL2L1* and *CSDA* exhibited the highest sensitivity (93%) and specificity (93%) amongst the 15 genes (Table 3). The majority of the 9 genes that were consistently expressed during and post rHuEpo in the MDS rHuEpo group, namely *ALAS2, BCL2L1, DCAF12, EPB42, GMPR, SELENBP1, SLC4A1, TMOD1* and *TRIM58* were associated with erythrocyte membrane structure and red blood cell metabolism; these 9 genes were in common with the 34 transcripts previously identified and validated [23], while 7 of the 9 genes were in common with the 15 genes showing high sensitivity and specificity. The major spliced mRNA isoform of the *ALAS2* has significant impact on erythroid heme biogenesis and hemogobin formation [33]. Erythrocyte survival is suppressed by the BH3 peptide through antagonizing Bcl-X(L) [34]. EPB42 deficiency causes hereditary spherocytosis, leading to chronic haemolytic anemia with abnormally shaped erythrocytes [35]. The *EPB42* may be involved in the regulation of erythrocyte shape and mechanical properties [36]. The *SELENBP1* is thought to play a role in rapid cell outgrowth by determining the direction of the outgrowth and the synthesis of actin filaments [37]. There are two structurally and functionally distinct versions of the protein encoded by the *SLC4A1* gene – the N-terminal 40 kDa cytoplasmic domain attaches to the red cell skeleton by binding ankyrin to maintain the structure of red blood cells and the C-terminal 50 kDa membrane domain is responsible for the transport of anions, by facilitating the exchange of chloride and bicarbonate across the plasma membrane of erythrocytes [38–40]. The TMOD1 protein also influences the structure of the erythrocyte membrane skeleton by regulating tropomyosin [41]. The *GMPR* gene is mapped to chromosome 6p23 [42] and maintains the intracellular balance of guanine and adenine nucleotides [43]. It was previously reported that the human red cell glucose-6-phosphate dehydrogenase is encoded by the chromosome 6- and chromosome X- encoded genes [44] but subsequently disproved by other studies [45, 46]. The *DCAF12* and *TRIM58* as well as *CSDA* genes are associated with terminal erythroid differentiation, red blood cell count and red blood cell interactome networks, respectively [47–49]. As stated previously, 5 genes (i.e. *PFN1*, *C13orf15*, *TSTA3*, *RPL41* and *TOMM40*) were identified by Varlet-Marie et al. using SAGE following administration of the erythropoiesis-stimulating agent (ESA) darbepoetin alpha and high doses and microdoses of rHuEpo [19]. However, none of these genes overlapped with the molecular signature of rHuEpo doping identified and validated in our previous and current studies. Differences in study design including the specific drugs, drug administration methods, sample size and detection platforms, are all likely factors contributing to these results, emphasising further the need for rigorous replication of any markers identified in a single study.

The identification of a similar panel of genes associated with the structure and function of red blood cells identified in the present study and those of other clinical groups (see ref. [33–49]) is encouraging and should assist the development of targeted therapy to treat patients with blood disorders. There are widespread applications utilising rHuEpo in clinical settings such as the treatment of patients with anaemia of chronic renal disease, improving quality of life in cancer patients and minimising the transfusion requirement [50], but not all patients respond effectively to treatment with rHuEpo. The transcriptomic markers identified in healthy individuals administered with rHuEpo in our previous and current studies represent useful targets to investigate the signal transduction pathways activated by erythropoietin and its receptor for improved therapeutic use of ESA and rHuEpo. With reference to anti-doping, the whole-blood trancriptomic markers discovered reflect closely the RET% changes over time (Fig. 3). This is consistent with the finding that a large amount of mRNA species in

whole blood originate from reticulocytes, evidenced by the separate clustering of whole blood samples treated with RNase H and leukocyte samples in microarray analysis [51]. New approaches that are able to interrogate the whole transcriptome with improved dynamic range for adequate quantification (e.g. RNA-seq) and preferably in leucocytes populations in a sufficient number of participants are warranted in future anti-doping transcriptomic studies and especially those involving the manipulation of red blood cells to improve further the detection of blood doping using molecular markers. Approaches enabling the interrogations of whole genome, transcriptome and metabolome have the increased capacity to measure rapidly and in the near future, inexpensively a large number of molecular signatures, which will collectively aid the decision making during identifying and differentiating numerous doping substances and methods by ABP experts when reviewing passports. For example, one of the transcripts validated in the present study is the *BPGM* gene (Table 2), encoding the 2–3 BPG. It is well known that blood doping can affect 2–3 BPG metabolism and is therefore closely monitored by the WADA ABP Expert Panel. In addition, the only subject evading detection by the haematological model of ABP in the present study (see Additional files 9 and 10, subject R) had a low mean cell volume and mean cell haemoglobin, indicating iron deficiency or a defect in iron metabolism and particularly that the subject participated in the present study is non-responsive to iron supplementation. However, 25 of the 41 genes included in the adaptive model analysis identified rHuEpo use in the same subject (see Additional file 11, subject 9). This illustrates the need for a holistic approach to drug detection; one based on the inclusion of a variety of parameters that provide information on related pathways and metabolism (e.g. iron, transferring and total iron binding capacity). Notably, the large-scale omics studies combined with the ABP are anticipated to be able to dramatically improve doping detection. It is also important to note that the potential for the validated transcripts in the present study to identify the rHuEpo use is high both when the transcripts are used alone (Tables 1 and 2; 24 genes exceeding a fold-change of 1.5 at 5% FDR post rHuEpo injections) or in combination with the current ABP (Table 3; 20 genes with a ROC area > 0.95), emphasising the need to understand erythropoiesis using a systems biology approach.

The *CD247* gene was significantly up-regulated exceeding a fold-change of 1.5 at 5% FDR immediately following exercise when compared to the baseline before and after rHuEpo/placebo, respectively; but this response rapidly subsided 30 min after exercise. The *CD247* encodes the T-cell receptor zeta, which is a component of the T-cell receptor-CD3 complex [52]. Low expression of the gene may relate to impaired immune response [53, 54]. This current observation is in line with previous studies indicating acute inflammatory response following acute resistance exercise [55, 56] and the subsequent anti-inflammatory response is to prevent the development of chronic inflammation [21, 57]. Six and fifteen genes were significantly down-regulated (6 is included in the 15) immediately following exercise before rHuEpo and after placebo, respectively (FDR < 0.05 and fold-change >1.5). Among these genes, only TRIM58 overlaps the 9 MDS genes. TRIM58 specifically expresses during late erythroid maturation, coinciding with enucleation and dynein loss [58]. The suppression of TRIM58 relates to reduced enucleation [58]. TRIM58 gene polymorphisms associate with the circulating erythrocyte size and number [48, 59]. Previous studies reported an increased number of leukocytes immediately following heavy exercise [21, 22], whereas the reduced level of TRIM58 following the repeated sprint test in the present study may reflect reduced proportions of reticulocytes in accordance with previous research [58]. This modified Wingate test did not reveal significant gene expression changes using the 35 transcripts when comparing before with after rHuEpo/placebo, or rHuEpo vs. placebo. These results, although preliminary, argue against exercise/training being a confounder as all gene expression alterations post strenuous exercise were restored after 30 min and well within the two hours stipulated by WADA before a blood sample can be obtained for anti-doping purposes [60]. Nevertheless, the explicit role of exercise needs to be further investigated not only in whole blood but also peripheral blood cells in order to better understand the molecular adaptations to both acute and chronic exercise (i.e. exercise training) and avoid confounding the analysis.

Previous studies have shown that short or prolonged residency at high altitude stimulates the secretion of Epo from the kidney [61–64]. For example, elevated levels of Epo are detected as early as 8 h after arrival at high altitude and reaches a peak 24 h after arrival [61]. These higher levels of Epo are also maintained throughout a period of high altitude exposure (e.g., ranging from 11 days to 4 weeks at approximately 3500 m or above [61, 64]). In the present ATS, 13 out of the 41 genes failed to show statistically significant changes at the transcriptional level following approximately 2-week training at moderate altitude in 12 elite runners given a F-test FDR > 0.05. Six of the remaining 28 genes that tended to be stimulated by altitude were in common with the 9 MDS genes (FDR > 0.059 with the fold-change varied from 1.39 to 1.63 during and after one week of altitude training; see Additional files 7 and 8). Among these 6 genes, SLC4A1 showed a tendency towards a 1.63-fold down-regulation (FDR = ~ 0.059) one

week after altitude training and TMOD1 a tendency for a 1.50-fold up-regulation (FDR = ~ 0.064) after ten days of altitude exposure; none of the other genes exceeded the 1.5 fold-change cut-off (FDR > 0.059) (see Additional file 7). In comparison, in the MDS rHuEpo group, DCAF12 showed a least 1.52-fold up-regulation (FDR = ~ 0.016) ten days after the first rHuEpo injection (i.e. During 2) and this level of expression was maintained for 4 weeks for all 6 genes (i.e. During 2–5, FDR < 0.05, Table 1). In addition, there was at least 1.50-fold down-regulation (FDR = ~ 0.025) in DCAF12 one week after the last rHuEpo injection (i.e. Post 1) with this level of change lasting throughout the 3 weeks for all 6 genes (i.e. Post 1–3, FDR < 0.05, Table 1), with majority of the 6 genes approaching or exceeding a 2 fold-change at the post rHuEpo stage (Table 1). There was also a more pronounced gene expression response in these 6 genes post rHuEpo administration compared to the placebo group (FDR < 0.05, Table 2), with an approximate 2.5 fold-change in ALAS2, SLC4A1 and TMOD1 2 and 3 weeks after the last injections (i.e. Post 2, 3, Table 2). As stated previously, the observed patterns of change in gene expression in MDS reflect closely the RET% changes over time (Fig. 3). In comparison, haematological analysis in the ATS elite runners revealed a small but significant decrease in HGB during altitude (approximately 10 days at altitude) (FDR = 0.036, see Additional file 2). No genes were differentially expressed in the control group or responded differently over time in the altitude group relative to the control group and this trend was in general agreement with the haematological results that showed no change over time despite exposure to altitude. The relatively long lasting effects and high magnitude of changes of the validated transcriptional markers of rHuEpo compared with the gene expression changes of the same markers (i.e. the 6 genes) in the ATS elite runners provide strong evidence in favour of applying such markers, alongside current anti-doping strategies to detect blood doping. The finding of no change in gene expression or in the measured haematological variables in the four elite rowers most likely reflects the limited exposure to moderate altitude, the small sample size and the typical inter-individual variation (see additional file 4). An altitude study with extended exposure (for 4 weeks at least) at high altitude (~ 3000 m) involving the analysis of both whole blood and peripheral blood for omics marker identification is required to better understand the molecular adaptations to altitude as compared to blood doping. Only then can the molecular response to altitude training be excluded, or if needed, integrated in the ABP along with other confounders to detect blood doping. Assessing and determining other confounding variables influencing the biological and analytical variability of the gene markers through altered erythropoiesis presents a critical next step to enhance the specificity and sensitivity of current ABP for unbiased detection of blood doping in conjunction with a systems biology approach.

Conclusions

In conclusion, several human whole-blood transcriptional signatures signifying predominantly altered red blood cell production were identified following rHuEpo injections ranging from high doses to microdoses. These findings support the use of molecular markers as potential biomarkers with an improved detection window and high sensitivity and specificity for developing the transcriptionally-enhanced ABP model for detecting blood doping. Collectively, the findings of the present study, interpreted in the context of the latest omics research, are encouraging and suggest a systems biology approach combining various omics signatures from genomics, transcriptomics, proteomics and metabolomics will inevitably provide a deeper understanding of the effects of erythropoietic stimulating agents on erythropoiesis with unparalleled potential to improve current drug detection strategies with particular reference to blood doping. However, continuous and rigorous efforts will be required to determine, accommodate and possibly eliminate other confounding effects on blood doping.

Additional files

Additional file 1: List of the 50 genes for the QuantiGene Plex Assay analysis in the MDS and ATS. (DOC 67 kb)

Additional file 2: Haemoglobin concentration (g·dL^{-1}), haematocrit (%), reticulocytes (%) and OFF-score in response to altitude training in the 11 ATS elite runners. Data is displayed by means with corresponding individual changes over time. B(−7), D(10), P(16), P(30) and P(42) represent pre, during, 48-h-, 1-week-, and 4-week-post altitude exposure, respectively. (PDF 7 kb)

Additional file 3: Haemoglobin concentration (g·dL^{-1}), haematocrit (%), reticulocytes (%) and OFF-score changes in the 7 ATS elite runners at the sea level. Data is displayed by means with corresponding individual changes over time. B(−7), P(30) and P(42) represent pre, 1-week-, and 4-week-post altitude exposure, respectively. (PDF 6 kb)

Additional file 4: Haemoglobin concentration (g·dL^{-1}), haematocrit (%), reticulocytes (%) and OFF-score in the four ATS elite rowers. Data is displayed by means with corresponding individual changes over time. B1(−14) and B2(−10): 14 and 10 days prior to 5-day simulated altitude, respectively; D1(5), D2(22), P1(29), P2(33) and P3(39): 5, 22, 29, 33 and 39 days relative to the first day of the simulated altitude prior to the natural altitude exposure. (PDF 7 kb)

Additional file 5: Individual expression of the *ALAS2, BCL2L1, DCAF12, EPB42, GMPR, SELENBP1, SLC4A1, TMOD1* and *TRIM58* genes over time in response to rHuEpo in the MDS, respectively. (PDF 23 kb)

Additional file 6: Genes differentially expressed immediately following the repeated sprint tests vs. baseline prior to and after the rHuEpo/placebo injection in the MDS, respectively. (XLS 21 kb)

Additional file 7: Summary of gene expression changes over time in response to altitude exposure in the 12 ATS elite runners for the 41 transcripts. (XLS 37 kb)

Additional file 8: Individual expression of the *ALAS2, BCL2L1, DCAF12, SLC4A1, TMOD1* and *TRIM58* genes over time in response to altitude

exposure in the 12 ATS elite runners, respectively. B(−7), D(10), P(16), P(21), P(30) and P(42) represent pre, during, 48-h-, 1-week-, 2-week- and 4-week-post altitude exposure, respectively. (PDF 9 kb)

Additional file 9: Adaptive model analysis and ROC curve in HGB concentration (g·L^{-1}) across 28 subjects (14 subjects × 2 trials) in the MDS. Red lines: individual limits as determined by the adaptive model for a specificity of 99%; blue line: actual test results. Subject A, D, F, G, J, K, M, P, R, S, U, X, Z and AA participated in the rHuEpo trial; subject B, C, E, H, I, L, N, O, Q, T, V, W, Y and BB participated in the placebo trial. X-axis: 13 time points and y-axis: HGB concentrations. (ZIP 88 kb)

Additional file 10: Adaptive model analysis and ROC curve in OFF-score across 28 subjects (14 subjects × 2 trials) in the MDS. Red lines: individual limits as determined by the adaptive model for a specificity of 99%; blue line: actual test results. Subject A, D, F. G, J, K, M, P, R, S, U, X, Z and AA participated in the rHuEpo trial; subject B, C, E, H, I, L, N, O, Q, T, V, W, X and AB participated in the placebo trial. X-axis: 13 time points and y-axis: OFF-scores. (ZIP 87 kb)

Additional file 11: Adaptive model analysis and ROC curve in gene expression changes of the 41 transcripts across 28 subjects (14 subjects × 2 trials) in the MDS. Red lines: individual limits as determined by the adaptive model for a specificity of 99%; blue line: actual test results. The first 1–14 graphs: subjects participated in the rHuEpo trial and following 15–28 graphs: the same subjects participated in the placebo trial. X-axis: 13 time points and y-axis: log2 gene expression. (ZIP 1076 kb)

Acknowledgements

The authors would like to thank all volunteers for their participation and cooperation as well as Mr. Mark Lees and Dr. Jun Wang for their help with data collection in the MDS.

Funding

The research and publication costs were funded by grants from the WADA (12C09YP and 13C28YP) and Affymetrix/Panomics.

Authors' contributions

YPP and JD study design; JD, MM and AK data collection; AM medical supervision; GW, BY, JS, NF, and MM molecular lab analysis; GW, PES, BY, JS and JDM statistical analysis; GW, PES, JS and YPP data interpretation; GW and PES illustration generation; GW, PES, JS and YPP drafting the manuscript; GW and YPP manuscript revision; all authors to comment, further edit and approve the manuscript.

Competing interests

The authors declare that they have no competing interests.

Author details

[1]Centre of Sports Medicine for Anti-Doping Research, University of Brighton, Eastbourne, UK. [2]Institute of Cardiovascular and Medical Sciences, College of Medical, Veterinary and Life Sciences, University of Glasgow, Glasgow, UK. [3]Brighton and Sussex Medical School, Brighton, UK. [4]Faculty of Medicine, University of Tartu, Tartu, Estonia. [5]Graduate School of Health and Sports Science, Juntendo University, Chiba, Japan. [6]School of Sport, Health and Applied Science, St Mary's University, Twickenham, London, UK. [7]Centre for Sports and Exercise, University of Edinburgh, Edinburgh, UK. [8]Affymetrix, Santa Clara, CA, USA. [9]BioKaizen Lab SA, Monthey, Switzerland. [10]Department of Movement, Human and Health Sciences, University of Rome "Foro Italico", Rome, Italy.

References

1. Sottas PE, Robinson N, Rabin O, Saugy M. The athlete biological passport. Clin Chem. 2011;57:969–76.

2. Lasne F, de Ceaurriz J. Recombinant erythropoietin in urine. Nature. 2000; 405:635.

3. Lundby C, Achman-Andersen NJ, Thomsen JJ, Norgaard AM, Robach P. Testing for recombinant human erythropoietin: problems associated with current anti-doping testing. J Appl Physiol. 2008;105:417–9.

4. Schwenke D. Improved detection of EPO in blood and urine based on novel velum SAR precast horizontal gels optimized for routine analysis. 2015; http://www.dyeagnostics.com/site/wp-content/uploads/2015/02/ Application_Note_Improvements-for-EPO-detection_Schwenke_2015.pdf. Accessed 22 Jan 2017.

5. Harmonization of Analysis and Reporting of Erythropoiesis-Stimulating Agents (ESAs) by Electrophoretic Techniques. 2014; https://www.wada-ama. org/sites/default/files/resources/files/WADA-TD2014EPO-Summary-Modifications-EN.PDF. Accessed 22 Jan 2017.

6. Parisotto R, Gore CJ, Emslie KR, Ashenden MJ, Brugnara C, Howe C, et al. A novel method utilising markers of altered erythropoiesis for the detection of recombinant human erythropoietin abuse in athletes. Haematologica. 2000; 85:564–72.

7. Sottas PE, Robinson N, Saugy M. The athlete's biological passport and indirect markers of blood doping. Handb Exp Pharmacol. 2010;195:305–26.

8. Lombardi G, Lanteri P, Colombini A, Lippi G, Banfi G. Stability of haematological parameters and its relevance on the athlete's biological passport model. Sports Med. 2011;41:1033–42.

9. Lippi G, Plebani M. Athlete's biological passport: to test or not to test? Clin Chem Lab Med. 2011;49:1393–5.

10. Sanchis-Gomar F, Martinez-Bello VE, Gomez-Cabrera MC, Viña J. Current limitations of the Athlete's biological passport use in sports. Clin Chem Lab Med. 2011;49:1413–6.

11. Ashenden M, Gough CE, Garnham A, Gore CJ, Sharpe K. Current markers of the athlete blood passport do not flag microdose EPO doping. Eur J Appl Physiol. 2011;111:2307–14.

12. Grover RF, Weil JV, Reeves JT. Cardiovascular adaptation to exercise at high altitude. Exerc Sport Sci Rev. 1986;14:269–302.

13. Mairbäurl H. Red blood cells in sports: effects of exercise and training on oxygen supply by red blood cells. Front Physiol. 2013;4:332.

14. Reichel C. OMICS-strategies and methods in the fight against doping. Forensic Sci Int. 2011;213:20–34.

15. Pitsiladis YP, Durussel J, Rabin O. An integrative 'omics' solution to the detection of recombinant human erythropoietin and blood doping. Br J Sports Med. 2014;48:856–61.

16. Naukkarinen J, Rissanen A, Kaprio J, Pietiläinen KH. Causes and consequences of obesity: the contribution of recent twin studies. Int J Obes. 2012;36:1017–24.

17. McCarthy MI. Genomics, type 2 diabetes, and obesity. N Engl J Med. 2010; 363:2339–50.

18. Vucic EA, Thu KL, Robison K, Rybaczyk LA, Chari R, Alvarez CE, et al. Translating cancer 'omics' to improved outcomes. Genome Res. 2012;22: 188–95.

19. Varlet-Marie E, Audran M, Ashenden M, Sicart MT, Piquemal D. Modification of gene expression : help to detect doping with erythropoiesis-stimulating agents. Am J Hematol. 2009;84:755–9.

20. Rupert JL. Transcriptional profiling: a potential anti-doping strategy. Scand J Med Sci Sports. 2009;19:753–63.

21. Connolly PH, Caiozzo VJ, Zaldivar F, Nemet D, Larson J, Hung SP, et al. Effects of exercise on gene expression in human peripheral blood mononuclear cells. J Appl Physiol (1985). 2004;97:1461–9.

22. Büttner P, Mosig S, Lechtermann A, Funke H, Mooren FC. Exercise affects the gene expression profiles of human white blood cells. J Appl Physiol (1985). 2007;102:26–36.

23. Durussel J, Haile DW, Mooses K, Daskalaki E, Beattie W, Mooses M, et al. The blood transcriptional signature of recombinant human erythropoietin administration and implications for anti-doping strategies. Physiol Genomics. 2016;48:202–9.

24. Davison RC, Wooles AL. Cycling. In: Winter EM, Jones AM, Davison RC, Bromley PD, Mercer TH, editors. Sport and exercise physiology testing guidelines: the British Association of Sport and Exercise Sciences Guide. Oxon: Routledge; 2006. p. 364.

25. WADA Athlete Biological Passport operating guidelines & compilation of required elements. https://www.wada-ama.org/sites/default/files/ resources/files/WADA-ABPOperating-Guidelines_v4.0-EN.pdf. Accessed 16 Oct 2017.

26. Bates D, Maechler M, Bolker B, Walker S. Fitting Linear Mixed-Effects Models Using lme4. J Stat Softw. 2015;67:1-48.

27. Bates D, Maechler M, Bolker B, Walker S. Fitting linear mixed-effects models using lme4. J Stat Softw. 2015;67:1-48.

28. De Rosario-Martinez H. Phia: post-hoc interaction analysis. R package version 0.2–0. 2015. http://CRAN.R-project.org/package=phia. Accessed 16 Oct 2017.

29. Smyth GK. Linear models and empirical bayes methods for assessing differential expression in microarray experiments. Stat Appl Genet Mol Biol. 2004;3:Article3.

30. Benjamini Y, Hochberg Y. Controlling the false discovery rate: a practical and powerful approach to multiple testing. J R Statist Soc B. 1995;57:289–300.

31. Sottas PE, Robinson N, Saugy M. A forensic approach to the interpretation of blood doping markers. Law Probability & Risk. 2008;7:191–210.

32. QuantiGene® Plex Assay. https://tools.thermofisher.com/content/sfs/ brochures/quantigene-assaysapplication-guide.pdf. Accessed 22 Jan 2017.

33. Cox TC, Sadlon TJ, Schwarz QP, Matthews CS, Wise PD, Cox LL, et al. The major splice variant of human 5-aminolevulinate synthase-2 contributes significantly to erythroid heme biosynthesis. Int J Biochem Cell Biol. 2004;36: 281–95.

34. Walsh M, Lutz RJ, Cotter TG, O'Connor R. Erythrocyte survival is promoted by plasma and suppressed by a Bak-derived BH3 peptide that interacts with membrane-associated Bcl-X(L). Blood. 2002;99:3439–48.

35. Kalfa TA, Connor JA, Begtrup AH. EPB42-related hereditary Spherocytosis. In: AM PRA, Ardinger HH, et al., editors. GeneReviews®. Seattle: University of Washington, Seattle; 2014.

36. Dahl KN, Parthasarathy R, Westhoff CM, Layton DM, Discher DE. Protein 4.2 is critical to CD47-membrane skeleton attachment in human red cells. Blood. 2004;103:1131–6.

37. Miyaguchi K. Localization of selenium-binding protein at the tips of rapidly extending protrusions. Histochem Cell Biol. 2004;121:371–6.

38. Low PS. Structure and function of the cytoplasmic domain of band 3: center of erythrocyte membrane-peripheral protein interactions. Biochim Biophys Acta. 1986;864:145–67.

39. Low PS, Zhang D, Bolin JT. Localization of mutations leading to altered cell shape and anion transport in the crystal structure of the Cytoplasmic domain of band 3. Blood Cells Mol Dis. 2001;27:81–4.

40. Reithmeier RA, Casey JR, Kalli AC, Sansom MS, Alguel Y, Iwata S. Band 3, the human red cell chloride/bicarbonate anion exchanger (AE1, SLC4A1), in a structural context. Biochim Biophys Acta. 2016;1858:1507–32.

41. Yao W, Sung LA. Erythrocyte tropomodulin isoforms with and without the N-terminal actin-binding domain. J Bio Chem. 2010;285:31408–17.

42. Murano I, Tsukahara M, Kajii T, Yoshida A. Mapping of the human guanosine monophosphate reductase gene (GMPR) to chromosome 6p23 by fluorescence in situ hybridization. Genomics. 1994;19:179–80.

43. Hedstrom L. The dynamic determinants of reaction specificity in the IMPDH/GMPR family of (β/α)8 barrel enzymes. Crit Rev Biochem Mol Biol. 2012;47:250–63.

44. Kanno H, Huang IY, Kan YW, Yoshida A. Two structural genes on different chromosomes are required for encoding the major subunit of human red cell glucose-6-phosphate dehydrogenase. Cell. 1989;58:595–606.

45. Beutler E, Gelbart T, Kuhl W. Human red cell glucose-6-phosphate dehydrogenase: all active enzyme has sequence predicted by the X chromosome-encoded cDNA. Cell. 1990;62:7–9.

46. Mason PJ, Bautista JM, Vulliamy TJ, Turner N, Luzzatto L. Human red cell glucose-6-phosphate dehydrogenase is encoded only on the X chromosome. Cell. 1990;62:9–10.

47. An X, Schulz VP, Li J, Wu K, Liu J, Xue F, et al. Global transcriptome analyses of human and murine terminal erythroid differentiation. Blood. 2014;123: 3466–77.

48. van der Harst P, Zhang W, Mateo Leach I, Rendon A, Verweij N, Sehmi J, et al. Seventy-five genetic loci influencing the human red blood cell. Nature. 2012;492:369–75.

49. D'Alessandro A, Righetti PG, Zolla L. The red blood cell proteome and Interactome: an update. J Proteome Res. 2010;9:144–63.

50. Ng T, Marx G, Littlewood T, Macdougall I. Recombinant erythropoietin in clinical practice. Postgrad Med J. 2003;79:367–76.

51. Feezor RJ, Baker HV, Mindrinos M, Hayden D, Tannahill CL, Brownstein BH, et al. Whole blood and leukocyte RNA isolation for gene expression analyses. Physiol Genomics. 2004;19:247–54.

52. Clevers H, Alarcon B, Wileman T, Terhorst C. The T cell receptor/CD3 complex: a dynamic protein ensemble. Annu Rev Immunol. 1988;6:629–62.

53. Christopoulos P, Dopfer EP, Malkovsky M, Esser PR, Schaefer HE, Marx A, et al. A novel thymoma-associated immunodeficiency with increased naive T cells and reduced CD247 expression. J Immunol. 2015;194:3045–53.

54. Bronstein-Sitton N, Cohen-Daniel L, Vaknin I, Ezernitchi AV, Leshem B, Halabi A, et al. Sustained exposure to bacterial antigen induces interferon-gamma-dependent T cell receptor zeta down-regulation and impaired T cell function. Nat Immunol. 2003;4:957–64.

55. Carlson LA, Tighe SW, Kenefick RW, Dragon J, Westcott NW, Leclair RJ. Changes in transcriptional output of human peripheral blood mononuclear cells following resistance exercise. Eur J Appl Physiol. 2011;111:2919–29.

56. Storey AG, Birch NP, Fan V, Smith HK. Stress responses to short-term intensified and reduced training in competitive weightlifters. Scand J Med Sci Sports. 2016;26:29–40.

57. Gjevestad GO, Holven KB, Ulven SM. Effects of exercise on gene expression of inflammatory markers in human peripheral blood cells: a systematic review. Curr Cardiovasc Risk Rep 2015;9:34.

58. Thom CS, Traxler EA, Khandros E, Nickas JM, Zhou OY, Lazarus JE, et al. Trim58 degrades Dynein and regulates terminal erythropoiesis. Dev Cell. 2014;30:688–700.

59. Kamatani Y, Matsuda K, Okada Y, Kubo M, Hosono N, Daigo Y, et al. Genome-wide association study of hematological and biochemical traits in a Japanese population. Nat Genet. 2010;42:210–5.

60. Athlete Biological Passport Operating Guidelines & Compilation of Required Elements. 2014; https://www.wada-ama.org/sites/default/files/resources/ files/wada_abp_operating_guidelines_2014_v5.0_en.pdf. Accessed 23 Jan 2017.

61. Basu M, Malhotra AS, Pal K, Prasad R, Kumar R, Prasad BA, et al. Erythropoietin levels in lowlanders and high-altitude natives at 3450 m. Aviat Space Environ Med. 2007;78:963–7.

62. Gunga HC, Röcker L, Behn C, Hildebrandt W, Koralewski E, Rich I, et al. Shift working in the Chilean Andes (> 3,600 m) and its influence on erythropoietin and the low-pressure system. J Appl Physiol (1985). 1996;81: 846–52.

63. Hudson JG, Bowen AL, Navia P, Rios-Dalenz J, Pollard AJ, Williams D, et al. The effect of high altitude on platelet counts, thrombopoietin and erythropoietin levels in young Bolivian airmen visiting the Andes. Int J Biometeorol. 1999;43:85–90.

64. Milledge JS, Cotes PM. Serum erythropoietin in humans at high altitude and its relation to plasma renin. J Appl Physiol (1985). 1985;59:360–4.

13

Genome-wide DNA methylation profiling reveals novel epigenetic signatures in squamous cell lung cancer

Yuan-Xiang Shi[1,2,3], Ying Wang[1,2], Xi Li[1,2,3], Wei Zhang[1,2,3], Hong-Hao Zhou[1,2,3], Ji-Ye Yin[1,2,3*] and Zhao-Qian Liu[1,2,3*]

Abstract

Background: Epigenetic alterations are strongly associated with the development of cancer. The aim of this study was to identify epigenetic pattern in squamous cell lung cancer (LUSC) on a genome-wide scale.

Results: Here we performed DNA methylation profiling on 24 LUSC and paired non-tumor lung (NTL) tissues by Illumina Human Methylation 450 K BeadArrays, and identified 5214 differentially methylated probes. By integrating DNA methylation and mRNA expression data, 449 aberrantly methylated genes accompanied with altered expression were identified. Ingenuity Pathway analysis highlighted these genes which were closely related to the carcinogenesis of LUSC, such as ERK family, NFKB signaling pathway, Hedgehog signaling pathway, providing new clues for understanding the molecular mechanisms of LUSC pathogenesis. To verify the results of high-throughput screening, we used 56 paired independent tissues for clinical validation by pyrosequencing. Subsequently, another 343 tumor tissues from the Cancer Genome Atlas (TCGA) database were utilized for further validation. Then, we identified a panel of DNA methylation biomarkers (CLDN1, TP63, TBX5, TCF21, ADHFE1 and HNF1B) in LUSC. Furthermore, we performed receiver operating characteristics (ROC) analysis to assess the performance of biomarkers individually, suggesting that they could be suitable as potential diagnostic biomarkers for LUSC. Moreover, hierarchical clustering analysis of the DNA methylation data identified two tumor subgroups, one of which showed increased DNA methylation.

Conclusions: Collectively, these results suggest that DNA methylation plays critical roles in lung tumorigenesis and may potentially be proposed as a diagnostic biomarker.

Keywords: Lung cancer, DNA methylation, Biomarker, Diagnosis, Epigenetics

Background

Lung cancer is the leading cause of cancer-related mortality throughout the world [1]. There are two main histological types of lung cancer, non-small cell lung cancer (NSCLC) and small cell lung cancer. NSCLC comprises three major histological subtypes: squamous cell carcinoma (LUSC), adenocarcinoma and large cell carcinoma [2]. Early diagnosis of cancer is one of the most important factors contributing to the successful and effective treatment. However, many patients are diagnosed with advanced lung cancer due to the asymptomatic nature of early stages and lack of effective screening modalities, resulting in a very low five-year survival rates for them. Therefore, it is essential to identify tumor specific molecular biomarkers for risk assessment and effective early screening.

Tumorigenesis involves a multi-step process, which is the result of the interactions of genetic, epigenetic and environmental factors. The change of these factors results in dysregulation of key oncogenes and tumor suppressor genes. Epigenetic mechanisms are heritable and reversible, including DNA methylation, histone modifications and chromatin organization. DNA methylation is a major epigenetic modification which leads to gene

* Correspondence: yinjiye@csu.edu.cn; liuzhaoqian63@126.com
[1]Department of Clinical Pharmacology, Xiangya Hospital, Central South University, Changsha 410008, People's Republic of China
Full list of author information is available at the end of the article

silencing at the transcriptional level. It is involved in some crucial biological processes, including proliferation, apoptosis, cell cycle, DNA repair, tumor invasion and metastasis [3]. Thus, identification of DNA methylation biomarkers has emerged as one of the most promising approaches to improve cancer diagnosis, it presents several advantages compared with other markers [4, 5]. Firstly, methylation changes in lung cancer appear to be early events and thus could be used to improve early detection of malignant tumors [6]. Additionally, the DNA methylation represents a very stable sign that can be detected in many different types of samples, including tumor tissues, cancer cells in body fluids [7, 8]. Most importantly, DNA methylation can be detected by a wide range of sensitive and cost efficient techniques even in samples with low tumor purity.

In previous studies, a variety of epigenetic biomarkers has been evaluated in lung cancer for early detection and prognosis prediction, however, most of them focused on a single gene. For example, P16, HOXA11 (Homeobox A11) and SOX17 (SRY-box 17) showed abnormal hypermethylation at their promoters, they were considered as biomarkers for lung cancer detection and prognosis prediction [9–11]. In recent years, many epigenetic biomarkers have been identified by using microarray [12]. However, they have either lacked clinical validation via large sample size or focused on a mix of lung cancer histologies, and thereby limited the ability to identify subtypes. Of note, LUSC and adenocarcinoma shows distinct differences in DNA methylation, expression profiles and lesion location, although they are similarly treated in clinical practice due to the largely unknown underlying molecular mechanisms [13]. Homogeneous treatment strategies have been traditionally implemented for the two fundamentally different subtypes in clinical practice, resulting in poor response to treatment. Therefore, a better understanding of their biological pattern is critical for finding subtype-specific diagnosis and treatment strategies [14]. The aim of this study is to identify epigenetic pattern in LUSC on a genome-wide scale.

Methods

All data analysis were performed using R (http://www.r-project.org/, version 2.15.0) and Bioconductor [15].

Patients and tissue collection

The study was approved by the Ethics Committee of Xiangya School of Medicine, Central South University. All the patients provided written informed consents in compliance with the code of ethics of the World Medical Association (Declaration of Helsinki) at the time of surgery for the donation of their tissue for this research. We also obtained the clinical research admission on the

Chinese Clinical Trial Registry and the registration number is ChiCTR-RCC-12002830 [16]. All fresh tissues were frozen in liquid nitrogen immediately after resection and stored at −80 °C. Their basic clinical characteristics were summarized in Table 1. In the current study, current smoker and current reformed smoker for ≤15 years were identified as smoker, whereas current reformed smoker for >15 years and never-smoker were defined as non-smoker.

Global methylation analysis

Genome-scale DNA methylation were analyzed by the Illumina Human Methylation 450 K BeadArrays according to manufacturer's instructions in the laboratory of CapitalBio Corporation (Beijing, China), which quantifies methylation levels (β-value) of 485,577 CpG-sites. Raw fluorescence intensity values were normalized by Illumina Genome Studio software. Normalized intensities were used to calculate β-values, which were calculated from mean methylated (M) and unmethylated (U) signal intensities for each locus of each sample using the formula $(\beta = (M)/(U + M + 100))$. All methylation data analysis was carried out by using R software (v2.1.5). First, we performed data quality control as following steps: 14,511 sites containing missing values were removed, 89,808 sites containing SNPs were removed, 10,245 sites on the X or Y chromosome were removed, 14 sites with P value greater

Table 1 Clinicopathological characteristics of patients for discovery and clinical validation cohorts

Clinical and pathological variables	Discovery cohort (N = 24)	Validation cohort (N = 56)
Age (years)		
< 60	13	29
≥ 60	11	27
Gender		
Male	22	54
Female	2	2
Smoking status		
Smoker	19	48
Non-smoker	5	8
Clinical stage		
I-II	14	28
III-IV	10	28
Differentiation		
Well	0	8
Moderate	16	27
Poor	8	21
Lymph node metastasis		
Yes	11	21
No	13	35

than 0.05 in at least 75% samples were removed, and finally 371,000 sites were retained from the original 485,577 sites. Secondly, site-level differential methylation analysis was performed: locus-by-locus analyses was conducted using the nonparametric Wilcoxon rank-sum test, and multiple comparisons correction was performed using Benjamini-Hochberg (BH) FDR from the package in R. Probes with FDR P-value <0.05 and β difference ≥ 0.2 were used to identify significantly differential DNA methylation, 5214 sites (1771 genes) were differentially methylated (Additional file 1: Figure S1).

Global gene expression analysis

Genome-scale mRNA expression profiles were detected by the Human 4 × 180 K expression microarray (Agilent Technologies, Santa Clara, California, USA). After strict data preprocessing and quality control, 32,205 sites were retained from the original sites. We analyzed differential expression using paired t-tests and Benjamini-Hochberg (BH) multiple comparisons correction. Corrected P-value <0.05 and absolute fold change >2 were used to identify significantly differential expressed mRNAs, and 3635 genes were differentially expressed.

Select the validation genes

In order to select the target genes, we designed our study into three steps. Firstly, to identify genes with the greatest changes, we further set up a fourfold cutoff to the average change in gene expression, the results show that 44 genes were coordinately hypermethylated and downregulated in tumors, and 26 genes were coordinately hypomethylated and up-regulated (Additional file 2: Table S3). Secondly, we looked at the literature one by one, looking for genes that were involved in the development of lung cancer and were not reported/reported less from the 70 negatively correlated genes. Finally, we selected several genes for clinical validation, and six genes (CLDN1, TP63, TBX5, TCF21, ADHFE1 and HNF1B) were identified.

Pyrosequencing analysis

Genomic DNA was extracted from samples by using QIAamp DNA Mini Kit (QIAGEN, Hilden, Germany), following the manufacturer's instructions. The genomic DNA was bisulfite-modified using an EpiTect Bisulfite Kit (QIAGEN, Hilden, Germany), according to the manufacturer's instruction. Primer design was carried out using the PyroMark Assay Design 2.0 software; one of the primers was biotinylated to enable capture by Streptavidin Sepharose (Additional file 2: Table S1). Bisulfite-treated DNA was amplified, followed by pyrosequencing using the Gold Q96 CDT Reagents (QIAGEN, Hilden, Germany).

Quantitative reverse transcription-polymerase chain reaction (qRT-PCR)

qRT-PCR was used to examine the mRNA expression as described previously [17]. Total RNA was extracted from samples with Trizol reagent (Takara, Dalian, China) and then reverse transcribed to cDNA using PrimeScriptTM RT-PCR Kit (Takara, Dalian, China). Real-time PCR was performed using SYBR® Premix DimerEraser™ (Perfect Real Time) (Takara, Dalian, China) in Roche LightCycler 480 II Real-Time PCR system (Roche Diagnostics Ltd., Rotkreuz, Switzerland). The data were calculated using the comparative cycle threshold (CT) (2-ΔΔCT) method. All primers were provided in Additional file 2: Table S1. The differences of mRNA expression level were compared by t test using SPSS 18.0 (SPSS Inc., Chicago, Illinois, USA).

Functional classification, the cancer genome atlas (TCGA) data and receiver operating characteristics (ROC) analysis

Gene Ontology analyses were performed by using the DAVID Functional Annotation Tool [18]. Gene network and pathway analyses were conducted by IPA (http://www.ingenuity.com). The NextBio database (http://www.nextbio.com) was used to analyze the overlap between our bioset and the other three most highly correlated NextBio biosets.

DNA methylation datasets in LUSC were downloaded from the Cancer Genome Atlas (TCGA) data portal (http://tcga-data.nci.nih.gov). We selected 343 tumor and 39 paired NTL samples, with both DNA methylation data and clinical features information available for performing the correlation analysis. Receiver operating curves were used to assess the predictive capacity of each marker. Area under the curve (AUC) was computed for each curve, and 95% confidence intervals (CI) were also estimated by bootstrapping with 1000 iterations.

Results

Genome-wide DNA methylation patterns in LUSC

In our study, a total of 24 LUSC and matched adjacent NTL tissues were analyzed, the strategy was diagrammatically outlined in Additional file 1: Figure S1. Single-CpG-site methylation levels are quantified by β, β ranges from zero (the CpG site is unmethylated) to one (the CpG site is fully methylated). Firstly, we investigated the overall distribution of methylation level in tumor versus NTL, the results showed a bimodal distribution of methylation (Additional file 3: Figure S2A). Normally, the methylation site can be grouped based on their positional context relative to closest CpG island (CGI) and the nearby transcripts (Additional file 3: Figure S2B). Thus, we further identified the methylation level distribution of probes located in five CpG island-based regions (CGIs, south and north shores, and south and north shelves) and six gene-based regions (TSS1500, TSS200, 5'-UTR, first exon, gene body, and 3'-UTR). As indicated in Additional file 3: Figure S2C, we found that most CpG sites in CGIs were hypomethylated as showed by a single peak with the β-value <0.2, while CpG sites in CGI shelf regions (both north and south) were hypermethylated as showed by a single peak with the β-value >0.6. In addition, CpG sites in CGI shore regions had variable methylation levels as indicated by a bimodal distribution, and this pattern is symmetric in the north and south shores of CGIs. In brief, the DNA methylation levels gradually increased with the CpG sites far away from CGIs. We further investigated that methylation patterns at gene context based on genomic content, the first exon and its upstream area (TSS1500, TSS200, 5'-UTR) are hypomethylated, while gene body and 3'-UTR are hypermethylated (Additional file 3: Figure S2D). We found that the CpG sites which closer to 3'-UTR have higher methylation levels. We also compared the methylation level between LUSC and NTL samples based on these groups, although the above distribution curves are similar with each other, our statistical analysis indicated that there were significant differences in LUSC versus NTL tissues (Additional file 2: Table S2).

Methylation differences in LUSC and matched NTL tissue

Then, we analyzed the methylation differences in LUSC and matched NTL tissues. After normalization, 371,000 probes from the methylation array were retained for analysis. Using the criteria of FDR p-value <0.05 and β difference ≥ 0.2, we identified 5214 probes (1771 gene) differentially methylated. Among them, 4001 probes (77%) were significantly hypermethylated, and 1213 probes (23%) were significantly hypomethylated in tumors (Fig. 1a). A two-dimensional hierarchical clustering analysis of the 5214 probes revealed a clear sorting of tumors and NTLs, indicating a substantial difference in DNA methylation profiles between the tumor and non-tumor samples (Fig. 1b). With these differentially methylated probes, we investigated their regional distribution in the gene context, CpG- island neighborhood and chromosome, respectively. The gene context regions of the hyper- or hypomethylated CpG sites were distributed similarly. As indicated in Fig. 1c, most of differentially methylated probes were located in the Gene body (29% in hypermethylated and 32% in hypomethylated). However, the CpG island-based regions of the significantly hyper- or hypomethylated CpG sites are distributed differently. 60% of the hypermethylated CpG sites are in CpG islands and that fewer are in the CpG shores (24%) and CpG shelves (4%). In contrast, just 5% of the hypomethylated CpG sites were in CpG islands, CpG shores

(14%) and CpG shelves (8%). In addition, chromosome location analysis showed that the majority of CpGs with differential methylation mapped to chromosome 2 and less in other chromosomes. We also added a distribution analysis of 371,000 probes in Additional file 4: Figure S4, and we compared the distribution of the differentially methylated probes and the overall probes in the genomic context. Compared to the overall distribution of all probes, the differentially methylated probes are distributed differently just in the CpG islands, 31% of the total probes located on the CpG island, 60% of the hypermethylated probes were located on the CpG island, and just 5% of the hypomethylated probes were located on the CpG island.

Identification of potentially functionally relevant DNA methylation changes in LUSC

To identify the potential functionally relevant methylation changes, we further selected 12 paired cancer and adjacent NTL tissue to detect the genome-scale mRNA expression profiles. Corrected P-value <0.05 and absolute fold change >2 were used to identify differentially expressed mRNAs, 3635 genes were identified to be differentially expressed. We performed an exploratory two-dimensional hierarchical clustering of the differentially expressed probes, the mRNA expression profiles of tumors and NTL resulted in separate clusters (Fig. 2a). After integrating analysis of differentially methylated genes (DMGs) and differentially expressed genes (DEGs), we identified 449 aberrantly methylated genes accompanied with altered expression. Of these, 184 genes were statistically significantly hypermethylated and down-regulated (41%), 72 genes (16%) were significantly hypomethylated and up-regulated, while 98 genes (22%) were significantly hypermethylated and up-regulated, 95 genes (21%) were significantly hypomethylated and down-regulated. To identify genes with the greatest changes, we further set up a fourfold cutoff to the average change in gene expression (Fig. 2b), the results show that 44 genes were coordinately hypermethylated and downregulated in tumors, and 26 genes were coordinately hypomethylated and up-regulated (Additional file 2: Table S3). We next asked whether the different groups of genes were associated with CpG islands or promoter regions methylation. As indicated in Fig. 2c, we found no statistically significant difference between groups whether or not the probes were located in the promoter region (P > 0.05). However, there were significant differences between groups with different locations of probes in CpG island (P < 0.01), hypermethylated genes were gathered at CpG island. To further investigate the relationships between DNA methylation and gene expression, we selected ten genes for verification. Scatter plot demonstrated that these probes showed an inverse

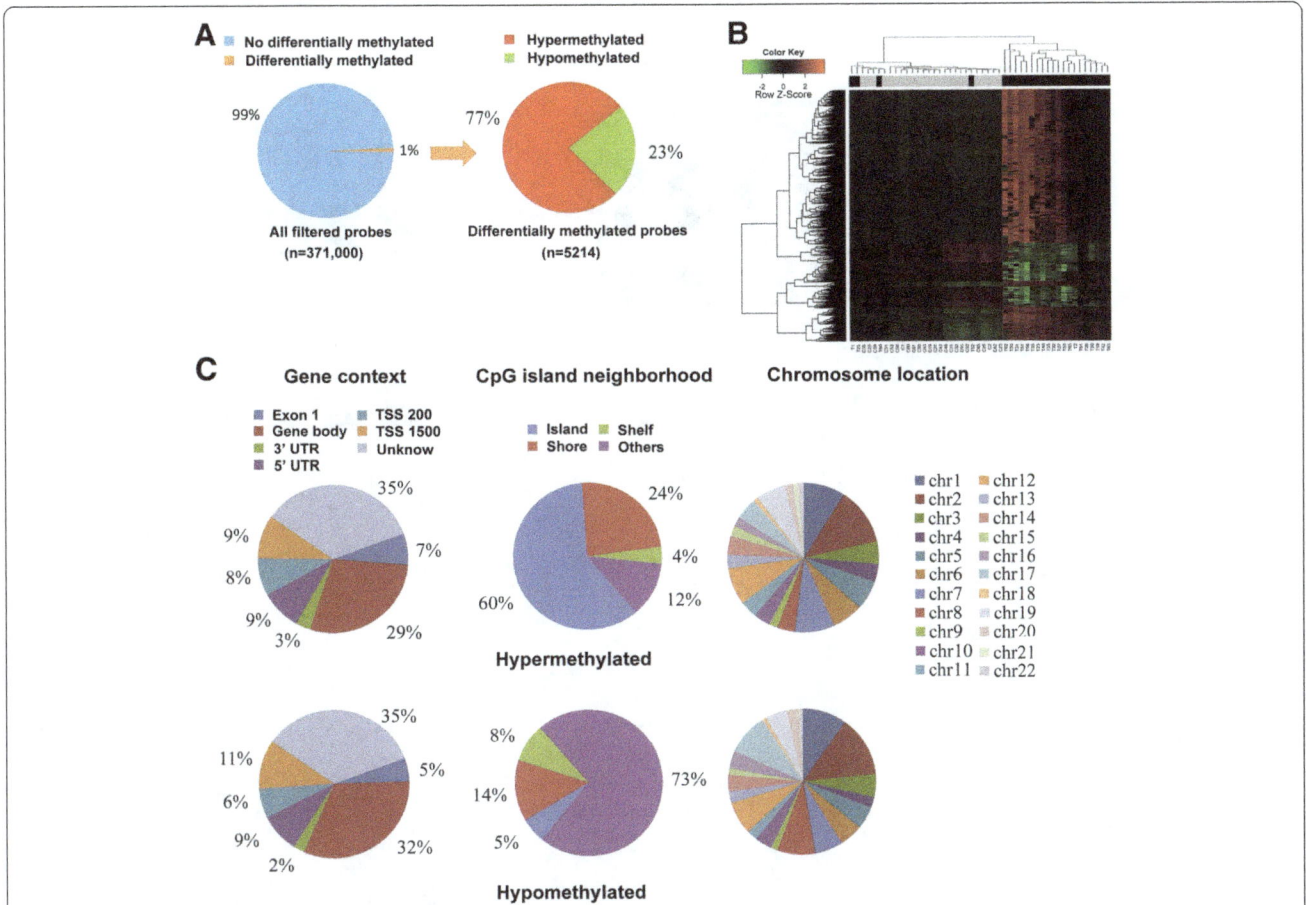

Fig. 1 Identification of DNA methylation differences between LUSC and NTL. **a** Pie charts showed the distribution of all filtered probes retained from the microarray, and revealed the methylation differences in LUSC and matched NTL tissues. **b** Two-dimensional hierarchical clustering was performed using the 5214 variable DNA methylation probes across all samples (*n* = 48). **c** The genomic distribution of differentially methylated probes in the gene context, CpG-site neighborhood and chromosome, respectively. TSS: transcription start site, UTR: untranslated region, Chr: chromosome

correlation of methylation with expression in tumor versus matched NTL, the Spearman correlation coefficient values for these ten genes were $r_{RAPGEFL1}$ = −0.829, r_{CLDN1} = −0.564, $r_{AKR1B10}$ = −0.709, r_{TP63} = −0.854, r_{TBX5} = −0.743, r_{TCF21} = −0.748, r_{ADHFE1} = −0.685, r_{GATA6} = −0.831, r_{GPR87} = −0.749, and r_{HNF1B} = −0.797, respectively (Fig. 2d).

To study biological functions of the 70 negatively correlated genes, Gene Ontology (GO) analysis was performed. In terms of the biological processes, most of the genes were related to development and adhesion. 7 of the top 10 categories of molecular function were related to protein binding, while cellular component mostly involved the plasma membrane and cell junction (Fig. 3a). Gene network analysis was further conducted using Ingenuity Pathways Analysis (IPA), we found that top two gene networks might be affected by the aberrant DNA methylation of the 256 negative correlation genes (Fig. 3b). Prominent in the first network were PP1 protein complex members, actin gene family, and NFKB signaling pathway members. The

second network was composed primarily of genes regulated by the ERK family, as well as the regulation of PIK3 complex members and Hedgehog signaling pathway members. Genes involved in the gene networks were associated with tissue morphology, organismal development, respiratory disease, cell death and survival.

Overlap analysis

To compare our study with other investigations, we employed the NextBio database (http://www.nextbio.com) to conduct the overlap analysis. The three most highly correlated NextBio biosets (LSCC, GSE30219, GSE19188) were selected. As indicated in Fig. 4a, of the 256 genes with inverse correlations, a total of 229 genes were significantly differentially expressed in our study and TCGA dataset (LSCC), most of the genes were expressed at the same direction, while 2 up regulated genes were down regulated in the LSCC dataset, and 4 down regulated genes were up regulated in LSCC dataset. To learn more about the overlap between our microarray and GEO

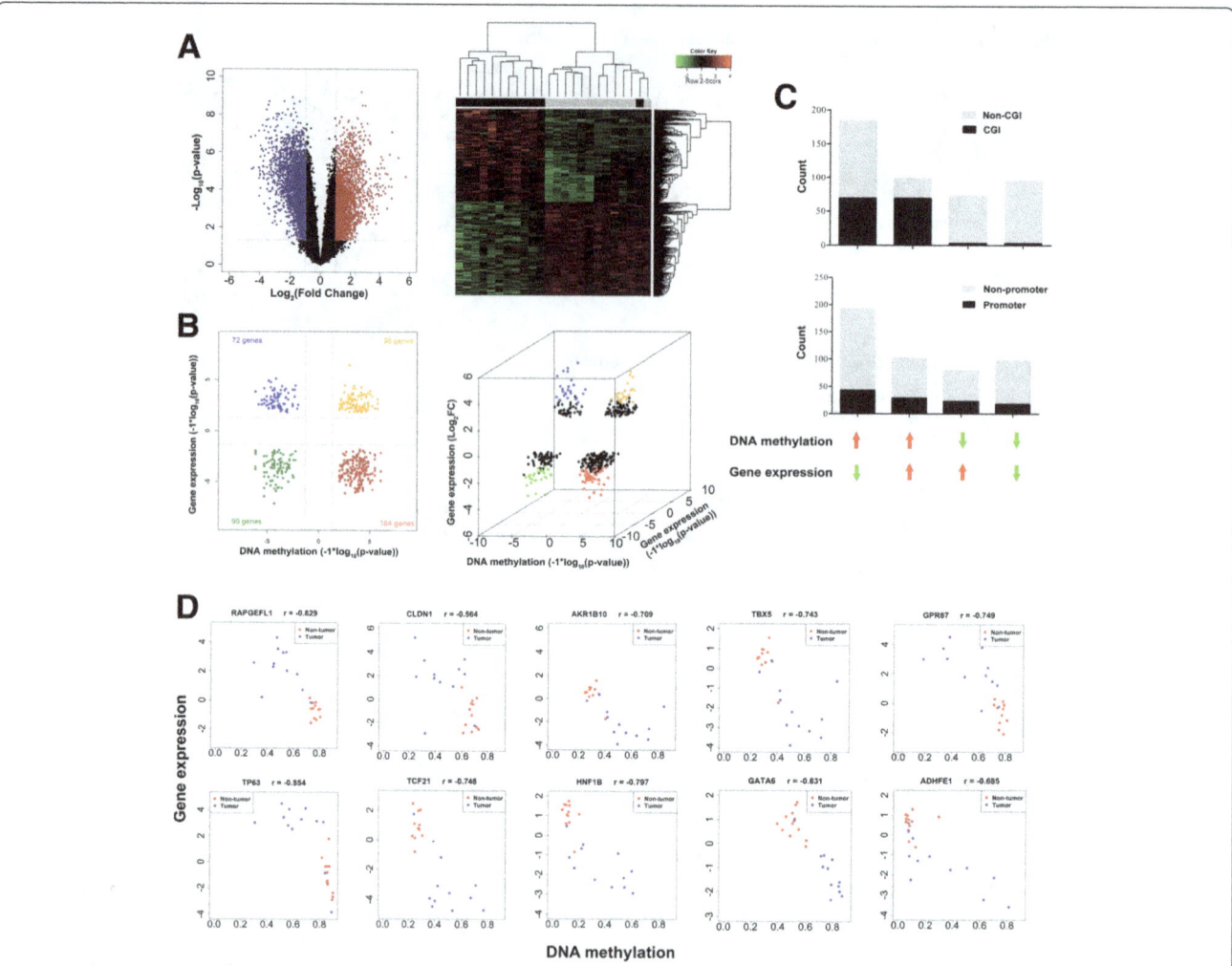

Fig. 2 Identification of genes showing coordinately changed DNA methylation and gene expression. **a** Volcano plot and two-dimensional hierarchical clustering of the differential mRNA expression analysis. Vertical dotted lines: fold change ≥2 or ≤2; Horizontal dotted line: the significance cutoff (FDR p-value = 0.05). Two-dimensional hierarchical clustering was performed using 4687 probes corresponding to 3635 genes across all samples (n = 24). **b** Starburst plot integrating differential DNA methylation and gene expression analyses. Vertical dotted lines: the significance cutoff (FDR p-value = 0.05); Horizontal dotted line: the significance cutoff (FDR p-value = 0.05). Three-dimensional starburst plot of 123 genes, integrating significant changes in DNA methylation (x-axis) and gene expression (y-axis), with a mean twofold or greater change in gene expression (z-axis). Indicated are genes that are hypermethylated and down-regulated in tumors (red); hypomethylated and up-regulated in tumors (blue); hypermethylated and up-regulated in tumors (orange); or hypomethylated and down-regulated in tumors (green). **c** Gene distribution in CpG islands and promoter region exhibiting hyper-or hypomethylation and up- or down-regulation. **d** Correlation plots of DNA methylation versus gene expression in tumors and normal tissues for selected genes. x-axis: DNA methylation level (β value), y-axis: mRNA expression level, r: correlation coefficient

database, we compared current study's differentially expressed genes with their results. As showed in Fig. 4b the two studies owned an overlap of 200 genes, just few genes were expressed at the different direction. Similar to this result, Fig. 4c showed that 196 genes were overlapped between our study and GSE19188 dataset. Taken together, there were 183 overlapping genes, highly consistent with the three previous studies on tumor and matched NTL (Fig. 4d).

Validation of the methylation biomarkers for LUSC diagnosis
To confirm our previous results, we selected six genes (CLDN1, TP63, TBX5, TCF21, ADHFE1 and HNF1B)

for further validation in another independent 56 paired LUSC and adjacent NTL tissues. The clinical characteristics of this cohort were summarized in Table 1. DNA methylation was detected by using pyrosequencing, mRNA expression was identified by using realtime PCR. As indicated in Fig. 5a and b, the results were consistent with our high-throughput analysis, and we found two hypomethylation and up-regulated expression genes, four hypermethylation and down-regulated expression genes. Next, receiver operating characteristics (ROC) analysis was performed to assess the diagnostic value of each individual biomarker to detect LUSC. Areas under

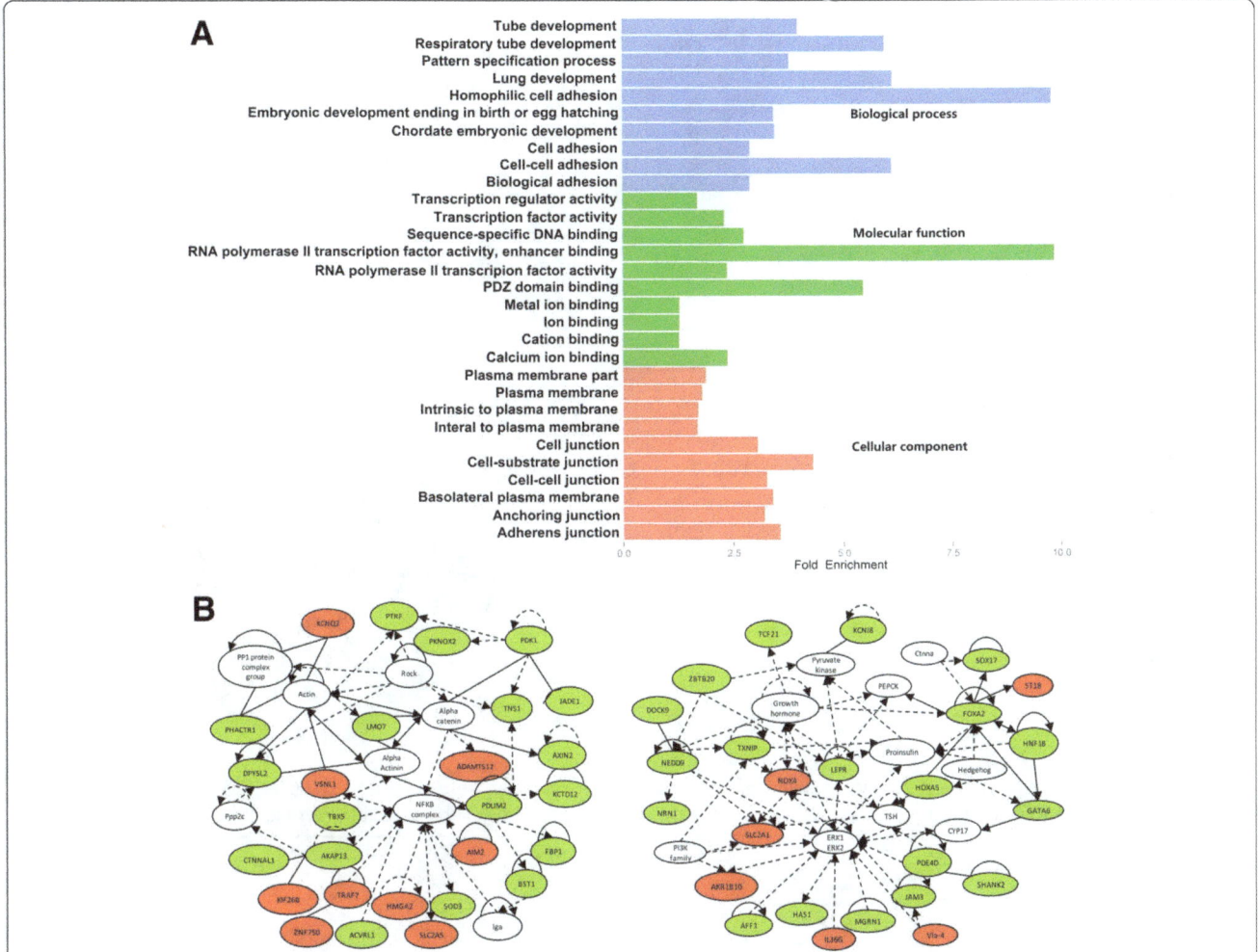

Fig. 3 GO and pathway analysis of significant DNA methylation changes associated with significant inverse gene expression changes. **a** The top ten significantly enriched GO categories were calculated. Blue: Biological process; Green: Molecular function; Red: Cellular component. **b** Gene networks identified through integrative pathways analysis of the negatively correlated genes. Red: the hypomethylated and up-regulated genes in tumor, Green: hypermethylated and down-regulated genes in tumors, Solid lines: direct interaction, Dashed lines: indirect interaction

the ROC curve (AUC) of tumor and NTL group were significantly different ($P < 0.01$) for all six genes with the following values $AUC_{CLDN1} = 0.836$, $AUC_{TP63} = 0.919$, $AUC_{TBX5} = 0.737$, $AUC_{TCF21} = 0.968$, $AUC_{ADHFE1} = 0.761$ and $AUC_{HNF1B} = 0.809$ (Fig. 5c). Considering that our study was limited by the number of patients, we expanded the sample size to further validation by using the Cancer Genome Atlas (TCGA) database. A total of 343 LUSC patients and 39 NTL tissue samples were selected (Additional file 2: Table S4). The methylation levels of the six selected genes were similar to those of our clinical validation cohort with significant differences between tumor and NTL (Additional file 5: Figure S3A), suggesting that the methylation statuses of the six selected biomarkers are a common feature for LUSC. Then, we performed ROC analysis to assess the performance of each individual biomarker to detect LUSC. Importantly, all the genes showed significant difference ($P < 0.01$) in AUC ($AUC_{CLDN1} = 0.919$,

$AUC_{TP63} = 0.958$, $AUC_{TBX5} = 0.984$, $AUC_{TCF21} = 0.985$, $AUC_{ADHFE1} = 0.852$ and $AUC_{HNF1B} = 0.908$), suggesting that they could be suitable as potential predictive biomarkers for LUSC diagnosis (Additional file 5: Figure S3B). Details of the CGs dinucleotides for these six genes are listed in Additional file 2: Table S5.

Subclassification of LUSC by methylation patterns

Finally, to explore the effect of clinical pathological features on DNA methylation, we performed correlation studies based on the stratification of clinical characteristics. Patients were divided into two groups according to each of the following five factors: age (<60 or ≥60 years old), smoking status (smokers or non-smokers), differentiation (poorly or moderately), complication (with or without) and TNM stages (I/II or III). The results showed that all factors mentioned above were highly associated with DNA methylation, respectively; and associations of each

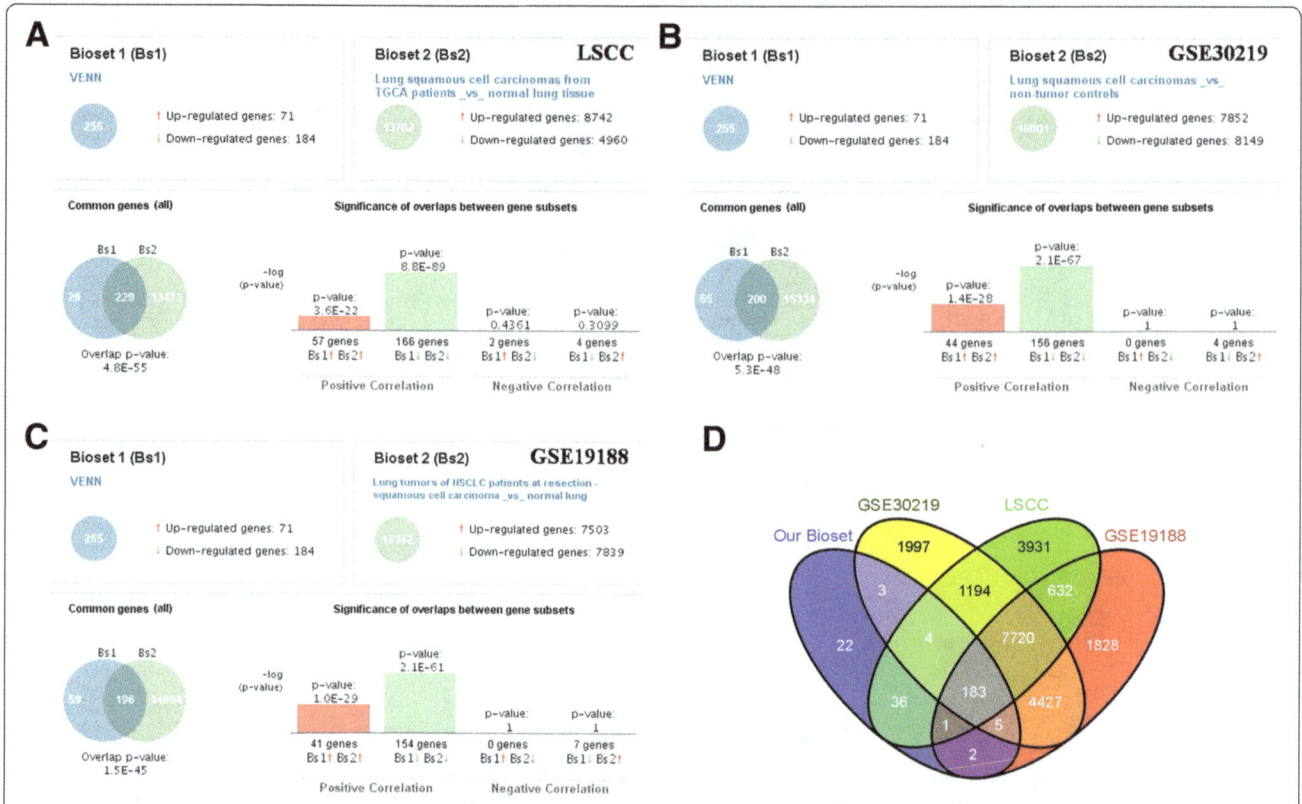

Fig. 4 Overlap analysis between our study and other studies. Three most highly correlated NextBio biosets: LSCC(**a**), GSE30219(**b**), GSE19188(**c**). **d** Venn diagram of NextBio analysis showing the overlap of our bioset with the three most highly correlated NextBio biosets

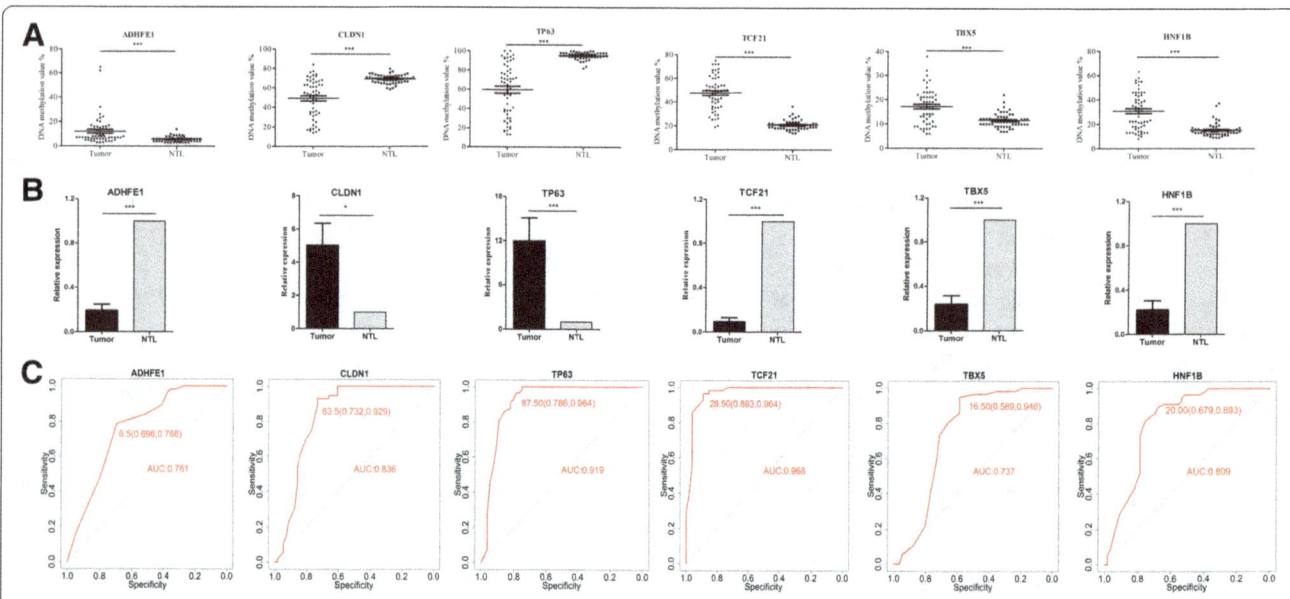

Fig. 5 Validation of selected methylation biomarkers. **a** Clinical validation of DNA methylation levels of selected genes in paired LUSC and adjacent NTL tissue by using pyrosequencing. **b** Clinical validation of mRNA expression of selected genes in paired LUSC and adjacent NTL tissue by using qRT-PCR. **c** ROC curves and area under the curve (AUC) for the candidate genes. Sensitivity, Specificity and the optimal cut-off values were marked in the figures. *** corresponds to $p < 0.001$; ** $p < 0.01$ and * $p < 0.05$

factor with DNA methlylation were more evident in NTL groups relative to tumor group (Fig. 6a). Then, to identify squamous cell lung cancer DNA methylation-based subclasses, we used the 5214 differentially methylated probes to perform an unsupervised hierarchical clustering. 24 tumors were clearly divided into two independent categories: Cluster1 ($n = 10$) and Cluster 2 ($n = 14$), Cluster 1 tumors were significantly hypomethylated as compared with Cluster 2 tumors (Fig. 6b). We also used the 2470

differentially methylated probes on the CpG island to perform an unsupervised hierarchical clustering, which suggested more significant differences between the two tumor subclusters (Fig. 6c).

Discussion

In the present study, we investigated the genome-wide DNA methylation patterns in 24 paired LUSC and adjacent NTL tissues by microarray, and identified 5214

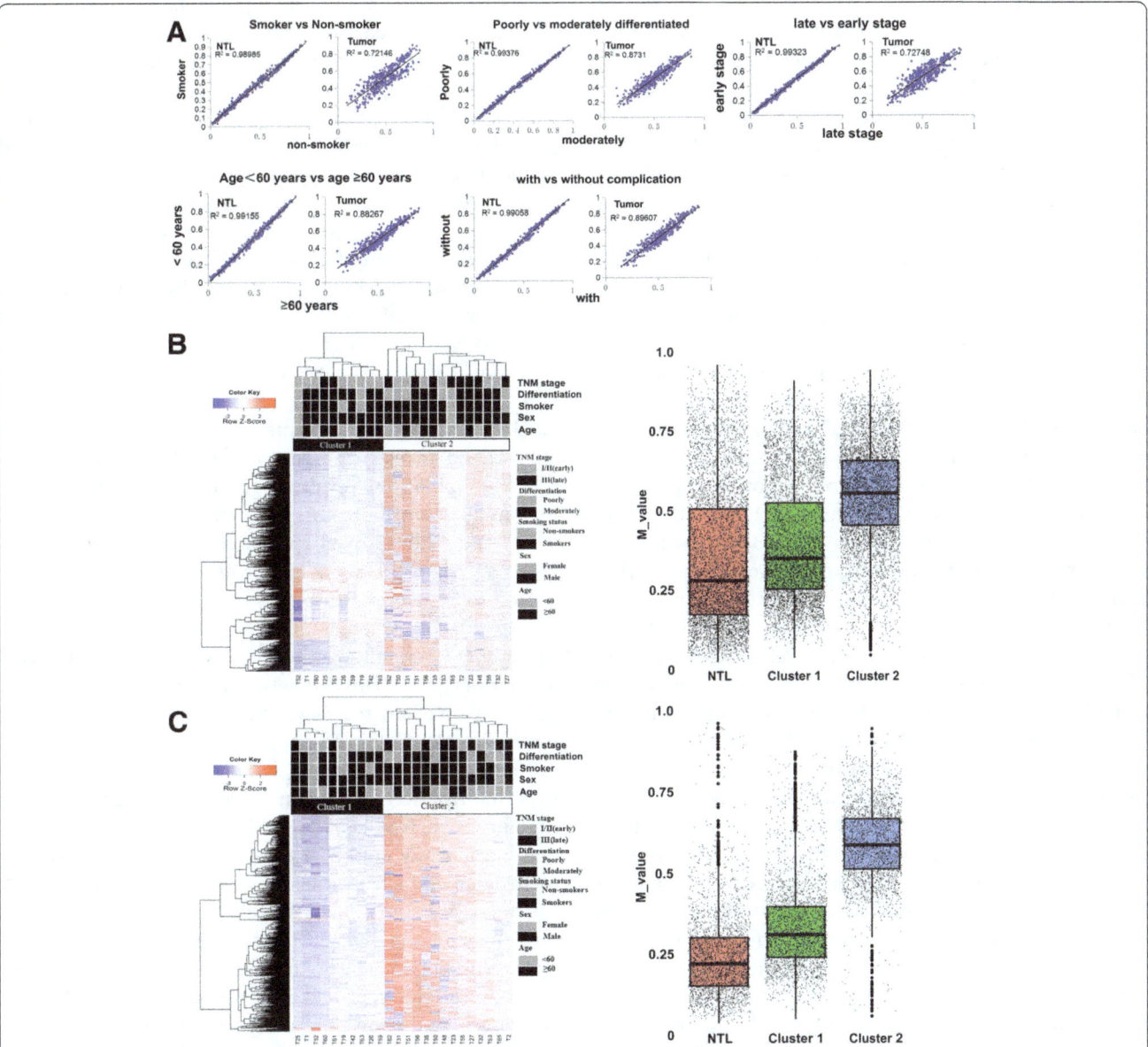

Fig. 6 Subgroup analysis of DNA methylation between LUSC and NTL. **a** Subgroup analysis of DNA methylation between LUSC and NTL according to smoking status, differentiation, TNM stage, age and complications. The correlation coefficient was given in the left corner. **b** Two-dimensional hierarchical clustering of all the significantly differently methylated probes in tumors was performed (n = 24). DNA methylation levels of Cluster 1, Custer 2 and NTL are shown using M-values. Note that methylation levels were significantly higher in Cluster 2 than in Cluster 1, although those of Cluster 1 were higher than those of NTL. **c** Two-dimensional hierarchical clustering of the 2470 differentially methylated probes (on the CpG island) was performed, two distinct clusters were identified. DNA methylation levels of Cluster 1, Custer 2 and NTL are shown using M-values. Note that methylation levels were significantly higher in Cluster 2 than in Cluster 1

probes showing significantly differential DNA methylation in cancer tissues. By integrating DNA methylation and mRNA expression data, 449 aberrantly methylated genes accompanied with altered expression were identified. GO analysis of these genes indicated that the most significantly related terms were development and binding. Pathway analysis highlighted many pathways which were closely related to the carcinogenesis of LUSC, such as ERK family, NFKB signaling pathway, Hedgehog signaling pathway, providing new clues for understanding the molecular mechanisms of LUSC pathogenesis. To verify the results of high-throughput screening, we used 56 paired independent tissues for clinical validation by pyrosequencing. Subsequently, another 343 tumor tissues from the TCGA were utilized for further validation. Then, we identified a panel of DNA methylation biomarkers in LUSC. Furthermore, we performed ROC analysis to assess the performance of each biomarker, results suggested that they could serve as potential predictive biomarkers for LUSC diagnosis. Finally, hierarchical clustering analysis of the DNA methylation data identified two tumor subgroups, one of which showed increased DNA methylation.

The pattern of DNA methylation in some certain types of cancers has been investigated including NSCLC [12]. DNA methylation analysis of cell-free blood samples has a substantial potential to serve as a minimally invasive tool for early diagnosis and clinical monitoring of cancer. Wielscher et al. found a model of four genes (HOXD10, PAX9, PTPRN2 and STAG3) that were able to differentiate lung cancer from controls, fibrotic ILD, and COPD [19]. NSCLC comprise multiple distinct biologic groups (such as epithelial-like NSCLCs and mesenchymal-like NSCLCs) with different prognoses. Walter et al. showed that patterns of DNA methylation can divide NSCLCs into these two phenotypically distinct subtypes of tumors and provide proof of principle that differences in DNA methylation can be used as a platform for predictive biomarker discovery and development [20]. Another research identified differentially methylated genes by comparing the global DNA methylation patterns between lung adenocarcinoma samples from smokers and nonsmokers. Their study provides an insightful perspective on smoking-associated DNA methylation and its role in tumorigenesis of the lung [21]. We investigated the genome-scale DNA methylation profile in LUSC and identified 5214 differentially methylated probes. Certain aberrantly methylated genes that were found in our study were also reported in previous other studies, which supported the result for each other. One of the previously reported methylated genes was *SOX17*, a canonical WNT antagonist previously shown functionally hypermethylated in breast, colorectal and lung cancers [22–24]. In our study, *SOX17* was also

a hypermethylated and down-regulated expression gene. Another example was *WIF1*, an extracellular antagonist that acts by binding to Wnt ligands. According to current findings, WIF1 promoter methylation was a frequent event as an epigenetic field manner and could be considered as a useful prognostic marker for adenocarcinoma patients [25]. In agreement with this report, our results also showed that WIF1 expression was down-regulated by promoter hypermethylation in LUSC. However, most of the differentially methylated genes identified in this study were novel. The top hypomethylated and up-regulated gene, aldo-keto reductase family 1 member B10 (AKR1B10), has been demonstrated previously to be specifically up-regulated in smoking-associated cancers such as squamous cell carcinoma and adenocarcinoma [26]. B-arrestin-1 (ARRB1), a scaffolding protein involved in the desensitization of signals arising from activated G-protein-coupled receptors, has been shown to play a role in invasion and proliferation of cancer cells, including nicotine-induced proliferation of NSCLC [27].

We selected several genes for clinical validation, and six genes (CLDN1, TP63, TBX5, TCF21, ADHFE1 and HNF1B) were identified. CLDN1 serves as an oncogene or a tumor suppressor in a tissue-specific manner. There is a significant correlation between down regulation of CLDN1 expression and methylation of its promoter CpG-island in estrogen receptor positive breast cancer [28, 29]. While our results suggested that hypomethylation might contribute to the upregulation of CLDN1 in LUSC and CLDN1 overexpression may play a role in the pathogenesis of LUSC. For the other genes, p63 was reported to be overexpressed in many tumors especially in LUSC [30–32], whereas adenocarcinoma and small cell carcinomas were almost all p63 low expression. TBX5 is a member of a phylogenetically conserved family of genes involved in the regulation of development, it is a novel functional tumor suppressor gene inactivated by promoter methylation in colon cancer [33]. TCF21 hypermethylation and reduced protein expression are ubiquitous in NSCLC [34]. TCF21 is expressed in normal lung airway epithelial cells, however, it is aberrantly methylated and silenced in the majority of head and neck squamous cell carcinomas and in NSCLC [35]. ADHFE1, a member of the group III metal dependent alcohol dehydrogenase family. The hypermethylation of ADHFE1 has recently been reported to be associated with colorectal cancer differentiation [36]. We performed receiver operating characteristics (ROC) analysis to assess the performance of each individual biomarker to detect LUSC. Our results suggested a strong diagnostic potential for these markers, and we hoped that they are potentially applicable in improving early LUSC diagnosis.

To further validate our findings, we asked if the findings of the three most highly correlated investigations (LSCC,

GSE30219, GSE19188) are consistent with our results. For the overlapped genes, our study and the previous investigations had high consistency with each other. The high reliability and reproducibility of the microarray technology in identifying the six genes are essential for its application in discovering the clinical biomarkers.

This study lays a foundation for the diagnosis, treatment and functional research of LUSC. However, there are some limitations in this study. We just examined the methylation of the target genes in tissue samples. In the future, we will also detect the methylation of these biomarkers in minimally/non-invasive samples.

Conclusions

In summary, we have identified and independently validated a powerful epigenetic signature of LUSC in tissue samples. We also described the clinicopathological characteristics of distinct molecular LUSC subgroups. The current study demonstrated that differences in genome-wide DNA methylation and gene expression patterns exist between LUSC and NTL. Our results suggested that DNA methylation plays critical roles in lung tumorigenesis and may potentially be proposed as a diagnostic biomarker.

Additional files

Additional file 1: Figure S1. Sketch and pipeline of the study design. (TIFF 2547 kb)

Additional file 2: Supplementary Materials. This docx file contains all supplementary tables. (DOCX 36 kb)

Additional file 3: Figure S2. Genomic context of CpG methylation. (**A**) The overall distribution of methylation sites in tumor versus NTL (**B**) A schematic diagram of CpGs depicts their genomic context relative to the nearest CpG island (top) or gene (bottom). (**C, D**) Density distribution of methylation probes in the CpG island-based regions and the gene-based regions. The x-axis is the mean β value in different regions. The y-axis is the signal density. The blue line is tumor, the red line is non-tumor. Transcription start site (TSS) 1500, TSS200, 5'untranslated region (UTR), and 3' UTR. (TIFF 2060 kb)

Additional file 4: Figure S4. The distribution of 371,000 probes in gene context, CpG-site neighborhood and chromosome, respectively. TSS: transcription start site, UTR: untranslated region, Chr: chromosome. (TIFF 9799 kb)

Additional file 5: Figure S3. Validation of the methylation biomarkers using 343 LUSC and 39 NTL tissue from TCGA database. (**A**) Validation of selected methylation biomarkers. ***corresponds to $P < 0.01$. (**B**) ROC curves and area under the curve (AUC) with 95% confidence intervals for the candidate genes. (TIFF 5194 kb)

Abbreviations

AUC: Area under the curve; CGI: CpG island; CI: Confidence intervals; DEGs: Differentially expressed genes; DMGs: Differentially methylated genes; GO: Gene Ontology; IPA: Ingenuity Pathways Analysis; LUSC: Squamous cell lung cancer; NSCLC: Non-small cell lung cancer; NTL: Non-tumor lung; ROC: Receiver operating characteristics; TCGA: The Cancer Genome Atlas; TSS: Transcription start site; UTR: Untranslated region

Acknowledgements
We thank all patients participated in this study.

Funding
This study was partially supported by the National High-tech R&D Program of China (863 Program) (2012AA02A517), National Natural Science Foundation of China (81,373,490, 81,573,508, 81,573,463), and Hunan Provincial Science and Technology Plan of China (2015TP1043).

Authors' contributions
Y.X.S., Z.Q.L. and J.Y.Y. conceived and designed the experiments, Y.X.S. performed the experiments, Y.X.S. and Y.W. contributed clinical samples and data, Y.X.S. and X.L. analyzed the data. Y.X.S. and J.Y.Y. wrote the manuscript. Y.X.S., J.Y.Y., Z.Q.L., W.Z. and H.H.Z. edited the manuscript. All authors read and approved the final manuscript.

Competing interests
The authors declare that they have no competing interest.

Author details
[1]Department of Clinical Pharmacology, Xiangya Hospital, Central South University, Changsha 410008, People's Republic of China. [2]Institute of Clinical Pharmacology, Central South University, Hunan Key Laboratory of Pharmacogenetics, Changsha 410078, People's Republic of China. [3]Hunan Province Cooperation Innovation Center for Molecular Target New Drug Study, Hengyang 421001, People's Republic of China.

References
1. Siegel RL, Miller KD, Jemal A. Cancer statistics, 2015. CA Cancer J Clin. 2015;65(1):5–29.
2. Liloglou T, Bediaga NG, Brown BR, Field JK, Davies MP. Epigenetic biomarkers in lung cancer. Cancer Lett. 2014;342(2):200–12.
3. Barros SP, Offenbacher S. Epigenetics: connecting environment and genotype to phenotype and disease. J Dent Res. 2009;88(5):400–8.
4. Heyn H, Esteller M. DNA methylation profiling in the clinic: applications and challenges. Nat Rev Genet. 2012;13(10):679–92.
5. Balgkouranidou I, Liloglou T, Lianidou ES. Lung cancer epigenetics: emerging biomarkers. Biomark Med. 2013;7(1):49–58.
6. Dai Z, Lakshmanan RR, Zhu WG, Smiraglia DJ, Rush LJ, Fruhwald MC, Brena RM, Li B, Wright FA, Ross P, et al. Global methylation profiling of lung cancer identifies novel methylated genes. Neoplasia. 2001;3(4):314–23.
7. Diaz-Lagares A, Mendez-Gonzalez J, Hervas D, Saigi M, Pajares MJ, Garcia D, Crujeiras AB, Pio R, Montuenga LM, Zulueta J, et al. A novel epigenetic signature for early diagnosis in lung cancer. Clin Cancer Res. 2016;22(13):3361–71.
8. Konecny M, Markus J, Waczulikova I, Dolesova L, Kozlova R, Repiska V, Novosadova H, Majer I. The value of SHOX2 methylation test in peripheral blood samples used for the differential diagnosis of lung cancer and other lung disorders. Neoplasma. 2016;63(2):246–53.
9. Xiao P, Chen JR, Zhou F, Lu CX, Yang Q, Tao GH, Tao YJ, Chen JL. Methylation of P16 in exhaled breath condensate for diagnosis of non-small cell lung cancer. Lung Cancer. 2014;83(1):56–60.
10. Hwang JA1, Lee BB, Kim Y, Park SE, Heo K, Hong SH, Kim YH, Han J, Shim YM, Lee YS, et al. HOXA11 hypermethylation is associated with progression of non-small cell lung cancer. Oncotarget. 2013;4(12):2317–25.
11. Balgkouranidou I, Chimonidou M, Milaki G, Tsaroucha E, Kakolyris S, Georgoulias V, Lianidou E. SOX17 promoter methylation in plasma circulating tumor DNA of patients with non-small cell lung cancer. Clin Chem Lab Med. 2016;54(8):1385–93.
12. Mullapudi N, Ye B, Suzuki M, Fazzari M, Han W, Shi MK, Marquardt G, Lin J, Wang T, Keller S, et al. Genome wide Methylome alterations in lung cancer. PLoS One. 2015;10(12):e0143826.

13. Gandara DR, Hammerman PS, Sos ML, Lara PN Jr, Hirsch FR. Squamous cell lung cancer: from tumor genomics to cancer therapeutics. Clin Cancer Res. 2015;21(10):2236–43.

14. Wang T, Zhang L, Tian P, Tian S. Identification of differentially-expressed genes between early-stage adenocarcinoma and squamous cell carcinoma lung cancer using meta-analysis methods. Oncol Lett. 2017;13(5):3314–22.

15. Zhang W, Spector TD, Deloukas P, Bell JT, Engelhardt BE. Predicting genome-wide DNA methylation using methylation marks, genomic position, and DNA regulatory elements. Genome Biol. 2015;16(1):14.

16. Wang Y, Qian C-Y, Li X-P, Zhang Y, He H, Wang J, Chen J, Cui J-J, Liu R, Zhou H, et al. Genome-scale long noncoding RNA expression pattern in squamous cell lung cancer. Sci Rep. 2015;5(1):11671.

17. Shi Y-X, Yin J-Y, Shen Y, Zhang W, Zhou H-H, Liu Z-Q. Genome-scale analysis identifies NEK2, DLGAP5 and ECT2 as promising diagnostic and prognostic biomarkers in human lung cancer. Sci Rep. 2017;7(1):8072.

18. Liu R, Guo CX, Zhou HH. Network-based approach to identify prognostic biomarkers for estrogen receptor-positive breast cancer treatment with tamoxifen. Cancer Biol Ther. 2015;16(2):317–24.

19. Wielscher M, Vierlinger K, Kegler U, Ziesche R, Gsur A, Weinhäusel A. Diagnostic performance of plasma DNA Methylation profiles in lung cancer, pulmonary fibrosis and COPD. EBioMedicine. 2015;2(8):929–36.

20. Walter K, Holcomb T, Januario T, Du P, Evangelista M, Kartha N, Iniguez L, Soriano R, Huw L, Stern H, et al. DNA Methylation profiling defines clinically relevant biological subsets of non-small cell lung cancer. Clin Cancer Res. 2012;18(8):2360–73.

21. Lu S, Tan Q, Wang G, Huang J, Ding Z, Luo Q, Mok T, Tao Q. Epigenomic analysis of lung adenocarcinoma reveals novel DNA methylation patterns associated with smoking. OncoTargets Therapy. 2013;1471

22. Zhang W, Glockner SC, Guo M, Machida EO, Wang DH, Easwaran H, Van Neste L, Herman JG, Schuebel KE, Watkins DN et al: Epigenetic inactivation of the canonical Wnt antagonist SRY-box containing gene 17 in colorectal cancer. Cancer Res 2008, 68(8):2764-2772.

23. Fu DY, Wang ZM, Li C, Wang BL, Shen ZZ, Huang W, Shao ZM. Sox17, the canonical Wnt antagonist, is epigenetically inactivated by promoter methylation in human breast cancer. Breast Cancer Res Treat. 2010;119(3):601–12.

24. Yin D, Jia Y, Yu Y, Brock MV, Herman JG, Guo M. SOX17 Methylation Inhibits Its Antagonism of Wnt Signaling Pathway in Lung Cancer. Discov Med. 2012;14(74):33–40.

25. Lee SM, Park JY, Kim DS. Wif1 hypermethylation as unfavorable prognosis of non-small cell lung cancers with EGFR mutation. Mol Cells. 2013;36(1):69–73.

26. Penning TM. AKR1B10: a new diagnostic marker of non–small cell lung carcinoma in smokers. Clin Cancer Res. 2005;

27. Perumal D, Pillai S, Nguyen J, Schaal C, Coppola D, Srikumar P. Chellappan: Nicotinic acetylcholine receptors induce c-Kit ligand/Stem Cell Factor and promote stemness in an ARRB1/β-arrestin-1 dependent manner in NSCLC. Oncotarget. 2014;5(21):10486–502.

28. Di Cello F, Cope L, Li H, Jeschke J, Wang W, Baylin SB, Zahnow CA: Methylation of the claudin 1 promoter is associated with loss of expression in estrogen receptor positive breast cancer. PLoS One 2013, 8(7):e68630.

29. Ogoshi K, Hashimoto S-i, Nakatani Y, Qu W, Oshima K, Tokunaga K, Sugano S, Hattori M, Morishita S, Matsushima K: Genome-wide profiling of DNA methylation in human cancer cells. Genomics 2011, 98(4):280-287.

30. Nobre AR, Albergaria A, Schmitt F. p40: a p63 isoform useful for lung cancer diagnosis - a review of the physiological and pathological role of p63. Acta Cytol. 2013;57(1):1–8.

31. Wang BY, Gil J, Kaufman D, Gan L, Kohtz DS, Burstein DE. p63 in pulmonary epithelium, pulmonary squamous neoplasms, and other pulmonary tumors. Hum Pathol. 2002;33(9):921–6.

32. Cancer Genome Atlas Research N. Comprehensive genomic characterization of squamous cell lung cancers. Nature. 2012;489(7417):519–25.

33. Yu J, Ma X, Cheung KF, Li X, Tian L, Wang S, Wu CW, Wu WK, He M, Wang M, et al. Epigenetic inactivation of T-box transcription factor 5, a novel tumor suppressor gene, is associated with colon cancer. Oncogene. 2010; 29(49):6464–74.

34. Richards KL, Zhang B, Sun M, Dong W, Churchill J, Bachinski LL, Wilson CD, Baggerly KA, Yin G, Hayes DN, et al. Methylation of the candidate biomarker TCF21 is very frequent across a spectrum of early-stage nonsmall cell lung cancers. Cancer. 2011;117(3):606–17.

35. Smith LT, Lin M, Brena RM, Lang JC, Schuller DE, Otterson GA, Morrison CD, Smiraglia DJ, Plass C. Epigenetic regulation of the tumor suppressor gene TCF21 on 6q23-q24 in lung and head and neck cancer. Proc Natl Acad Sci. 2006;103(4):982–7.

36. Moon JW, Lee SK, Lee YW, Lee JO, Kim N, Lee HJ, Seo JS, Kim J, Kim HS, Park SH. Alcohol induces cell proliferation via hypermethylation of ADHFE1 in colorectal cancer cells. BMC Cancer. 2014;14:377.

14

Weighted gene co-expression network analysis of expression data of monozygotic twins identifies specific modules and hub genes related to BMI

Weijing Wang[1], Wenjie Jiang[1], Lin Hou[2], Haiping Duan[1,3], Yili Wu[1], Chunsheng Xu[1,3,4], Qihua Tan[5,6], Shuxia Li[6] and Dongfeng Zhang[1*] (ORCID)

Abstract

Background: The therapeutic management of obesity is challenging, hence further elucidating the underlying mechanisms of obesity development and identifying new diagnostic biomarkers and therapeutic targets are urgent and necessary. Here, we performed differential gene expression analysis and weighted gene co-expression network analysis (WGCNA) to identify significant genes and specific modules related to BMI based on gene expression profile data of 7 discordant monozygotic twins.

Results: In the differential gene expression analysis, it appeared that 32 differentially expressed genes (DEGs) were with a trend of up-regulation in twins with higher BMI when compared to their siblings. Categories of positive regulation of nitric-oxide synthase biosynthetic process, positive regulation of NF-kappa B import into nucleus, and peroxidase activity were significantly enriched within GO database and NF-kappa B signaling pathway within KEGG database. DEGs of *NAMPT*, *TLR9*, *PTGS2*, *HBD*, and *PCSK1N* might be associated with obesity. In the WGCNA, among the total 20 distinct co-expression modules identified, coral1 module (68 genes) had the strongest positive correlation with BMI ($r = 0.56$, $P = 0.04$) and disease status ($r = 0.56$, $P = 0.04$). Categories of positive regulation of phospholipase activity, high-density lipoprotein particle clearance, chylomicron remnant clearance, reverse cholesterol transport, intermediate-density lipoprotein particle, chylomicron, low-density lipoprotein particle, very-low-density lipoprotein particle, voltage-gated potassium channel complex, cholesterol transporter activity, and neuropeptide hormone activity were significantly enriched within GO database for this module. And alcoholism and cell adhesion molecules pathways were significantly enriched within KEGG database. Several hub genes, such as *GAL*, *ASB9*, *NPPB*, *TBX2*, *IL17C*, *APOE*, *ABCG4*, and *APOC2* were also identified. The module eigengene of saddlebrown module (212 genes) was also significantly correlated with BMI ($r = 0.56$, $P = 0.04$), and hub genes of *KCNN1* and *AQP10* were differentially expressed.

Conclusion: We identified significant genes and specific modules potentially related to BMI based on the gene expression profile data of monozygotic twins. The findings may help further elucidate the underlying mechanisms of obesity development and provide novel insights to research potential gene biomarkers and signaling pathways for obesity treatment. Further analysis and validation of the findings reported here are important and necessary when more sample size is acquired.

Keywords: BMI, Differentially expressed genes, Gene module, Hub gene, Monozygotic twins, Obesity, WGCNA

---

* Correspondence: zhangdf1961@126.com
[1]Department of Epidemiology and Health Statistics, Public Health College, Qingdao University, No. 38 Dengzhou Road, Shibei District, Qingdao 266021, Shandong Province, People's Republic of China
Full list of author information is available at the end of the article

Background

Obesity, as a complex disorder mediated by the interplay between genetic and environmental factors [1], has been a public health and policy problem due to its prevalence, costs, and health effects [2]. The therapeutic management of obesity includes lifestyle changes, medications, and surgery. However, the treatment of obesity is challenging because of diverse patient conditions, prolonged and chronic nature of disease, difficulty of maintaining dieting and physical exercise frequently [3–5], limited effectiveness and side effects of the medication [6], and high cost and risk of complications of surgery [7]. Other efforts are focused in the development of novel therapeutics, yet the effectiveness requires to be tested and confirmed [8–10] and the safety requires to be assessed [11]. Therefore, for the purpose of identifying new diagnostic biomarkers and therapeutic targets and developing novel therapeutic strategies which not only produce sufficient weight loss but also lack side effects, further elucidating the molecular mechanisms underlying obesity development is necessary and urgent.

Recently, gene expression profiling analysis has yielded insights into the measurement of alterations in genetic expression patterns, and has facilitated the identification of differentially expressed genes (DEGs) being crucial to obesity. In a study conducted by Roque DR, et al., obesity related genes, such as *LPL*, *IRS-1*, *IGFBP4*, and *IGFBP7*, etc., were found to be upregulated with increasing BMI among endometrial cancer patients [12]. The study of Gruchala-Niedoszytko, M, et al. also found a series of genes (*PI3*, *LOC100008589*, *RPS6KA3*, *LOC441763*, *IFIT1*, and *LOC100133565*) with a different expression that may be related to an increased BMI [13]. And genes of *PGC1-α*, *FNCD5*, and *FGF*, which play roles in adipose tissue development and function, were abundantly expressed in subcutaneous, visceral, and epigastric adipose tissues of extreme obesity patients based on gene expression profiling [14]. However, due to the gene expression profiling analysis merely focused on the effect of individual genes and transcripts, without regard to their correlated patterns of expression and the effect of networks of genes, it may fail to detect important biological pathways or gene-gene interactions related to obesity.

Weighted gene co-expression network analysis (WGCNA) is a systems biology method for analyzing the correlation patterns of large and high-dimensional gene expression data sets [15]. It can be used to find modules of highly correlated genes, correlate module eigengenes (MEs) to external sample traits, calculate module membership (MM) and gene significance (GS), and find intramodular hub genes, etc. WGCNA has yielded novel insights into the molecular aspects to identify candidate biomarkers or therapeutic targets. At present, it has been increasingly applied to analyze various gene expression profiles of hepatocellular carcinoma

[16], pneumocyte senescence induced by thoracic irradiation [17], psoriasis [18], severe asthma [19], coronary artery disease [20], and lung cancer [21], etc. Although widely being employed, the WGCNA has, to our knowledge, not yet been applied to analyze the expression profiles of BMI-discordant monozygotic twin pairs.

While monozygotic twins are characteristic of the genetic similarity and rearing-environment sharing, they show phenotypic discordance for certain complex traits and diseases. Thus, the discordant monozygotic model is becoming a popular and powerful tool for identifying non-genetic contributions to a phenotype variation including subtle difference in gene expression not mediated by cis- or trans-eQTL effects, and for linking environmental exposure to differential epigenetic regulation while controlling for individual genetic make-up [22–24]. Therefore, to reveal the potential molecular mechanisms of obesity, we performed both differential gene expression analysis and WGCNA to analyze the expression profiles of BMI-discordant monozygotic twin pairs. The potentially important DEGs were identified, and the modules correlated with external traits and the hub genes related to BMI were determined. The results may help further elucidate the underlying mechanisms of obesity and provide novel insights to research potential gene biomarkers and signaling pathways for the treatment of obesity.

Methods

Subjects recruitment

The sampling of monozygotic twins was based on the Qingdao Twin Registry at the Qingdao Center for Disease Control and Prevention [25]. Twins were recruited to a clinical investigation after sampling randomly through residence registry and the local disease control network (2012–2013). Written informed consent was obtained from all subjects. We excluded subjects (i) being pregnant or breastfeeding, (ii) undergoing diabetes, (iii) undergoing cardiovascular disease, and (iv) taking any medications within 1 month before participation, and incomplete twin pairs were dropped. The zygosity of twin pairs was determined by DNA testing using 16 short tandem repeat DNA markers. Finally, a total of 7 BMI-discordant monozygotic twin pairs with median age of 52 years (range: 43–65 years) were identified.

For each subject, we took three anthropometric measurements following standard procedures with at least one-minute interval and calculated the mean of these three measurements. Height was measured to the nearest centimeter using a vertical scale with a horizontal moving headboard. And body weight was measured to the nearest 0.1 kg using a standing beam scale. Then BMI was calculated as weight (kg) divided by the square of height (m). Besides, BMI was classified into three classes: Class I,

$18.5 \leq$ BMI < 24 kg/m^2, normal; Class II, $24 \leq$ BMI < 28 kg/m^2, overweight; and Class III, BMI ≥ 28 kg/m^2, obesity. Blood sample was kept frozen at -80 °C for 6 months before sending to routine laboratory testing.

RNA library construction and sequencing and quality control

After total messenger RNA (mRNA) being extracted from whole blood by using TRIzol reagent (Invitrogen, San Diego, USA), the RNA concentration and purity were tested with NanoDrop 2000 Spectrophotometer (Termo Fisher Scientifc, Wilmington, USA) and the RNA integrity was measured with RNA Nano 6000 Assay Kit of Agilent Bioanalyzer 2100 system (Agilent Technologies, Santa Clara, USA).

Then the high-quality RNA was sent to Biomarker Technologies Corporation (Biomarker Technologies Corporation, Beijing, China) for further analysis. The RNA-Seq libraries were constructed with NEBNext UltraTM RNA Library Prep Kit for Illumina (New England Biolabs, Ipswich, USA) following the manufacturer's recommendations as follows: purifying mRNA with NEBNext Poly (A) mRNA Magnetic Isolation Module, randomly fragmenting isolated mRNA, synthesizing and purifying double-stranded cDNAs, selecting fragment sizes using Agencourt AMPure XP system, and obtaining cDNA library by PCR enrichment. At last, we sequenced the prepared cDNA library using the Illumina HiSeq 2500.

To obtain high-quality clean data (Q30 > 85%), quality control for the raw sequencing data was performed by removing reads containing adapter sequences, unknown nucleotides >5%, and low-quality reads. After mapping to the human genome by TopHat2 [26], we estimated the gene expression levels with fragments per kilobase of exon per million fragments mapped (FPKM) value by Cufflinks software [27].

Differential expression analysis

In the differential expression analysis between the 7 BMI-discordant twin pairs using EBSeq [28], the Benjamini-Hochberg method corrected P-value, i.e., False discovery rate (FDR), was estimated to circumvent false positive results which occurred in the multiple tests [29, 30]. The fold change (FC) of the expression values between twins was also calculated. Then DEGs were defined as those met the criteria of |log2FC| > 1and FDR < 0.01.

Weighted gene co-expression network analysis (WGCNA)

The WGCNA package in R is a comprehensive collection of R functions for performing various aspects of weighted correlation network analysis [15, 31]. Based on the expression profiles of 7 monozygotic twin pairs, the network construction, module detection, module and gene selection, calculations of topological properties, visualization, and interfacing with external software package were conducted following the tutorials provided.

Modules identification

Briefly, after calculating Pearson correlations between each gene pair, we established a weighted adjacency matrix by raising the co-expression similarity to a power $\beta = 29$. Subsequently, we constructed the topological overlap matrix (TOM) using correlation expression values [32–34]. Then each TOM was used as input for hierarchical clustering analysis [35], and gene modules (i.e. clusters of genes with high topological overlap) was detected by using a dynamic tree cutting algorithm (deep split = 2, cut height = 0.27). The co-expression module structure was visualized by heatmap plots of topological overlap in the gene network. Relationships among modules were summarized by a hierarchical clustering dendrogram of the eigengenes and by a heatmap plot of the corresponding eigengene network.

Relating modules to external traits

To identify modules that were significantly associated with the traits of interest-BMI, BMI classes, and disease status (obesity versus non-obesity), we correlated the MEs (i.e. the first principle component of a module) [36] with external traits and searched the most significant associations.

Hub genes analysis

The MM was defined as the correlation of gene expression profile with ME. And the GS measure was defined as (the absolute value of) the correlation between gene and external traits. Genes with highest MM and highest GS in modules of interest were natural candidates for further research [37–40]. Thus, the intramodular hub genes were chosen by external traits based GS > 0.2 and MM > 0.8 with a threshold of P-value <0.05 [41]. The gene-gene interaction network was constructed and visualized using VisANT 5.0 [42].

Functional annotation and enrichment analysis

Genes identified in the differential expression analysis and in module of interest in WGCNA were annotated by utilizing BLAST software within the following databases: NCBI nonredundant protein sequences (NR) [43], Clusters of Orthologous Groups (COG) [44], KOG [45], Kyoto Encyclopedia of Genes and Genomes (KEGG) [46, 47], and Gene Ontology (GO) [48, 49]. Subsequently, we drew histogram by mapping GO function of genes in modules of interest to the corresponding secondary features on the background of all genes' GO annotation. The Pearson Chi-Square test was applied to indicate significant relationships between the two input datasets if all the expected counts were greater than or equal to 5 for 2 × 2 matrixes. And the Fisher's exact test was applied if one of the expected counts was less than 5. Then we implemented GO enrichment analysis based on a hypergeometric test [50] and calculated a Fisher's Exact P-Value which was then corrected by Benjamini-Hochberg method. Besides, the KEGG pathways enrichment analysis was conducted by applying the KEGG Orthology-Based Annotation System (KOBAS) utilizing a hypergeometric test [51]. The P-value <0.05 was used as the enrichment cut-off criterion.

Results

Differential expression analysis

A total of 7 BMI-discordant monozygotic twin pairs with median age of 52 years were included for the gene expression profiling analysis (Table 1). The extracted cDNA samples from twins were subjected to sequencing using an Illumina HiSeq2500 platform. The Q30 of each sample was not less than 92.26% and the mapped rate ranged from 87.62% to 93.18% (Additional file 1: Table S1). Under the threshold of |log2FC| > 1 and FDR < 0.01, a range from 360 to 1116 DEGs were identified between co-twin pairs (Table 1). It appeared that 32 DEGs were with a trend of up-regulation in at least three of twins with higher BMI when compared to their siblings (Additional file 2: Table S2). Of these, three genes were found up-regulated in 4 twin pairs, and the others were found up-regulated in 3 twin pairs.

Table 1 The characteristics of the BMI-discordant monozygotic twin pairs (43–65 years) and summary of differentially expressed genes

Subject ID	Height, m	weight, kg	BMI, kg/m²	DEG Set	All DEGs	Up-regulated	Down-regulated
E01	1.56	72	29.6				
E02	1.54	62	26.1	E02_vs_E01	462	418	44
E03	1.62	89	34.1				
E04	1.63	73	27.5	E04_vs_E03	1116	579	537
E05	1.65	53	19.5				
E06	1.6	65.3	25.5	E05_vs_E06	656	356	300
E07	1.73	67.7	22.6				
E08	1.72	75.9	25.7	E07_vs_E08	576	426	150
E09	1.67	81.4	29.2				
E10	1.66	67.2	24.5	E10_vs_E09	360	163	197
E11	1.7	55.9	19.3				
E12	1.7	71.2	24.6	E11_vs_E12	625	187	438
E13	1.55	63	26.2				
E14	1.56	71	29.2	E13_vs_E14	661	426	235

Note: DEG: differentially expressed gene; Up-regulated: the number of up-regulated genes; Down-regulated: the number of down-regulated genes

As the summarized results of enrichment analysis within GO and KEGG databases shown (Table 2), several potentially important findings emerged (Corrected P-value < 0.05), including positive regulation of nitric-oxide synthase biosynthetic process (P = 5.34E-03), positive regulation of NF-kappa B import into nucleus (P = 1.04E-02), peroxidase activity (P = 6.82E-03), and NF-kappa B signaling pathway (P = 4.49E-02). Genes of NAMPT, TLR9, PTGS2, HBD, and BCL2L1 were involved in these significant findings. In addition, PCSK1N gene might also be associated with obesity. We compared previously implicated BMI-related gene expression differences in study of Homuth, G, et al. [52] with ours to validate the findings further. This comparison revealed consistency for positive BMI-associated expression differences, including HBD, XK, SELENBP1, SNCA, LAS2, PLEK2, GLRX5, TMOD1, SLC4A1, BCL2L1, TRIM58, DCAF12, NFIX, BSG, PLVAP, and PCSK1N.

WGCNA

Modules identification

WGCNA was applied to investigate gene sets that were related to traits of interest-BMI, BMI classes, and disease status using the gene expression data of 7 monozygotic twin pairs. After using a dynamic tree cutting algorithm, a total of 20 distinct co-expression modules containing 48 to 9274 genes per module were identified, and 1912 uncorrelated genes were assigned into a grey module which was ignored in the following study (Fig. 1, and Additional file 3: Table S3). The heatmap plot of topological overlap in the gene network is depicted (Fig. 2).

Relating modules to external traits

To understand the physiologic significance of the modules, we correlated the 20 MEs with traits of interest and searched for the most significant associations. According to the heatmap of module-trait correlation (Fig. 3), genes clustered in coral1 module (68 genes) had the strongest positive correlation with BMI (r = 0.56, P = 0.04) and disease status (r = 0.56, P = 0.04), whereas statistically non-significant correlation was found with BMI classes (r = 0.51, P = 0.06). Nevertheless, the ME of saddlebrown module (212 genes) was only significantly correlated with BMI (r = 0.56, P = 0.04). Thus, we would mainly consider coral1 module in the following because this module may indicate external traits more accurately. None of the other modules had a significant association with external traits.

Relationships among modules

To study the relationships among modules and determine their correlation with trait of BMI, we correlated the MEs. The eigengene network using a dendrogram and a heatmap

Table 2 The results of GO and KEGG pathway enrichment analysis for differentially expressed genes with a trend of up-regulation

Category	Term	Gene symbol	Corrected P-value
Gene Ontology term–Biological Process	Porphyrin-containing compound biosynthetic process (GO:0006779)	SPTB; ANK1	1.04E-03
Gene Ontology term–Biological Process	Decidualization (GO:0046697)	BSG; PTGS2	4.53E-03
Gene Ontology term–Biological Process	Positive regulation of nitric-oxide synthase biosynthetic process (GO:0051770)	NAMPT; TLR9	5.34E-03
Gene Ontology term–Biological Process	Positive regulation of NF-kappa B import into nucleus (GO:0042346)	PTGS2; TLR9	1.04E-02
Gene Ontology term–Biological Process	Embryo implantation (GO:0007566)	PTGS2; BSG	2.92E-02
Gene Ontology term–Biological Process	Adult locomotory behavior (GO:0008344)	TMOD1; SNCA	4.20E-02
Gene Ontology term–Cellular Component	Cortical cytoskeleton (GO:0030863)	TMOD1; SLC4A1; ANK1	1.88E-04
Gene Ontology term–Cellular Component	Spectrin-associated cytoskeleton (GO:0014731)	ANK1; SPTB	1.30E-03
Gene Ontology term–Cellular Component	Basolateral plasma membrane (GO:0016323)	ANK1; SLC4A1; TLR9	3.62E-02
Gene Ontology term–Molecular Function	Ankyrin binding (GO:0030506)	SLC4A1; SPTB	5.97E-03
Gene Ontology term–Molecular Function	Peroxidase activity (GO:0004601)	PTGS2; HBD	6.82E-03
Gene Ontology term–Molecular Function	Structural constituent of cytoskeleton (GO:0005200)	ANK1; TUBB2A; SPTB	9.49E-03
Gene Ontology term–Molecular Function	Kinesin binding (GO:0019894)	SNCA; KLC3	1.43E-02
KEGG pathway	Small cell lung cancer (ko05222)	PTGS2; BCL2L1	3.10E-02
KEGG pathway	NF-kappa B signaling pathway (ko04064)	PTGS2; BCL2L1	4.49E-02

plot are depicted in Fig. 4. The dendrogram (Fig. 4a) indicated that the coral1 and saddlebrown modules were highly correlated, and trait of BMI fell within the meta-module grouping together the two modules. The heatmap plot (Fig. 4b) showed the detailed eigengenes adjacencies of all modules and trait of BMI.

Functional annotation and enrichment analysis for coral1 module

In order to provide an interpretation of the biological mechanism associated with the genes clustered in module of interest–coral1, we conducted functional annotation (Additional file 4: Table S4) and enrichment analysis.

Three main annotated categories-biological process, cellular component, and molecular function were obtained in GO database (Fig. 5, and Additional file 5: Table S5). The proportion of genes in coral1 module increased significantly in certain subgroups, including single-organism process ($P = 0.024$), multicellar organismal process ($P = 0.004$), developmental process ($P = 0.009$), localization ($P = 0.002$), signaling ($P = 0.005$), extracellar region ($P = 0.009$), and transporter activity ($P = 0.023$).

As the summarized results of enrichment analysis within GO database shown (Table 3), several potentially important findings emerged (Corrected P-value < 0.05). In the biological processes, categories of positive regulation of phospholipase activity ($P = 2.91E-03$), high-density lipoprotein particle clearance ($P = 4.36E-03$), chylomicron remnant clearance ($P = 4.36E-03$), reverse cholesterol transport ($P = 2.24E-02$), and positive regulation of axon extension ($P = 2.61E-02$) were significantly enriched. Among the 6 enrichment categories in the cellular component, intermediate-density lipoprotein particle ($P = 2.74E-03$), chylomicron ($P = 9.97E-03$), low-density lipoprotein particle ($P = 1.89E-02$), and very-low-density lipoprotein particle ($P = 2.74E-02$) were related to lipid transport and metabolism. And categorie of voltage-gated potassium channel complex ($P = 1.43E-02$) may be potentially involved in the regulating energy homeostasis. In the molecular function, the categories of cholesterol transporter activity ($P = 9.72E-03$) and neuropeptide hormone activity ($P = 1.98E-02$) should also be highlighted.

As shown in Additional file 6: Table S6, the KEGG annotation results were classified according to KEGG pathway classification. Two pathways of alcoholism and cell adhesion molecules (CAMs) were significantly enriched in KEGG database (Table 3). And the COG function classification results are shown in Additional file 7: Table S7.

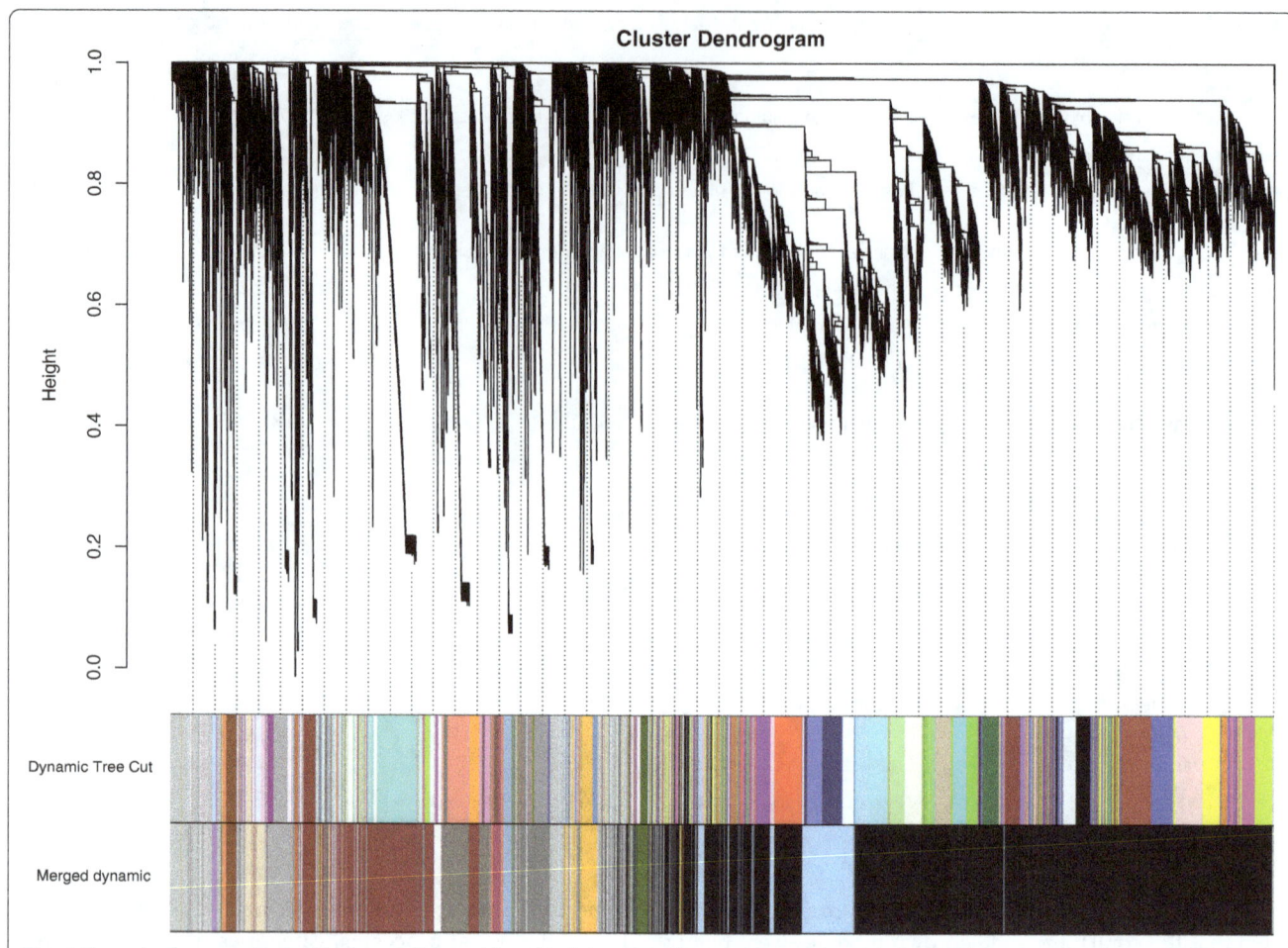

Fig. 1 Gene dendrogram obtained by average linkage hierarchical clustering. The color row underneath the dendrogram shows the assigned original module and the merged module

Hub genes analysis in coral1 module

Figure 6 shows the scatterplots of GS for traits of BMI, BMI classes, and disease status versus MM in coral1 module. MM and GS for BMI (Fig. 6a), BMI classes (Fig. 6b), and disease status (Fig. 6c) exhibited very significant positive correlations, implying that the most important (central) elements of coral1 module also tended to be highly correlated with these external traits. The identified 21 hub genes (Additional file 8: Table S8) included *GAL, ASB9, KCNT1, NPPB, TBX2, KCNK15, IL17C, APOE, LBX1, LRRC38, LINGO1, ABCG4, LCN15, RFLNA, SOX18, C1orf146, APOC2, PRSS29P, LOC102724223, C7orf71,* and *IGKV1D-17.* The visualized plot of the gene-gene interaction network in coral1 module is shown in Fig. 7.

The 21 hub genes were involved in several enriched functional items (Table 3), including high-density lipoprotein particle clearance, chylomicron remnant clearance, phospholipid efflux, reverse cholesterol transport, chylomicron, voltage-gated potassium channel complex, very-low-density lipoprotein particle, and cholesterol

transporter activity, most of which were associated with lipid transport and metabolism. None of the 68 genes in coral1 module was identified as DEGs.

Saddlebrown module

In the functional annotation analysis within GO database, the proportion of genes in saddlebrown module increased in subgroups of extracellular region ($P = 0.002$) and extracellular region part ($P = 0.019$) (Additional file 9: Figure S1, and Additional file 10: Table S9). The categories of structural constituent of eye lens ($P = 1.14$E-02) and troponin T binding ($P = 2.75$E-02) were significantly enriched in the molecular function, whereas no categories were significantly enriched in biological process and cellular component. The results of KEGG pathway classification and COG function classification are shown in Additional file 11: Table S10 and Additional file 12: Table S11, respectively. BMI based GS and MM exhibited a very significant correlation in saddlebrown module (Additional file 13: Figure S2), and hub genes of *KCNN1, CNN1,* and *AQP10* were identified as DEGs.

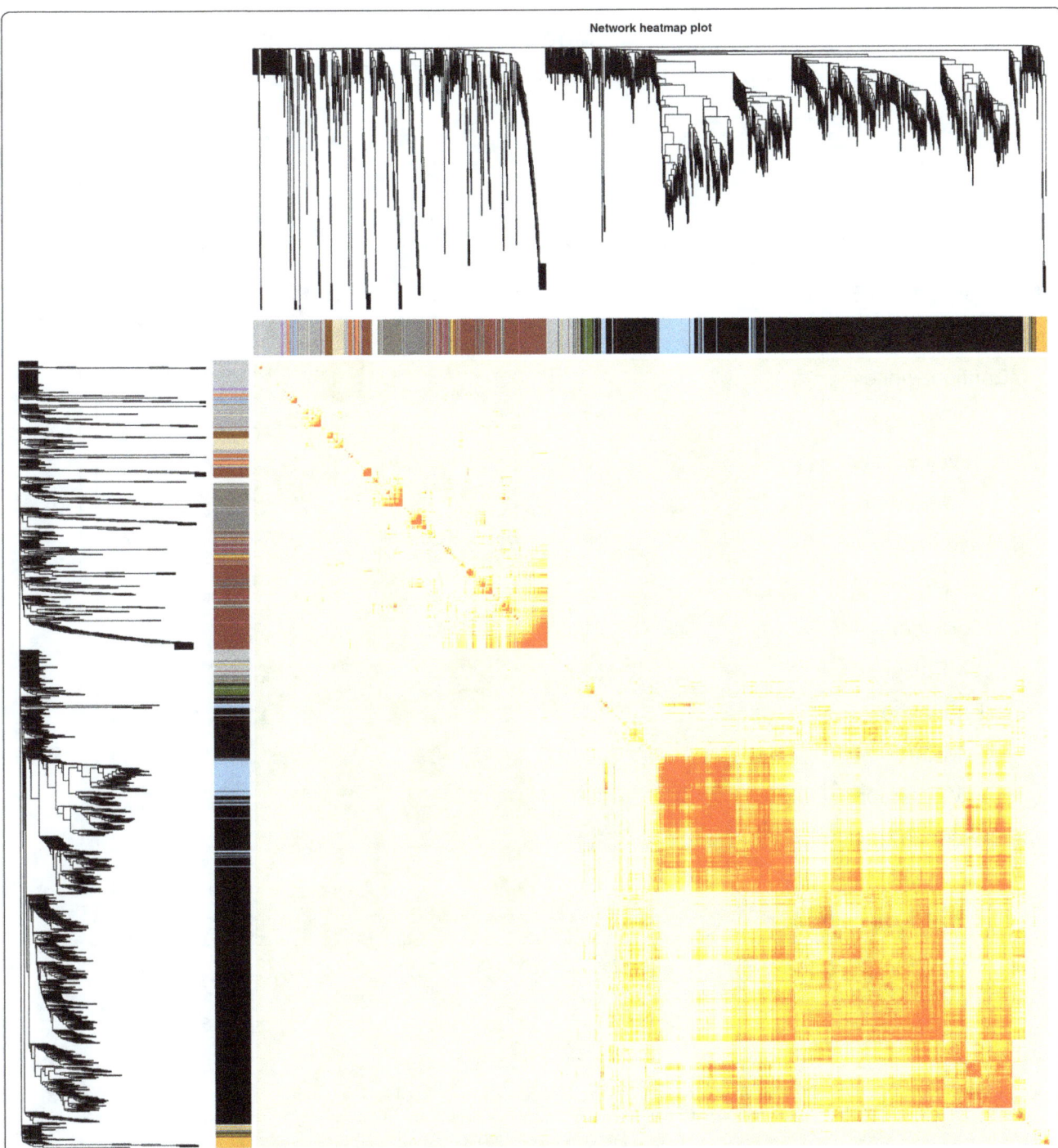

Fig. 2 Heatmap plot of topological overlap in the gene network. In the heatmap, each row and column corresponds to a gene, light color denotes low topological overlap, and progressively darker red denotes higher topological overlap. Darker squares along the diagonal correspond to modules. The gene dendrogram and module assignment are shown along the left and top

Discussion

In the differential expression analysis based on the gene expression data of 7 BMI-discordant monozygotic twin pairs, we identified 32 genes with a trend of up-regulation in twins with higher BMI when compared to their siblings. Several potentially important enrichment findings emerged, including positive regulation of nitric-oxide synthase biosynthetic process, positive regulation of NF-kappa B import into nucleus, peroxidase activity, and NF-kappa B signaling pathway (Table 2). And up-regulated genes-*NAMPT*, *TLR9*, *PTGS2*, *HBD*, and *PCSK1N* might be associated

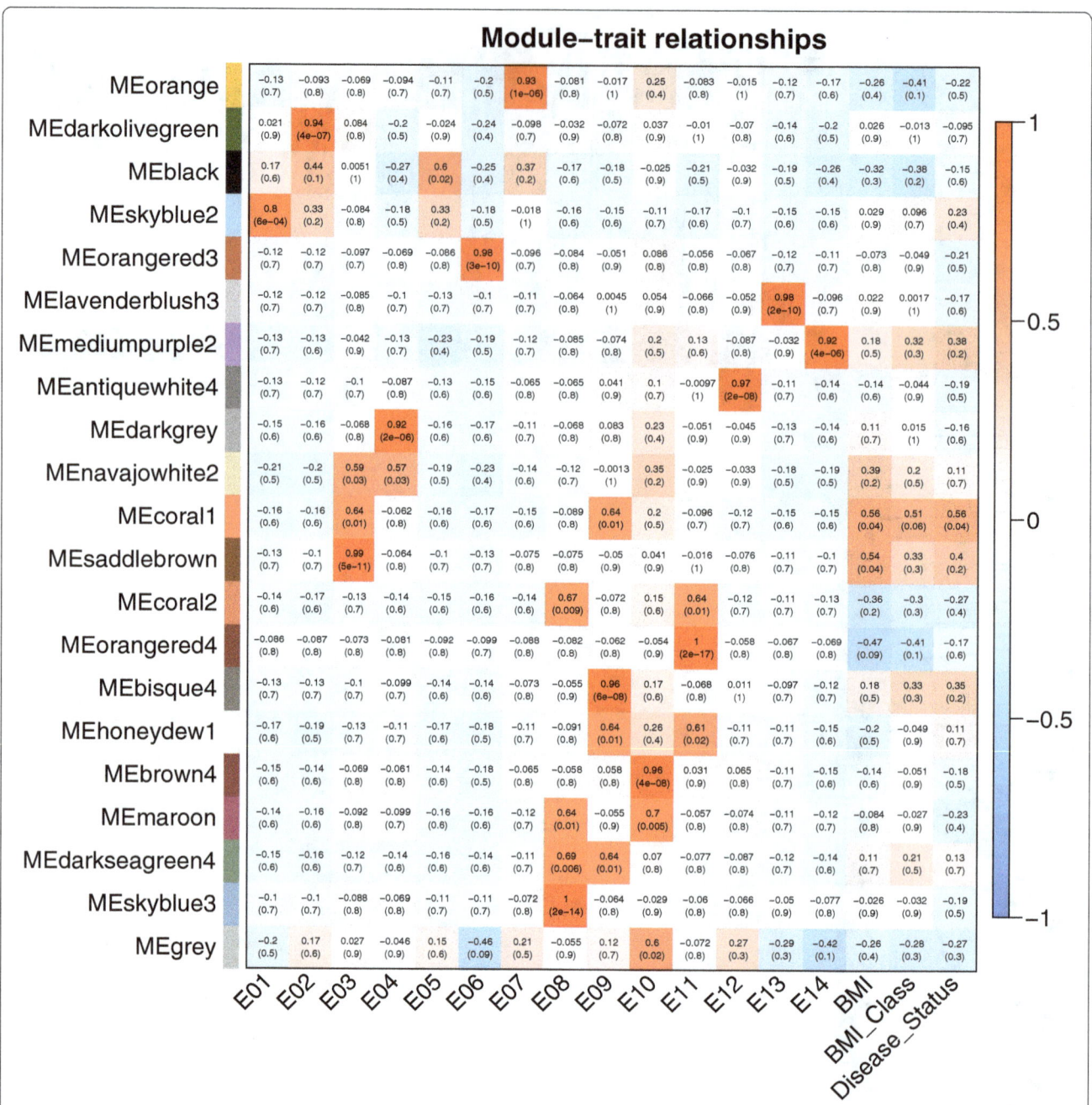

Fig. 3 Relationships of consensus module eigengenes and external traits. Each row in the table corresponds to a consensus module, and each column to a sample or trait. Numbers in the table report the correlations of the corresponding module eigengenes and traits, with the *P*-values printed below the correlations in parentheses. The table is color coded by correlation according to the color legend

with obesity risk. In addition, we also applied WGCNA to quantitatively analyze the interconnectedness of gene expression data and assessed the importance of genes within the networks. Among the 20 distinct co-expression modules identified, genes clustered in coral1 module had the strongest positive correlation with BMI and disease status (Fig. 3), indicating that the highly co-expressed genes in this module had potential biological significance. Functional enrichment analysis revealed several significant enrichments

of BMI-related categories for coral1 module. Importantly, several hub genes were strongly related to lipid transport and metabolism (Table 3) and may be particularly valuable for identifying the candidate biomarkers and therapeutic targets for obesity assessed by BMI. Besides, the ME of saddlebrown module was also significantly correlated with BMI (Fig. 3) and 3 hub genes were identified as DEGs.

Obesity is a complex disease under the control of both genetic and environmental factors through the interface

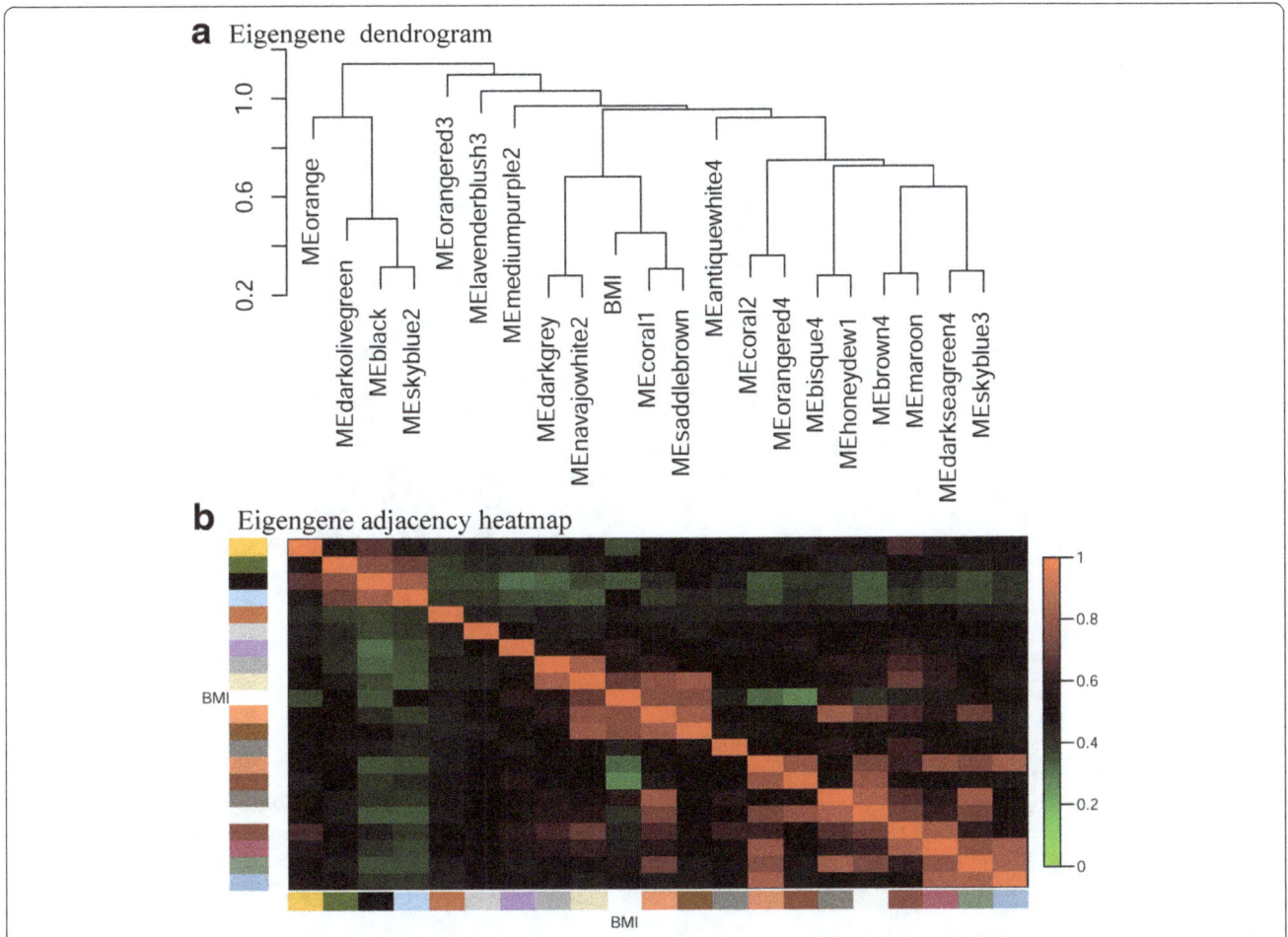

Fig. 4 Relationships among modules. **a** Hierarchical clustering of module eigengenes that summarize the modules found in the clustering analysis. Branches of the dendrogram (the meta-modules) group together eigengenes that are positively correlated. **b** Heatmap plot of the adjacencies in the eigengene network including the trait of interest-BMI. Each row and column in the heatmap corresponds to one module eigengene (labeled by color) or BMI. In the heatmap, red represents high adjacency, while blue color represents low adjacency. Squares of red color along the diagonal are the meta-modules

of epigenetics, where different combinations of genetic and epigenetic variations can lead to a common phenotype. Considering this, we would not necessarily expect each of the 7 BMI-discordant monozygotic twin pairs to present exactly the same series of gene aberrations, and the stringency of the criterion on the commonality of gene changes was relaxed. In the differential expression analysis, it appeared that 32 genes were with a trend of up-regulation in at least three of twins with higher BMI when compared to their siblings (Additional file 2: Table S2). Besides, four potentially important enrichment findings emerged and 5 up-regulated genes associated with obesity risk were identified as follows (Table 2).

Nitric oxide (NO), whose production is mostly through the action of the nitric oxide synthase (NOS) family of enzymes, is emerging as a central regulator of energy metabolism and body composition. The isoform of inducible nitric oxide synthase (iNOS)-derived NO can promote

insulin resistance and inflammation in key peripheral tissues such as liver, skeletal muscle, and adipose tissue. In addition, iNOS may affect glucose homeostasis. Thus, the iNOS isoform appears to promote deleterious changes in metabolism [53]. Considering this, the two up-regulated genes (*NAMPT* and *TRL9*) involved in the category of positive regulation of nitric-oxide synthase biosynthetic process should be considered notably (Table 2). And it is indicated that the protein of *NAMPT* gene can directly activate pathways leading to iNOS induction [54].

It has been revealed that a characteristic feature of obesity linking it to insulin resistance is the presence of chronic low-grade inflammation which is indicative of activation of the innate immune system. The IKK/NF-κB pathway is a well-known inflammatory signaling pathway involved in the pathogenesis of obesity [55, 56], and the two genes-*PTGS2* and *TLR9* involved in this enrichment term should also be focused (Table 2). In addition, the

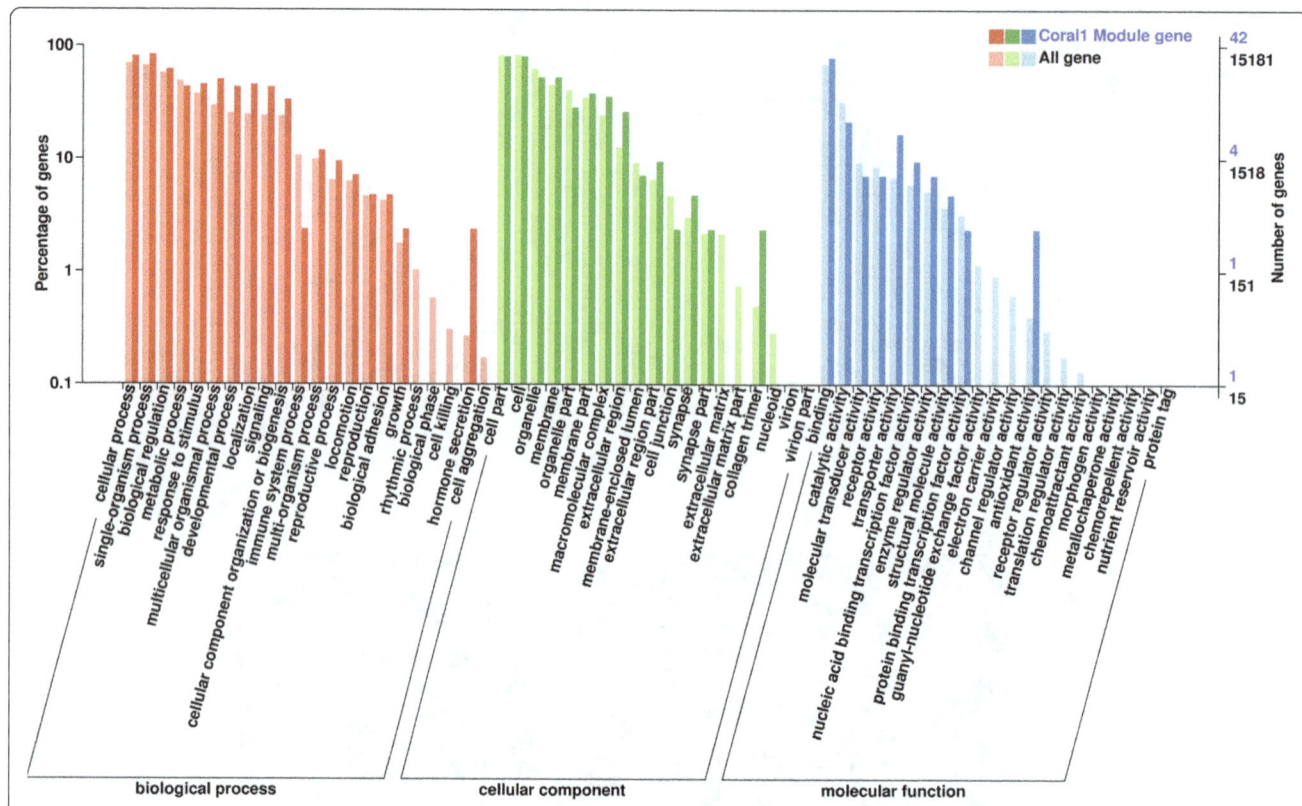

Fig. 5 GO classification in coral1 module. Annotation statistics of genes in the secondary node of GO. The horizontal axis shows secondary nodes of three categories in GO. The vertical axis displays the percentage of annotated genes versus the total gene number. The left columns display annotation information of the total genes and the right columns represent annotation information of the genes clustered in coral1 module

Table 3 The results of GO and KEGG pathway enrichment analysis for genes clustered in coral1 module

Category	Term	Gene Symbol	Corrected P-value
Gene Ontology term–Biological Process	Positive regulation of phospholipase activity (GO:0010518)	CYR61; APOC2*	2.91E-03
Gene Ontology term–Biological Process	High-density lipoprotein particle clearance (GO:0034384)	APOC2*; APOE*	4.36E-03
Gene Ontology term–Biological Process	Chylomicron remnant clearance (GO:0034382)	APOC2*; APOE*	4.36E-03
Gene Ontology term–Biological Process	Phospholipid efflux (GO:0033700)	APOE*; APOC2*	1.30E-02
Gene Ontology term–Biological Process	Reverse cholesterol transport (GO:0043691)	APOC2*; APOE*	2.24E-02
Gene Ontology term–Biological Process	Positive regulation of axon extension (GO:0045773)	APOE*; RAB25	2.61E-02
Gene Ontology term–Cellular Component	Intermediate-density lipoprotein particle (GO:0034363)	KCNJ1; KCNT1*; KCNA1	2.74E-03
Gene Ontology term–Cellular Component	Chylomicron (GO:0042627)	APOE*; APOC2*	9.97E-03
Gene Ontology term–Cellular Component	Voltage-gated potassium channel complex (GO:0008076)	TBX2*; LBX1*; SOX18*	1.43E-02
Gene Ontology term–Cellular Component	Low-density lipoprotein particle (GO:0034362)	TRH; GAL*	1.89E-02
Gene Ontology term–Cellular Component	Very-low-density lipoprotein particle (GO:0034361)	APOC2*; APOE*	2.74E-02
Gene Ontology term–Cellular Component	Dendrite (GO:0030425)	MYO1A; NKD2	3.63E-02
Gene Ontology term–Molecular Function	Cholesterol transporter activity (GO:0017127)	ABCG4*; APOE*	9.72E-03
Gene Ontology term–Molecular Function	Neuropeptide hormone activity (GO:0005184)	GAL*; TRH	1.98E-02
KEGG pathway	Alcoholism (ko05034)	SHC2; GRIN2C; HIST3H2BB	1.79E-02
KEGG pathway	Cell adhesion molecules (CAMs) (ko04514)	SDC1; CADM3; CLDN6	2.00E-02

Note: * represents the hub genes in coral1 module

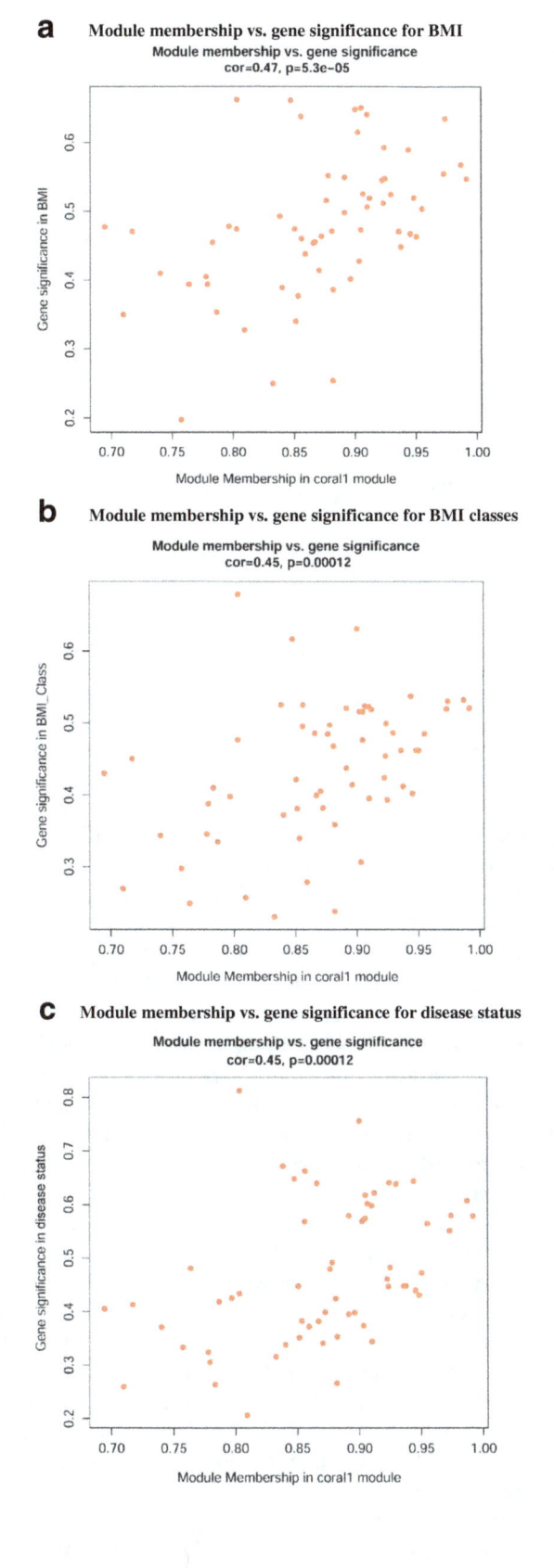

a Module membership vs. gene significance for BMI

b Module membership vs. gene significance for BMI classes

c Module membership vs. gene significance for disease status

Fig. 6 Scatterplots of gene significance (GS) for external traits versus module membership (MM) in the coral1 module. MM and GS for BMI, BMI classes, and disease status exhibit very significant correlations, implying that the most important (central) elements of coral1 module also tend to be highly correlated with these external traits. **a** Module membership vs. gene significance for BMI; (**b**) Module membership vs. gene significance for BMI classes; and (**c**). Module membership vs. gene significance for disease status

protein encoded by *PTGS2* gene is indicated to be linked with energy homeostasis and metabolic processes based on a cohort of children presenting with syndromic obesity [57]. Even though both these two genes were enriched to the NF-kappa B signaling pathway in KEGG database (Table 2), *PTGS2* gene was involved in the inflammation process while *BCL2L1* gene might be related to survival process.

In mammals, the peroxidases comprise 8 glutathione peroxidases (GPx1–GPx8) so far identified. Too much data regarding the association between obesity and GPx1, GPx3, GPx4, and GPx7 has been reported [58]. Thus, the two genes of *PTGS2* and *HBD* could be regard as the candidates for further research (Table 2).

Moreover, SNPs in or near *PCSK1* loci may also contribute to obesity risk [59, 60]. The associations with BMI for other DEGs should be explored further.

To validate the identified DEGs further, we compared previously implicated BMI-related gene expression differences [52] with ours. This comparison revealed consistency for positive BMI-associated expression differences including *HBD*, *XK*, *SELENBP1*, *SNCA*, *LAS2*, *PLEK2*, *GLRX5*, *TMOD1*, *SLC4A1*, *BCL2L1*, *TRIM58*, *DCAF12*, *NFIX*, *BSG*, *PLVAP*, and *PCSK1N*. Two consistent genes (*SNCA* and *DCAF12*) were also revealed when compared with the BMI-related genes by the Data-driven Expression Prioritized Integration for Complex Traits (DEPICT) method.

In the WGCNA, the proportion of genes in coral1 module increased significantly in subgroups of developmental process, signaling, extracellar region, and transporter activity (Fig. 5, and Additional file 5: Table S5), indicating that these functions may be associated with metabolism and accelerated growth and development of obesity individuals. Notably, GO enrichment analysis also provided more significant results with more biological meanings as follows (Table 3).

Obese subjects frequently suffer atherogenic dyslipidemia which is commonly manifested as elevated plasma free fatty acids, triglycerides (TG) and very low-density lipoprotein (VLDL) levels, decreased high-density lipoprotein cholesterol (HDL-C) levels, and abnormal low-density lipoprotein cholesterol (LDL-C) [61, 62]. Our study suggested that categories involved in lipid metabolism and transport were significantly enriched within the coral1 module, including high-density lipoprotein

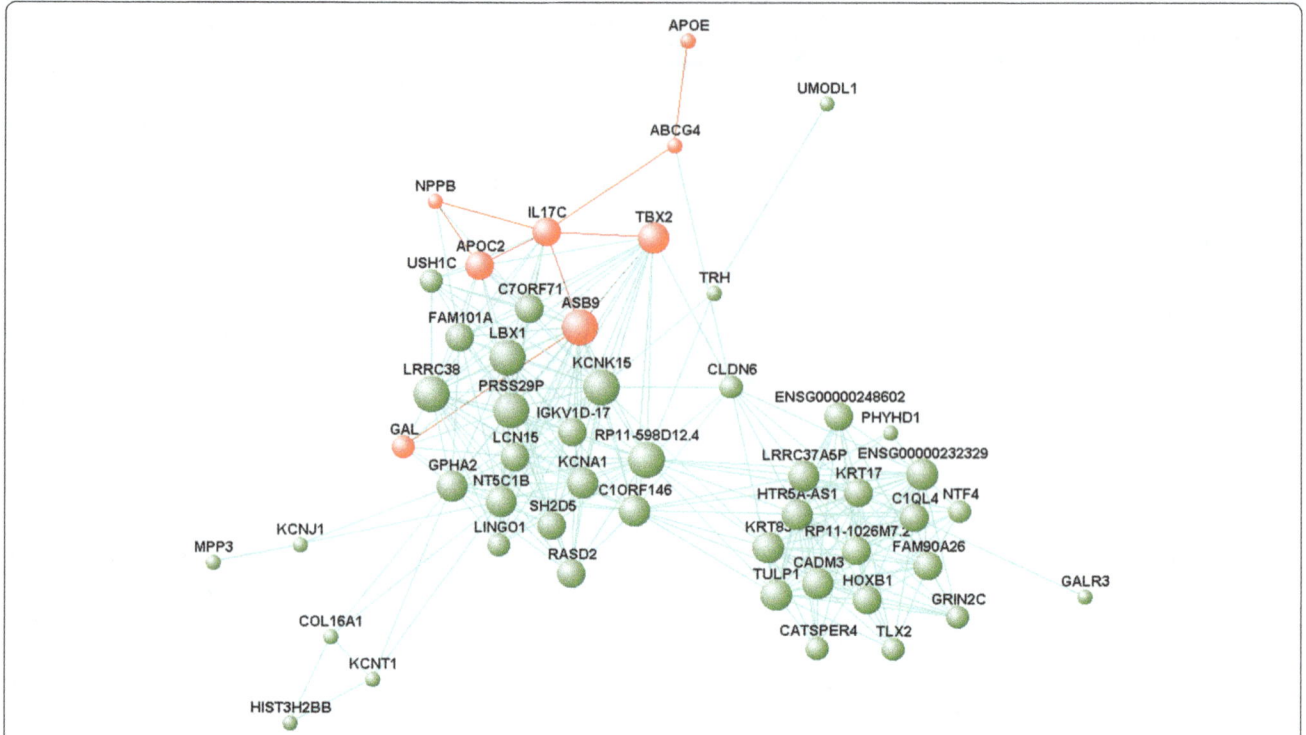

Fig. 7 Interaction of gene co-expression patterns by VisANT 5.0 in the coral1 module. The node size and edge number are proportional to degree and connection strength, respectively. Eight red nodes indicate the hub genes potentially related to BMI in the coral1 module. Among the 8 genes, *GAL*, *APOE*, *APOC2*, and *NPPB* have been demonstrated to be associated with obesity and the others would be associated with obesity as the related works suggested

particle clearance, reverse cholesterol transport, low-density lipoprotein particle, cholesterol transporter activity, chylomicron, chylomicron remnant clearance, very-low-density lipoprotein particle, and intermediate-density lipoprotein particle (Table 3).

Category of positive regulation of phospholipase activity was also enriched in the coral1 module (Table 3). Phospholipids were identified as potential biomarkers for obesity [63]. And it was reported that members of the phospholipase A2 (PLA2) family of enzymes, such as PLA2G1B [64], PLA2G5, and PLA2G2E [65], can serve a distinct role in generating active lipid metabolites, which can promote inflammatory metabolic diseases including obesity [66, 67]. In addition, AdPLA enzyme in white adipose tissue can function as a regulator of lipolysis through increasing prostaglandin E2 (PGE2) formation and decreasing intracellular cAMP [68].

The hypothalamic peptides, such as neuropeptide Y (NPY) and melanin concentrating hormone (MCH) [69], and the peripheral neuropeptides, such as hormone leptin [70], play important roles in regulating food intake and maintaining energy balance [71]. Normally, a dynamic equilibrium exists between orexigenic peptides and anorexigenic peptides [70]. And after receiving stimulus information of neural signal, hormone signal, and metabolites, etc., the hypothalamus appestat maintains the

dynamic equilibrium of energy by neuro-humoral response. Therefore, the enriched category of neuropeptide hormone activity may also exert a significant meaning regarding obesity (Table 3).

Another significant GO enriched category was voltage-gated potassium channel (Kv) complex (Table 3). Studies had suggested the relation of subtype-Kv1.3 to insulin sensitivity and the participation of Kv1.3 in regulating energy homeostasis and body weight [72, 73]. Hence, Kv1.3 may be a putative and promising pharmacological target for the treatment of obesity, type II diabetes mellitus and related metabolic diseases [72].

Two pathways of alcoholism and cell adhesion molecules (CAMs) were significantly enriched in KEGG database (Table 3). A growing body of literatures indicate that overlapping central pathways may be involved with uncontrolled eating and excessive ethanol drinking [74]. And emerging link between familial alcoholism risk and obesity in women and possibly in men is identified in recent years [75]. Furthermore, some genetic variants are associated with both alcohol dependence and obesity [76, 77]. Therefore, the genes involved in the alcoholism pathway may be used as potential links between alcoholism and obesity, and as promising targets for controlling ethanol abuse and food intake. As for the CAMs, a review concluded that anthropometric indicators, body composition and eating pattern positively modulate the subclinical inflammation of obesity through reducing CAMs and chemokines [78]. Moreover, a recent study also identified the relationship of adiposity to several CAMs [79].

We visualized the gene-gene interaction network in coral1 module to obtain an insight on the hidden mechanisms (Fig. 7), and a total of 21 hub genes were identified (Additional file 8: Table S8). The hub genes were involved in various gene families and might serve as candidates for additional mechanistic studies and therapeutic interventions. Hub genes of *GAL*, *APOE*, *APOC2*, and *NPPB* have been demonstrated to be associated with obesity as follows: (I) *GAL*: Galanin peptides, as the protein for *GAL*, is undoubtedly involved in the regulation of food intake and body weight. It has been identified that both central galanin and peripheral galanin can affect appetite, food intake and body weight of animals, and the latter can also affect gastrointestinal motility and brown adipose tissue activity [80, 81]. Particularly, newly discovered galanin-like peptide (GALP) may play a role in boosting appetite, body weight, and obesity [80, 82]. Overall, galanin and its receptors may serve as a novel anti-obesity strategy in the future. (II) *APOE*: ApoE synthesized by adipocyte is a polymorphic glycoprotein in humans, and is a major constituent of HDL, VLDL, and remnant lipoproteins (RLPs). ApoE was identified playing an important role in the development of obesity and insulin resistance in experimental mouse models [83], and the mutation in *APOE* was involved in lipid metabolism [84] and lipid levels [85] in population studies. Moreover, an equally vital role in adipocyte triglyceride accumulation and VLDL-induced adipogenesis was summarized [86]. Overall, it has been identified that *APOE* expression serves as a key peripheral contributor to the development of obesity and related metabolic dysfunctions [83, 87]. (III) *APOC2*: Apolipoprotein C-II (ApoC-II), as a constituent of chylomicrons, VLDL, LDL and HDL, is a cofactor for lipoprotein lipase, which can hydrolyze TG. The gene *APOC2* mutation can result in hypertriglyceridemia, which is one of the main characteristics of obese subjects [88]. Besides, an excess of ApoC-II is related to increase of triglyceride-rich particles and alterations in HDL particle distribution [89]. (IV) *NPPB*: A growing body of evidence indicates that the natriuretic peptides (NPs) system holds the potential to be amenable to therapeutical intervention against obesity. Vila, G, et al. demonstrate that B-Type Natriuretic Peptide (BNP) plays an important role in reducing circulating ghrelin concentrations, decreasing hunger, and increasing feeling of satiety in healthy individuals [90]. Moreover, the function of enhancing lipolysis and energy expenditure, and modulating adipokine release and food intake is also identified [91]. In addition, one recent review emphasized the ability of NPs to regulate body weight and energy homeostasis by driving lipolysis, facilitating beiging of adipose tissues, and promoting lipid oxidation and mitochondrial respiration [92]. Moreover, another review drew the similar conclusion [93].

Although there was no strong indication that *TBX2*, *ASB9*, *IL17C*, and *ABCG4* were the causal variant of obesity in the population, studies showed that these genes may also be part of the multifactorial etiology of this complex condition as follows: (I) *TBX2*: The results of a prospective cohort on the associations of menarche-related genetic variants with pubertal growth in adolescents indicated that SNPs (rs757608) near *TBX2* is associated with a rapid weight gain [94]. (II) *ASB9*: It indicated that overexpression of *ASB9* can induce ubiquitination of ubiquitous mitochondrial creatine kinase (uMtCK) [95] in a specific, SOCS box-independent manner [96]. The intracellular creatine kinase (CK) system may be involved in the storage of fat and the development of obesity [97–99]. Besides, one cross-sectional study recently provided further evidence that CK may play a role in the pathophysiology of obesity and serve as a marker to identify individuals at risk for obesity [100]. (III) *IL17C*: Obesity, in some sense, is considered to be an inflammatory predisposition. And interleukin-17 (IL-17) may impact adipose tissue due to the association with induction of tissue inflammation. Particularly, the potential implications of IL-17 in relation to obesity has been consolidated by Ahmed, M and Gaffen, SL [101]. And one study also suggested a linear negative association between IL-17 and visceral adipose tissue thickness [102]. However, the exact role of IL-17C in obesity remains to be explored. (IV) *ABCG4*: An additional hub gene that should be further investigated is *ABCG4*, one member of the *ABCG* family. Studies indicated that ABCG4 promotes cholesterol efflux from cells to HDL [103, 104].

Even though no sufficient studies showed the association of two genes of *KCNT1* and *LBX1* with obesity, the results of functional annotation and enrichment analysis indicated that they were involved in intermediate-density lipoprotein particle and voltage-gated potassium channel complex, respectively. Thus, they may also be regarded as the targets for etiology research of obesity. Other hub genes in coral1 module were of unknown function in terms of obesity currently, whereas they may also be interesting potential candidates to be future researched and validated.

Among the hub genes in saddlebrown module, *KCNN1*, *CNN1*, and *AQP10* were up-regulated with increasing BMI in twins. (I) *KCNN1*: The lipotoxicity in morbid obesity can gradually impair insulin action in the liver and muscle, aggravating insulin resistance [105], and Ye, J proposed an energy-based concept of insulin resistance, in which insulin resistance is a result of energy surplus in cells [106]. The protein of gene *KCNN1*–small conductance calcium-activated potassium channel protein 1, can serve as a key regulator of excitability and endocrine function in beta cells [107]. (II) *AQP10*: Aquaglyceroporins, such as AQP10, represent novel additional pathways for the transport of glycerol in human adipocytes [108], and the deregulation in the expression of aquaglyceroporins in adipose tissue is associated with human obesity [109, 110].

Several strengths must be noticed in our study. First, gene expression levels may be under the effect of subjects' genetic background, gender, age, and environmental exposures as well as by some experimental variables related to clinical sampling, processing, and data analysis. However, the discordant monozygotic model, which is characteristic of the genetic similarity and rearing-environment sharing, is becoming a popular and powerful tool for identifying non-genetic contributions to a phenotype variation including difference in gene expression. Hence, our results of WGCNA, based on the expression data generated from BMI-discordant monozygotic model, may be more credible. Another strength of our study was that the WGCNA provides information on the correlated patterns of expression and the effect of networks of genes, which is useful for detecting important biological pathways or gene-gene interactions related to obesity. Specifically, a set of genes sharing similar functions and correlated to one another in coral1 and saddlebrown modules were identified in our study, some of which have already been verified to play efficient roles in obesity.

Nevertheless, our study has potential limitations as well. First, our study was with small sample size and limited statistical power resulting from the challenges of identifying and recruiting qualified monozygotic twins discordant for BMI. The BMI-discordant monozygotic model, however, helps to mitigate confounding factors associated with genetic polymorphisms in studies of unrelated human subjects and to identify non-genetic contributions to a phenotype variation including difference in gene expression. Besides, we had identified significant genes and specific modules potentially related to BMI. It's still important and necessary to validate our findings when more sample size is acquired. Second, we couldn't validate our results with an external and independent dataset because of lacking public BMI-discordant monozygotic dataset with adequate size currently. However, we compared previously implicated BMI-related gene expression differences with our findings, and 16 consistent positive BMI-associated findings were revealed. Third, some genes may be involved in multiple processes/functions which require different gene sets. However, it was

difficult to characterize such gene interactions because of the impossibility of forming overlapping modules by WGCNA. Fourth, as in any other studies based on microarray technology, changes in protein levels may not reflect similar changes in mRNA levels accurately because posttranslational modification also acts importantly in controlling biological processes. Hence, it may be necessary to validate our results by other techniques. And fifth, in saddlebrown module, the 3 hub genes also identified as DEGs were differentially expressed in just one twin pair. More studies are needed to confirm these results.

Conclusions

In summary, we identified 32 DEGs with a trend of upregulation in twins with higher BMI when compared to their siblings in the differential expression analysis and determined one module most positively correlated with BMI and several hub genes in the WGCNA. The potentially significant genes and pathways correlated with BMI identified in our analysis may help further elucidate the molecular mechanisms underlying obesity development and provide novel insights regarding future prognostic and therapeutic approaches. Further analysis and validation of the candidate biomarkers of obesity reported here are necessary, including those that have not yet been definitely identified.

Additional files

Additional file 1: Table S1. Summary of the sequencing reads and the mapped results for the 7 monozygotic twin pairs (DOCX 17 kb)

Additional file 2: Table S2. The summary of differentially expressed genes with a trend of up-regulation in BMI-discordant monozygotic twin pairs and the corresponding functional annotation results (XLSX 23 kb)

Additional file 3: Table S3. Module assignments for genes in network of WGCNA (XLSX 302 kb)

Additional file 4: Table S4. The result for functional annotation within databases of NR, COG, KOG, KEGG, and GO for genes clustered in coral1 module (XLSX 24 kb)

Additional file 5: Table S5. GO classification of genes in coral1 module (XLSX 13 kb)

Additional file 6: Table S6. KEGG categories of genes in coral1 module (XLSX 12 kb)

Additional file 7: Table S7. COG function classification in coral1 module (XLSX 12 kb)

Additional file 8: Table S8. The hub genes found by the criterion of BMI based GS > 0.2 and MM > 0.8 with a threshold of *P*-value <0.05 in coral1 module (DOCX 16 kb)

Additional file 9: Figure S1. GO classification in saddlebrown module. Annotation statistics of genes in the secondary node of GO. The horizontal axis shows secondary nodes of three categories in GO. The vertical axis displays the percentage of annotated genes versus the total gene number. The left columns display annotation information of the total genes and the right columns represent annotation information of the genes clustered in saddlebrown module. (TIFF 777 kb)

Additional file 10: Table S9. GO classification of genes in saddlebrown module (XLSX 14 kb)

Additional file 11: Table S10. KEGG categories of genes in saddlebrown module (XLSX 18 kb)

Additional file 12: Table S11. COG function classification in saddlebrown module (XLSX 15 kb)

Additional file 13: Figure S2. Scatterplots of BMI based gene significance (GS) versus module membership (MM) in the saddlebrown module. GS for BMI and MM exhibit a very significant correlation, implying that the most important (central) elements of saddlebrown module also tend to be highly correlated with BMI trait (TIFF 168 kb)

Abbreviations
BMI: Body mass index; DEGs: Differentially expressed genes; FC: Fold change; FDR: False discovery rate; GO: Gene Ontology; GS: Gene significance; KEGG: Kyoto Encyclopedia of Genes and Genomes; MM: Module membership; WGCNA: Weight gene co-expression network analysis

Acknowledgements
Not applicable.

Funding
This study was supported by the grants from the Natural Science Foundation of China (81773506), the Entrepreneurial Innovation Talents Project of Qingdao City (13-CX-3), and the EFSD/CDS/Lilly Programme award (2013).

Authors' contributions
DFZ and WJW contributed to the conception and design. HPD and CSX organized the collection of samples and phenotypes. WJW and WJJ contributed to sample data and sequencing data management. QHT and SXL analyzed the sequencing data and WJW, YLW, and LH interpreted the analysis results. WJW, DFZ and WJJ drafted the manuscript, HPD, YLW, and CSX were involved in the discussion, and QHT, SXL, DFZ, and LH revised it. All the authors read the manuscript and gave the final approval of the version to be published. All the authors agreed to be accountable for all aspects of the work.

Competing interests
The authors declare that they have no competing interests.

Author details
[1]Department of Epidemiology and Health Statistics, Public Health College, Qingdao University, No. 38 Dengzhou Road, Shibei District, Qingdao 266021, Shandong Province, People's Republic of China. [2]Department of Biochemistry, Medical College, Qingdao University, No. 38 Dengzhou Road, Shibei District, Qingdao 266021, Shandong Province, People's Republic of China. [3]Qingdao Municipal Center for Disease Control and Prevention, No. 175 Shandong Road, Shibei District, Qingdao 266033, Shandong Province, People's Republic of China. [4]Qingdao Institute of Preventive Medicine, No. 175 Shandong Road, Shibei District, Qingdao 266033, Shandong Province, People's Republic of China. [5]Epidemiology, Biostatistics and Bio-demography, Institute of Public Health, University of Southern Denmark, DK-5000 Odense C, Denmark. [6]Human Genetics, Institute of Clinical Research, University of Southern Denmark, DK-5000 Odense C, Denmark.

References
1. Paquot N, De Flines J, Rorive M. Obesity: a model of complex interactions between genetics and environment. Revue medicale de Liege. 2012;67:332–6.
2. U.S. Department of Health and Human Services. The surgeon General's call to action to prevent and decrease overweight and obesity. Rockville: U.S. Department of Health and Human Services, Public Health Service, Office of the Surgeon General; 2001.
3. Strychar I. Diet in the management of weight loss. CMAJ. 2006;174:56–63.
4. Shick SM, Wing RR, Klem ML, McGuire MT, Hill JO, Seagle H. Persons successful at long-term weight loss and maintenance continue to consume a low-energy, low-fat diet. J Am Diet Assoc. 1998;98:408–13.
5. Tate DF, Jeffery RW, Sherwood NE, Wing RR. Long-term weight losses associated with prescription of higher physical activity goals. Are higher levels of physical activity protective against weight regain? Am J Clin Nutr. 2007;85:954–9.
6. Rucker D, Padwal R, Li SK, Curioni C, Lau DC. Long term pharmacotherapy for obesity and overweight: updated meta-analysis. BMJ. 2007;335:1194–9.
7. Colquitt JL, Pickett K, Loveman E, Frampton GK. Surgery for weight loss in adults. Cochrane Database Syst Rev. 2014. Art. No.: CD003641. doi:10.1002/14651858.CD003641.pub4.
8. Jayasinghe TN, Chiavaroli V, Holland DJ, Cutfield WS, O'Sullivan JM. The new era of treatment for obesity and metabolic disorders: evidence and expectations for gut microbiome transplantation. Front Cell Infect Microbiol. 2016;6:15.
9. Fanelli RD, Andrew BD. Is Endoluminal bariatric therapy a new paradigm of treatment for obesity? Clin Gastroenterol H. 2016;14:507–15.
10. Sukhdev S, Bhupender S, Singh KS. Pharmacotherapy & Surgical Interventions Available for obesity management and importance of pancreatic lipase inhibitory Phytomolecules as safer anti-obesity therapeutics. Mini Rev Med Chem. 2017;17:371–9.
11. Halpern B, Mancini MC. Safety assessment of combination therapies in the treatment of obesity: focus on naltrexone/bupropion extended release and phentermine-topiramate extended release. Expert Opin Drug Saf. 2017;16:27–39.
12. Roque DR, Makowski L, Chen TH, Rashid N, Hayes DN, Bae-Jump V. Association between differential gene expression and body mass index among endometrial cancers from the cancer genome atlas project. Gynecol Oncol. 2016;142:317–22.
13. Gruchala-Niedoszytko M, Niedoszytko M, Sanjabi B, van der Vlies P, Niedoszytko P, Jassem E, et al. Analysis of the differences in whole-genome expression related to asthma and obesity. Pol Arch Med Wewn. 2015;125:722–30.
14. Gerhard GS, Styer AM, Strodel WE, Roesch SL, Yavorek A, Carey DJ, et al. Gene expression profiling in subcutaneous, visceral and epigastric adipose tissues of patients with extreme obesity. Int J Obesity. 2014;38:371–8.
15. Langfelder P, Horvath S. WGCNA: an R package for weighted correlation network analysis. BMC Bioinform. 2008;9:559.
16. Zhang J, Baddoo M, Han C, Strong MJ, Cvitanovic J, Moroz K, et al. Gene network analysis reveals a novel 22-gene signature of carbon metabolism in hepatocellular carcinoma. Oncotarget. 2016;7:49232–45.
17. Xing Y, Zhang J, Lu L, Li D, Wang Y, Huang S, et al. Identification of hub genes of pneumocyte senescence induced by thoracic irradiation using weighted gene coexpression network analysis. Mol Med Rep. 2016;13:107–16.
18. Sundarrajan S, Arumugam M. Weighted gene co-expression based biomarker discovery for psoriasis detection. Gene. 2016;593:225–34.
19. Modena BD, Bleecker ER, Busse WW, Erzurum SC, Gaston BM, Jarjour NN, et al. Gene Expression Correlated to Severe Asthma Characteristics Reveals Heterogeneous Mechanisms of Severe Disease. Am J Respir Crit Care Med. 2016; 195:1449-63.
20. Liu J, Jing L, Tu X. Weighted gene co-expression network analysis identifies specific modules and hub genes related to coronary artery disease. BMC Cardiovasc Disord. 2016;16:54.
21. Guo Y, Xing Y. Weighted gene co-expression network analysis of pneumocytes under exposure to a carcinogenic dose of chloroprene. Life Sci. 2016;151:339–47.
22. Castillo-Fernandez JE, Spector TD, Bell JT. Epigenetics of discordant monozygotic twins: implications for disease. Genome Med. 2014;6:60.

23. Tan Q, Christiansen L, von Bornemann Hjelmborg J, Christensen K. Twin methodology in epigenetic studies. J Exp Biol. 2015;218:134–9.

24. Zhang D, Li S, Tan Q, Pang Z. Twin-based DNA methylation analysis takes the center stage of studies of human complex diseases. J Genet Genomics. 2012;39:581–6.

25. Pang Z, Ning F, Unger J, Johnson CA, Wang S, Guo Q, et al. The Qingdao twin registry: a focus on chronic disease research. Twin Res Hum Genet. 2006;9:758–62.

26. Kim D, Pertea G, Trapnell C, Pimentel H, Kelley R, Salzberg SL. TopHat2: accurate alignment of transcriptomes in the presence of insertions, deletions and gene fusions. Genome Biol. 2013;14:R36.

27. Trapnell C, Williams BA, Pertea G, Mortazavi A, Kwan G, van Baren MJ, et al. Transcript assembly and quantification by RNA-Seq reveals unannotated transcripts and isoform switching during cell differentiation. Nat Biotechnol. 2010;28:511–5.

28. Leng N, Dawson JA, Thomson JA, Ruotti V, Rissman AI, Smits BM, et al. EBSeq: an empirical Bayes hierarchical model for inference in RNA-seq experiments. Bioinformatics. 2013;29:1035–43.

29. Benjamini Y, Hochberg Y. Controlling the false discovery rate - a practical and powerful approach to multiple testing. J Roy Stat Soc B. 1995;57:289–300.

30. Benjamini Y. Discovering the false discovery rate. J R Stat Soc B. 2010;72:405–16.

31. Langfelder P, Horvath S. Fast R functions for robust correlations and hierarchical clustering. J Stat Softw. 2012;46

32. Zhang B, Horvath S. A general framework for weighted gene co-expression network analysis. Stat Appl Genet Mol Biol. 2005;4:Article17.

33. Yip AM, Horvath S. Gene network interconnectedness and the generalized topological overlap measure. BMC Bioinform. 2007;8:22.

34. Li A, Horvath S. Network neighborhood analysis with the multi-node topological overlap measure. Bioinformatics. 2007;23:222–31.

35. Ravasz E, Somera AL, Mongru DA, Oltvai ZN, Barabasi AL. Hierarchical organization of modularity in metabolic networks. Science. 2002;297: 1551–5.

36. Langfelder P, Horvath S. Eigengene networks for studying the relationships between co-expression modules. BMC Syst Biol. 2007;1:54.

37. Horvath S, Zhang B, Carlson M, Lu KV, Zhu S, Felciano RM, et al. Analysis of oncogenic signaling networks in glioblastoma identifies ASPM as a molecular target. Proc Natl Acad Sci U S A. 2006;103:17402–7.

38. Ghazalpour A, Doss S, Zhang B, Wang S, Plaisier C, Castellanos R, et al. Integrating genetic and network analysis to characterize genes related to mouse weight. PLoS Genet. 2006;e130:2.

39. Fuller TF, Ghazalpour A, Aten JE, Drake TA, Lusis AJ, Horvath S. Weighted gene coexpression network analysis strategies applied to mouse weight. Mamm Genome. 2007;18:463–72.

40. Oldham MC, Horvath S, Geschwind DH. Conservation and evolution of gene coexpression networks in human and chimpanzee brains. Proc Natl Acad Sci U S A. 2006;103:17973–8.

41. Horvath S, Dong J. Geometric interpretation of gene coexpression network analysis. PLoS Comput Biol. 2008;4:e1000117.

42. VisANT 5.0. Visual analyses of metabolic networks in cells and ecosystems. http://visant.bu.edu. Accessed 19 Apr 2017.

43. Deng YY, Li JQ, Wu SF, Zhu YP, Chen YW, He FC. Integrated nr database in protein annotation system and its localization. Comput Eng. 2006;32:71–4.

44. Tatusov RL, Galperin MY, Natale DA, Koonin EV. The COG database: a tool for genome-scale analysis of protein functions and evolution. Nucleic Acids Res. 2000;28:33–6.

45. Koonin EV, Fedorova ND, Jackson JD, Jacobs AR, Krylov DM, Makarova KS, et al. A comprehensive evolutionary classification of proteins encoded in complete eukaryotic genomes. Genome Biol. 2004;5:R7.

46. Kanehisa M, Goto S, Kawashima S, Okuno Y, Hattori M. The KEGG resource for deciphering the genome. Nucleic Acids Res. 2004;32:D277–80.

47. Kanehisa M, Goto S. KEGG: kyoto encyclopedia of genes and genomes. Nucleic Acids Res. 2000;28:27–30.

48. Tweedie S, Ashburner M, Falls K, Leyland P, McQuilton P, Marygold S, et al. FlyBase: enhancing drosophila gene ontology annotations. Nucleic Acids Res. 2009;37:D555–9.

49. Ashburner M, Ball CA, Blake JA, Botstein D, Butler H, Cherry JM, et al. Gene ontology: tool for the unification of biology. The gene ontology consortium. Nat Genet. 2000;25:25–9.

50. Falcon S. Gentleman R. Hypergeometric testing used for gene set enrichment analysis. Bioconductor case studies. New York: Springer New York; 2008. p. 207–20.

51. Xie C, Mao X, Huang J, Ding Y, Wu J, Dong S, et al. KOBAS 2.0: a web server for annotation and identification of enriched pathways and diseases. Nucleic Acids Res. 2011;39:W316–22.

52. Homuth G, Wahl S, Muller C, Schurmann C, Mader U, Blankenberg S, et al. Extensive alterations of the whole-blood transcriptome are associated with body mass index: results of an mRNA profiling study involving two large population-based cohorts. BMC Med Genet. 2015;8:65.

53. Sansbury BE, Hill BG. Regulation of obesity and insulin resistance by nitric oxide. Free Radic Biol Med. 2014;73:383–99.

54. Romacho T, Azcutia V, Vazquez-Bella M, Matesanz N, Cercas E, Nevado J, et al. Extracellular PBEF/NAMPT/visfatin activates pro-inflammatory signalling in human vascular smooth muscle cells through nicotinamide phosphoribosyltransferase activity. Diabetologia. 2009;52:2455–63.

55. Kalin S, Heppner FL, Bechmann I, Prinz M, Tschop MH, Yi CX. Hypothalamic innate immune reaction in obesity. Nat Rev Endocrinol. 2015;11:339–51.

56. Ringseis R, Eder K, Mooren FC, Kruger K. Metabolic signals and innate immune activation in obesity and exercise. Exerc Immunol Rev. 2015;21: 58–68.

57. Vuillaume ML, Naudion S, Banneau G, Diene G, Cartault A, Cailley D, et al. New candidate loci identified by array-CGH in a cohort of 100 children presenting with syndromic obesity. Am J Med Genet A. 2014;164a:1965–75.

58. Picklo MJ, Long EK, Vomhof-DeKrey EE. Glutathionyl systems and metabolic dysfunction in obesity. Nutr Rev. 2015;73:858–68.

59. Hsiao TJ, Hwang Y, Chang HM, Lin E. Association of the rs6235 variant in the proprotein convertase subtilisin/kexin type 1 (PCSK1) gene with obesity and related traits in a Taiwanese population. Gene. 2014;533:32–7.

60. Rouskas K, Kouvatsi A, Paletas K, Papazoglou D, Tsapas A, Lobbens S, et al. Common variants in FTO, MC4R, TMEM18, PRL, AIF1, and PCSK1 show evidence of association with adult obesity in the Greek population. Obesity (Silver Spring). 2012;20:389–95.

61. Schmidt AM. The growing problem of obesity: mechanisms, consequences, and therapeutic approaches. Arterioscler Thromb Vasc Biol. 2015;35:e19–23.

62. Jung UJ, Choi MS. Obesity and its metabolic complications: the role of adipokines and the relationship between obesity, inflammation, insulin resistance, dyslipidemia and nonalcoholic fatty liver disease. Int J Mol Sci. 2014;15:6184–223.

63. Rauschert S, Uhl O, Koletzko B, Hellmuth C. Metabolomic biomarkers for obesity in humans: a short review. Ann Nutr Metab. 2014;64:314–24.

64. Cash JG, Kuhel DG, Goodin C, Hui DY. Pancreatic acinar cell-specific overexpression of group 1B phospholipase A2 exacerbates diet-induced obesity and insulin resistance in mice. Int J Obes. 2011;35:877–81.

65. Sato H, Taketomi Y, Ushida A, Isogai Y, Kojima T, Hirabayashi T, et al. The adipocyte-inducible secreted phospholipases PLA2G5 and PLA2G2E play distinct roles in obesity. Cell Metab. 2014;20:119–32.

66. Garces F, Lopez F, Nino C, Fernandez A, Chacin L, Hurt-Camejo E, et al. High plasma phospholipase A2 activity, inflammation markers, and LDL alterations in obesity with or without type 2 diabetes. Obesity (Silver Spring). 2010;18:2023–9.

67. Hui DY. Phospholipase a(2) enzymes in metabolic and cardiovascular diseases. Curr Opin Lipidol. 2012;23:235–40.

68. Wolf G. Adipose-specific phospholipase as regulator of adiposity. Nutr Rev. 2009;67:551–4.

69. Baltatzi M, Hatzitolios A, Tziomalos K, Iliadis F, Zamboulis C. Neuropeptide Y and alpha-melanocyte-stimulating hormone: interaction in obesity and possible role in the development of hypertension. Int J Clin Pract. 2008;62: 1432–40.

70. Jeanrenaud B, Rohner-Jeanrenaud F. Effects of neuropeptides and leptin on nutrient partitioning: dysregulations in obesity. Annu Rev Med. 2001; 52:339–51.

71. Arora S. Anubhuti. Role of neuropeptides in appetite regulation and obesity–a review. Neuropeptides. 2006;40:375–401.

72. Perez-Verdaguer M, Capera J, Serrano-Novillo C, Estadella I, Sastre D, Felipe A. The voltage-gated potassium channel Kv1.3 is a promising multitherapeutic target against human pathologies. Expert Opin Ther Targets. 2016;20:577–91.

73. Xu J, Koni PA, Wang P, Li G, Kaczmarek L, Wu Y, et al. The voltage-gated potassium channel Kv1.3 regulates energy homeostasis and body weight. Hum Mol Genet. 2003;12:551–9.

74. Thiele TE, Navarro M, Sparta DR, Fee JR, Knapp DJ, Cubero I. Alcoholism and obesity: overlapping neuropeptide pathways? Neuropeptides. 2003; 37:321–37.

75. Grucza RA, Krueger RF, Racette SB, Norberg KE, Hipp PR, Bierut LJ. The emerging link between alcoholism risk and obesity in the United States. Arch Gen Psychiatry. 2010;67:1301–8.

76. Wang L, Liu X, Luo X, Zeng M, Zuo L, Wang KS. Genetic variants in the fat mass- and obesity-associated (FTO) gene are associated with alcohol dependence. J Mol Neurosci. 2013;51:416–24.

77. Wang KS, Zuo L, Pan Y, Xie C, Luo X. Genetic variants in the CPNE5 gene are associated with alcohol dependence and obesity in Caucasian populations. J Psychiatr Res. 2015;71:1–7.

78. Adrielle Lima Vieira R, Nascimento de Freitas R, Volp AC. Adhesion molecules and chemokines; relation to anthropometric, body composition, biochemical and dietary variables. Nutr Hosp. 2014;30:223–36.

79. Christoph MJ, Allison MA, Pankow JS, Decker PA, Kirsch PS, Tsai MY, et al. Impact of adiposity on cellular adhesion: the multi-ethnic study of atherosclerosis (MESA). Obesity (Silver Spring). 2016;24:223–30.

80. Fang P, Yu M, Guo L, Bo P, Zhang Z, Shi M. Galanin and its receptors: a novel strategy for appetite control and obesity therapy. Peptides. 2012;36: 331–9.

81. Leibowitz SF. Regulation and effects of hypothalamic galanin: relation to dietary fat, alcohol ingestion, circulating lipids and energy homeostasis. Neuropeptides. 2005;39:327–32.

82. Gundlach AL. Galanin/GALP and galanin receptors: role in central control of feeding, body weight/obesity and reproduction? Eur J Pharmacol. 2002;440: 255–68.

83. Kypreos KE, Karagiannides I, Fotiadou EH, Karavia EA, Brinkmeier MS, Giakoumi SM, et al. Mechanisms of obesity and related pathologies: role of apolipoprotein E in the development of obesity. FEBS J. 2009;276:5720–8.

84. Srivastava A, Mittal B, Prakash J, Srivastava P, Srivastava N. Analysis of MC4R rs17782313, POMC rs1042571, APOE-Hha1 and AGRP rs3412352 genetic variants with susceptibility to obesity risk in north Indians. Ann Hum Biol. 2016;43:285–8.

85. Breitling C, Gross A, Buttner P, Weise S, Schleinitz D, Kiess W, et al. Genetic contribution of variants near SORT1 and APOE on LDL cholesterol independent of obesity in children. PLoS One. 2015;10:e0138064.

86. Li YH, Liu L. Apolipoprotein E synthesized by adipocyte and apolipoprotein E carried on lipoproteins modulate adipocyte triglyceride content. Lipids Health Dis. 2014;13:136.

87. Wu CL, Zhao SP, Yu BL. Intracellular role of exchangeable apolipoproteins in energy homeostasis, obesity and non-alcoholic fatty liver disease. Biol Rev Camb Philos Soc. 2015;90:367–76.

88. Ramasamy I. Update on the molecular biology of dyslipidemias. Clin Chim Acta. 2016;454:143–85.

89. Kei AA, Filippatos TD, Tsimihodimos V, Elisaf MS. A review of the role of apolipoprotein C-II in lipoprotein metabolism and cardiovascular disease. Metabolism. 2012;61:906–21.

90. Vila G, Grimm G, Resl M, Heinisch B, Einwallner E, Esterbauer H, et al. B-type natriuretic peptide modulates ghrelin, hunger, and satiety in healthy men. Diabetes. 2012;61:2592–6.

91. Gruden G, Landi A, Bruno G. Natriuretic peptides, heart, and adipose tissue: new findings and future developments for diabetes research. Diabetes Care. 2014;37:2899–908.

92. Palmer BF, Clegg DJ. An emerging role of Natriuretic peptides: igniting the fat furnace to fuel and warm the heart. Mayo Clin Proc. 2015;90:1666–78.

93. Schlueter N, de Sterke A, Willmes DM, Spranger J, Jordan J, Birkenfeld AL. Metabolic actions of natriuretic peptides and therapeutic potential in the metabolic syndrome. Pharmacol Ther. 2014;144:12–27.

94. Tu W, Wagner EK, Eckert GJ, Yu Z, Hannon T, Pratt JH, et al. Associations between menarche-related genetic variants and pubertal growth in male and female adolescents. J Adolesc Health. 2015;56:66–72.

95. Kwon S, Kim D, Rhee JW, Park JA, Kim DW, Kim DS, et al. ASB9 interacts with ubiquitous mitochondrial creatine kinase and inhibits mitochondrial function. BMC Biol. 2010;8:23.

96. Debrincat MA, Zhang JG, Willson TA, Silke J, Connolly LM, Simpson RJ, et al. Ankyrin repeat and suppressors of cytokine signaling box protein asb-9 targets creatine kinase B for degradation. J Biol Chem. 2007;282:4728–37.

97. Sun G, Ukkola O, Rankinen T, Joanisse DR, Bouchard C. Skeletal muscle characteristics predict body fat gain in response to overfeeding in never-obese young men. Metabolism. 2002;51:451–6.

98. Oudman I, Clark JF, Brewster LM. The effect of the creatine analogue beta-guanidinopropionic acid on energy metabolism: a systematic review. PLoS One. 2013;8:e52879.

99. Haan YC, van Montfrans GA, Brewster LM. The high creatine kinase phenotype is hypertension- and obesity-prone. J Clin Hypertens (Greenwich). 2015;17:322.

100. Haan YC, Oudman I, Diemer FS, Karamat FA, van Valkengoed IG, van Montfrans GA, et al. Creatine kinase as a marker of obesity in a multi-ethnic population. Mol Cell Endocrinol. 2017;442:24–31.

101. Ahmed M, Gaffen SL. IL-17 in obesity and adipogenesis. Cytokine Growth Factor Rev. 2010;21:449–53.

102. Zizza A, Guido M, Grima P. Interleukin-17 regulates visceral obesity in HIV-1-infected patients. HIV Med. 2012;13:574–7.

103. Wang N, Lan D, Chen W, Matsuura F, Tall AR. ATP-binding cassette transporters G1 and G4 mediate cellular cholesterol efflux to high-density lipoproteins. Proc Natl Acad Sci U S A. 2004;101:9774–9.

104. Kusuhara H, Sugiyama Y. ATP-binding cassette, subfamily G (ABCG family). Pflugers Arch. 2007;453:735–44.

105. Mitrou P, Raptis SA, Dimitriadis G. Insulin action in morbid obesity: a focus on muscle and adipose tissue. Hormones (Athens). 2013;12:201–13.

106. Ye J. Mechanisms of insulin resistance in obesity. Front Med. 2013;7:14–24.

107. Andres MA, Baptista NC, Efird JT, Ogata KK, Bellinger FP, Zeyda T. Depletion of SK1 channel subunits leads to constitutive insulin secretion. FEBS Lett. 2009;583:369–76.

108. Laforenza U, Scaffino MF, Gastaldi G. Aquaporin-10 represents an alternative pathway for glycerol efflux from human adipocytes. PLoS One. 2013;8: e54474.

109. Madeira A, Moura TF, Soveral G. Aquaglyceroporins: implications in adipose biology and obesity. Cell Mol Life Sci. 2015;72:759–71.

110. Mendez-Gimenez L, Rodriguez A, Balaguer I, Fruhbeck G. Role of aquaglyceroporins and caveolins in energy and metabolic homeostasis. Mol Cell Endocrinol. 2014;397:78–92.

15

Genotypic and phenotypic features of all Spanish patients with McArdle disease: a 2016 update

Alfredo Santalla[1,2,†], Gisela Nogales-Gadea[3,5*,†], Alberto Blázquez Encinar[2,5,6], Irene Vieitez[7], Adrian González-Quintana[2,5], Pablo Serrano-Lorenzo[2,5], Inés García Consuegra[2,5,6], Sara Asensio[2,5], Alfonsina Ballester-Lopez[3,5], Guillem Pintos-Morell[3,4,5], Jaume Coll-Cantí[3,5,8], Helios Pareja-Galeano[2,9], Jorge Díez-Bermejo[2,9], Margarita Pérez[9], Antoni L. Andreu[10], Tomàs Pinós[5,10], Joaquín Arenas[2,5], Miguel A. Martín[2,5,6] and Alejandro Lucia[2,9]

Abstract

Background: We recently described the genotype/phenotype features of all Spanish patients diagnosed with McArdle disease as of January 2011 ($n = 239$, prevalence of ~1/167,000) (*J Neurol Neurosurg Psychiatry* 2012;83:322–8). Several caveats were however identified suggesting that the prevalence of the disease is actually higher.

Methods: We have now updated main genotype/phenotype data, as well as potential associations within/between them, of all Spanish individuals currently diagnosed with McArdle disease (December 2016).

Results: Ninety-four new patients (all Caucasian) have been diagnosed, yielding a prevalence of ~1/139,543 individuals. Around 55% of the mutated alleles have the commonest *PYGM* pathogenic mutation p.R50X, whereas p.W798R and p.G205S account for 10 and 9% of the allelic variants, respectively. Seven new mutations were identified: p.H35R, p.R70C, p.R94Q, p.L132WfsX163, p.Q176P, p.R576Q, and c.244-3_244-2CA. Almost all patients show exercise intolerance, the second wind phenomenon and high serum creatine kinase activity. There is, however, heterogeneity in clinical severity, with 8% of patients being asymptomatic during normal daily life, and 21% showing limitations during daily activities and fixed muscle weakness. A major remaining challenge is one of diagnosis, which is often delayed until the third decade of life in 72% of new patients despite the vast majority (86%) reporting symptoms before 20 years. An important development is the growing proportion of those reporting a 4-year improvement in disease severity (now 34%) and following an active lifestyle (50%). Physically active patients are more likely to report an improvement after a 4-year period in the clinical course of the disease than their inactive peers (odds ratio: 13.98; 95% confidence interval: 5.6, 34.9; $p < 0.001$). Peak oxygen uptake is also higher in the former (20.7 ± 6.0 vs. 16.8 ± 5.3 mL/kg/min, $p = 0.0013$). Finally, there is no association between *PYGM* genotype and phenotype manifestation of the disease.

(Continued on next page)

* Correspondence: gnogales@igtp.cat
[†] Equal contributors
[3] Grup de Recerca en Malalties Neuromusculars i Neuropediatriques, Department of Neurosciences, Institut d'Investigació en Ciències de la Salut Germans Trias i Pujol, Universitat Autònoma de Barcelona, Camí de les Escoles, s/n 08916 (Barcelona), Badalona, Spain
[5] Centre for Biomedical Network Research on Rare Diseases (CIBERER), Instituto de Salud Carlos III, Madrid, Spain
Full list of author information is available at the end of the article

(Continued from previous page)

Conclusions: The reported prevalence of McArdle disease grows exponentially despite frequent, long delays in genetic diagnosis, suggesting that many patients remain undiagnosed. Until a genetic cure is available (which is not predicted in the near future), current epidemiologic data support that adoption of an active lifestyle is the best medicine for these patients.

Keywords: McArdle disease, Spanish patients, Genotype, Phenotype, Glycogenosis type V

Background

Glycogenosis type V [glycogen storage disease type V (GSD V), McArdle disease or *myophosphorylase* deficiency; OMIM® 232,600] is an autosomal recessive disease of carbohydrate metabolism. It is caused by inherited pathogenic mutations in the gene Phosphorylase, Glycogen, Muscle (*PYGM*), encoding the muscle-specific isoform of glycogen phosphorylase, *myophosphorylase*, which catalyzes the breakdown of glycogen into glucose-1-phosphate in this tissue [1]. This myopathy is arguably the prototype of exercise intolerance, which typically consists of acute crises of early fatigue and contractures, occasionally accompanied by rhabdomyolysis and myoglobinuria [1]. The current best diagnostic tool for McArdle disease is genetic testing to determine whether patients are homozygous or alternatively compound heterozygous for pathogenic *PYGM* mutations [1]. Yet the so-called 'second wind' phenomenon, that is, marked improvement in tolerance to dynamic exercise (*eg*, bicycling at a constant, submaximal wattage) after 6–10 min of exertion, with subsequent disappearance of previous tachycardia, is a unique characteristic of patients with McArdle disease that is easily measurable [2]. Thus, laboratory assessment of this second wind can be used to support (or discard) the presence of McArdle disease before eventual genetic diagnostic confirmation.

Epidemiologic data available on McArdle disease are relatively scarce and usually limited in sample size [3–7]. We recently described the main genotype and phenotype features of all Spanish patients diagnosed with McArdle disease, as of January 2011 [8]. According to our prior study, reporting on the largest series of McArdle patients published to date (*n* = 239), the prevalence of the disease was ~1/167,000 Spanish individuals of Caucasian descent. Several caveats were however identified that led us to believe that the actual prevalence of the disease might be higher [8]. A number of patients are likely to remain undiagnosed owing to the rarity of the disease (which is still not well known by many clinicians) or to the mildness of the symptoms in some cases (with no actual interference with daily living activities). Further, many paediatricians are probably unaware of the fact that McArdle disease is to be considered, to a large extent, a paediatric condition, which should expedite diagnosis. In fact, only 4% of Spanish patients were genetically diagnosed during the first decade of life (despite 58% of the total reporting onset of symptoms during childhood), and 47% had not been diagnosed until 30+ years of age [8]. Our previous observations concur with those of recent population estimates by De Castro and colleagues using next-generation sequencing (NGS) of the *PYGM* gene [9]. These authors suggested that the currently accepted prevalence of McArdle disease in Americans of European descendent (~1/100,000) [10] is an underestimate, with the actual disease prevalence being at least 2-fold higher, and thus ≥3-fold higher than the prevalence we recently reported for Spanish patients [8].

The diagnostic protocol followed by the National Health System, where the blood of candidate patients is routinely sent to each of 3 'reference' centres for genetic analysis (*Hospital 12 de Octubre*, Madrid; *Hospital Val d'Hebron*, Barcelona; and *Hospital Meixoeiro*, Vigo), makes it relatively easy to gather data on Spanish McArdle patients. Further, an increasing number of patients with exercise intolerance are referred to our exercise physiology facilities (*Universidad Europea de Madrid* or *Universidad Pablo Olavide*, Seville), which allows us to assess the second wind. Thus, the aim of this study was to update the main genotype and phenotype characteristics, as well as potential associations within/between them, of all Spanish individuals who are currently diagnosed with McArdle disease.

Methods

We have used a cross-sectional design to perform an update of the main *PYGM* genotype and phenotype data [clinical and laboratory variables (muscle biopsy when available, serum creatine kinase (CK) activity), exercise capacity] of all Spanish individuals diagnosed with McArdle disease, as of December 7th, 2016. The study also has a prospective element, as we have followed-up the 4-year progression of the clinical severity of the disease (see below) in a sub-cohort of 151 patients (*ie*, *n* = 89 from 2006 to 2010, already reported by us [8], and *n* = 62 from 2011 and onwards). The study was approved by the local ethics committees and followed the tenets of the Declaration of Helsinki, 1961.

PYGM genotyping

Mutant *PYGM* alleles were identified in patients' blood samples using SNaPShot mini-sequencing (ThermoFisher) or polymerase chain reaction and restriction fragment length polymorphism methods [11], followed by: Sanger sequencing of the entire coding region and intron/exon boundaries of the *PYGM* gene [12], or the use of a NGS customised gene panel on a PGM-IonTorrent platform (ThermoFisher), consisting of 35 genes, including *PYGM*, associated with metabolic myopathies. In some cases, analysis of muscle or blood mRNA/cDNA was needed to demonstrate the molecular pathogenicity of a presumed mutant allele, particularly when an alteration of the splicing mechanism was suspected [13, 14].

Phenotype data

We recorded from clinical histories [or from personal interview with (and direct evaluation of) the patients (in those visiting the aforementioned exercise physiology laboratories for functional evaluation)] data on comorbidities, exercise intolerance, self-reported second wind phenomenon, permanent muscle weakness [15, 16], basal serum CK activity after 1+ days with no exercise (last result available), muscle biopsy results (corroborating lack of staining for *myophosphorylase* and no *myophosphorylase* activity in biochemical analyses), and clinical severity class following the classification originally reported by Martinuzzi et al. [5].

Since autumn 2010, we (AS, AL, MP) have interviewed 151 patients on the progression of their disease within the previous 4-year period to ascertain: (i) improvement (change to a lower severity class in the aforementioned Martinuzzi's scale), (ii) worsening (change to a higher severity class), or (iii) constant (no change). On the same day of the interview, patients were also asked about their physical activity (PA) habits and were classified as physically active if they followed international PA guidelines, that is, doing ≥150 min/week of moderate-vigorous PA (walking/brisk walking, bicycling, swimming) [17, 18].

Patient's peak oxygen uptake (VO$_2$peak) was determined during cycle-ergometry [19] or treadmill testing until exhaustion (in children) as reported elsewhere [19], and the second wind diagnostic test was performed following the methodology reported by Vissing and Haller [2] using consistently the same equipment (metabolic cart and cycle-ergometer) and under the supervision of the same researchers (AS, AL, MP).

Statistical analysis

Descriptive data are expressed as frequencies (%) and mean ± standard deviation. We compared phenotype data between genders with the χ^2 test or unpaired Student's t test. To determine whether the clinical condition of patients deteriorates with aging, we compared patients' age between the different severity groups with 1-factor analysis of variance (and with Tukey's test for post hoc comparisons), and between patients showing an improvement in disease progression vs. those showing worsening/ no change (with unpaired Student's t test).

We also calculated the odds ratio (OR) and 95% confidence interval (CI) to determine the association: between *PYGM* genotype and disease phenotype/progression, on the one hand, and between PA levels and disease progression, on the other. All statistical analyses were performed using the PASW (v.18.0 for WINDOWS, Chicago) and the level of significance was set at 0.05.

Results

Compared with the first report on all diagnosed Spanish patients as of January 2011 [8], 94 new patients of Spanish nationality (all of Caucasian descent) have been diagnosed with McArdle disease in only 5 years, to sum a total of 333 patients (183 male). Three patients (all males) included in the previous report [8] have died since January 2011, all due to cardiovascular disease, at the age of 56, 67 and 89 years. As such, the prevalence of McArdle disease in the Spanish (Caucasian) population is now ~1/139,543 persons living.

Table 1 shows the *PYGM* mutational spectrum of Spanish patients, which in essence has not changed in the last years. The genetic analysis showed that ~55% of the mutated alleles harbour the commonest Caucasian stop codon mutation p.R50X, 10% the missense p.W798R mutation (a virtually Spanish private mutation) and 9% correspond to the p.G205S (a relatively common Caucasian mutation). The *PYGM* exons containing more mutations are, in decreasing order, exons 1, 18, 17, 15 and 12. No mutations were found in exons 6–8. Yet, 7 novel *PYGM* mutations were identified: 5 missense mutations, p.H35R, p.R70C, p.R94Q, p.Q176P, p.R576Q, 1 frameshift predicting a premature stop codon (p.L132WfsX163, and 1 splice-site microdeletion mutation c.244-3_244-2CA (Additional files 1 and 2). Of note, all the patients carrying a new mutation exhibited the second wind phenomenon, as assessed by us in the laboratory, and consequently the functional diagnosis of McArdle disease was also proven in these cases. We identified the 2 mutant *PYGM* alleles in all but 3 people (99.1% of total). Although only 1 mutant allele has been identified in these 3 individuals, we also consider them to be patients with McArdle disease because they also experienced the second wind phenomenon in laboratory assessment. The main reason for not having yet detected the second pathogenic mutation in these patients is simply lack of time because diagnosis analyses started only 2 months ago (October 2016).

Phenotype data, and how they compare to our previous report [8], are shown in Table 2. They do not essentially differ between sexes (data not shown, $p > 0.1$ for most comparisons). The mean age of the cohort has now slightly increased since the first report (by 3 years), simply reflecting the aging of the previous 239 patients, who still account for the majority of the cohort. The main clinical features of McArdle disease have remained essentially unchanged with regard to our previous report, except for a slight decrease

Table 1 *PYGM* mutations identified in all Spanish McArdle patients (*N* = 333)

Type of mutation	N	%
p.R50X (c.148C > T) / p.R50X (c.148C > T)	114	34.2%
p.R50X (c.148C > T) / p.W798R (c.2392 T > C)	29	8.7%
p.G205S (c.613G > A) / p.G205S (c.613G > A)	20	6.0%
p.W798R (c.2392 T > C) / p.W798R (c.2392 T > C)	16	4.8%
p.R50X (c.148C > T) / p.G205S (c.613G > A)	14	4.2%
p.R50X (c.148C > T) / p.K754fsX49 (c.2262delA)	8	2.4%
p.C784X (c.2352C > A) / p.R94W (c.280C > T)	5	1.5%
p.R50X (c.148C > T) / p.R94W (c.280C > T)	6	1.8%
p.R50X (c.148C > T) / p.R602W (c.1804C > T)	3	0.9%
p.R50X (c.148C > T) / p.A660D (c.1979C > A)	3	0.9%
p.R50X (c.148C > T) / p.E383K (c.1147G > A)	3	0.9%
p.G205S (c.613G > A) / c.1768 + 1G > A	2	0.6%
p.R50X (c.148C > T) / c.1768 + 1G > A	4	1.2%
p.R50X (c.148C > T) / p.A365V (c.1094C > T)	3	0.9%
p.R50X (c.148C > T) / p.A55GfsX21 (c.163_167delGCTCT)	2	0.6%
p.R50X (c.148C > T) / p.A704V (c.2111C > T)	4	1.2%
p.R50X (c.148C > T) / p.D534fsX5 (c.1601delA)	2	0.6%
p.R50X (c.148C > T) / p.L5VfsX22 (c.13_14delCT)	3	0.9%
p.R50X (c.148C > T) / p.R194W (c.580C > T)	2	0.6%
p.R50X (c.148C > T) / p.R715W (c.2143C > T)	3	0.9%
p.R50X (c.148C > T) / p.T488 N (c.1463 > A) + p.K215 K (c.645G > A)	2	0.6%
p.R576X (c.1726C > T) / p.G136AfsX159 (c.407G > A)	2	0.6%
p.R771PfsX33 (c.2310_2311dupCC) / p.R771PfsX33 (c.2310_2311dupCC)	2	0.6%
p.W388SfsX34 (c.1162_1169delTGGCCGGTinsA)/ p.K754fsX49 (c.2262delA)	2	0.6%
p.W798R (c.2392 T > C) / p.K215 K (c.645G > A)	2	0.6%
p.K609 K (c.1827 G > A) / p.K609 K (c.1827 G > A)	1	0.3%
p. Y733X (c.2199C > G) + p.Y733X (c.2199C > G)	1	0.3%
p.A55GfsX21 (c.163_167delGCTCT) / p.A55GfsX21 (c.163_167delGCTCT)	1	0.3%
p.A660D (c.1979C > A) / p.A660D (c.1979C > A)	1	0.3%
p.E125X (c.373G > T) / p.E125X (c.373G > T)	1	0.3%
p.G174D (c.521G > A) / p.K609 K (c.1287G > A)	1	0.3%
p.G205S (c.613G > A) / p.A365V (c.1094C > T)	1	0.3%
p.G205S (c.613G > A) / p.I83F (c.247A > T)	1	0.3%
p.G205S (c.613G > A) / p.Q176_M177insVQ (c.529-8 g > a)	1	0.3%
p.I83F (c.247A > T)/ p.R94W (c.280C > T)	1	0.3%
p.K754NfsX49 (c.2262delA) / c.2380-1G > A	1	0.3%
p.K754NfsX49 (c.2262delA) / p.K754NfsX49 (c.2262delA)	3	0.9%
p.L116P (c.347 T > C) / p.L116P (c.347 T > C)	1	0.3%
p.L587P (c.1760 T > C) / p.A660D (c.1730 G > A)	2	0.6%
p.L5VfsX22 (c.13_14delCT) / p.K754fsx49 (c.2262delA)	1	0.3%

Table 1 *PYGM* mutations identified in all Spanish McArdle patients (*N* = 333) *(Continued)*

Type of mutation	N	%
p.M442 K (c.1325 T > A) / p.M442 K (c.1325 T > A)	1	0.3%
p.N134KfsX161 (c.402delC) / p.R491AfsX7 (c.1470dupG)	1	0.3%
p.Q577R (c.1730 A > G) / p.A660D (c.1730A > G)	1	0.3%
p.R194W (c.580C > T) + p.E797VfsX18 (c.2385_2386delAA)/ p.R194W (c.580C > T) + p.E797VfsX18 (c.2385_2386delAA)	1	0.3%
p.R50X (c.148C > T) / [a]	4	1.2%
p.R50X (c.148C > T) / c.(1969 + 214)_(2177 + 369)de	1	0.3%
p.R50X (c.148C > T) / c.1827 G > A	1	0.3%
p.R50X (c.148C > T) / c.855 + 5G > A	1	0.3%
p.R50X (c.148C > T) / p.E349K (c.1045G > A)	1	0.3%
p.R50X (c.148C > T) / p.G455R (c.1363G > C)	1	0.3%
p.R50X (c.148C > T) / p.G695R (c.2083G > A)	2	0.6%
p.R50X (c.148C > T) / p.K215 K (c.645G > A)	1	0.3%
p.R50X (c.148C > T) / p.L587P (c.1760 T > C)	1	0.3%
p.R50X (c.148C > T) / p.L5VfsX22 (c.13_14delCT) + p.R324G	2	0.6%
p.R50X (c.148C > T) / p.N685Y (c.2053A > T)	1	0.3%
p.R50X (c.148C > T) / p.Q734HfsX7 (c.211_217dupCGCAGCA)	1	0.3%
p.R50X (c.148C > T) / p.Q755X (c.2263C > T)	1	0.3%
p.R50X (c.148C > T) / p.R576X (c.1726C > T)	1	0.3%
p.R50X (c.148C > T) / p.T488 N (c.1463C > A)	1	0.3%
p.R50X (c.148C > T) / p.T692KfsX30 (c.2075_2076delCCinsAAA)	1	0.3%
p.R50X (c.148C > T) / p.W388SfsX34 (c.1162_1169delTGGCCGGTinsA)	1	0.3%
p.R50X (c.148C > T) / R715W (c.2143C > T)	1	0.3%
p.R50X (c.148C > T) / p.V456 M (c.1366G > A)	1	0.3%
p.R50X (c.148C > T) / p.L354P (c.1061 T > C)	1	0.3%
p.W798R (c.2392 T > C) / p.R590H (c.1769G > A)	1	0.3%
p.Y574X (c.1722 T > G) / p.K609 K (c.1827G > A)	1	0.3%
p.L5VfsX22 (c.13_14delCT) / p.L5VfsX22 (c.13_14delCT)	4	1.2%
p.R50X (c.148C > T) / p.E27AfsX50 (c.78_79delTG)	2	0.6%
p.R50X (c.148C > T) / p.L116P (c.347 T > C)	1	0.3%
c.1092-1G > T / c.2444-3_244-2delCA	1	0.3%
p.R50X (c.148C > T) / p.K609 K (c.1287G > A)	1	0.3%
p.R491Afs (c.1470dupG) / [a]	1	0.3%
p.W388SfsX421 (c.1162_1169delTGGCCGGT)/ p.W388SfsX421 (c.1162_1169delTGGCCGGT)	1	0.3%
p.G205S (c.613G > A) / p.R590H (c.1769G > A)	1	0.3%
p.W798R (c.2392 T > C) / [a]	1	0.3%
p.R50X (c.148C > T) / p.R490W (c.1468C > T)	1	0.3%
p.R50X (c.148C > T) / p.V456 M (c.1366G > A)	1	0.3%
p.W798R (c.2392 T > C) / c.212_218dup (p.Q73HfsX)	1	0.3%
c.2262delA (p.K754Nfs) / c.244-3_244-2delCA	1	0.3%

Table 1 PYGM mutations identified in all Spanish McArdle patients (N = 333) (Continued)

Type of mutation	N	%
p.K754fsX49 (c.2262delA) / c.773-2A > T	1	0.3%
p.Q734HfsX7 (c.211_217dupCGCAGCA) / p.Q734HfsX7 (c.211_217dupCGCAGCA)	1	0.3%
p.R50X (c.148C > T) / p.R576Q	1	0.3%
p.R50X (c.148C > T) / p.L132WfsX153 (c.393delG)	1	0.3%
p.R50X (c.148C > T) / p.Q176P (c.527A > C)	1	0.3%
p.R491Afs (c.1470dupG) / p.R491Afs (c.1470dupG)	1	0.3%
p.R94W (c.280C > T) / p.R94W (c.280C > T)	1	0.3%
p.G135R (c.403G > A) / p.R70C	1	0.3%
c.244-3_244-2delCA / c.1093-1G > T	1	0.3%
p.R50X (c.148C > T) / p.H35R (c.104A > G)	1	0.3%
p.W798R (c.2392 T > C) / p.A365V (c.1094C > T)	1	0.3%

[a]Unidentified mutation in one allele

in the proportion of patients with fixed muscle weakness (25% → 21%) [8]. Thus, virtually all patients have exercise intolerance, which has been accompanied by repeated episodes of myoglobinuria (that patients typically refer to as 'dark urine') in half of the cohort. In addition, the vast majority of patients are able to report the second wind phenomenon, which they typically refer to as the ability to resume exercise with attenuated fatigue, tachycardia and myalgia after they take a quick rest. Importantly, this phenomenon was easily measurable in all patients (see Additional file 3 for a representative example), except in 1 child (aged 12 years), which also concurs with our previous observation that, as opposed to adults, the second wind may not be easily detectable in some of the youngest patients [20].

A laboratory feature that characterises the disease and can further assist in its diagnosis is the typically high levels of the muscle damage marker, serum CK activity, which is above reference limits in nearly all patients and above 2000 U/L in ~2 thirds of them, and is also in accordance with our previous report [8]. Other characteristics also remain essentially unchanged, such as frequency distribution among severity classes [at least when grouping the 2 highest severity classes (2 and 3) together], and with the same percentage (8%) of patients who are virtually asymptomatic during normal daily living (ie, belonging to class 0, which denotes exercise intolerance only during strenuous activities or sports participation and absence of myoglobinuria episodes). Also, association with comorbidities induced directly or indirectly by the disease appears low, with very few cases of chronic renal failure and a prevalence of major noncommunicable diseases (diabetes, cardiovascular disease, cancer) that does not appear to differ from that expected for the adult Spanish population. Further, 1 patient has reached remarkable longevity

Table 2 Main phenotype data in all Spanish McArdle patients (n = 333)

	N with data	Result (men + women)	Main change with regards to previous data (n = 239 patients) [1]
Gender		55% male	↔
Age, years (mean ± SD, range)	333	48 ± 19 (12, 99)[a]	↑
BMI, kg/m^2 (mean ± SD, range)	132	24.7 ± 4.8 (16, 43)	↔
Familial consaguinity (%)	188	17%	↑
Symptomatic father (%)[a]	201	3%	↔
Symptomatic mother (%)[a]	201	6%	↔
Symptoms' onset (%)	235		
1st decade		66%	↑
2nd decade		20%	↓
3rd decade		5%	↔
≥4th decade		9%	↔
Genetic diagnosis (%)	275		
1st decade		5%	↔
2nd decade		23%	↔
3rd decade		23%	↔
≥4th decade		49%	↔
Exercise intolerance (%)	272	99.6%	↔
Second wind, self-reported (%)	164	91.5%	↑
Second wind, laboratory-determined (%)	119	99.2%	↔
Fixed muscle weakness (%)	240	21%	↓
Recurrent episodes of myoglobinuria (%)	240	51%	↔
Disease severity (%)[b]	240		
Class 0		8%	↔
Class 1		41%	↔
Class 2		30%	↑
Class 3		21%	↓
Disease progression (%)	151		
Improvement		34%	↑↑
Worsening		28%	↔
Constant		35%	↓↓

Table 2 Main phenotype data in all Spanish McArdle patients (n = 333) (Continued)

	N with data	Result (men + women)	Main change with regards to previous data (n = 239 patients) [1]
Acute renal failure (%)	173	6%	↔
Chronic renal failure (%)	171	1%	↔
Comorbidities (%)			
Diabetes[c]	140	9%	↔
CAD	136	9%	↔
Hypertension	135	11%	↔
Cancer	134	1%	↔
Obesity	127	10%	↔
COPD	131	1%	↔
Serum CK (%) >200 U/L, >1000 U/L	179	98%, 68%	↔, ↓
Biopsy diagnosis (%)	205	100%	↔
Physical activity data (% active)	120	50%	↑↑
VO$_2$peak, mLO$_2$/ kg/min (mean ± SD, range)	120	19.9 ± 6.6 (5.9, 41.5)	↔

Data on 3 patients (all males) who died recently (all after the 6th decade of life, due to cardiovascular disease) are included
Symbols: [a] 'symptomatic' refers to having the main symptomatic features of McArdle disease (i.e., exercise intolerance with or without myoglobinuria or self-reported second wind)
[b]disease severity class following the classification originally reported by Martinuzzi et al. [2]: 0 = asymptomatic or virtually asymptomatic (mild exercise intolerance, but no functional limitation in any daily life activity); 1 = exercise intolerance, contractures, myalgia, and limitation of acute strenuous exercise, and occasionally in daily life activities; no record of myoglobinuria, no muscle wasting or weakness; 2 = same as 1, plus recurrent exertional myoglobinuria, moderate restriction in exercise, and limitation in daily life activities; 3 = same as 2, plus fixed muscle weakness, with or without wasting and severe limitations on exercise and most daily life activities
[c]diabetes diagnosed based on a glucose tolerance test. Abbreviations: BMI, body mass index; CAD, coronary artery disease; COPD, chronic obstructive pulmonary disease; VO$_2$peak, peak oxygen uptake

(99 years), supporting the overall benign nature of the disease compared with other glycogenoses.

An important problem that persists with regards to the first series [8] is that, despite most patients reporting onset of exercise intolerance symptoms since childhood, typically in physical education classes and in the school playground, genetic diagnosis has been delayed until much later in life. Among the 94 new patients, genetic diagnosis has not been available until the age of 20+ years in 75% of cases, despite the vast majority (90%) reporting symptoms before that age. Yet a major difference compared with our previous report is an increase in the proportion of patients (i) reporting a 4-year improvement in disease severity (21% → 34%) at the expense of those showing no change (51% → 35%); and (ii) adopting an active lifestyle in recent

times (32% → 50%). Indeed, those patients who are physically active are 14-fold more likely to report an improvement after a 4-year period in the clinical course of the disease compared with their inactive peers (OR: 13.98; 95%CI: 5.6, 34.9; $p < 0.001$). In addition, a key fitness and health indicator, VO$_2$peak, is significantly ($p = 0.001$) higher in physically active patients (20.7 ± 6.0) than in their inactive referents (16.8 ± 5.3 mL/kg/min).

Age has a detrimental effect on several phenotypic features of the disease. The mean age (57 ± 19 years) of those patients in the highest severity class 3 (that includes presence of fixed muscle weakness) is higher than in those in the lower severity classes 1 (46 ± 19 years, $p = 0.007$) and 2 (41 ± 17 years, $p < 0.001$). Finally, *PYGM* genotype is not significantly associated with any of the phenotype data reported in Table 2 after controlling for sex and age (data not shown). Likewise, *PYGM* is not associated with any phenotype data within the subset of patients reporting a 4-year improvement in disease severity, with a similar proportion of *PYGM* alleles harbouring a missense mutation in this subgroup compared with the whole patient cohort (i.e., ~30% of total number of alleles in both cases).

Discussion

This is the largest series of patients' data that is available to date on McArdle disease and as such can provide corroborative or novel insights on this disorder. Despite heterogeneity, mainly in terms of *PYGM* genotype but also of disease severity [with almost 1 in 10 patients being fundamentally asymptomatic in daily living (*ie*, belonging to Martinuzzi's class 0) vs. 2 in 10 being clearly limited in daily life activities and having fixed muscle weakness (class 3)], several features of the disorder, mainly pertaining to phenotype and laboratory data, are common to the vast majority of patients. This should raise the suspicion of the presence of McArdle disease until genetic confirmation is achieved. These common features include intolerance to strenuous exercise coupled with the second wind phenomenon in almost all patients, as well as basal hyper-CK-emia.

The *PYGM* genotype data shown here are in overall agreement with previous reports on smaller Caucasian cohorts [3, 4, 6, 7, 11, 21–23] with p.R50X/p.R50X and p.R50X/p.W798R combinations accounting for 43% of all *PYGM* genotypes and p.R50X being by far the commonest pathogenic genetic variant. In addition, we have identified 7 novel mutations that must be added to the list of pathogenic variants causing McArdle disease [24]. That heterogeneity of *PYGM* genotypes does not account for heterogeneity in the clinical manifestation of the disease is in agreement with previous studies showing no genotype-phenotype correlation [4, 6, 22]. In fact, in our series, biochemical analysis of muscle biopsies consistently showed null *myophosphorylase* activity and most reported *PYGM* mutations have functional consequences, with many actually resulting in no gene transcript levels owing to a protective intracellular mechanism, the so-called nonsense-mediated decay, which degrades transcripts that contain premature termination codons [25]. By contrast, regular PA is likely the main modulator of the phenotype manifestation of the disease, which explains the heterogeneous presentation of the disorder among patients despite all having the same defect at the muscle molecular level, that is, complete inability to metabolize glycogen stores.

A promising novel result is that a growing number of Spanish patients (around one half of them) are now adopting a relatively active lifestyle, which we believe is especially important when considering the strong, positive association between regular PA and a favourable progression of clinical symptoms. These results support and extend the findings from our previous report [8]. This recent tendency reflects, at least in part, the fact that patients are now following our recommendations [notably, during educational talks given by us (AL, AS, GNG, TP) to the patients in each yearly meeting of the Spanish Association of Glycogenosis Patients] to perform low-moderate intensity PA regularly. Furthermore, VO$_2$peak was significantly higher in the physically active patients. Indeed, 8 physically active adult patients (vs. only 2 inactive adult patients) have a VO$_{2\text{peak}}$ $\geq$ 8 METs (where 1 MET = resting metabolic rate or 3.5 mL O$_2$/kg/min for most humans). This is also an important finding because 8 METs is the minimum threshold for optimal health, above which the risk for cardiovascular mortality in adults is significantly reduced compared with lower values [26]. Our findings on PA also support previous studies showing the benefits of regular, low-moderate intensity exercise training (bicycling, brisk walking) during 8 [19] to 14 weeks to increase the VO$_{2\text{peak}}$ of McArdle patients [27].

We believe an important challenge identified by us here as well as previously [8] is the fact that diagnosis of the disease is usually delayed until adulthood despite symptoms frequently occurring since childhood. Misdiagnosis might also be as a potential cause of delayed diagnosis of McArdle disease [28]. This undoubtedly calls for a better characterisation and monitoring of the disease from childhood and adolescence. The possibility of diagnosing McArdle disease should be clearly mentioned in the specific educational training of paediatricians because early diagnosis would favour implementation of regular PA habits (with carbohydrate ingestion and gradual warm-up before strenuous activities) since early childhood (coupled with a diet rich in complex carbohydrates to ensure constant availability of glucose to working muscles) [29]. In this regard, all affected children report problems during physical education classes because the latter usually involve strenuous exercises, such as running, which is particularly painful and fatiguing for all of them compared to their healthy peers. Early diagnosis is particularly important when considering that practice of regular, low-moderate PA usually tracks from childhood to adolescence and from adolescence into adulthood, thereby laying the foundation for a healthy lifestyle over life [30]. As for diagnostic tools, we believe that objective assessment of the second wind followed by genetic analyses using blood samples (quick screening initially for the three most prevalent mutations, p.R50X, p.W798R and p.G205S, which account, alone or in combination, for ~74% of all *PYGM* mutant alleles, and then, whenever needed, searching for further mutations by Sanger sequencing or NGS of the *PYGM* gene) is the most efficient strategy, with no need for performing unpleasant muscle biopsies unless it is a requirement to identify the genetic alteration or to validate the molecular consequences of novel mutations in the *PYGM* gene. Other tests that were traditionally implemented, such as electromyography recordings or the ischaemic forearm test are probably not useful anymore. A question that remains unanswered however is pregnancy outcomes in all women, especially which is the best way of delivery, whether vaginal or caesarean.

Conclusions

The reported prevalence of McArdle disease seems to grow considerably despite genetic diagnosis being frequently delayed until adulthood and onset of symptoms occurring since childhood. Thus, awareness of this disease and monitorisation is still insufficient, especially among paediatricians. Until a genetic cure is available (which is not predicted in the near future), the current epidemiologic data support that adoption of an active lifestyle is the best medicine for these patients.

Acknowledgements
We thank the Spanish Association of Glycogenosis for their help and their support; without their contribution this article would not be possible. We thank Dr. Kenneth McCreath for editorial support.

Funding
Publication of the article funded by Cátedra ASISA-UEM (2016/UEM41). The research of Alejandro Lucia, Tomàs Pinós, Joaquin Arenas, Gisela Nogales-Gadea and Miguel Angel Martín in the field of McArdle disease and muscle diseases is funded by the Fondo de Investigaciones Sanitarias ISCIII (FIS, grant numbers PI15/00558, PI12/00914, PI16/01492, PI13/00855 PI14/00903, PI15/01756 and PI15/00431) and cofinanced by FEDER. Gisela Nogales-Gadea is supported by a Miguel Servet research contract (ISCIII CD14/00032 and FEDER) and by AFM Telethon Trampoline Grant 21108. The funding bodies had no role in the design of the study and collection, analysis, and interpretation of data. Alfonsina Ballester-Lopez is funded by an FI Agaur fellowship. Helios Pareja-Galeano is supported by the Cátedra ASISA-UEM (2016/UEM41).

About this supplement
This article has been published as part of *BMC Genomics* Volume 18 Supplement 8, 2017: Proceedings of the 34th FIMS World Sports Medicine Congress. The full contents of the supplement are available online at https://bmcgenomics.biomedcentral.com/articles/supplements/volume-18-supplement-8.

Authors' contributions

AL, AS and GN-G obtained patient data, analysed and interpreted the patient data and wrote the manuscript. ABE, IV, AGQ, PSL, IGC, SA, AB-L, TP, and MAM obtained and analysed genotype data of patients from Spain. GP-M, JC-C, HP-G, MP, ALA, JA and TP were in charge of data collection and analysis. All authors read and approved the final manuscript.

Competing interests

The authors declare that they have no competing interests.

Author details

[1]Universidad Pablo de Olavide, Sevilla, Spain. [2]Instituto de Investigación Hospital 12 de Octubre (i+12), Madrid, Spain. [3]Grup de Recerca en Malalties Neuromusculars i Neuropediatriques, Department of Neurosciences, Institut d'Investigació en Ciències de la Salut Germans Trias i Pujol, Universitat Autònoma de Barcelona, Camí de les Escoles, s/n 08916 (Barcelona), Badalona, Spain. [4]Servicio de Pediatría, Hospital Universitari Germans Trias i Pujol, Badalona, Spain. [5]Centre for Biomedical Network Research on Rare Diseases (CIBERER), Instituto de Salud Carlos III, Madrid, Spain. [6]Laboratorio de Enfermedades Mitocondriales y Neuromusculares, Hospital 12 de Octubre, Madrid, Spain. [7]Rare Diseases and Pediatric Medicine Group, Galicia Sur Health Research Institute, Complexo Hospitalario Universitario de Vigo (CHUVI), SERGAS, Vigo, Spain. [8]Servicio de Neurología, Hospital Universitari Germans Trias i Pujol, Badalona, Spain. [9]Universidad Europea de Madrid, Madrid, Spain. [10]Departament de Patologia Mitocondrial i Neuromuscular, Hospital Universitari Vall d'Hebron, Institut de Recerca (VHIR), Universitat Autónoma de Barcelona, Barcelona, Spain.

References

1. Lucia A, Nogales-Gadea G, Perez M, Martin MA, Andreu AL, Arenas J. McArdle disease: what do neurologists need to know? Nat Clin Pract Neurol. 2008;4(10):568–77.
2. Vissing J, Haller RG. A diagnostic cycle test for McArdle's disease. Ann Neurol. 2003;54(4):539–42.
3. Bruno C, Cassandrini D, Martinuzzi A, Toscano A, Moggio M, Morandi L, Servidei S, Mongini T, Angelini C, Musumeci O, et al. McArdle disease: the mutation spectrum of PYGM in a large Italian cohort. Hum Mutat. 2006;27(7):718.
4. Martin MA, Rubio JC, Buchbinder J, Fernandez-Hojas R, del Hoyo P, Teijeira S, Gamez J, Navarro C, Fernandez JM, Cabello A, et al. Molecular heterogeneity of myophosphorylase deficiency (McArdle's disease): a genotype-phenotype correlation study. Ann Neurol. 2001;50(5):574–81.
5. Martinuzzi A, Sartori E, Fanin M, Nascimbeni A, Valente L, Angelini C, Siciliano G, Mongini T, Tonin P, Tomelleri G, et al. Phenotype modulators in myophosphorylase deficiency. Ann Neurol. 2003;53(4):497–502.
6. Quinlivan R, Buckley J, James M, Twist A, Ball S, Duno M, Vissing J, Bruno C, Cassandrini D, Roberts M, et al. McArdle disease: a clinical review. J Neurol Neurosurg Psychiatry. 2010;81(11):1182–8.
7. Vieitez I, Teijeira S, Fernandez JM, San Millan B, Miranda S, Ortolano S, Louis S, Laforet P, Navarro C. Molecular and clinical study of McArdle's disease in a cohort of 123 European patients. Identification of 20 novel mutations. Neuromuscul Disord. 2011;21(12):817–23.
8. Lucia A, Ruiz JR, Santalla A, Nogales-Gadea G, Rubio JC, Garcia-Consuegra I, Cabello A, Perez M, Teijeira S, Vieitez I, et al. Genotypic and phenotypic features of McArdle disease: insights from the Spanish national registry. J Neurol Neurosurg Psychiatry. 2012;83(3):322–8.
9. De Castro M, Johnston J, Biesecker L. Determining the prevalence of McArdle disease from gene frequency by analysis of next-generation sequencing data. Genet Med. 2015;17(12):1002–6.
10. Haller RG. Treatment of McArdle disease. Arch Neurol. 2000;57(7):923–4.
11. Rubio JC, Garcia-Consuegra I, Nogales-Gadea G, Blazquez A, Cabello A, Lucia A, Andreu AL, Arenas J, Martin MA. A proposed molecular diagnostic flowchart for myophosphorylase deficiency (McArdle disease) in blood samples from Spanish patients. Hum Mutat. 2007;28(2):203–4.
12. Kubisch C, Wicklein EM, Jentsch TJ. Molecular diagnosis of McArdle disease: revised genomic structure of the myophosphorylase gene and identification of a novel mutation. Hum Mutat. 1998;12(1):27–32.
13. Garcia-Consuegra I, Rubio JC, Nogales-Gadea G, Bautista J, Jimenez S, Cabello A, Lucia A, Andreu AL, Arenas J, Martin MA. Novel mutations in patients with McArdle disease by analysis of skeletal muscle mRNA. J Med Genet. 2009;46(3):198–202.
14. Garcia-Consuegra I, Blazquez A, Rubio JC, Arenas J, Ballester-Lopez A, Gonzalez-Quintana A, Andreu AL, Pinos T, Coll-Canti J, Lucia A, et al. Taking advantage of an old concept, "illegitimate transcription", for a proposed novel method of genetic diagnosis of McArdle disease. Genet Med. 2016;18(11):1128–35.
15. Dimauro S, Tsujino S. Nonlysosomal glycogenoses. In: Engel AG, Franzini-Armstrong C, editors. Myology. New York: McGraw-Hill; 1995. p. 1554–6.
16. Nadaj-Pakleza AA, Vincitorio CM, Laforet P, Eymard B, Dion E, Teijeira S, Vietez I, Jeanpierre M, Navarro C, Stojkovic T. Permanent muscle weakness in McArdle disease. Muscle Nerve. 2009;40(3):350–7.
17. WHO. Global recommendations on physical activity for health. In: Switzerland; 2010.
18. Physical activity guidelines for Americans. The Oklahoma nurse. 2008;53(4):25.
19. Mate-Munoz JL, Moran M, Perez M, Chamorro-Vina C, Gomez-Gallego F, Santiago C, Chicharro L, Foster C, Nogales-Gadea G, Rubio JC, et al. Favorable responses to acute and chronic exercise in McArdle patients. Clin J Sport Med. 2007;17(4):297–303.
20. Perez M, Ruiz JR, Fernandez Del Valle M, Nogales-Gadea G, Andreu AL, Arenas J, Lucia A. The second wind phenomenon in very young McArdle's patients. Neuromuscul Disord. 2009;19(6):403–5.
21. Aquaron R, Berge-Lefranc JL, Pellissier JF, Montfort MF, Mayan M, Figarella-Branger D, Coquet M, Serratrice G, Pouget J. Molecular characterization of myophosphorylase deficiency (McArdle disease) in 34 patients from southern France: identification of 10 new mutations. Absence of genotype-phenotype correlation. Neuromuscul Disord. 2007;17(3):235–41.
22. Deschauer M, Morgenroth A, Joshi PR, Glaser D, Chinnery PF, Aasly J, Schreiber H, Knape M, Zierz S, Vorgerd M. Analysis of spectrum and frequencies of mutations in McArdle disease. Identification of 13 novel mutations. J Neurol. 2007;254(6):797–802.
23. Rubio JC, Gomez-Gallego F, Santiago C, Garcia-Consuegra I, Perez M, Barriopedro MI, Andreu AL, Martin MA, Arenas J, Lucia A. Genotype modulators of clinical severity in McArdle disease. Neurosci Lett. 2007;422(3):217–22.
24. Nogales-Gadea G, Brull A, Santalla A, Andreu AL, Arenas J, Martin MA, Lucia A, de Luna N, Pinos T. McArdle disease: update of reported mutations and polymorphisms in the PYGM gene. Hum Mutat. 2015; 36(7):669–78.
25. Nogales-Gadea G, Rubio JC, Fernandez-Cadenas I, Garcia-Consuegra I, Lucia A, Cabello A, Garcia-Arumi E, Arenas J, Andreu AL, Martin MA. Expression of the muscle glycogen phosphorylase gene in patients with McArdle disease: the role of nonsense-mediated mRNA decay. Hum Mutat. 2008;29(2):277–83.
26. Kodama S, Saito K, Tanaka S, Maki M, Yachi Y, Asumi M, Sugawara A, Totsuka K, Shimano H, Ohashi Y, et al. Cardiorespiratory fitness as a quantitative predictor of all-cause mortality and cardiovascular events in healthy men and women: a meta-analysis. JAMA. 2009;301(19):2024–35.
27. Haller R, Wyrick P, Taivassalo T, Vissing J. Aerobic conditioning: an effective therapy in McArdle's disease. Ann Neurol. 2006;59(6):922–8.
28. Scalco RS, Chatfield S, Junejo MH, Booth S, Pattni J, Godfrey R, Quinlivan R, McArdle Disease Misdiagnosed as Meningitis. Am J Case Rep. 2016;17:905–8.
29. Perez M, Foster C, Gonzalez-Freire M, Arenas J, Lucia A. One-year follow-up in a child with McArdle disease: exercise is medicine. Pediatr Neurol. 2008; 38(2):133–6.
30. Malina RM. Tracking of physical activity and physical fitness across the lifespan. Res Q Exerc Sport. 1996;67(3 Suppl):S48–57.

PERMISSIONS

All chapters in this book were first published in GENOMICS, by BioMed Central; hereby published with permission under the Creative Commons Attribution License or equivalent. Every chapter published in this book has been scrutinized by our experts. Their significance has been extensively debated. The topics covered herein carry significant findings which will fuel the growth of the discipline. They may even be implemented as practical applications or may be referred to as a beginning point for another development.

The contributors of this book come from diverse backgrounds, making this book a truly international effort. This book will bring forth new frontiers with its revolutionizing research information and detailed analysis of the nascent developments around the world.

We would like to thank all the contributing authors for lending their expertise to make the book truly unique. They have played a crucial role in the development of this book. Without their invaluable contributions this book wouldn't have been possible. They have made vital efforts to compile up to date information on the varied aspects of this subject to make this book a valuable addition to the collection of many professionals and students.

This book was conceptualized with the vision of imparting up-to-date information and advanced data in this field. To ensure the same, a matchless editorial board was set up. Every individual on the board went through rigorous rounds of assessment to prove their worth. After which they invested a large part of their time researching and compiling the most relevant data for our readers.

The editorial board has been involved in producing this book since its inception. They have spent rigorous hours researching and exploring the diverse topics which have resulted in the successful publishing of this book. They have passed on their knowledge of decades through this book. To expedite this challenging task, the publisher supported the team at every step. A small team of assistant editors was also appointed to further simplify the editing procedure and attain best results for the readers.

Apart from the editorial board, the designing team has also invested a significant amount of their time in understanding the subject and creating the most relevant covers. They scrutinized every image to scout for the most suitable representation of the subject and create an appropriate cover for the book.

The publishing team has been an ardent support to the editorial, designing and production team. Their endless efforts to recruit the best for this project, has resulted in the accomplishment of this book. They are a veteran in the field of academics and their pool of knowledge is as vast as their experience in printing. Their expertise and guidance has proved useful at every step. Their uncompromising quality standards have made this book an exceptional effort. Their encouragement from time to time has been an inspiration for everyone.

The publisher and the editorial board hope that this book will prove to be a valuable piece of knowledge for researchers, students, practitioners and scholars across the globe.

LIST OF CONTRIBUTORS

Shane M. Heffernan
MMU Sports Genomics Laboratory, Manchester Metropolitan Universit Crewe, Manchester,UK
School of Public Health, Physiotherapy and Sports Science, University College Dublin, Dublin 4, Ireland

Liam P. Kilduff, Mark A. Bennett
A-STEM, College of Engineering, Swansea University, Swansea, UK

Robert M. Erskine
Research Institute for Sport & Exercise Sciences, Liverpool John Moores University, Liverpool, UK
Institute of Sport, Exercise and Health,University College London, London, UK

Stephen H. Day, Georgina K. Stebbings
MMU Sports Genomics Laboratory, Manchester Metropolitan University, Crewe, Manchester, UK

Christian J. Cook
A-STEM, College of Engineering, Swansea University, Swansea, UK
5School of Sport, Health and Exercise Sciences, Bangor University, Bangor, UK

Stuart M. Raleigh
Centre for Physical Activity and Chronic Disease, Institute of Health and Wellbeing, University of Northampton, Northampton, UK

Guan Wang, Yannis P. Pitsiladis
FIMS Reference Collaborating Centre of Sports Medicine for Anti-Doping Research, University of Brighton, Brighton, UK

Malcolm Collins
Division of Exercise Science and Sports Medicine, Department of Human Biology, University of Cape Town (UCT), Cape Town, South Africa

Alun G. Williams
MMU Sports Genomics Laboratory, Manchester Metropolitan University, Crewe, Manchester, UK
Institute of Sport, Exercise and Health, University College London, London, UK

L. M. Vera, C. Metochis, J. F. Taylor, M. Clarkson, H. Migaud, D. R. Tocher
Institute of Aquaculture, Faculty of Natural Sciences, University of Stirling,FK94LA, Stirling, Scotland, UK

K. H. Skjærven
National Institute of Nutrition and Seafood Research (NIFES), Nordnes, 5817 Bergen, Norway

Ruoyu Zhang, Yiqin Wang, Zhenglong Gu
Division of Nutritional Sciences, Cornell University, Ithaca, NY 14853, USA

Kaixiong Ye
Department of Biological Statistics and Computational Biology, Cornell University, Ithaca, NY 14853, USA

Martin Picard
Department of Psychiatry, Division of Behavioral Medicine, Department of Neurology and Columbia Translational Neuroscience Initiative, Columbia Aging Center, Columbia University Medical Center, New York, NY 10032, USA

Marten Michaelis, Joachim M. Weitzel
Institute of Reproductive Biology, University of Rostock, Rostock, Germany
Leibniz Institute for Farm Animal Biology (FBN),Institute of Reproductive Biology, FBN Dummerstorf, Wilhelm-Stahl-Allee 2,18196, Dummerstorf, Germany

Alexander Sobczak, Jennifer Schoen
Institute of Reproductive Biology, University of Rostock, Rostock, Germany

Dirk Koczan
Institute of Immunology, University of Rostock, Rostock, Germany

Martina Langhammer, Norbert Reinsch
Institute of Genetics and Biometry, Leibniz Institute for Farm Animal Biology (FBN), Dummerstorf, Germany

Prathima P. Thirugnanasambandam
Queensland Alliance for Agriculture and Food Innovation, The University of Queensland, St. Lucia, QLD 4072, Australia
ICAR - Sugarcane Breeding Institute, Coimbatore, Tamil Nadu, India

Nam V. Hoang
Queensland Alliance for Agriculture and Food Innovation, The University of Queensland, St. Lucia, QLD 4072, Australia
College of Agriculture and Forestry, Hue University, Hue, Vietnam

Agnelo Furtado
Queensland Alliance for Agriculture and Food Innovation, The University of Queensland, St. Lucia, QLD 4072, Australia

Frederick C. Botha
Sugar Research Australia, Indooroopilly, QLD 4068, Australia

Robert J. Henry
Queensland Alliance for Agriculture and Food Innovation, The University of Queensland, St. Lucia, QLD 4072, Australia
The University of Queensland, Room 2.245, Level 2, The John Hay Building, Queensland Biosciences Precinct [#80], 306 Carmody Road, St Lucia, QLD 4072, Australia

Julien G. Levy, J. Creighton Miller Jr., Elizabeth A. Pierson
Department of Horticultural Sciences, Texas A&M University, College Station, TX 77843, USA

Azucena Mendoza, Cecilia Tamborindeguy
Department of Entomology, Texas A&M University, College Station, TX 77843, USA

Nguyen-Phuong Pham, Séverine Layec, Eric Dugat-Bony, Françoise Irlinger, Christophe Monnet
UMR GMPA, AgroParisTech, INRA, Université Paris-Saclay, 78850 Thiverval-Grignon, France

Marie Vidal
US 1426, GeT-PlaGe, Genotoul, INRA, 31326 Castanet-Tolosan, France

Zachary N. Harris
Missouri State University, Biology Department, 901 S. National Ave, Springfield, MO, USA

Present address: Saint Louis University, Department of Biology, 1 N.Grand Blvd, Saint Louis, MO, USA

Laszlo G. Kovacs
Missouri State University, Biology Department, 901 S. National Ave,Springfield, MO, USA

Jason P. Londo
United States Department of Agriculture, Agricultural Research Service, Grape Genetics Research Unit, 630 W. North Street, Geneva, NY, USA

Corey L. Campbell, Laura B. Dickson, Saul Lozano-Fuentes, William C. Black
Department of Microbiology, Immunology and Pathology, Colorado State University, Campus Delivery 1692, Fort Collins, CO 80523, USA

Punita Juneja, Francis M. Jiggins
Department of Genetics, University of Cambridge, Downing Street, Cambridge CB2 3EH, UK

Helen McCormick, Jennifer E. Cropley, Eleni Giannoulatou, Catherine M. Suter
Victor Chang Cardiac Research Institute, 405 Liverpool Street, Darlinghurst, NSW 2010, Australia
St Vincents Clinical School, Faculty of Medicine,University of New South Wales, Kensington, NSW 2052, Australia

Paul E. Young, Suzy S. J. Hur
Victor Chang Cardiac Research Institute, 405 Liverpool Street, Darlinghurst, NSW 2010, Australia

Keith Booher, Hunter Chung
Zymo Research, Murphy Ave, Irvine, CA 92614, USA

T. Maroilley, G. Lemonnier, J. Lecardonnel, Y. Ramayo-Caldas, C. Rogel-Gaillard, J. Estellé
GABI, INRA, AgroParisTech, Université Paris-Saclay, 78350 Jouy-en-Josas, France

D. Esquerré
GenPhySE, INRA, INPT, ENVT, Université de Toulouse, 31326 Castanet-Tolosan, France

M. J. Mercat
IFIP - Institut du porc/BIOPORC, La Motte au Vicomte, BP 35104, 35651 Le Rheu, France

Guan Wang
Centre of Sports Medicine for Anti-Doping Research, University of Brighton, Eastbourne, UK

Department of Movement, Human and Health Sciences, University of Rome "Foro Italico", Rome, Italy

Jérôme Durussel
Institute of Cardiovascular and Medical Sciences, College of Medical, Veterinary and Life Sciences, University of Glasgow, Glasgow, UK

Jonathan Shurlock
Brighton and Sussex Medical School, Brighton, UK

Martin Mooses
Faculty of Medicine,University of Tartu, Tartu, Estonia

Noriyuki Fuku
Graduate School of Health and Sports Science, Juntendo University, Chiba, Japan

Georgie Bruinvels, Charles Pedlar, Richard Burden
School of Sport, Health and Applied Science, St Mary's University, Twickenham, London, UK

Andrew Murray
Centre for Sports and Exercise, University of Edinburgh, Edinburgh, UK

Brendan Yee
Affymetrix,Santa Clara, CA, USA. 9BioKaizen Lab SA, Monthey, Switzerland

Anne Keenan, John D. McClure
Institute of Cardiovascular and Medical Sciences, College of Medical, Veterinary and Life Sciences, University of Glasgow, Glasgow, UK

Pierre-Edouard Sottas
BioKaizen Lab SA, Monthey, Switzerland

Yannis P. Pitsiladis
Centre of Sports Medicine for Anti-Doping Research, University of Brighton, Eastbourne
Department of Movement, Human and Health Sciences, University of Rome "Foro Italico", Rome, Italy

Yuan-Xiang Shi, Wei Zhang Hong-Hao, Zhou Ji-Ye Yin and Zhao-Qian Liu
Department of Clinical Pharmacology, Xiangya Hospital, Central South University, Changsha 410008, People's Republic of China

Institute of Clinical Pharmacology, Central South University, Hunan Key Laboratory of Pharmacogenetics, Changsha 410078, People's Republic of China
Hunan Province Cooperation Innovation Center for Molecular Target New Drug Study, Hengyang 421001, People's Republic of China

Ying Wang
Department of Clinical Pharmacology, Xiangya Hospital, Central South University, Changsha 410008, People's Republic of China
Institute of Clinical Pharmacology, Central South University, Hunan Key Laboratory of Pharmacogenetics, Changsha 410078, People's Republic of China

Weijing Wang, Wenjie Jiang,Yili Wu and Dongfeng Zhang
Department of Epidemiology and Health Statistics, Public Health College, Qingdao University, No. 38 Dengzhou Road, Shibei District, Qingdao 266021,Shandong Province, People's Republic of China

Lin Hou
Department of Biochemistry, Medical College, Qingdao University, No. 38 Dengzhou Road, Shibei District, Qingdao 266021, Shandong Province, People's Republic of China

Haiping Duan
Department of Epidemiology and Health Statistics, Public Health College, Qingdao University, No. 38 Dengzhou Road, Shibei District, Qingdao 266021,Shandong Province, People's Republic of China
Qingdao Municipal Center for Disease Control and Prevention, No. 175 Shandong Road, Shibei District, Qingdao 266033, Shandong Province, People's Republic of China

Chunsheng Xu
Department of Epidemiology and Health Statistics, Public Health College, Qingdao University, No. 38 Dengzhou Road, Shibei District, Qingdao 266021, Shandong Province, People's Republic of China
Qingdao Municipal Center for Disease Control and Prevention, No. 175 Shandong Road, Shibei District, Qingdao 266033, Shandong Province, People's Republic of China

Qingdao Institute of Preventive Medicine, No.175 Shandong Road, Shibei District, Qingdao 266033, Shandong Province, People's Republic of China

Qihua Tan
Epidemiology, Biostatistics and Bio-demography, Institute of Public Health, University of Southern Denmark, DK-5000 Odense C, Denmark
Human Genetics, Institute of Clinical Research, University of Southern Denmark, DK-5000 Odense C, Denmark

Shuxia Li
Human Genetics, Institute of Clinical Research, University of Southern Denmark, DK-5000 Odense C, Denmark

Alfredo Santalla
Universidad Pablo de Olavide, Sevilla, Spain
Instituto de Investigación Hospital 12 de Octubre (i+12), Madrid, Spain

Gisela Nogales-Gadea, Alfonsina Ballester-Lopez
Grup de Recerca en Malalties Neuromusculars i Neuropediatriques, Department of Neurosciences, Institut d'Investigació en Ciències de la Salut Germans Trias i Pujol, Universitat Autònoma de Barcelona, Camí de les Escoles, s/n 08916 (Barcelona), Badalona, Spain

Alberto Blázquez Encinar, Inés García Consuegra, Miguel A. Martín
Instituto de Investigación Hospital 12 de Octubre (i+12), Madrid, Spain
Centre for Biomedical Network Research on Rare Diseases (CIBERER), Instituto de Salud Carlos III, Madrid, Spain
Laboratorio de Enfermedades Mitocondriales y Neuromusculares, Hospital 12 de Octubre, Madrid, Spain

Irene Vieitez
Rare Diseases and Pediatric Medicine Group, Galicia Sur Health Research Institute, Complexo Hospitalario Universitario de Vigo (CHUVI), SERGAS, Vigo, Spain

Adrian González-Quintana, Pablo Serrano-Lorenzo, Sara Asensio, Joaquín Arenas
Instituto de Investigación Hospital 12 de Octubre (i+12), Madrid, Spain

Centre for Biomedical Network Research on Rare Diseases (CIBERER), Instituto de Salud Carlos III, Madrid, Spain.

Guillem Pintos-Morell
Grup de Recerca en Malalties Neuromusculars i Neuropediatriques, Department of Neurosciences, Institut d'Investigació en Ciències de la Salut Germans Trias i Pujol, Universitat Autònoma de Barcelona, Camí de les Escoles, s/n 08916 (Barcelona), Badalona, Spain
Servicio de Pediatría, Hospital Universitari Germans Trias i Pujol, Badalona, Spain
Centre for Biomedical Network Research on Rare Diseases(CIBERER), Instituto de Salud Carlos III, Madrid, Spain

Jaume Coll-Cantí
Grup de Recerca en Malalties Neuromusculars i Neuropediatriques, Department of Neurosciences, Institut d'Investigació en Ciències de la Salut Germans Trias i Pujol, Universitat Autònoma de Barcelona, Camí de les Escoles, s/n 08916 (Barcelona),Badalona, Spain
Centre for Biomedical Network Research on Rare Diseases (CIBERER), Instituto de Salud Carlos III, Madrid, Spain
Servicio de Neurología, Hospital Universitari Germans Trias i Pujol,Badalona, Spain

Helios Pareja-Galeano, Jorge Díez-Bermejo and Alejandro Lucia
Instituto de Investigación Hospital 12 de Octubre (i+12), Madrid, Spain
Universidad Europea de Madrid, Madrid, Spain

Margarita Pérez
Universidad Europea de Madrid, Madrid, Spain

Antoni L. Andreu
Departament de Patologia Mitocondrial i Neuromuscular, Hospital Universitari Vall d'Hebron, Institut de Recerca (VHIR), Universitat Autónoma de Barcelona, Barcelona, Spain.

Tomàs Pinós
Centre for Biomedical Network Research on Rare Diseases(CIBERER), Instituto de Salud Carlos III, Madrid, Spain
Departament de Patologia Mitocondrial i Neuro-muscular, Hospital Universitari Vall d'Hebron, Institut de Recerca (VHIR), Universitat Autónoma de Barcelona, Barcelona, Spain.

Index

www.ingramcontent.com/pod-product-compliance
Lightning Source LLC
Chambersburg PA
CBHW080406190526
45161CB00003B/148